THE POTENTIAL EFFECTS OF GLOBAL CLIMATE CHANGE ON THE UNITED STATES

THE POTENTIAL EFFECTS OF GLOBAL CLIMATE CHANGE ON THE UNITED STATES

Editors

Joel B. Smith and Dennis A. Tirpak
United States Environmental Protection Agency
Office of Policy, Planning, and Evaluation
Office of Research and Development
Washington, DC

HEMISPHERE PUBLISHING CORPORATION
A member of the Taylor & Francis Group

New York Washington Philadelphia London

DISCLAIMER: This material was circulated for review and comment and does not necessarily reflect the official position of the U.S. Environmental Protection Agency. Mention of trade names does not constitute an endorsement.

THE POTENTIAL EFFECTS OF GLOBAL CLIMATE CHANGE ON THE UNITED STATES

1 2 3 4 5 6 7 8 9 0 E B E B 9 8 7 6 5 4 3 2 1 0

Cover design by Renee Winfield.
A CIP catalog record for this book is available from the British Library.

Library of Congress Cataloging-in-publication Data

The Potential effects of global climate change on the United States / editors, Joel B. Smith and Dennis A. Tirpak.
p. cm.
"United States Environmental Protection Agency, Office of Policy, Planning, and Evaluation; Office of Research and Development."

1. Climatic changes. 2.Climatic changes—United States. 3. Global warming—United States. 4. Greenhouse effect, Atmospheric—United States. 5. United States—Climate. I. Smith, Joel B. II. Tirpak, Dennis A. III. United States. Environmental Protection Agency. Office of Policy, Planning, and Evaluation. IV. United States. Environmental Protection Agency. Office of Research and Development.
QC981.8.C5P6832 1990
551.6973—dc20 90-30124
ISBN 1-56032-071-0 CIP

CONTENTS

ACKNOWLEDGEMENTS

This report was made possible because of the hundreds of people who participated in workshops, conducted research projects, reviewed draft manuscripts, and contributed ideas that shaped the final product. They shared a common belief that an objective analysis of global climate change could be undertaken despite the uncertainties in scientific information. We are grateful for their support and encouragement. It was the difference that sustained us through this effort. In particular, we wish to acknowledge the authors who organized and integrated the following chapters.

The following people contributed to this report:

Chapter 1:	Introduction	Joel B. Smith
Chapter 2:	Global Climate Change	Dennis A. Tirpak
Chapter 3:	Methodology	Joel B. Smith
Chapter 4:	California	George A. King Robert L. DeVelice Ronald P. Neilson Robert C. Worrest
Chapter 5:	Great Lakes	Joel B. Smith
Chapter 6:	Southeast	James G. Titus
Chapter 7:	Great Plains	Cynthia Rosenzweig William E. Riebsame
Chapter 8:	Water Resources	Mark W. Mugler Michael C. Rubino
Chapter 9:	Sea Level Rise	James G. Titus
Chapter 10:	Agriculture	Cynthia Rosenzweig Margaret M. Daniel
Chapter 11:	Forests	Jack K. Winjum Ronald P. Neilson
Chapter 12:	Biological Diversity	Lauretta M. Burke Ross A. Kiester
Chapter 13:	Air Quality	Joseph J. Bufalini Peter L. Finkelstein Eugene C. Durman
Chapter 14:	Health	Janice A. Longstreth
Chapter 15:	Urban Infrastructure	Ted R. Miller
Chapter 16:	Electricity Demand	Kenneth P. Linder
Chapter 17:	Variability	Linda O. Mearns

Chapter 18:	Research Needs	Anthony Janetos
Chapter 19:	Policy	James G. Titus

Scenario data were assembled and distributed by Roy Jenne. We wish to thank J. Hansen, S. Manabe, and M. Schlesinger for providing us with the results from their GCM runs. Special thanks are also necessary to Joan O'Callaghan and Karen Swetlow for editing; Roberta Wedge for assistance on production of the report; and Margaret Daniel and Michael Greene for research and administrative assistance.

CHAPTER 1

INTRODUCTION

Since the beginning of the Industrial Revolution, emissions of greenhouse gases have increased. Fossil fuel burning, which releases CO_2, CO, N_2O, and other pollutants, has expanded many times over. Changes in agriculture have led to increased emissions of CH_4 and N_2O. Population growth has contributed to deforestation in many areas of the globe, which in turn has affected the global carbon cycle. Atmospheric concentrations of tropospheric ozone and CFCs have also increased, primarily due to industrial activity.

Scientists studying the effect of increasing concentrations of greenhouse gases have tried to predict how quickly and to what extent atmospheric temperature may rise. For analytic convenience, the convention has been to examine the implications of a CO_2 doubling. In 1979, the National Academy of Sciences concluded that such a doubling would lead to an increase of 1.5 to 4.5°C in global air temperatures. Since then, other researchers have examined the increase in all greenhouse gases and have concluded that greenhouse gas increase equivalent to CO_2 doubling will occur as early as the 2030s, with commensurate global warming lagging by several decades.

The earth's atmosphere has undergone many cycles of warming and cooling in the past. Paleoclimatologists have estimated that at the glacial maximum of the last ice age, which was about 18,000 years ago, the earth was approximately 5°C cooler than at present. This is generally attributed to changes in orbital characteristics combined with lower trace gas concentrations and different climate feedbacks. Two things are different about current greenhouse warming. First, it will raise temperatures higher than the planet has experienced in the last 100,000 years. Second, past climate changes of comparable magnitude have generally occurred over tens of thousands of years. The greenhouse effect is expected to raise atmospheric temperatures several degrees in less than a century.

CONGRESSIONAL REQUEST FOR REPORTS

The significant implications of the greenhouse effect have been the subject of discussion within the scientific community for the past three decades. In recent years members of the Congress have held hearings and began to explain the implications for public policy. This interest was accentuated during a series of hearings held in June 1986 by the Subcommittee on Pollution of the Senate Environment and Public Works Committee. Following the hearings, members of the Senate Environment and Public Works

Committee sent a formal request to the EPA Administrator, asking the Agency to undertake two studies on climate change due to the greenhouse effect.

> One of the studies we are requesting should examine the potential health and environmental effects of climate change. This study should include, but not be limited to the potential impacts on agriculture, forests, wetlands, human health, rivers, lakes, and estuaries, as well as other ecosystems and societal impacts. This study should be designed to include original analyses, to identify and fill in where important research gaps exist, and to solicit the opinions of knowledgeable people throughout the country through a process of public hearings and meetings.

The Congress also requested that EPA prepare a study on policy options to stabilize current levels of atmospheric greenhouse gas concentrations. This study analyzes policy options for limiting gas concentrations including energy efficiency, alternative technologies, reforestation options, CFC reductions, and other options for limiting CH_4 and N_2O. Congress requested the studies in the Fiscal Year 1987 Continuing Resolution.

GOALS OF THIS REPORT

This report builds on the past contributions of many scientists throughout the world. Most notably, the reports by the National Academy of Sciences (1979, 1983, 1987), the World Meterological Organization, United Nations Environment Programme, and the International Council of Scientific Unions (1985), Scope 29 (1986), Department of Energy (1985). It is an attempt to identify some of the sensitivities, direction and magnitude, linkages, regional differences, national impacts, policy implications, and uncertainties among the effects of global climate warming, but it is not a prediction. We hope it will provide useful information to climate modelers and effects researchers. We also hope that officials, at all levels of government will be encouraged to examine the implications of climate change on long-term policies. Since this is the first study of this type, we expect that a great deal more research, analysis, and planning will be needed in the future. We do not pretend to have all the answers.

This report is intended to achieve the following goals of identifying:

Sensitivities

Since climate change on a regional level is uncertain, we cannot predict regional effects. However, we can identify the sensitivities of systems to climate change. Our goal is to use a variety of scenarios to determine what climatic variables are important in causing impacts and how sensitive systems are to changes in these variables.

Direction and Magnitude

Since the scenarios do not encompass all possible combinations of climate change due to increased greenhouse gases, the results do not represent the entire range of possible effects. Yet, the results from various scenarios help define the direction and magnitude of effects. First, we examine them to see if a direction of change (e.g. more water, lower crop yields) is evident. Second, we attempt to determine if the magnitude of change is significant. Yet, results outside the bounds of our results cannot be ruled out at this time.

Linkages

Individual environmental systems will not be affected by climate change in isolation. Water resources, for example, may be affected not only by changes in water supply, but also by changes in demand for water for such purposes as irrigation. Wildlife may be directly affected by changes in climate and indirectly by changes in habitat. This report attempts to identify linkages among effects, quantitatively, where possible, and qualitatively elsewhere.

Regional Differences

Effects were examined in several regions to broaden the analysis and to allow for comparison of effects on similar systems in different regions. We do not attempt to identify "winners and losers" since all regions are likely to change. The effects are simply likely to occur in different sectors and at different rates.

National Impacts

While regional case studies are the major method used in this report, several issues have been analyzed on national scales. National analyses can build on regional studies to provide more comprehensive information on the sensitivities of systems to climate change. In some cases, national analyses gives us some insights into the total cost to the United States over the next 50 to 100 years.

Policy Implications

In their current decisions on the use of natural resources and related systems, States, municipalities, industries, and businesses are assuming that climate will not change. An examination of the sensitivities of systems to climate change may indicate that they should reconsider some of their long-term decisions. This report discusses some policy implications of climate change, but it does lay out a prescription policy agenda.

Uncertainties

There are many uncertainties about the rate of warming and specific changes in regional weather patterns. There are also uncertainties about how ecological and other systems will be affected by climate change. These uncertainties are reflected in the models used to estimate climate change and impacts. This report attempts to clearly state these limitations.

Guidance for Research

This effects analysis should provide climate modelers with information about how general circulation models could be improved. It should also help define research needs for future analysis of the potential impacts of climate change.

Fundamentally, these goals center on the identification of important issues and state-of-the-science investigations in each environmental system. Because each component of science and policy development is at an early stage, the goals of the report are to develop insights and estimates of the ranges of possible future effects and to use that information for identifying where the policies and research programs of EPA and other agencies should be reexamined.

STRUCTURE OF THE ANALYSIS

Key Systems

This report focuses on several systems that may be particularly affected by climate change.

Sea Level Rise
Water Resources
Agriculture

Forests
Biodiversity and Wildlife
Energy Demand
Air Pollution
Health

Regional Case Studies

Analysis of the regional impacts of climate change on various resources help display the diversity of possible effects of climate change across the United States and make the possible effects of global climatic change easier to understand. Analysis of regional impacts also makes it easier to identify and analyze the interrelationship of effects and possible responses. For example, rising sea level and reduced river flow would have combined impacts on estuarine salinity. Changes in agriculture could lead to increased demand for irrigation, further reducing available water supplies. Furthermore, the identification of combined effects on a regional scale aids in the analysis of potential adaptation strategies.

Four regional case studies were selected: the Southeast, the Great Lakes, California, and the southern Great Plains. These regions were picked for several reasons. Each of these regions is important for economics, social, and environmental reasons, and each offers some unique current characteristics that makes it an interesting example of the range of possible environmental issues that will most likely need to be considered. The Southeast depends heavily on forestry and agriculture, and has extensive and fragile wetlands and coastal ecosystems. The Great Lakes are the dominant natural resource in their region, supplying freshwater, fishery resources, and a pathway for shipping and transportation, and provide a natural laboratory for environmental issues that affect both the United States and Canada. California already must manage its water supplies carefully, and its agricultural industry provides many crops for the United States, a large share of the international market, and is among the most productive agricultural regions in the world. The Great Plains is one of the largest producers of grain crops in the world. Although these regions are diverse, they do not encompass the entire range of regional differences in the United States. The analysis of effects in these regions does not cover all potential impacts in the United States.

National Studies

The effects on a number of systems were quantitatively analyzed on a national scale. National agricultural markets were analyzed with respect to their sensitivities to changes in yield derived from our agricultural models. Options for adapting to a sea level rise were examined on a national scale, as were the possibilities of health impacts. Forestry, water management, air quality, and biodiversity issues were explored by analyzing the results of several of the regional case studies with a broader perspective. In each case, the national-level analyses provide an additional level of qualitative integration that a purely regional analysis could not. The structure of the regional and national studies is displayed in Figure 1-1.

ANALYTIC APPROACHES

The climate scenarios we used were based on outputs from General Circulation Models (see Chapter 3). Where possible, we tried to obtain quantitative estimates of effects. However, the development of quantitative estimates was constrained by the availability of well-documented models that included some interaction of the particular effect in question and climatic variability. We obtained information on sensitivities by reviewing the literature and by gathering expert judgement. The approach of using existing models, all of which were originally constructed for other purposes, makes the interpretation of results instructive, but somewhat limited with respect to the full range of climatically relevant questions that could be asked.

PROCESS

We used a seven-stage process to define the scope of this report, select the projects, write the chapters, and review the results.

Step 1: Initial Scoping of the Report

This stage immediately followed the request from the Senate Environment and Public Works Committee. We agreed on using the regional case study approach and the four regions to be investigated, and on using climatic scenarios. We also decided not to attempt to analyze environmental effects outside of the United States in this report. Our rationale for thisdecision was based on available time, funds, and the lack of suitable models that would be immediately accessible to us.

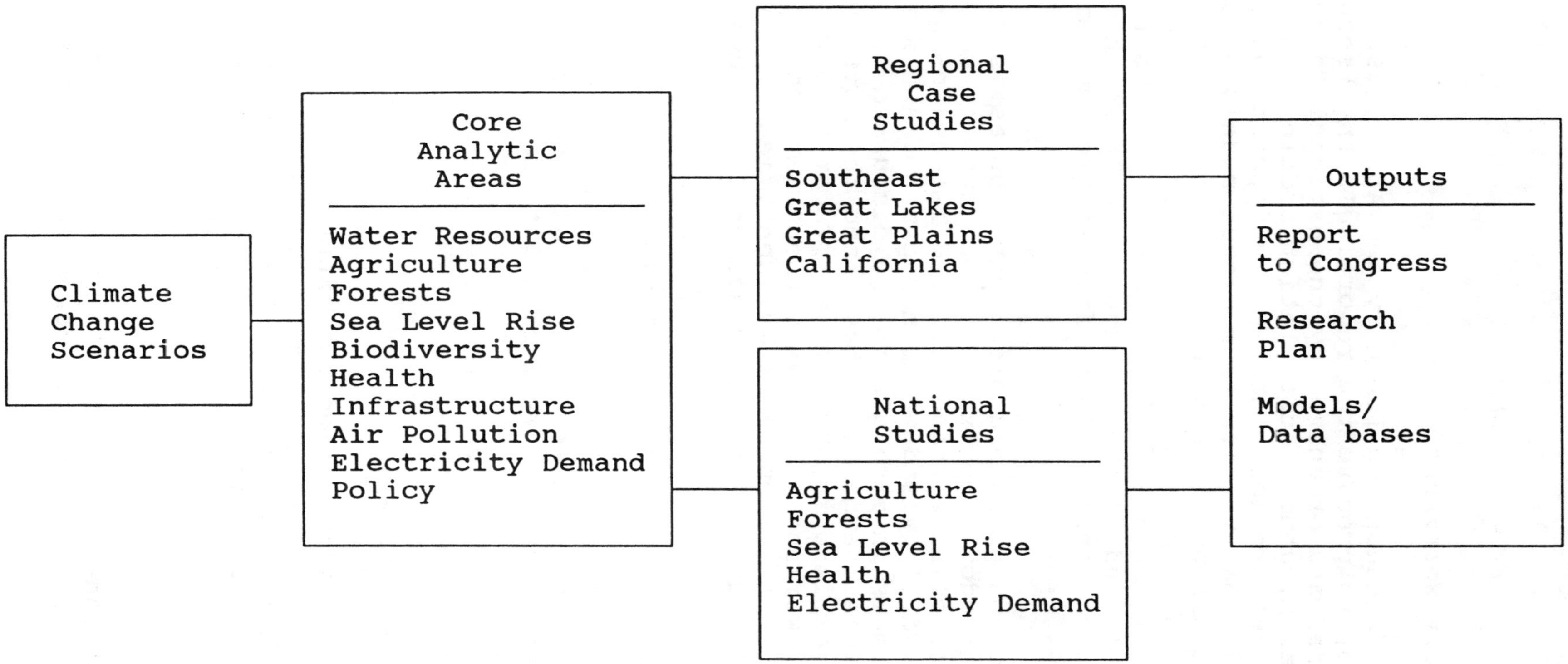

Figure 1-1. Elements of Effects Report

Step 2: Preparatory Workshops

We held two workshops in February and April, 1987 in Boulder, Colorado, to prepare the report. In the February workshop, sponsored and organized by the National Center for atmospheric Research, general circulation modelers convened to discuss some of the problems inherent in attempting to understand the regional results from global models. several major topics were discussed from the standpoint of how the results from GCMs should be used in impact studies. A list of variables that would be available for use by effects researchers was produced at the end of the workshop. In addition, several potential studies on aspects of the frequency of extreme weather events were identified.

The April workshop was organized with the assistance of the University of Colorado. Approximately one hundred scientists explored the major climate change related issues in agriculture, forest effects, hydrology and water management, and sea-level rise. Working groups in each discipline discussed the potential impacts that climate change might have, and the most important uncertainties to explore in order to arrive at better predictions. The working groups were then rearranged into geographically oriented groups. They identified a series of studies that would address the major scientific issues in each region.

Step 3: Identification of Potential Projects

From the lists identified in the two Boulder workshops, and additional studies on urban and regional air quality subsequently identified internally by EPA, we arrived at list of investigators from whom we would solicit proposals. The decision to solicit proposals was based primarily on the potential coverage of environmental issues in each region.

Step 4: Reviews of Proposals

At least one intramural and two extramural reviewers examined each proposal. All comments were responded to and proposals modified and appropriate. EPA used a combination of cooperative agreements, existing contracts, and interagency agreements to fund projects for this report.

Step 5: Planning and Integration

All the researchers met with EPA staff in October 1987 to discuss scenarios, goals, and approaches for the studies.

Researchers discussed integration of projects within regions as well as the commonality of approaches within disciplines.

Step 6: Preliminary Project Review

In April 1988, EPA assembled panels of scientists to provide a preliminary review of most of the agriculture, forestry, and hydrology projects. The principal investigators of the appropriate projects were asked to present their work orally and in written drafts. EPA project managers used the comments from the review panels to make mid-course corrections in the conduct of a few projects, and as a guide to interpreting the results of individual projects.

Step 7: Project and Report Peer Review

At least two to three reviewers examined the final reports from all principal investigators before the EPA project managers accepted them. During this time, EPA staff on the report project team wrote the overviews that are reflected in this final report. In October 1988, a special subcommittee of EPA's Science Advisory Board was convened and asked to review the entire Environmental Effects Report. Following the SAB's written review, the EPA project team responded to comments and produced the final version of the Environmental Effects Report. The draft of the report was simultaneously sent to other Federal agencies for review and comment, and these comments were also taken into account in the final version.

STRUCTURE OF THIS REPORT

This report is divided into several sections. Section I consists of Chapter II on trends in emissions of greenhouse gases and potential impacts on climate; and Chapter III on the choice of scenarios and effects modelling. Section II contains results from the regional case studies, with each chapter devoted to different regions. Each regional chapter describes the climate sensitive stems in the region, reviews previous studies on impacts of climate change on the region, describes the structure of regional studies for the report, discusses regional climate change scenarios, reviews the design, results, and limitations of the studies and discusses the broader socio-economic and policy implications of climate change for the region. The regional chapters include relevant regional results from national studies. In Section III, the results of national analyses are presented. Each chapter covers a different system. They include an overview of relevant regional studies and presentation of results from national analyses. Each chapter discusses the current state of issue, reviews previous literature on climate change and the issue, discusses studies used for this report, presents national results from regional and national studies, and discusses broader socio-economic and policy implications. The design and limitations for each study are presented only once - in a regional chapter if it is a regional study or in a national chapter if it is a national study. Not all regionally relevant results are presented in the appropriate regional chapters. Results for health and livestock are presented only in the Health and Agriculture chapters. Section IV includes conclusion chapters. It begins with an overview of studies on climate variability. Chapter XVIII discusses where future research on climate change effects should go and Chapter XIX discusses policy implication and recommendations.

This report is meant as an overview of the individual studies. Those studies are printed in appendix volumes. In the overview, the studies are referenced by the author's name or names in parentheses. Previous published work is reference by the author's name and year of publication.

RELATIONSHIP TO CURRENT NATIONAL AND INTERNATIONAL ACTIVITIES

National Research and Policy Activities

The Global Climate Protection Act of 1987 requested EPA to develop a national policy on global climate change and to prepare an assessment of scientific information. The very scope of this issue suggests that this request can only be fulfilled in cooperation with other federal agencies, hence, EPA is working with these agencies to formulate a process to fullfill this request. The scientific assessment will be conducted in coordination with NASA, NOAA, NSF, and other agencies. To the extent possible, this scientific assessment will also be developed on an international basis and should be available in 1990.

The development of a national policy will be coordinated with the Department of Energy and other natural resource departments. The goal will be to build on this report and others under development by Federal agencies to identify the adoptive policies and other measures that may be appropriate to deal with this issue. An interim report on a national policy is scheduled for development by the end of 1989. The nature of this issue suggests that a continuous review of domestic policy will be required for many years into the future.

Chapter 1

International Activities

In 1987, the United Nations Environment Programme (UNEP) and the World Meterological Organization (WMO) were asked by member governments to establish an Intergovernmental Panel on Climate Change (IPCC) for the specific purpose of reviewing the scientific information and potential response strategies. The World Meterological Programme has primary responsibility for the World Climate Research Programme and the United Nations Environment Programme has responsibility for the World Climate Impacts Programme. The UNEP was the primary international agency responsible for negotiations leading up to the Montreal Protocol to Protect the Ozone Layer. As this report is being developed, the agenda for the IPCC has not yet been agreed upon by participating governments. The first meeting is scheduled for November 1988. Nevertheless, it is expected that the IPCC will be the primary forum for multilateral discussions between governments on this issue.

Other governments and international agencies are also examining this issue. Conferences are currently planned in 1989 by Italy, Canada, and the Netherlands. The United States has bilateral activities with the Soviet Union and China. The Organization for Economic Cooperation and Development (OECD) and the International Energy Agency are examining their potential contributions.

CHAPTER 2

GLOBAL CLIMATE CHANGE

The last two decades have witnessed a substantial increase in interest in the greenhouse theory. Once the subject of purely scientific curiosity during the last half of the nineteen century (Tyndall, 1863), the greenhouse process is now generally recognized as essential to the existence of life on earth in its current forms. Water vapor and carbon dioxide as primary trace gases were the first to receive attention. The scientific community developed a hierarchy of models with increasingly complex representations of the atmosphere, oceans, and terrestrial biosphere to represent the earth's climate and understand the greenhouse effect.

During the last decade, several developments have broadened the focus of scientific investigations on this global warming. The first was the finding that other trace gases--e.g., the chlorofluorocarbons (CFCs), methane ($_CH_4$), and nitrous oxide (N_2O)--could have a greater temperature effect on a molecule-per-molecule basis than CO_2. The second was the increasing recognition that these other trace gases could substantially alter the chemical composition and temperature of the atmosphere. Stratospheric ozone changes recently became the subject of intense laboratory and field experiments as scientists have struggled to understand the Antarctic ozone. Finally, it was recognized that the growth rates of these other trace gases are such that their combined growth could result in the equivalent doubling of CO_2 by approximately 2030 -- a time much sooner than previously assumed.

These discoveries lead the climate research community to deal with a series of challenges with a new urgency, including the chemistry of the stratosphere and the troposphere, the radiation balance of the planet, the importance of climate feedback mechanisms, and the role of the oceans in global climate. In addition, a variety of studies have been assembled to review current scientific information on the greenhouse effect, including reports by the National Academy of Sciences (NAS, 1979, 1983, 1987) and the World Meterological Organization (WMO, 1986a; 1986b). All these studies support the hypothesis that increasing concentrations of trace gases will lead to a global warming.

This chapter reviews and synthesizes the results of these prior efforts. Together, these sections are intended to provide a broad overview of the major issues affecting our knowledge of the greenhouse effect. The EPA report (EPA, 1989) on policy options for stabilizing the atmosphere provides a more detailed treatment of the factors affecting the earth's climate.

GREENHOUSE THEORY

The greenhouse effect may be described as the process by cwhich the biosphere balances the earth's energy budget. The main components of this process are incoming radiation from the sun, some of which is reflected back to space, and the outgoing longwave (infrared) radiation at the top of the atmosphere (Figure 2-1). The trace gases, mainly composed of CO_2, and water vapor, allow solar radiation to pass through the atmosphere to the earth's surface. Infrared radiation from the earth's surface is partly trapped by the greenhouse gases which are mainly CO_2 and water vapor. Scientists estimate that without the pretense of greenhouse gases, the earth would be approximately 58°C colder. The relative transparency of the gases to solar radiation and the trapping of infrared radiation is commonly referred to as the greenhouse effect.

When concentrations of radiatively important greenhouse gases increase, less longwave infrared passes back to space, and the global energy budget shown in Figure 2-1 becomes imbalanced. That is, less energy leaves the atmosphere than enters it. To maintain a balance with incoming solar radiation,--i.e., a net radiative flux at the top of the atmosphere equal to zero,--the lower part of the atmosphere (the troposphere) and the surface warm, increasing the emission of infrared radiation into space, until a new balance is achieved. Hence, as more greenhouse gases increase in the atmosphere, more infrared radiation is trapped, and the greater greenhouse effect. One very important factor in this greenhouse process is the altitude of the greenhouse gases. In general, gases absorbing solar radiation in the lower troposphere tend to warm the earth's surface, while gases absorbing radiation in the stratosphere tend to cool it.

THE ROLE OF TRACE GASES

It is now recognized emissions of many greenhouse gases are increasing and that these gases influence the chemistry of the stratosphere and troposphere in many ways. (A comprehensive discussion of greenhouse gases can be found in WMO, 1985; Ramanathan, 1985; Wang et al., 1986; and EPA, 1988). The radiatively important gases include water vapor, carbon dioxide,

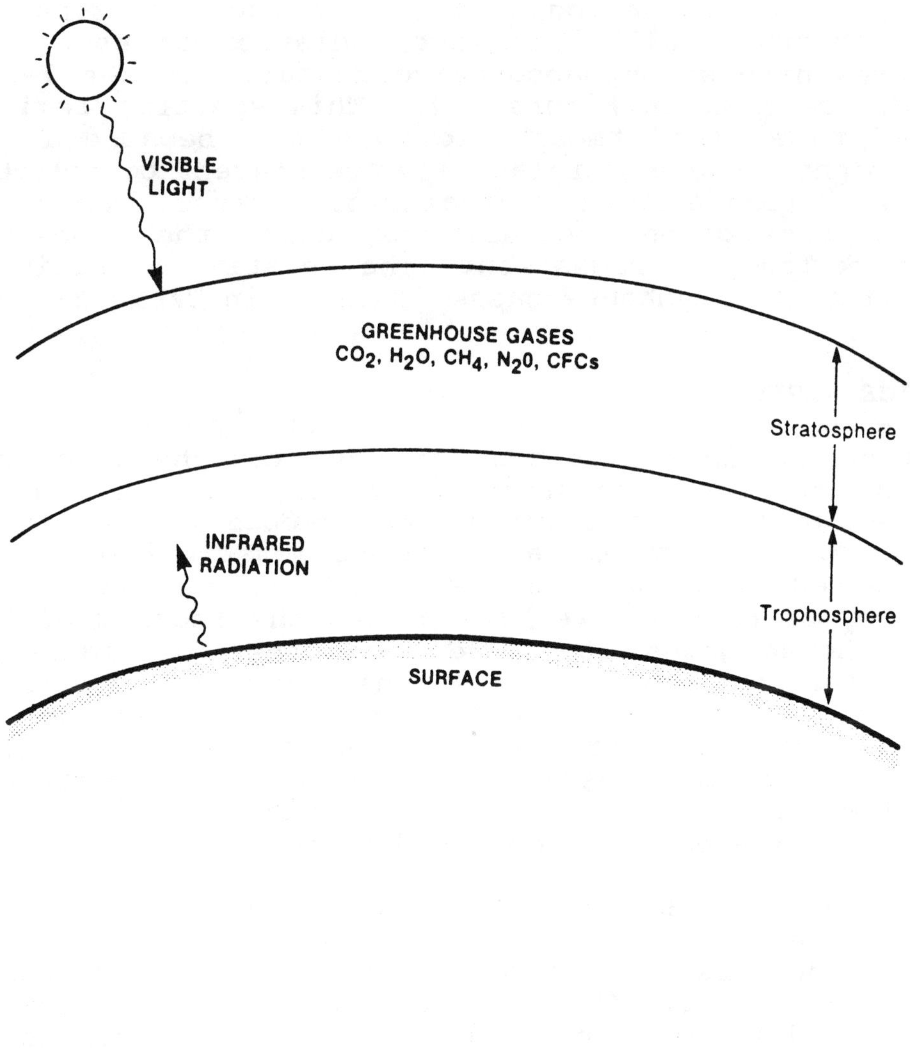

Figure 2-1. Global energy balance and the greenhouse effect.

Source: World Meteorological Organization (1985).

methane, chlorofluorocarbons, nitrous oxide, and tropospheric ozone. Other gases that are important because of their direct radiative effect or because of their role in the chemistry of the atmosphere include chlorocarbons, hydrocarbons, aldehydes, fluorinated and brominated species, carbon monoxide, nitrogen oxides, and compounds of sulfur.

The lifetimes, dominant sources, and sinks of greenhouse gases are shown in Table 2-1. Once emitted into the atmosphere, these greenhouse gases will be present for more than 100 years. In addition, they absorb radiation in parts of the electromagnetic spectrum that generally allow longwave radiation to escape into space. The gases have strong absorptive features in the 7-13 M spectral band, as shown in Figure 2-2. This spectral region is often referred to as the atmospheric "window," because in this region the atmosphere there is relatively transparent to radiation. Pollutants with strong absorptive features, however, can be very effective greenhouse gases, because they dirty the atmospheric window and block longwave radiation. The relative effectiveness of fifteen different greenhouse gases is shown in Table 2-2.

TRENDS IN TRACE GASES

Information on the composition of the atmosphere generally comes from either direct measurements, mostly made in the last several decades, or by measurement of gas trapped in Greenland or Antarctic ice. In some cases, attempts to measure old gas have even been directed at exotic sources, such as ancient-Egyptian tombs. For easy comparison, we have placed the records of three important greenhouse gases alongside each other in Figure 2-3. While not matching perfectly, the instrument and ice core records generally appear to overlap. In some cases, such as with CH_4, concentrations appear to mirror the general growth in global populations, while in other cases, such as with CO_2, concentrations appear to follow the direct growth in industrial activities. The following is a brief summary of observed trends.

Carbon Dioxide (CO_2) - The most reliable recent measurements of CO_2 were initiated by Keeling in 1958, when concentrations on Mauna Loa, Hawaii, were found to be 315 ppm. This compares to a number of ice core studies that generally place the preindustrial concentrations in the range of 270-290 ppm. The mean growth of CO_2 for the period 1850-1958 was about 4ppm/decade, while the growth in recent

Table 2-1. Estimates of the Abundance of Trace Chemicals in the Global Atmosphere of Years 1980 and 2030.
Source: Ramanathan, Cicerone, Singh and Hiehl (1985)

Chemical Group	Chemical Formula	Dominant Source*	Dominant Sink*	Estimated Average Residence Time (τj)	Year 1980 Global Average Mixing Ratio (ppb)+	Year 2030 probable global average concentration: Probable Value	Year 2030 probable global average concentration: Probable Value	Remarks (also see text for details)
Carbon Dioxide	CO_2	N.A	O	2	$339x10^3$	$450x10_3$	--	Based on a 2.4% per year increase in anthropogenic CO_2 release rates over the next 50 years
Nitrogen Compounds	N_2O	N.A	S(UV)	120	300	375	350 - 450	Combustion and fertilizer sources
	NH_3	N.A	T	0.01	<1	--	--	Concentration variable and poorly characterized
	$(NO+NO_2)$	N.A	T(OH)	0.001	0.05	0.05	0.05 - 0.1	Concentration variable and poorly characterized
Sulfur Compounds	CSO	N.A	T(O.OH)?	I(?)	0.52	0.52	--	Sources and sinks largely unknown
	CS_2	N.A	T	I(?)	<0.005	<0.005	--	Sources uncharacterized
	SO_2	A(?)	T(OH)	0.001	0.1	0.1	0.1 - 0.2	Given the short lifetime the global presence of SO_2 is unexplained
	H_2S	N	T(OH)	0.001	<0.05	<0.5	--	
Fully Fluorinated Species	CF_4(F14)	A	I	>500	0.07	0.24	0.2 - 0.31	Aluminum industry a major source
	C_2F_6(F116)	A	I	>500	0.004	0.02	0.01 - 0.04	Aluminum industry a major source
	SF_6	A	I	>500	0.001	0.003	0.002- 0.05	
Chlorofluorocarbons	$CClF_3$(F13)	A	S(UV).I	400	0.007	0.06	0.04 - 0.1	All chlorofluorocarbons are of exclusive man-made origin. A number of regulatory actions are pending. The nature of regulatory actions are pending. The nature of regulations and their effectiveness would greatly affect the growth of these chemicals over the next 50 years
	CCl_2F_2(F12)	A	S(UV)	110	0.28	1.8	0.9 - 3.5	
	$CHClF_2$(F22)	A	T(OH)	20	0.06	0.9	0.4 - 1.9	
	CCl_3F(F11)	A	S(UV)	65	0.18	1.1	0.5 - 2.0	
	CF_3CF_2Cl(F115)	A	S(UV)	380	0.005	0.04	0.02 - 0.1	
	$CClF_2CClF_2$(F114)	A	S(UV)	180	0.015	0.14	0.06 - 0.3	
	CCl_2FCClF_2(F113)	A	S(UV)	90	0.025	0.17	0.08 - 0.3	
Chlorocarbons	CH_3Cl	N(O)	T(OH)	1.5	0.6	0.6	0.6 - 0.7	Dominant natural chlorine carrier of oceanic origin
	CH_2Cl_2	A	T(OH)	0.6	0.03	0.2	0.1 - 0.3	A popular reactive but non-toxic solvent
	$CHCl_3$	A	T(OH)	0.6	0.01	0.03	0.02 - 0.1	Used for manufacture of F22; many secondary sources also exist
	CCl_4	A	S(UH)	25-50	0.13	0.3	0.2 - 0.4	Used in manufacture of fluorocarbons; many other applications as well
	CH_2ClCH_2Cl	A	T(OH)	0.4	0.03	0.1	0.06 - 0.3	A major chemical intermediate (global production=10 tg/yr); possibly toxic

Table 2-1. Estimates of the Abundance of Trace Chemicals in the Global Atmosphere of Years 1980 and 2030.
Source: Ramanathan, Cicerone, Singh and Hiehl (1985) (continued)

Chemical Group	Chemical Formula	Dominant Source*	Dominant Sink*	Estimated Average Residence Time (τj)	Year 1980 Global Average Mixing Ratio (ppb)+	Year 2030 probable global average concentration: Probable Value	Year 2030 probable global average concentration: Probable Value	Remarks (also see text for details)
	CH_3CCl_3	A	T(OH)	8.0	0.14	1.5	0.7 - 3.7	Non-toxic, largely uncontrolled degreasing solvent
	C_2HCl_3	A	T(OH)	0.02	0.005	0.01	0.005- 0.02	Possibly toxic, declining markets because of substitution to CH_3CCl_3
	C_2Cl_4	A	T(OH)	0.5	0.3	0.07	0.03 - 0.2	Possibly toxic; moderate growth due to substitution to CH_3CCl_3
Brominated and iodated species	CH_3Br	N	T(OH)	1.7	0.01	0.01	0.01 - 0.02	Major natural bromine carrier
	$CBrF_3$(FI3BI)	A	S(UV)	110	0.001	0.005	0.003- 0.01	Fire extinguisher
	CH_2BrCH_2Br	A	T(OH)	0.4	0.002	0.002	0.001- 0.01	Major gasoline additive for lead scavenging; also a fumigant
	CH_2I	N	T(UV)	0.02	0.002	0.002	--	Exclusively of oceanic origin
Hydrocarbons CO, H_2	CH_4	N	T(OH)	5-10	1650	2340	1850 - 3300	A trend showing increase over the last 2 years has been identified
	C_2H_6	N	T(OH)	0.3	0.8	0.8	0.8 - 1.2	Predominantly of auto exhaust origin
	C_2H_2	A	T(OH)	0.3	0.06	0.1	0.06 - 0.16	No trend has been identified to date
	C_3H_8	N.	T(OH)	0.03	0.05	0.05	0.05 - 0.1	No trend has been identified to date
	CO	N.A	T(OH)	0.3	90	115	90 - 160	No trend has been identified to date
	H_2	N.A	T(SL.OH)	2	560	760	560 - 1140	
Ozone	O_3 (Tropospheric)	N	T(UV, SL,O)	0.1-0.3	F(Z)**	12.5%		A small trend appears to exist but data are insufficient
Aldehydes	HCHO	N	T(OH,UV)	0.001	0.2	0.2	--	Secondary products of hydrocarbon oxidation
	CH_3CHO	N	T(OH,UV)	0.001	0.02	0.02	--	1980 concentration estimated from theory

*N - Natural; A - Anthropogenic; O - Oceanic; S - Stratosphere; UV - Ultraviolet Photolysis; T - Troposphere; OH - Hydroxyl radical removal; I - Ionospheric and extreme UV and electron capture removal; SL - Soil sink

+These concentrations are integrated averages; for chemicals with lifetimes of 10 years or less, significant latitudinal gradients can be expected in the troposphere; for chemicals with extremely short lifetimes (0.001-0.3 years) vertical gradients may also be encountered

**Varies from 25 ppbv at the surface to about 70 ppbv at 9 km. The concentration was increased uniformly by the same percentage from the surface to 9 km.

***These values are not used in the present assessment.

Source: Ramanathan, Cicerone, Singh, and Kiehl (1985)

Table 2-2. Radiative Forcing for a Uniform Increase in Trace Gases

Compound	Radiative Forcing (°C/ppb)
CO_2	.000004
CH_4	.0001
N_2O	.001
CFC-11	.07
CFC-12	.08
CFC-13	.10
Halon 1301	.10
F-116	.08
CC14	.05
CHC13	.04
F-14	.04
HCFC-22	.03
CH2 C12	.02
CH3 CC13	.01
C2 H2	.01
SO2	.01

Source: Adapted from Ramanathan et al. (1985).

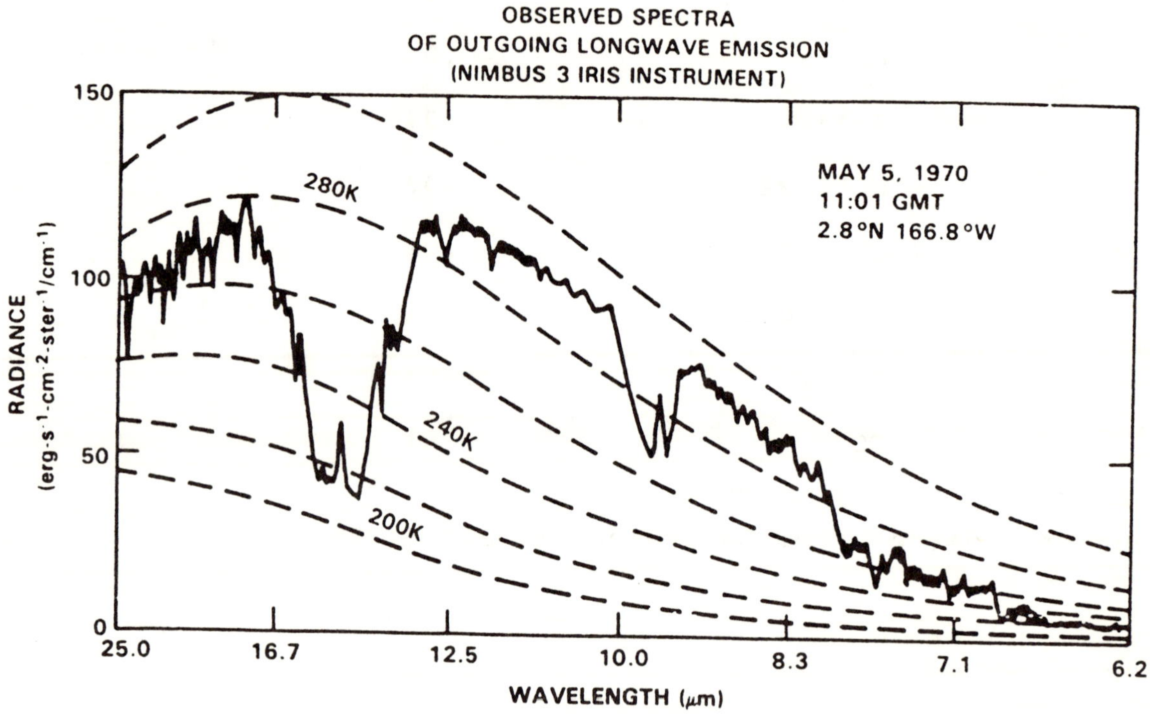

Figure 2-2. Sample spectra from the IRIS instrument on board NIMBUS-3 satellite. The dashed lines indicate the effective radiation temperature for different wave lengths.

Source: Hanel et al. (1972).

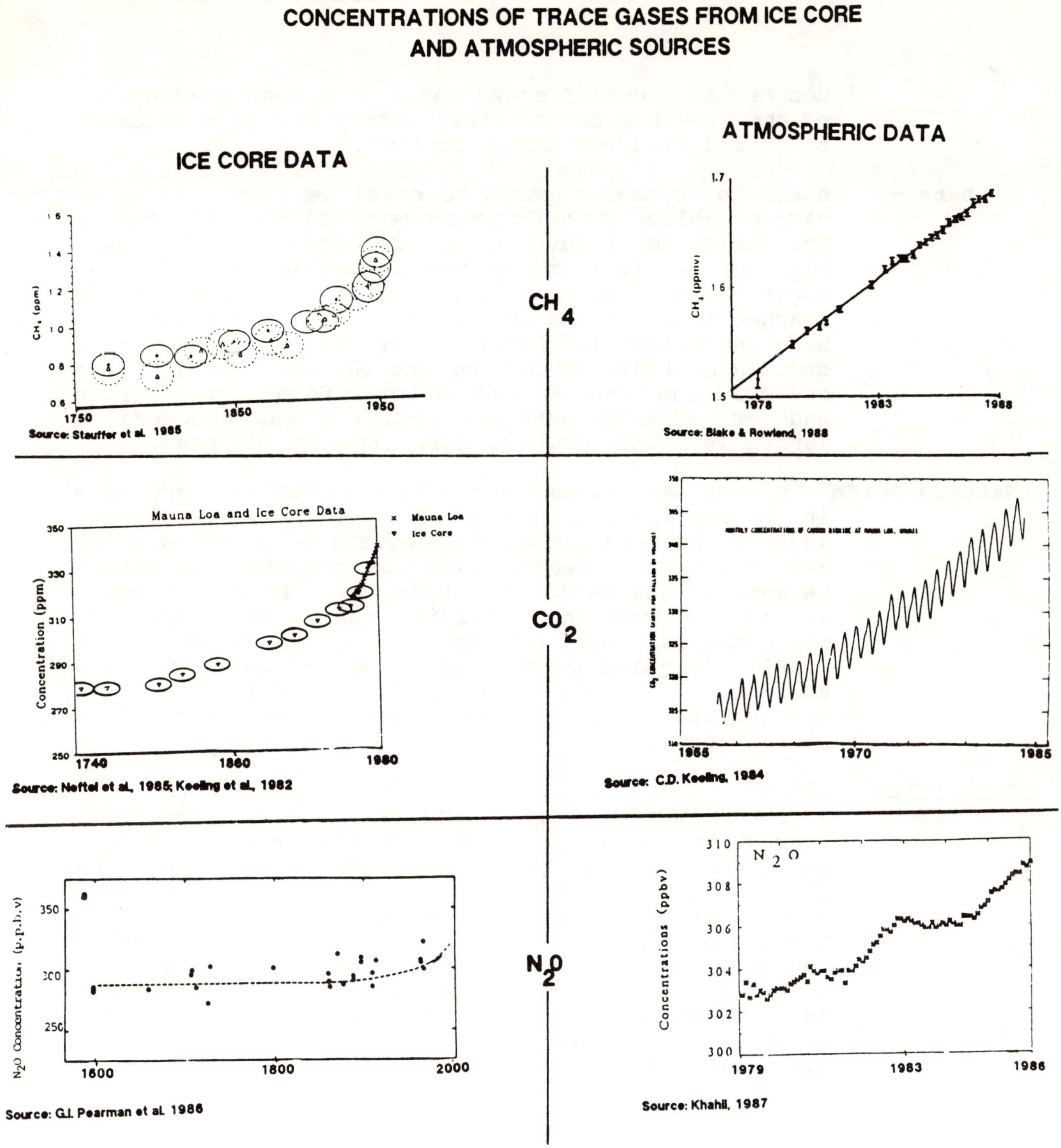

Figure 2-3. Greenhouse gas trends in ice cores and atmospheric instrument data.

decades is about 15 ppm/decade. The near quadrupling of the growth rate is mainly attributed to combustion of fossil fuel and deforestation.

Methane - (CH_4) Atmospheric measurements taken since 1979 have established that the concentration of methane is increasing at a rate of approximately 1% per year, although earlier measurements begun in 1965 had suggested a rate of 1.8% per year. Analyses of air trapped in polar ice show that levels of methane have been increasing for the last several centuries (Craig and Chou, 1982; Rasmusson and Khalil, 1984). The sources of methane include enteric fermentation, rice paddies, biomass burning, coal and gas production, landfills, wetlands, and other natural sources.

Nitrous Oxide - Recent measurement suggests that N_2O is increasing in the atmosphere at a rate of 0.2-0.3% per year (WMO, 1986). The mean global concentration of N_2O is about 300 ppb, with very little geographic variations because of the gas's long lifetime. There currently are no observations providing quantitative data on preindustrial N_2O concentrations, although Weiss (1981) estimated concentration of 281-291 ppb. The sources of N_2O include oceans, fossil fuel and biomass combustion, agricultural fertilizers and land disturbances.

Chlorofluorocarbons - (CFCs) Global measurements of halocarbons (CFC-11($CFCl_3$), CFC-12(CF_2Cl_2), CCl_4 and CH_3CCl_3) since 1978 suggest that each gas is increasing with time. Growth rates range from 5% per year for CFC-11 and CFC-12 to 7% per year for CH_3CCl_3 and 1% per year for CCl_4. Since chlorofluorocarbon production began in the twentieth century, there are no preindustrial measurements for comparison. The less abundant chlorine and fluorine compounds are also increasing in the atmosphere, although less is known about changes in atmospheric concentrations. An extensive discussion of chlorofluorocarbon sources including refrigerants, aerosols, sprays, insulating material and solvents is contained in EPA (1987).

Ozone - An important greenhouse gas/ozone, plays a major role in the chemistry of the stratosphere and the troposphere. Current understanding of atmospheric chemistry suggests that hydrocarbons, carbon monoxide, and methane act to increase the abundance of tropospheric ozone by suppressing ohydroxyl radicals (Thompson, 1986). There are data and theories that suggest that ozone in the free troposphere is increasing with time, perhaps at a rate of 0.5 to 1% per year (Logan, 1982; Angell and Kovshover, 1983). Ramanathan estimated that perhaps ozone increased by about 25% between 1880 and 1980. Extensive analysis with stratospheric models suggest that CFCs act to deplete stratospheric ozone. Hence, there are counterbalancing global trends and important differences between the stratosphere and the troposphere, as well as differences between regions (WMO, 1986). The net effect on climate may be a warming due to the greater greenhouse effect of tropospheric ozone, (Ramanathan and Dickensen, 1979), but important research questions remain to be answered.

Other Gases - Carbon Monoxide (CO_2) Nitrogen Oxides (NO_x)

Other gases play an important role in the balance of the earth's radiation. In particular, carbon monoxide (CO) and nitrogen oxides (NO_x) affect both methane and tropospheric ozone. One study by Rasmussen and Khalil, (1984), indicated CO increases of about 5% per year. Results by Dorgashina et al. (1984) suggest a 1-2% per year increase. Important sources of CO include fossil fuel combustions, oxidation of natural hydrocarbons and CH_4, and burning of savanna, agricultural land, and forests There is no direct evidence to define the history of CO over the past centuries, but a number of factors, such as the north/south gradient, suggest that an increase by a factor of two may have been possible since prehistoric times (WMO, 1986). Measurement of nitrogen species, NO_x and ammonia in the troposphere are few in remote regions. These species have very short lifetimes (0.5 to 5 days). While global levels of NO_x are certainly higher than in preindustrial times, the total impact of man's activities on NO_x remains unclear.

RADIATIVE FORCING AND CLIMATE SENSITIVITY

The biochemical and physical processes that change the fluxes of radiative energy into and out of the surface-atmosphere system are generally referred to as radiative-forcing factors. If the process tends to warm the earth-atmosphere system, the forcing is deemed to be positive. The greenhouse gases discussed in the previous section exert a positive radiative forcing, as do high clouds. If the process tends to cool the earth-atmosphere system, the forcing is deemed to be negative. Examples of negative radiative forcing include stratospheric aerosols from volcanos and low clouds. The potential radiative effect of various climate-forcing factors based on model analysis by Hansen as modified in WMO, 1986, is shown in Figure 2-4.

At present the direct climate forcing from increasing greenhouse gases is not large enough to completely dominate other forcing factors, such as stratospheric aerosols and solar irradiance (the total energy from the sun) (Hansen, 1987a). For example, solar irradiance was observed to decrease by approximately 0.09% between 1979 and 1985. This negative forcing is somewhat larger than the positive forcing from CO_2 increases during the same Figure period. An even larger negative forcing occurred during 1982-84 as a result of stratospheric aerosols generated by the volcano El Chichón. Volcanos can eject sulfuric acid aerosols and other particles high into the stratosphere. Ramanathan (1988) estimated that a major eruption such as El Chichón could contribute cooling that would nearly equal the trace gas warming for the period 1975-85. These climate forcings can mask the effect of man-induced greenhouse gases for periods ranging from a few years to decades. But over a long period, such natural factors as volcanoes should have no lasting effect.

Other climate forcing, such as changes in tropospheric aerosols or surface albedo are also potentially important, particularly on a regional basis. Examples of changing aerosols abundance include arctic haze, desert aerosols, and urban/rural aerosols. Changes in albedo variations may be associated with large-scale changes in land use (deforestation) or the latitudinal extent of sea ice and snow cover as global warming occurs.

ROLE OF ATMOSPHERIC MODELS

Scientists attempt to understand how the earth-atmosphere system works by building mathematical models of different complexity. The simplest approach that allows an understanding of the radiative and chemical processes is a one-dimensional (1-D)

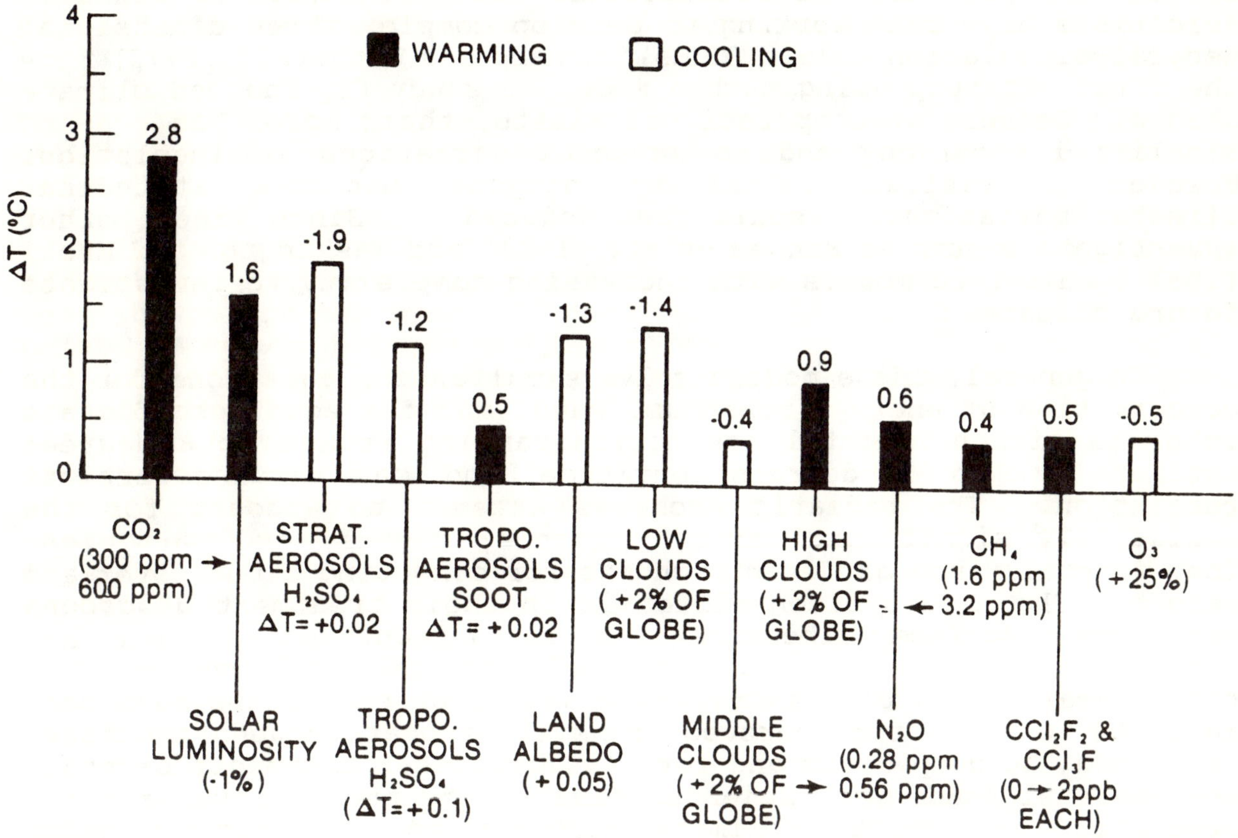

Figure 2-4. One dimensional radiative-convective model estimates of surface temperature effects of various global radiative perturbations. All of the results, except for CH_4, and the figure are adapted from Hansen et al., 1981. The CH_4 result was obtained as follows: The radiative forcing of doubled CH_4 as given in WMO (1982) is multiplied by the climate sensitivity of Hansen et al. (1981) model to obtain T_s.

model. This type of model allows no horizontal variations and, hence, is only useful for obtaining globally averaged results or results over some particular location. Nevertheless, it is an important tool that allows investigators to quickly conduct experiments that serve as a basis for more complex models.

Conducting analyses of the type discussed in this report necessitates having climate data sets for temperature, winds, precipitation, and other variables for different regions of the country and globe. The development of such data requires that the chemical and dynamic processes of the entire earth-atmosphere system be represented as realistically as practical. Atmospheric scientists have been working to develop complex three-dimensional general-circulation models (GCMs). Manabe and Wetherald (1975) made the first attempt, using such a model to study CO_2-induced climate change. Because of computational limits, their model had simplified geography and no seasonal variations in insolation. However, it yielded preliminary insights into how latitudinal climate variations could be deduced. Since then other investigators such as Hansen et al. (1984) and Washington and Meehl (1983), have used models with increasing complexity to investigate future climate.

In general, these models solve simultaneous equations for the conservation of energy, momentum, mass, and the equation of state on grids with horizontal resolution ranging from 4 to 8 degrees latitude by 5 to 10 degrees longitude, and with varying vertical resolution. The radiation schemes attempt to account for the radiatively significant gases, aerosols, clouds, and particles. They generally use different schemes for computing cloud cover and height. The models often differ in in their treatment of ground hydrology, surface albedo (which is a function of ice and terrestrial vegetation), and diurnal and seasonal cycles. In recent years, more complex representations of the oceans have been included to allow for improved transfer of heat between the mixed layer and the deep ocean and for the horizontal transport of heat. The National Academy of Sciences (1987) attempted to evaluate the probability of certain climate features as a result of model analyses. The results shown in Table 2-3 suggest that such features as increased polar warming and increased global precipitation are probable in a doubled CO_2 world. However,

Table 2-3. Summary of Scientific Confidence

Large Stratospheric Cooling (virtually certain). Reduced ozone concentrations in the upper stratosphere will lead to reduced absorption of solar ultraviolet radiation and therefore less heating. Increases in the stratospheric concentration of carbon dioxide and other radiatively active trace gases will increase the radiation of heat from the stratosphere. The combination of decreased heating and increased cooling will lead to a major lowering of temperatures in the upper stratosphere.

Global-Mean Surface Warming (very probable). For an equivalent doubling of atmospheric carbon dioxide, the long-term global-mean surface warming is expected to be in the range of 1.5 to 4.5°C.

Global-Mean Precipitation Increase (very probable). Increased heating of the surface will lead to increased evaporation and, therefore, to greater global mean precipitation. Some individual regions might well experience decreases in rainfall.

Reduction of Sea Ice (very probable). As the climate warms, total sea ice is expected to be reduced.

Polar Winter Surface Warming (very probable). As the sea ice boundary is shifted poleward enhanced surface warming in winter polar regions is likley. Warming of the polar surface air may be as much as 3 times the global mean warming.

Summer Continental Dryness/Warming (likely in the long term). Several studies have predicted a marked long-term drying of the soil moisture over some mid-latitude interior continental regions during summer. This dryness is mainly caused by an earlier termination of snowmelt and rainy periods, and an earlier onset of the spring-to-summer reduction of soil wetness.

High-Latitude Precipitation Increase (probable). As the climate warms, the increased poleward penetration of warm, moist air should increase the average annual precipitation in high latitudes.

Rise in Global Mean Sea Level (probable). A rise in mean sea level is generally expected due to thermal expansion of sea water in the warmer future climate. Far less certain is the contribution due to melting or calving of land ice.

Source: Modified from NAS (1987) (The original text should be consulted to understand the context completely.)

improved understanding of ocean circulation processes and the coupled ocean-atmospheric system is necessary before the climate system can be completely understood.

For comparative purposes, the global surface air temperature changes induced by seven GCMs for a doubled CO_2 concentration are presented in Table 2-4. The first four models show changes ranging from 1.3 to 3.0°C with the lower values generally associated with simple ocean models with zero heat capacity and no horizontal or vertical heat transport. The studies with higher values were conducted with fixed-depth oceans that have heat capacity and prescribed horizontal heat transports. This range of results is reflected in the National Academy of Sciences' 1987 report, which reaffirmed that a 1.5 to 4.5°C increase in global temperatures is likely under doubled CO_2 conditions. A more recent review by Dickenson (1986); recommended a 1.5 to 5.5°C increase, and a paper by Lashof (1988) suggests that when all biogeochemical feedbacks are considered a temperature sensitivity of as much as 8 to 10°C may be possible.

For comparative purposes, the global surface air temperature changes induced by seven GCMs for a doubled CO_2 concentration are presented in Table 2-4. The first four models show changes ranging from 1.3 to 3.0°C with the lower values generally associated with simple ocean models with zero heat capacity and no horizontal or vertical heat transport. The studies with higher values were conducted with fixed-depth oceans that have heat capacity and prescribed horizontal heat transports. This range of results is reflected in the National Academy of Sciences' 1987 report, which reaffirmed that a 1.5 to 4.5°C increase in global temperatures is likely under doubled CO_2 conditions. A more recent review by Dickenson (1986); rcommended a 1.5 to 5.5°C increase, and a paper by Lashof (1988) suggests that when all biogeochemical feedbacks are considered a temperature sensitivity of as much as 8 to 10°C may be possible.

Uncertainties in General Circulation Models

The introduction of greenhouse gases into the atmosphere causes the earth's system to seek a new equilibrium in several ways. Often called climate feedbacks, these responses include changes in water vapor, ice cover albedo, clouds, and the oceans. Among these, the water vapor feedback is the most important, since it amplifies surface warming by about 50 to 80 percent. In general, the increase in trapped infrared radiation warms the troposphere and the oceans, causing moisture to evaporate, which further enhances greenhouse warming. Another important effect is the ice-snow albedo feedback. This process focuses on the melting

Table 2-4. Surface Air Temperature Change Induced by a Doubled CO_2 Concentration as Simulated by Selected General Circulation Models

Study	T_s (K)
Manabe & Wetherald (1975)	2.9
Manabe & Wetherald (1980)	3.0
Schlesinger (1983)	2.0
Washington & Meehl (1983)	1.3
Washington & Meehl (1984)	3.5
Hansen et al. (1984)	4.2[2]
Manabe & Wetherald	4.0

Source: World Meteorological Organization (1986).

of ice and snow and the changes in absorbed solar radiation, which amplify initial warming. This feedback amplifies global warming by only 10 to 20 percent, but near the sea-ice margins, the warming can be two to five times larger. Therefore, this process is an important contributor to the larger temperature increases predicted by general circulation models for polar regions.

Cloud Feedbacks

The increased water vapor from the oceans should also alter cloud distributions and height. The exact nature of future cloud changes -- i.e., whether more clouds will form and at what altitude -- remains unclear. Somerville and Remer (1984) have proposed a process by which clouds could reduce warming by increasing their optical thickness. Recent observations by the International Satellite Cloud Climatology project, which seeks to collect global data on clouds, suggest that warmer tropical regions typically have more cirrus clouds that trap greater amounts of infrared radiation.

After water vapor, clouds play the next most important role in the energy balance of the earth-atmosphere system. They could either enhance or partly negate greenhouse warming, but are unlikely to completely reverse the effect of greenhouse gases. While the temperature sensitivity of current GCMs increases with the introduction of interactive clouds, we have much to learn about cloud processes before we can make reliable conclusions about the nature of cloud feedback processes.

Ocean Processes - The Timing of Global Warming

The oceans play an important role in the climate change process in several ways. First, they absorb CO_2, with most estimates suggesting that approximately 50 percent of past CO_2 emissions are sequestered in the ocean. And second, they absorb heat into their deeper layers, thereby delaying global warming. A more thorough discussion of the CO_2 and the oceans can be found in EPA (1989).

A key policy question asked by decision makers is, How soon will global warming take place? To estimate the rates of future warming, Hansen (1984,) used a one-dimensional radiative convective model coupled to an ocean box-diffusion model. The scenarios used for trace gas growth are generally the same as those discussed in other chapters of this report. The results shown in Figure 2-5 suggest approximately a 30-to 40-year delay in the full magnitude of global warming. Although response times as rapid as 10 years may be possible for climate sensitivities as low as 2°C, and as slow as a 100 years for a sensitivity of 5°C. Hansen describes the

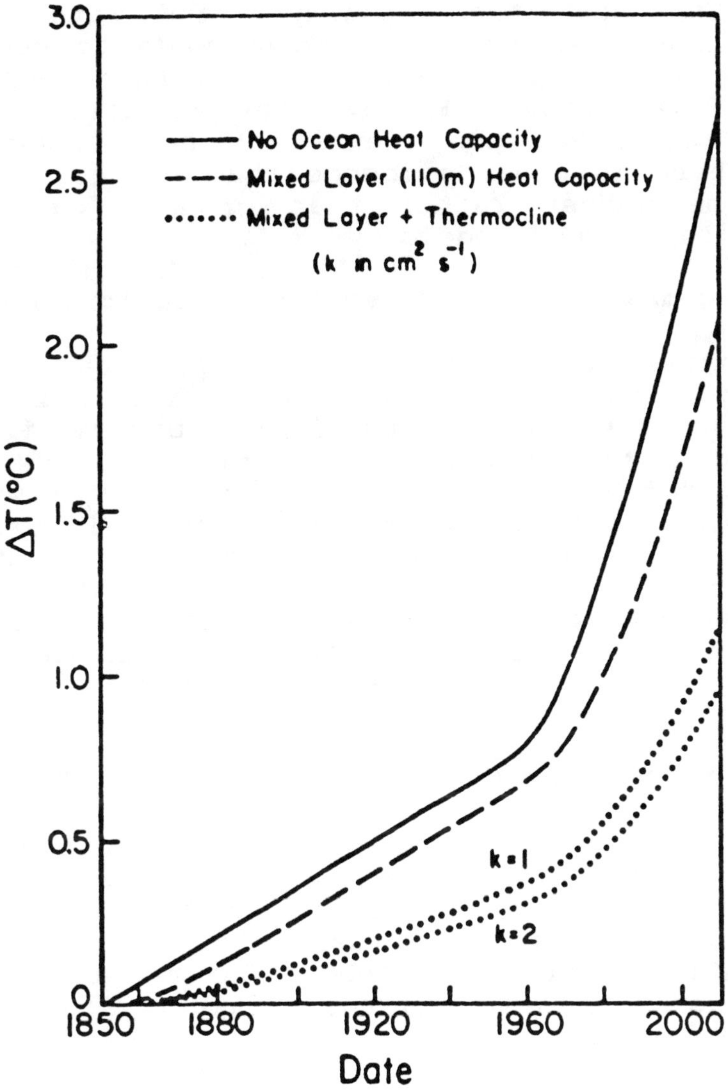

Figure 2-5. Transient estimates of global warming.

Source: Hansen (1984).

growing gap between realized warming and equilibrium warming, and suggests that the warming already built into the system, but not yet evident, may need to be considered in setting national policies.

Models such as this can neither portray complex connections and diverse processes that govern the exchange of heat in the oceans, nor represent the topography and water densities in different regions. Attempts are being made to model the ocean differently or more completely (e.g., Hoffert and Flannery, 1985; Sarmiento and Bryan, 1982). But even these models will not capture the potential surprises the oceans may hold. Broecker (1987) argues that there is clear evidence that changes in the earth's climate can be sudden rather than gradual, as suggested by experiments with current models.

Regional Intercomparisons of General Circulations Models

The Department of Energy recently sponsored a study of the results produced by four different general circulation models (DOE, 1988). The study examined results of both surface air temperatures and precipitation for both current climate and doubled CO_2. The study points out that the models often agree well when comparing seasonal or annual averages over large areas (Figures 2-6 and 2-7), but do not agree as well when the spatial extent is reduced. (See Table 2-5) Collectively, the results point out important differences between models, and suggest caution when using model results for estimating climate impacts. (See Chapter 3 for further discussion of the use of models and scenarios.)

EVIDENCE of CLIMATE CHANGE

A key question often asked is whether there is evidence to suggest that global climate is already changing. A comprehensive response to this question requires the examination of several climate variables (e.g., temperature, wind, and precipitation) at different locations around the globe. The most comprehensive data set used to measure the state of the climate system is global mean surface air temperatures.

Several investigators used meterological data to examine temperature change during the last century (Bradley et al., 1985). The results by Hansen and Lebedeff (1987) indicate a global warming of about 0.5 to 0.7°C in the past century, with warming of similar magnitude in both hemispheres (Figure 2-8). Hansen and Lebedeff also found that eliminating urban stations to account for possible heat island effects reduced these values by about 0.1°C. Jones et

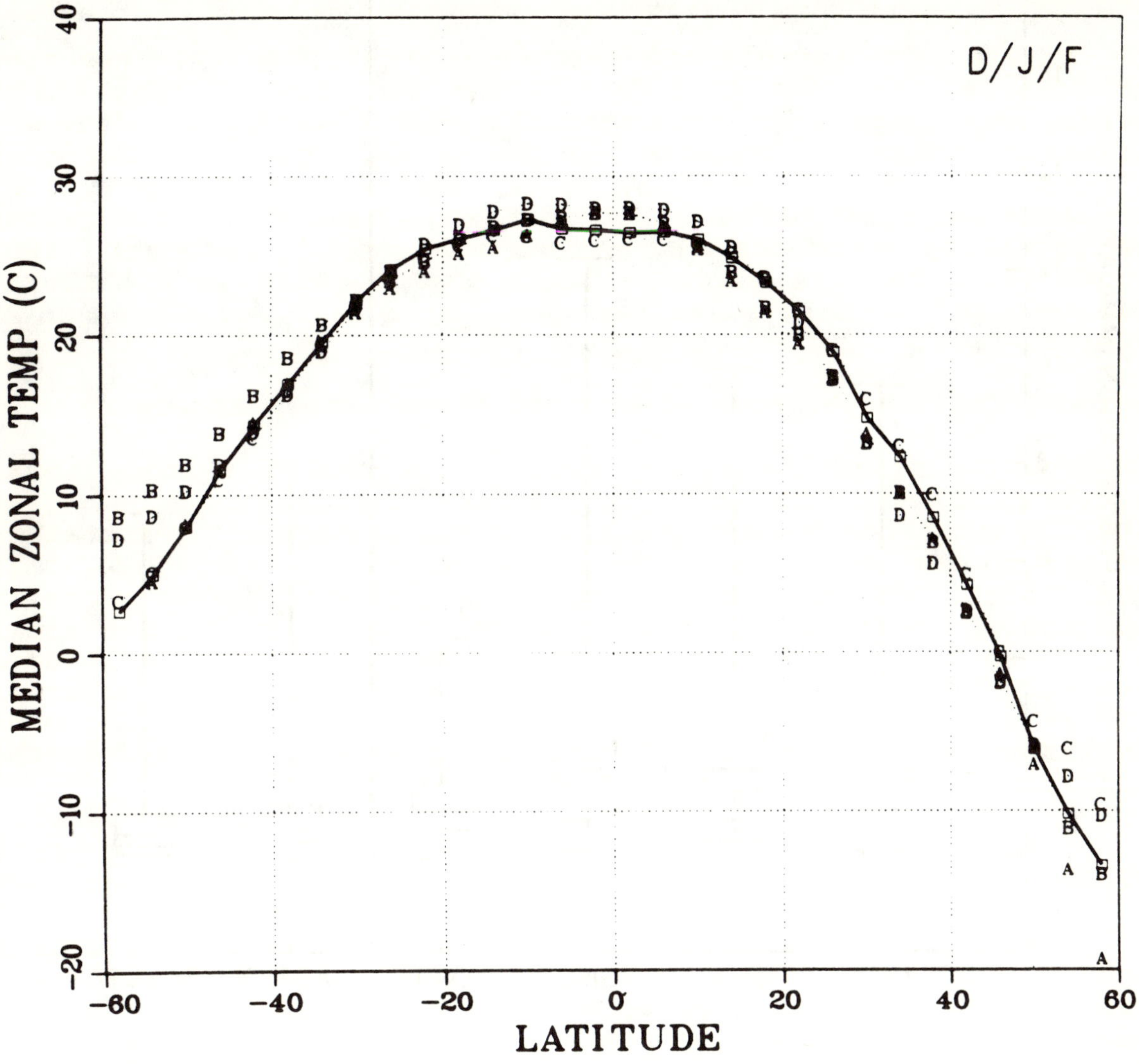

Figure 2-6. Zonal average December, January and February control temperatures for four GCMs and historical data (solid line) A=CCM, B=GFDL, C=GISS, D=OSU.

Source: DOE Report, "Regional Intercomparisons of General Circulation Model Predictions and Historical Climate Data." April, 1988. S. Grotch.

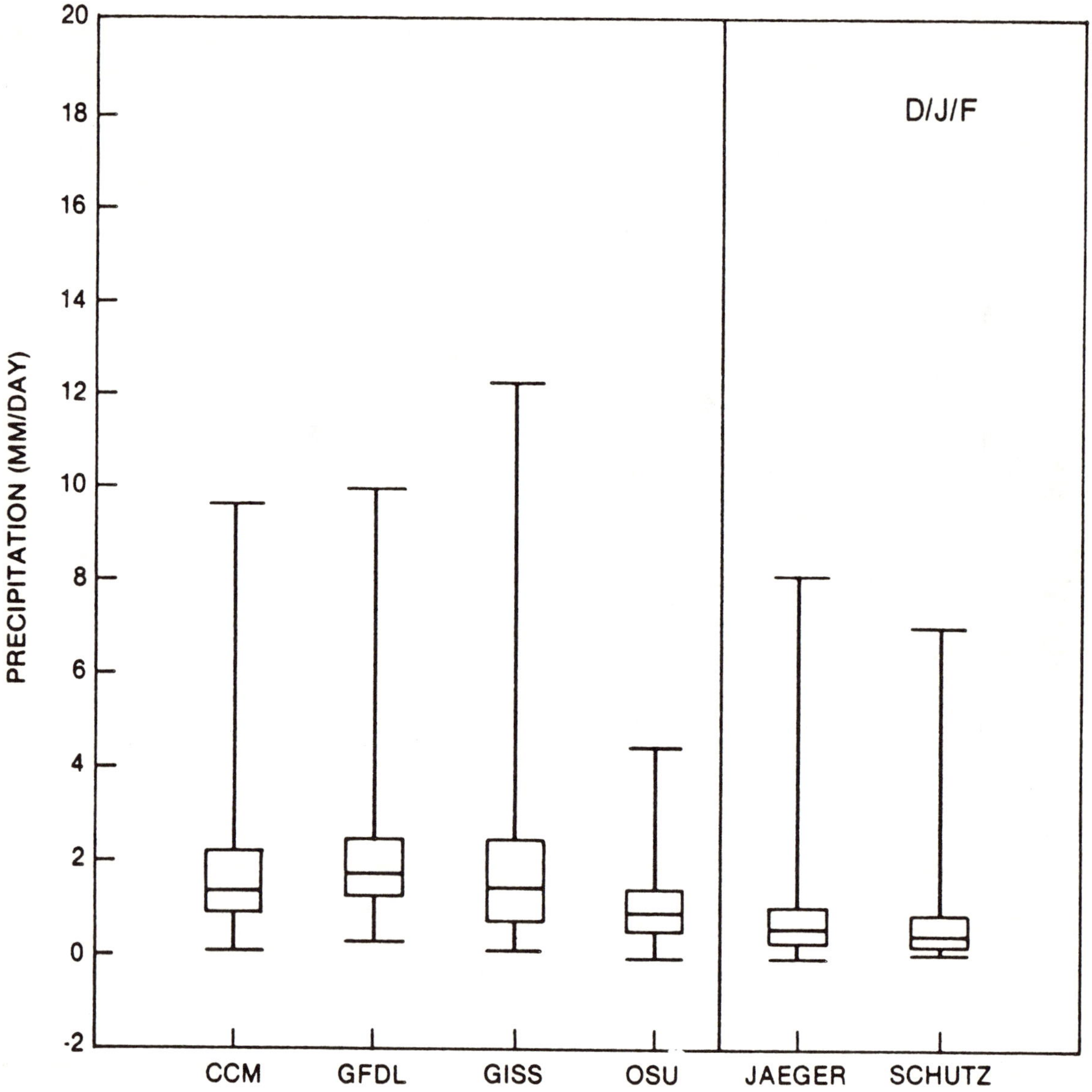

Figure 2-7. Seasonal control precipitation (mm/day) over NH land areas as predicted by the GCMs and the Jaeger and Schutz-Gates Historical Gridded Data Sets for December, January and February.

Source: DOE Report, "Regional Intercomparisons of General Circulation Model Predictions and Historical Climate Data." April, 1988. S. Grotch.

Table 2-5. Seasonal Comparison of Observed and Modeled Control Temperature for Different Geographical Regions

Variable and Model	Domain of Comparisons Global	North America	Contiguous U.S.	Midwestern U.S.
	December-January-February			
Observed Median Temperature (°C)	8.5	-5.8	0.9	-1.5
Difference in Median Temperatures (GCM - Observation)				
CCM	1.6	0.3	-2.1	-0.5
GFDL	1.5	1.8	-0.8	-1.3
GISS	0.8	-0.5	0.0	1.1
OSU	0.3	0.5	-0.6	-1.0
	June-July-August			
Observed Median Temperature (°C)	13.9	18.9	23.0	23.0
Difference in Median Temperatures (GCM - Observation)				
CCM	1.3	6.0	6.3	6.8
GFDL	-0.2	0.6	0.1	3.7
GISS	0.4	-3.1	-4.5	-4.8
OSU	-0.6	-2.2	-2.2	-1.6

Source: DOE Report, "Regional Intercomparisons of General Circulation Model Predictions and Historical Climate Data" April, 1988. S. Grotch.

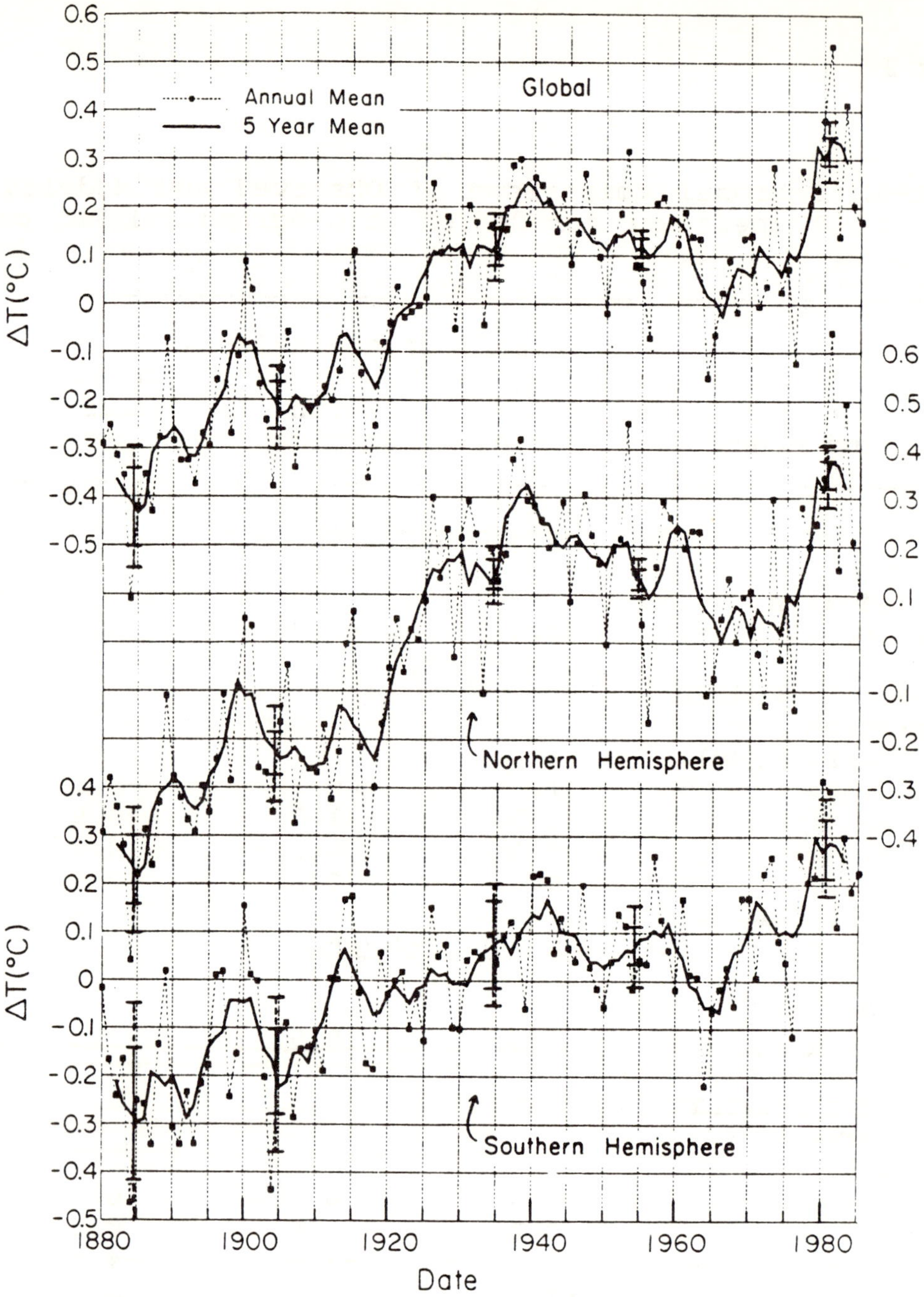

Figure 2-8. Global and hemispheric surface air temperature change estimated from meteorological station records. The northern hemisphere scale is on the right. The 5-year running mean is the linear average for the 5 years centered on the plotted year. The uncertainty bars (95% confidence limits) are based on the error analysis in Section 5; the inner bars refer to the 5-year mean and the outer bars to the annual mean.

Source: J. Hansen, and S. Lebedeff. "Global Trends of Measured Surface Air Temperature." J. Geophys. Res. vol. 92, November 20, 1987:13,351.

al. (1982) found similar trends for sea surface temperatures beginning in 1861 and noted that the warmest three years have occurred in the 1980s. (The very warm year of 1987 subsequently suggests that the warmest four years have occurred in the 1980s). Jones et al.'s results show little trend in the nineteenth century, marked warming to 1940, relatively steady conditions to the mid-1970s, and a subsequent rapid warming up to the present.

Temperature changes in the upper atmosphere were reported by Angell (1983). More recently, NASA (1988) examined satellite, radiosonde, and rocketsonde data on temperatures in the stratosphere at various altitudes. Radiosonde analysis from 1970 to the present show no significant trends in temperature below 30 mb, except in equatorial and Antarctic regions. During 1979-1986, while accounting for the El Chichón volcano effects, global temperatures show a decline between 1.75°C at 0.5 mb (45-55km) to 1°C below 5 mb. These actual temperature trends are less than current model predictions, but are in the same direction predicted by theory.

Lachenbruch and Marshall (1986) examined temperature changes in the arctic region by conducting temperature measurements in permafrost in northern Alaska. The advantage of permafrost records is that they represent a systematic record of recent history and a relatively smooth signal of temperature change. Their results indicate a variable but widespread warming of the permafrost in the range of 2 to 4 °C during the last few decades to a century. While more extensive measurements are needed to assess changes in this region, the results appear to be generally consistent with model predictions that suggest more rapid warming in polar regions.

Finally, there is remarkable evidence, based on ice core records that document the long-term changes in temperatures and their relationship to atmospheric CO_2 concentrations. The Vostok ice core records by Barnola et al. (1987) provide direct evidence of these trends over the last 160,000 years. The record in Figure 2-9 shows two very large increases in temperature and CO_2. One occurred approximately 15,000 years ago, corresponding to a jump from approximately 220 ppm CO_2, and another occurred nearly 140 thousand years ago. The high CO_2 level of approximately 280 ppm, corresponds to the preindustrial CO_2 concentrations that prevailed until 100-300 years ago. The low CO_2 concentrations generally correspond to the glacial conditions that prevailed for most of the last 100,000 years. While it is difficult to assign a cause for these past changes, it is reasonable to conclude that, given current CO_2 concentrations of 345 ppm, global temperatures will soon equal or exceed the maximum temperatures of more than 100,000 years ago.

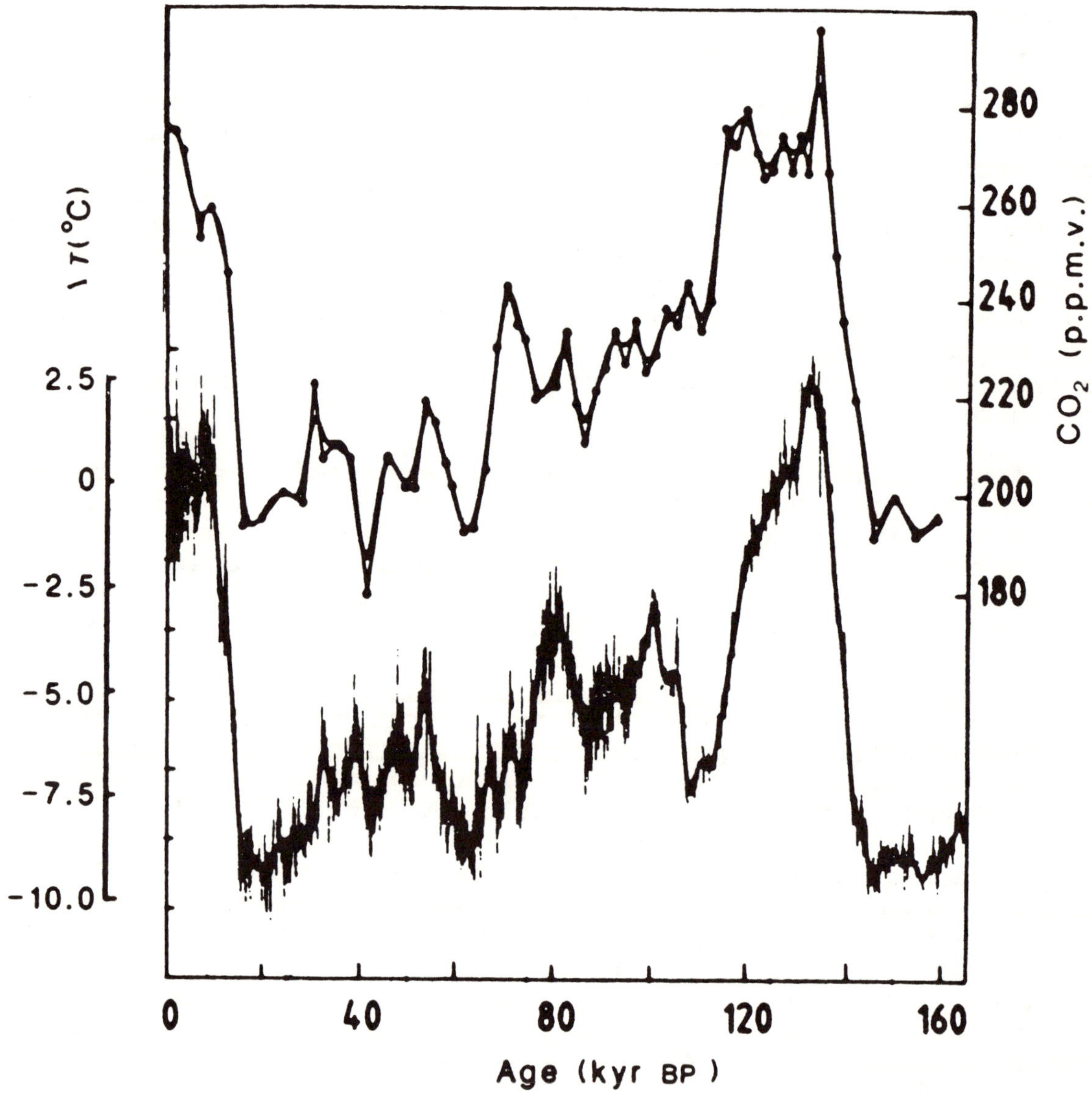

Figure 2-9. Carbon dioxide levels and temperatures over the last 160,000 years. From Vostok 5 Ice Core.

Source: Barnola et al. (1987).

REFERENCES

Angell, J.K. and Korshover, J. 1983. Global Variations in total ozone and layer - mean ozone: An update through 1981, Journal of Climate and Applied Meteorology, 22, 1611-1627.

Angell, J.K. and Korshover. 1983. Global temperature variations in the troposphere and stratosphere 1985-1982, Monthly Weather Review 111, 901-921.

Barnola, J.M., Raymond, D., Korotkevich, Y.S., Lorius, C. 1987. Vostok ice core provides 160,000 year record of atmospheric CO_2 Nature 32, 408-414.

Blake, Donald R. and F. Sherwood Rowland. 1988. Continuing Worldwide Increase in Tropospheric Methane, 1978 to 1987. Science 239:1129-1131.

Bradley, R.S., Kelly, P.M., Jones, P.D., Goodess, C.M., and Diaz A.F. 1988. A Climatic data bank for northern hemisphere land areas 1985-1980. DOE Tech. Report 117. (DOE/EV/10739-2). Dept. of Energy, Washington, D.C.

Broecker, W. 1987. "Unpleasant surprises in the Greenhouse?" Nature 328, 123-126.

Craig H. and Chou C.C. 1982. Methane: The Record in Polar Ice Cores, Geophysical Research Letters, 9, 1221-1224.

Department of Energy. 1988. Regional Intercomparisons of General Circulation Model Predictions and Historical Climate Data. DOE/NBB-0084.

Dickenson, R.E. 1986. The climate system and modeling of future climate in The Greehouse Effect, Climate Change and Ecosystems, eds. B. Bolin, B.A. Doos, J. Jager and R.A. Warrick, Wiley Press, New York, pp. 206-270.

Dvoryashina, Y.V., Dianov, V.I. and Yurganov, L.N. 1984. On the variations of atmospheric total column carbon monoxide abundance for 1970-1982, Physical Atmosphere Oceans, 20,40-47.

Environmental Protection Agency. 1987. Assessing the Risks of Greenhouse gases that Can Modify the Stratosphere. EPA 400-1-87/001C Vol. I-VI. Washington, D.C.

Hansen, J., Fung, I., Lacis, A., Lebedeff, S., Rind, D., Ruedy, R., Russell, G., and Stone, P. 1987. Prediction of Near Term Climate Evolution: What Can We Tell Decision Makers Now? First North American Conference on Preparing for Climate Change, Washington, D.C.

Hansen, J., and Lebedeff, S. 1987. Global Trends of Measured Surface Air Temperature, Journal of Geophysical Research, 92, 13,345-372.

Hansen, J. Lasis, A., Rind, D., Russell, G., Stone, P., Fung, I., Ruedy, R., and Lerner, J. 1984. Climate Sensitivity: Analyses of Feedback mechanisms in Climate Processes and Climate Sensitivity, Maurice Ewing Series, 5. eds. J. Hansen and T.Takahasi. American Geophysical Union, p. 368, Washington, D.C.

Hansen, J. Russell, G., Rind, D., Stone, P., Lacis, A., Lebedeff, S., Ruedy, R., and Travis, L. 1983. Efficient three-dimensional global models for climate studies: Model I and II Monthly Weather Review 111, 609-662.

Hoffert, M.I., and Flannery, B.B. 1985. Model Projections of the Equilibrium Climatic Response to Increased Carbon Dioxide, in MacCracken, M.C., and F.M. Luther Projecting the Climatic Effects of Increasing Carbon Dioxide. DOE/ER-0237, U.S. Dept. of Energy, Washington, D.C.

Jones, P.D., Wigley, T.M. and Wright, P.B. 1986. "Global Temperature Variations between 1861 and 1984." Nature 322, 430-434.

Jones, P.D., Wigley, T., and Kelly, P.M. 1982. "Variations in Surface Air Temperatures: 1 Northern Hemisphere, 1881-1980, Monthly Weather Review 110, 59-70.

Keeling, C., R. Bacastow, and T. Whorf. 1982. Measurement of the concentration of carbon dioxide at Mauna Loa Observatory, Hawaii. In W. Clark, ed., Carbon Dioxide Review: 1982. pp. 377-384.

Keeling, Charles D. and Alane F. Carter. 1984. Seasonal, Latitudinal, and Secular Variations in the Abundance and Isotopic Ratios of Atmospheric CO2. Results From Oceanographic Cruises in the Tropical Pacific Ocean. Journal of Geophysical Research 9:4615-4628.

Khalil, M., and R. Rasmussen. 1983. Increase and seasonal cycles in the atmospheric concentration of nitrous oxide(N2O). Tellus 35B:161-169.

Lachenbruch, A.H. and Marshall B.V. 1986. Changing Climate: Geothermal Evidence from Permafrost in the Alaska Arctic Science 234, 689-696.

Lashof, D.A. 1988. The Dynamic Greenhouse: Feedback Processes that May Influence Future Concentrations of Atmospheric Trace Gases, Submitted to Journal of Climate Change.

Logan, J.A. 1982. Paper presented at Second Sumposium on the Composition of the Nonurban Troposphere, Williamburg, Va.

Manabe, S., and Wetherald, R.T. 1975. The effects of doubling CO_2 - concentrations on the climate of a general circulation model. Journal of Atmospheric Sciences 32, 3-15.

National Academy of Sciences (NAS). 1979. Carbon Dioxide and Climate: A Second Assessment, National Academy Press, Washington, D.C.

National Academy of Sciences (NAS). 1979. Carbon Dioxide and Climate: A Scientific Assessment, National Academy Press, Washington, D.C.

National Academy of Science (NAS). 1987. Current Issues in Atmospheric change National Academy Press, Washington, D.C.

National Aeronatics and Space Administration. 1988. Ozone Trends Panel Report.

Neftel, A., E. Moor, H. Oeschger, B. Stauffer. 1985. Evidence from polar ice cores for the increase in atmospheric CO2 in the past two centuries. Nature 315: 45-47.

Pearman, G., D. Etherridge, F. de Silva, and P. Fraser. 1986. Evidence of Changing concentrations of Atmospheric CO2, N2O, and CH4 from Air Bubbles in Antartic Ice. Nature, vol.320:248-250.

Ramanathan, V. 1988. The Greenhouse Theory of Climate Change: A Test by an Inadvertent Global Experiment. Science 240, 293-299.

Ramanathan, V. and Dickinson, R.E. 1979. The role of stratospheric ozone in the zonal and seasonal radiative energy balance of the earth-troposphere system. Journal of Atmospheric Sciences, 36, 1084-1104.

Ramanathan, V., R.J. Cicerone, H.B. Singh, and J.T. Kiehl. 1985. "Trace Gas Trends and Their Potential Role in Climate Change." Journal of Geophysical Research, 90(D3): 5557-5566.

Rasmussen, R., Khalil M.A.K. 1984. Atmospheric Methane in Recent and Ancient Atmospheres: Concentrations, Trends and Interhemispheric Gradients, Journal of Geophysical Research, 89 11599-11605.

Sarmiento, J., and Bryan, K. 1982. "An Ocean Transport Model for the North Atlantic." Journal of Geophysical Research 87, 9694-9698.

Somerville, R.C.J. and L.A. Remer. 1984. "Cloud Optical Thickness Feedbacks in the CO_2 Climate Problem." Journal of Geophysical Research 89, 9668-9672.

Stauffer, B., G. Fischer, A. Neftel, and H. Oeschger. 1985. Increase in Atmospheric Methane Recorded in an Antarctic Ice Core. Science 229:1386-1388.

Stone, P. 1985. Climate Response Times: Dependence on Climate Sensitivity and Ocean Mixing Science 229, 857-859.

Thompson, A.M., Cicerone, R.J. 1986. Possible Perturbations to Atmospheric CO, CH_4, OH. Journal of Geophysical Research, 91, 10,853-10,864.

Tyndall, J. 1961. "On the Absorption and Radiation of Heat by Gases and Vapours, and on the Physical Coxection of Radiation, Absorption and Conduction, "Philosophical Magazine and Journal of Science 22(146): 169-194.

Wang, W.C., Wuebbles, D.J., Washington, W.M., Isaacs, R.G., Molnar, G. 1986. Trace Gas Trends and Other Potential Perturbations to Global Climate. Review of Geophysics 24(1): 110-140.

Washington, W.M., and Meihl, G.A. 1983. General circulation model experiments on the climate efforts due to a doubling and a quadrupling of CO_2 concentrations, Journal of Geophysical Research 88, 6600-6610.

Weiss, R.F. 1981. The Temporal and Spatial Distribution of Tropospheric Nitrous Oxide, Journal of Geophysical Research, 86, 7185-7195.

World Meterological Organization (WMO). 1986. Atmospheric Ozone 1985. Assessment of Our Understanding of the Processes Controlling Its Present Distribution and Change, Global Ozone Research and Monitoring Project--Report No. 16, WMO, Geneva, Switzerland.

CHAPTER 3

METHODOLOGY

The basic components in analyzing the potential effects of global warming include assumptions of how climate will change and approaches to estimate effects that will occur as a result of those changes.

SCENARIOS OF CLIMATE CHANGE

Need for Scenarios

At this time, scientists are unable to reliably predict how the greenhouse effect may change regional climate. Unknown factors include the amount that temperature will rise in certain regions, whether rainfall will increase or decrease, how winds may shift, or whether storm frequency may change.

Scenarios can help illustrate events that may occur in the future. Scenarios may be used to identify possible effects of climate change and to evaluate responses to those effects. To incorporate uncertainties surrounding regional climate change, regional scenarios should include a variety of potential climatic changes consistent with the state of knowledge regarding global warming. By analyzing many scenarios, scientists may be able to determine the direction and relative magnitude of change. It is important to realize that scenarios are not predictions, and their probabilities are not known.

By using scenarios as inputs into models of ecological and manmade systems, the range of possible climate-induced effects within the systems may be determined. For example, climate scenarios can be used in a hydrologic model of a river basin to examine how the river's flow may change. These models of the relationship between climate and particular systems will be referred to as "effects models."

Scenario Components

To assess the potential effects of global climate change, regional scenarios of such change should have the following characteristics:

1. The scenarios should be consistent with global warming caused by anthropogenic increase in greenhouse gas emissions. A doubling of CO_2 concentrations in the

atmosphere is thought to possibly increase global temperatures by approximately 1.5 to 4.5°C (3 to 8°F). The regional temperature change may be higher or lower as long as it is internally consistent with the global range.

2. The scenarios must include a sufficient number of meteorologic variables to meet the requirements for running effects models. Some models of the relationship between climate and a system only need temperature and precipitation, while others also need solar radiation, humidity, winds, and other variables.

3. The meteorologic variables should be internally consistent. While a scenario is not a prediction, it should at least be feasible. The laws of physics limit how meteorologic variables may change in relationship to each other. For example, if global temperatures increase, global precipitation must also rise.

4. The scenarios should provide meteorologic variables on a daily basis. Many of the effects models used in this study need daily meteorologic inputs.

5. Finally, the scenarios should illustrate what climate would look like on a spatial scale fine enough for effects analysis. Many effects models consider changes in individual stands of trees or farm fields. To run them, scenarios must illustrate how climate may change locally.

Types of Scenarios

Two basic questions must be considered in analyzing the potential impacts of the greenhouse effect: What would be the effects of a large climate change in the future? How quickly will the effects become apparent over time? The first question asks what the world will be like in the future; the second is about the speed of change.

One way of examining the first question is to use scenarios of an equilibrium future climate. Equilibrium is defined as climate that has reached a steady state. The equilibrium climate should be significantly different from today's climate.

A drawback of an equilibrium scenario is that it occurs at an arbitrary point in the future and assumes that the climate has reached a stable level corresponding with the higher concentrations of greenhouse gases. It does not indicate how climate may change between now and the equilibrium condition. Thus, equilibrium scenarios are not useful for determining the sensitivity of systems to small climate change (for example, a 1°C warming) or for indicating how soon the effects of global warming may be seen.

To help identify sensitivities and give a sense of when effects may occur, this study uses transient scenarios of climate change. A transient scenario is a scenario of how climate may change over time. For this study, it would be helpful to have a transient scenario starting in current conditions and ending when the equilibrium climate is reached. The options for creating scenarios and the scenarios chosen for this study are discussed in this chapter.

A number of options exist for creating regional scenarios of global warming. These options include the following:

1. arbitrary changes in climate,

2. analogue warming, and

3. use of General Circulation Models.

<u>Arbitrary Changes</u>

A simple way of constructing a scenario is to assume that climate variables change by some arbitrary amount. For example, one could assume that temperature increases by 2 or 4°C, or that rainfall rises or falls by 10%. Such scenarios are relatively easy to use and can help to identify the sensitivities of systems to changes in different variables. To determine how sensitive a system is to temperature alone, one could hold other variables at current climate levels and change temperature by arbitrary amounts.

A major drawback to using scenarios with arbitrary changes is that they may not be realistic, since temperature, evaporation, precipitation, wind, and other variables will most likely change. A combination of unrealistic meteorological changes may yield an unrealistic effect. Scenarios with arbitrary changes may be useful for determining sensitivities to particular variables, but not for determining the possible magnitudes of effects.

Analogue Warming

Many climatologists have advocated the use of historic warming periods as an analogue of how a future warming may affect regional climates (Vinnikov 1987). The instrumental weather record can be used by comparing a cool decade on record, such as the 1880s, with a warm decade, such as the 1930s (Wigley, 1987).

Paleoclimatic data may also be incorporated into an analogue warming scenario. Paleoclimatologists have determined how rainfall and temperature patterns on a broad regional scale differed in the past. The changes associated with past climates that were warmer than now may be used as an analogue warming scenario. For example, 6,000 years ago, the temperatures were about 1°C warmer.

The advantage of using an analogue is that it gives a realistic sense of how regional and local weather patterns change as global climate warms. The 1880s and 1930s daily historic data could be used to determine geographic and variability differences in climate on a local scale.

However, analogues have several drawbacks. First, they are not consistent with the range of global warming now thought likely under the greenhouse effect: 1.5 to 4.5°C. The warmest period of the last 125,000 years was 1°C warmer than the present temperature. In addition, the cause of the past warmings was not only a change in the concentration of greenhouse gases, but also shifts in the Earth's axis, which caused different regional distribution of climate than would be associated with changes in radiative forcing. Second, historic records do not provide enough detail to conduct comprehensive analysis of the 1°C warming. Paleoclimatic records only indicate broad regional patterns of change for a few variables, such as temperature, rainfall, and solar radiation. We can not discern local, daily, or interannual climate from these records. Even using the 1930s presents some problems. Daily records are available only for temperature and rainfall. Some effects models need more variables, such as wind or radiation. Furthermore, the number of weather stations with 1930s data is limited, which could present calibration problems.

General Circulation Models (GCMs)

GCMs are dynamic models that simulate the physical process of the atmosphere and oceans to estimate global climate. These models have been developed over two decades and require extensive computations to run. They can be run to estimate current climates and climates under different compositions of greenhouse gases. The GCMs are often used to simulate climate following a doubling of carbon dioxide levels, otherwise referred to as doubled CO_2. Output is given in regional grid boxes.

The climate change estimated by the GCMs is consistent with the effective doubling of CO_2 concentrations of the atmosphere. The effective doubling of CO_2 means that the combined radiative forcing of the increase in concentrations of all greenhouse gases has the same radiative forcing as doubling CO_2 levels.

The effective doubling of CO_2 concentrations will occur around the year 2030 if current emissions trends continue. The climate change caused by an effective doubling of CO_2 may be delayed by 30-40 years or longer.* Estimates of climate change caused by this effective doubling of CO_2 are referred to as "2xCO_2 scenarios."

GCMs have several advantages over the other approaches for creating scenarios. First, the models are used to estimate how global climate may change in response to increased concentrations of greenhouse gases. Thus, regional outputs are consistent with a global warming associated with 2xCO_2. Second, the estimates of climate variables (for example, rainfall, temperature, and humidity levels) are internally consistent. Third, GCMs estimate outputs for many meteorologic variables (including wind, radiation, cloud cover, and soil moisture) providing enough input for effects models. Fourth, GCMs simulate climate variability, in some cases on a daily basis.

One of the major disadvantages of GCMs is their low spatial resolution. GCMs give outputs in grid boxes that vary in size from 4 by 5° to as much as 8 by 10°. Figure 3-1 shows the grid boxes from the Goddard Institute for Space Studies (GISS) model overlaid on a map of the United States. The grid boxes are 8 by 10° and are larger than the land area of France (Mitchell, 1988). Within each area, the actual climate may be quite variable. For example, the weather in southern Washington State may be quite different from the weather in northern California, but both are in the same grid box. The models, however, do not account for variations within each grid box. Instead, they estimate average climatic conditions for the entire grid box. So, for any

*This estimate does not account for the effect of the Montreal Protocol on emissions of chlorofluorocarbons (CFCs). If implemented, the protocol will slow the rate of CFC emissions, delaying the date of "effective doubling."

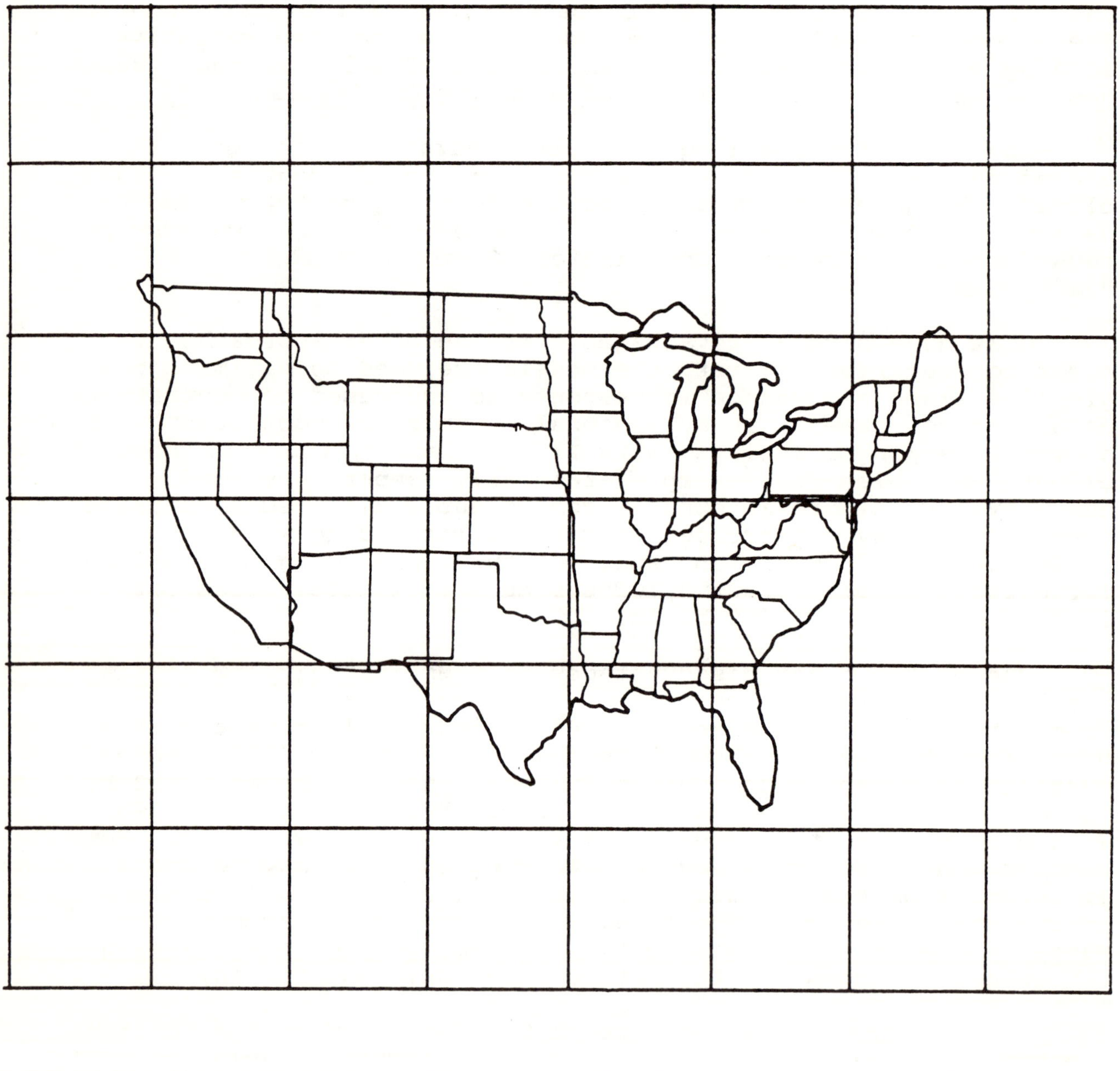

Figure 3-1. GISS Model of the United States.

simulated day, they provide a single value for temperature, for rainfall, and for other variables for the entire area of the box.

A second disadvantage is that the models do not satisfactorily simulate the oceans, which lowers confidence in the GCMs' ability to simulate regional climate and climate variability. GCMs disagree on regional climate, with two models often showing opposite changes in some variables (such as precipitation) for the same location. (Model disagreement on regional climate change is the subject of a recent study by Grotch, 1988.) A low level of confidence also exists regarding model estimates of variability, on both a daily and an interannual basis. (See the Climate Variability chapter of this report for a discussion of GCM estimates regarding changes in variability.) In sum, a relatively greater degree of confidence exists regarding estimates of average and global climate changes than regarding regional climate and variability estimates.

Choice of $2xCO_2$ Scenario

GCM outputs are employed as a basis for constructing the scenarios to be used in our report because they produce regional climate estimates consistent with doubled CO_2 concentrations. Yet, GCMs are relatively new tools that need a great degree of refinement. Their results must be applied with caution. A conservative approach requires using only the outputs from the GCMs for which the atmospheric modelers have expressed the greatest confidence, specifically the average monthly changes in each grid box.

The scenarios described in this chapter are a hybrid between GCM outputs and historic weather data. The estimates of average monthly change in temperature, precipitation, and other weather variables are used in a grid box from the GCMs. Model simulations of current conditions are compared with simulations of $2xCO_2$ conditions. The ratios of $(2xCO_2)/(1xCO_2)$ are multiplied by historic weather conditions at weather stations in a grid box. Parry et al. (1987) used this approach in an analysis of impacts of climate change on agriculture. Thus, if a grid box is 2°C warmer under the GCM scenario, all stations in that grid are assumed to be 2°C warmer. The effect of this is to keep variation from station to station within a grid the same as in the historic base period. Furthermore, interannual and daily variability remain the same. If it rains 10 days in a month, in the scenario it also rains 10 days in the month, and the amount of rainfall is adjusted by the GCM output.

The years 1951-80 were chosen as the base period to which average $2xCO_2$ changes were applied. Several decades of data allow for a great deal of interannual variability, giving a wide range of warm, cold, wet, and dry years. Since the data are from the most recent decades, they are the most complete historic data available.

GCMs Used

To obtain a range of scenarios, output from three GCMs is used:

- o Goddard Institute for Space Studies (GISS) (Hansen et al., 1988),
- o Geophysical Fluid Dynamics Laboratory (GFDL) (Wetherald and Manabe 1988), and
- o Oregon State University (OSU) (Schlesinger and Zhao, 1988).

The average seasonal temperature and precipitation for the United States gridpoints for each model are displayed in Figure 3.2. All three models yield an average warming over the United States, with OSU giving 3°C, GISS 4.3°C, and GFDL 5.1°C. The seasonal patterns are different, with GISS having a larger warming in winter and fall, GFDL having the highest temperature change in the spring, and OSU having little seasonal variability. All three models estimate that annual precipitation would increase. GISS and OSU estimate that annual precipitation would rise by respectively, 73 mm (2.92 inches) and 62 mm (2.48 inches), while GFDL has a rainfall increase of only 33 mm (1.31 inches). The first two models have precipitation increases in all four seasons, while GFDL has a summer decline in rainfall. In sum, OSU appears to be the "mildest" scenario, with the lowest temperature rise and largest increase in precipitation. GFDL appears to be the most "extreme," with the highest temperature rise, the smallest increase in precipitation, and a summer decrease in rainfall.

Limitations

A major limitation of the $2xCO_2$ scenarios used for this study is the lack of temporal and spatial variability. By applying average monthly changes to the historic data set, it is assumed that the daily and interannual patterns of climate remain the same. This assumption is probably unrealistic, since a change in average conditions will probably lead to a change in

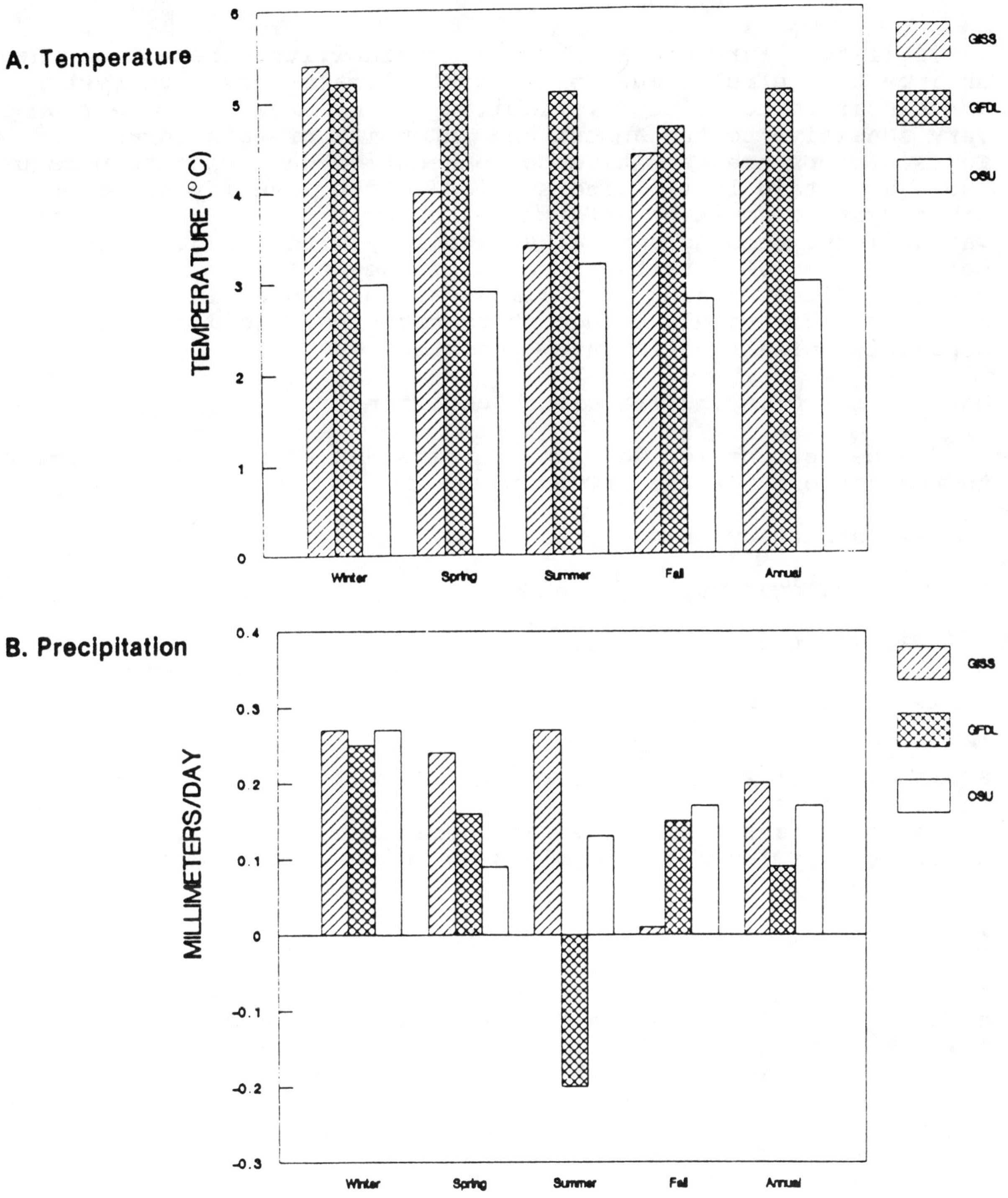

Figure 3-2. Average changes in temperature (°C) and precipitation (mm/day) over lower 48 States grid boxes.

variability. Furthermore, holding variability constant can have an impact on effects analysis. Most climate-sensitive systems are sensitive to climate variability. For example, riverflow is very sensitive to the amount and intensity of rainstorms. Certain crops are sensitive to consecutive days with temperatures above a certain level. The studies do not identify how these and other systems could be affected by changes in temporal climate variability. Holding spatial variability within a grid box constant also affects the results of the analyses done for this report. Climate change may also lead to changes in wind patterns, which could change storm patterns, cloud distribution, deposition of air pollutants, and other systems.

Options for Creating Transient Scenarios

The options for developing transient scenarios are similar to the options for the 2xCO_2 scenarios:

1. arbitrary changes,

2. analogue warming, and

3. GCM transient runs.

<u>Arbitrary Changes</u>

Although the manner in which a system responds to an arbitrary 1 or 2°C temperature warming and arbitrary changes in other variables could be examined, the problems of physical inconsistent assumptions pertain here also. In addition, the arbitrary warming scenario gives no indication of when the warming may occur.

<u>Analogue Warming</u>

Wigley (1987) has suggested using analogues as scenarios for climates that may occur in the next several decades. He notes that the warming from the late 19th century to 1940 was about 0.4°C, which may approximate the transient warming over the next two decades. The problem here is that climate may change faster in the future than in the early 20th century. (The average decadal warming may be as much as 0.5°C, rather than the 0.1°C identified for earlier years.) Furthermore, the analogue only takes one as far as a 0.5°C -- or in the case of paleoclimatic records, a 1°C -- warming. It does not indicate what happens in decades after the 0.5-1.0°C level is reached. In addition, the analogue may not represent the regional distribution of climate associated with greenhouse forcing.

GCM Transient Run

GISS has modeled how global climate may change as concentrations of greenhouse gases gradually rise over the next century. This is called the transient run. GISS has modeled climate change under several assumptions of trace gas growth. The transient runs start in 1958 with the atmospheric concentrations of greenhouse gases that existed at that time. The concentrations of the gases and equivalent radiative forcing are then increased from 1958 until an arbitrary point in the future according to assumptions regarding trace gas growth. The GISS transient run yields daily climate estimates from 1958 until that arbitrary point.

For example, one of the transient scenarios, which is known as GISS A, assumes that trace gas concentrations continue to increase at historic rates and net greenhouse forcing increases exponentially. The scenario is run from 1958 to 2062. The end of the transient corresponds with a global warming equivalent to that of the equilibrium climate from the doubled CO_2 run. GISS B assumes a decreasing trace gas concentration growth rate such that the annual increase in climate forcing stays the same (Hansen et al., 1988). GISS B stops in 2029.

Since the GCMs are used to produce this transient run, the advantages and disadvantage of using this approach are the same as those described in the discussion of doubled CO_2 scenarios. In addition, the timing of the changes estimated by the GCMs is complicated by the uncertainties regarding the growth of greenhouse gas emissions and the roles of the oceans and clouds in delaying climate changes.

Choice of Transient Scenario

This study uses transient scenarios based on the GISS transient run because, of all of the different approaches, only this study is internally consistent and allows examination of the entire range of climate change between current conditions and $2xCO_2$ climate.

In creating the transient scenario, an approach similar to that used for the $2xCO_2$ scenario is employed. Since relatively little confidence exists in the GCM's estimates of changes in interannual and daily variability, the monthly means are calculated for each decade of the transient. This process gives average decadal temperature, precipitation, and other changes. The average decadal temperature changes in GISS A and B for the United States are shown in Figure 3-3.

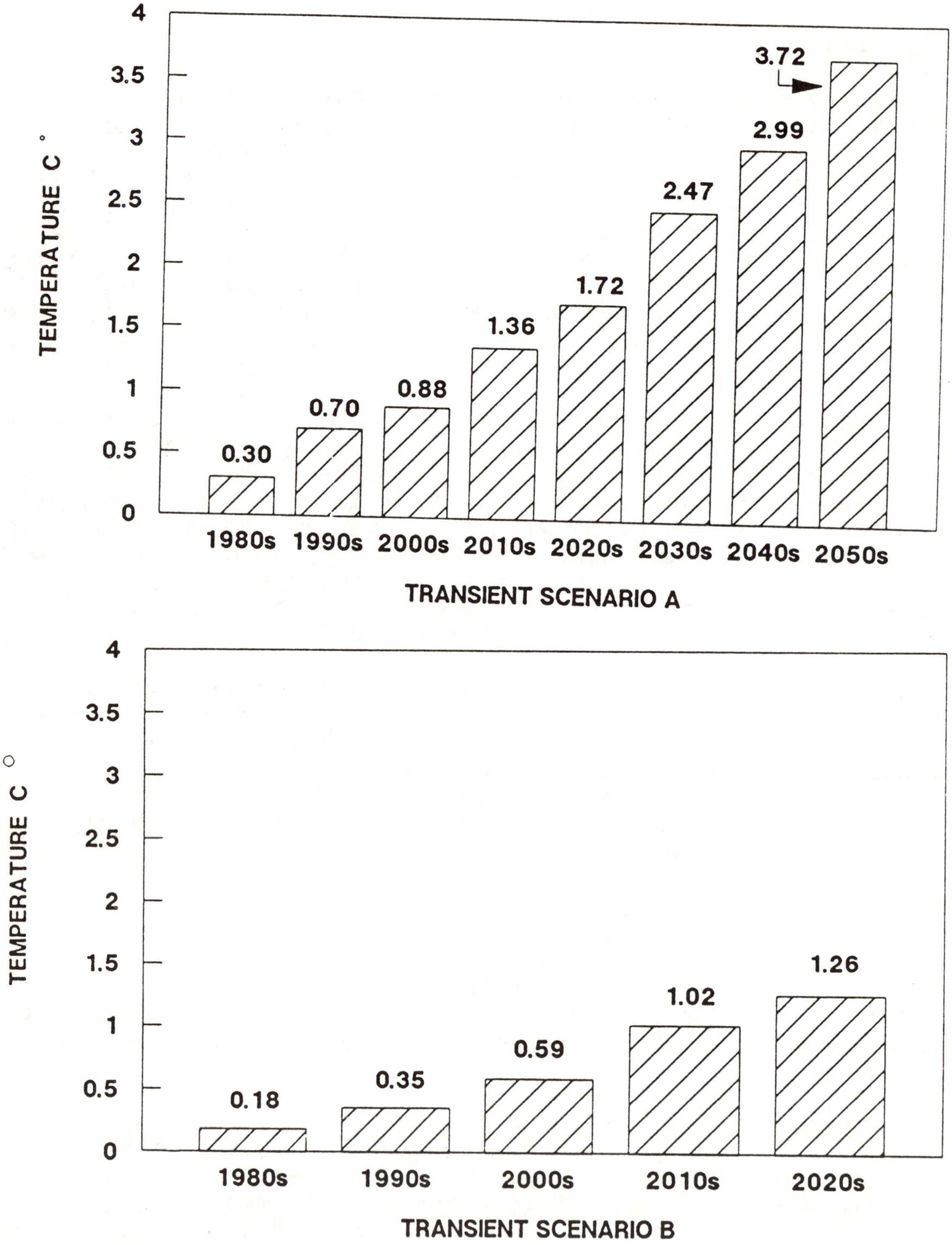

Figure 3-3. GISS transients "A" and "B" average temperature change for lower 48 States grid points

As in the $2xCO_2$ scenario, the average meteorologic changes from the transient are combined with a historic time series. What is different from the $2xCO_2$ scenario is that a gradual change in temperature and other variables is mixed with a historic time series with its own variability. This can produce a "sawtooth" effect.

In this study, the historic time series 1951-80 is used, and the transient monthly statistics are applied to the time series. The procedure for creating the transient scenario is to first linearly interpolate between decadal means. This smoothes out sharp decadal changes from the transient GCM results. The monthly underlying climatic change, starting with the decade of the 1980s in the GISS run to the actual 1950s data, is then added. The statistics from the 1990s in the GISS run are added to the 1960s, and those from the 2000s are added to the 1970s. Then, the monthly averages from the 2010s are combined with 1950s actual data, and so on, until the end of the transient is reached. Although there is a cooling from the 1950s to the 1960s, followed by a warming in the 1970s, the underlying warming of the transient, which is 3.7°C by the middle of the 2050s in GISS A, is much greater than the variability in the base period. This is shown in Figure 3-4, which displays the smoothed decadal averages from GISS A, the 1951-80 decadal averages for the United States, and the combination of the two to create one of the transient scenarios.

Limitations

Since the transient scenarios are also derived from GCMs, the same limitations concerning temporal and spatial variability pertain as in the $2xCO_2$ scenario. An additional limitation in the transient scenario is the rate of change. The GISS transient runs assume a gradual rate of change in temperature. This is complicated by the addition of the 1951-80 base period, which yields short-term oscillations and a sawtooth effect. Broecker (1987) has shown that past climatic changes may have been abrupt. A sudden warming could mean that significant effects happen sooner and more suddenly than the results of the transient analysis used in this study indicate.

Other Scenarios

In a few cases, researchers used meteorologic data from the 1930s as an analogue scenario. This scenario was used to provide additional information on the sensitivity of systems to climate change. In a few other cases, researchers only examined

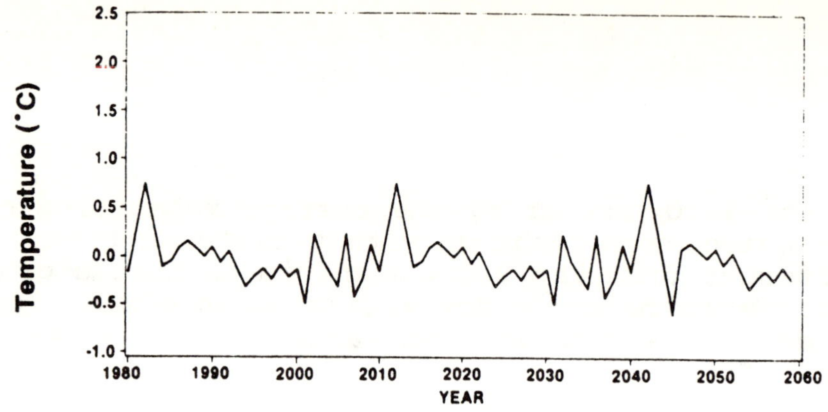

A. 1951 - 1980

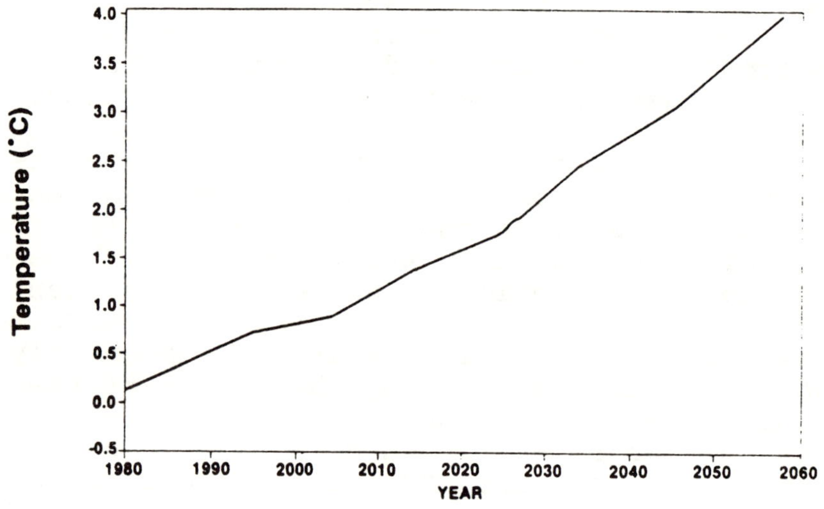

B. SMOOTHED GISS "A"

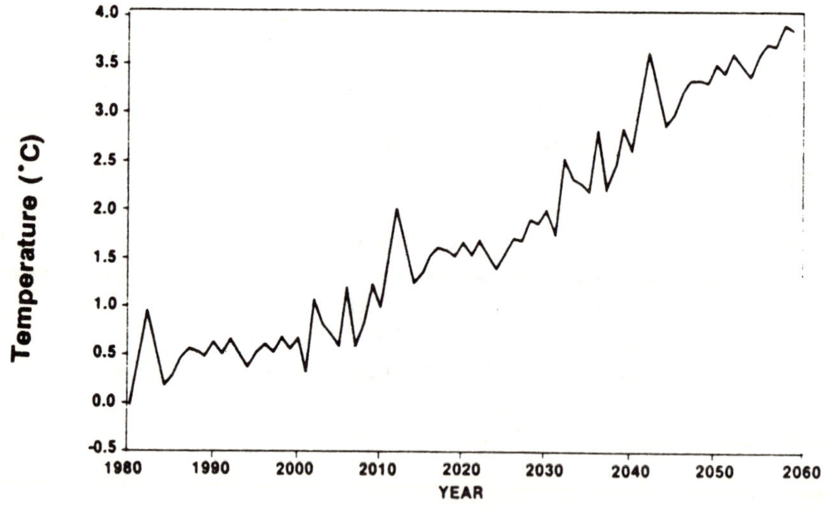

C. TRANSIENT SCENARIO, SMOOTHED GISS "A"
AND 1951 - 1980 REPEATED

Figure 3-4. Transient scenarios.

paleoclimatic records. In these cases, the goal was to determine how a system responded to past climate change.

Unlike the climate scenarios, the alternative sea level rise scenarios were not based solely on the differences between various general circulation models. Instead, they were based on the range of estimates that previous studies have projected for the year 2100. (Hoffman et al. 1983, 1986; Meier et al. 1985; Revelle 1983; Thomas 1986) which have generally considered alternative rates of greenhouse gas emissions, climate sensitivity ranging from 1.5 to 4.5°C for a CO_2 doubling; and uncertainties regarding ocean expansion and glacial melting. Estimates for the year 2100 generally range from 50-200cm.

This report uses three scenarios for the year 2100: 50, 100, and 200 cm, the current trend of 12 cm/century. Because most studies have not reported estimates for the intermediate years, we followed the convention of a recent NRC report (Dean et al.) and interpolated sea level rise using a parabola, as illustrated in figure 3-0. Because various coastal areas are also sinking (and in a few cases rising), relative sea level rise at specific locations was estimated by adding current local subsidence trends.

EFFECTS ANALYSES

In this study, the preferred approach for analyzing potential impacts of climate change was to develop quantitative estimates. Most researchers estimated impacts by running models that simulate the relationship of weather and the relevant system. The climate scenarios were used as inputs into the models. Since the researchers only had several months to do the analysis, they either used "off-the-shelf" models or analytic techniques. In many cases, existing models were calibrated to new sites. This lack of time also limited the gathering of new data to a few studies.

A drawback of using quantitative models of systems to estimate sensitivities is that the models are applied to climates for which they were not developed. The models estimate relationships between weather on record and the system. This relationship is then extrapolated to an unprecedented climate. It is possible that in the new climate situation, the statistical relationship may be different due to crossing of a threshold or some other reason. With the drawbacks of quantitative models, the current statistical relationships are the best basis for quantitatively estimating sensitivities.

For the most part, researchers analyzed the potential effects of climate change on systems as they currently exist. Although these changes may be quite substantial, potential changes in populations, GNP, technology, and other factors were not considered. In some cases, researchers ran additional scenarios with assumptions about technological and other changes. In addition, potential responses to climate change were considered in some, but not all, cases. For these and many other reasons, the results should only be interpreted as indicative of the sensitivity of current systems to climate, not as a prediction of what the effects will be.

In some situations, quantitative models of the relationship between climate and a particular system did not exist. In those cases, other approaches were used to try to identify sensitivities. Some researchers examined how systems responded to analogue warmings. In other cases, expert judgment was used. This consisted of literature reviews to assemble information on sensitivities as they appear in the literature, and workshops and interviews to poll experts on how they thought systems would respond to global warming.

REFERENCES

Broecker, W.S. 1987. Unpleasant surprises in the greenhouse? Nature 328:123-126.

Grotch, S.L. 1988. Regional Intercomparisons of General Circulation Model Predictions and Historical Climate Data. Prepared for United States Department of Energy. TR041.

Hansen, J., I. Fung, A. Lacis, D. Rind, G. Russell, S. Lebedeff, R. Ruedy, and P. Stone. 1988. Global climate changes as forecast by the GISS 3-D model. Journal of Geophysical Research (In press.)

Hansen, J., G. Russell, D. Rind, P. Stone, A. Lacis, S. Lebedeff, R. Ruedy, and L. Travis. 1983. Efficient three-dimensional global models for climate studies: models I and II. Monthly Weather Review 3(4):609-622.

Hoffman, J.S., D. Keyes, and J.G. Titus. 1983. Projecting future sea level rise. Washington, DC: Government Printing Office.

Hoffman, J.S., J. Wells, and J.G. Titus. 1986. Future global warming and sea level rise. In Iceland coastal and river symposium. ed. G. Sigbjarnarson. Regkjavid/Iceland: National Energy Authority

Meier, M.F., et al. 1985. Glaciers, ice sheets, and sea level. Washington, D.C.: National Academy Press.

Mitchell, J.F.B. 1988. Local effects of greenhouse gases. Nature 332:399-400.

Parry, M., T. Carter, N. Konijin, and J. Lockwood et al. 1987. The Impact of Climatic Variations on Agriculture: Introduction to the IIASA/UNEP Case Studies in Semi-Arid Regions. International Institute for Applied Systems Analysis.

Revelle, R. 1983. Probable future changes in sea level resulting from increased atmospheric carbon dioxide. In Changing Climate. Washington, DC: National Academy Press.

Schlesinger, M. and Z. Zhao. 1988. "Seasonal Climate Changes Induced by Doubled CO_2 or Simulated by the OSU Atmospheric GCM/Mixed-Layer Ocean Model" Oregon State University Climate Research Institute Report. Carvallis, Oregon.

Thomas, R.H. 1986. In Titus, J.G. 1986. Effects of changes in stratosphere ozone and global climate. EPA and UNEP: Washington, DC.

Vinnikov, K.Y., and N.A. Lemeshko. 1987. Soil Moisture Content and Runoff in the USSR Territory With Global Warming. Journal of Meteorology and Hydrology No. 12.

Wetherald R.T. and S. Manabe. 1988. "Cloud Feedback Process in a GCM" Journal of Atmospheric Sciences. April, 1988

Wigley, T.M.L. 1987. Climate Scenarios. Prepared for the European Workshop in Interrelated Bioclimate and Land-Use Changes. National Center for Atmospheric Research. NCAR 3142-86/3.

CHAPTER 4

POTENTIAL IMPACTS OF CLIMATIC WARMING ON CALIFORNIA

FINDINGS

Global warming could cause higher winter and lower spring runoff in California and increasing the difficulty of meeting water supply needs. It could also increase the salinity in San Francisco Bay and the Delta and the relative abundance of marine species in the Bay; degrade water quality in alpine lakes; raise ambient ozone levels; and increase electricity demand. Changes in agriculture are uncertain.

Water Resources

- Higher temperatures would lead to higher winter runoff from the mountains surrounding the Central Valley, because less precipitation would fall as snow, and the snowpack would melt earlier. Runoff in the late spring and summer would consequently be reduced.

- As a result, the amount and reliability of water supply from reservoirs in the Central Valley basin would decrease. Annual water deliveries from the State Water Project could be reduced by 7 to 16 percent. Even if operating rules are changed, current reservoirs would not have the capacity to store the heavier winter runoff and at the same time retain flood control capabilities.

- The water area and volume of the San Francisco Bay estuary could increase due to a one-meter (40-inch) sea level rise. If the Delta and Bay levees are maintained, the increase in area and volume would be, respectively, 300 km^2 and 1.2 billion m^3 (about 30 and 15 percent). If the levees fail, the area and volume would, respectively, triple and double, an increase of 2,400 km^2 and 6.8 billion m^3.

- Saline (brackish) water might migrate inland between 4 km (2.5 miles) and 10 km (6 miles) if sea level rises 1 meter, if the levees fail, and if there is no erosion of tidal channels. Freshwater releases into the Delta might have to be doubled in order to repel saline water near the major freshwater pumping facilities.

Wetlands and Fisheries

- The wetlands in San Francisco Bay estuary will gradually be inundated as sea level rises. The amount of wetland lost will be a function of the rate of sea level rise and whether shorelines are protected. If sea level rises 1 meter by 2100, the rate of rise will be greater than wetland vertical accretion by the middle of the next century. If sea level rises 2 to 3 meters by 2100, wetlands will start to be inundated early in the next century.

- If salinity increases within the San Francisco Bay estuary, wetland vegetation will shift from brackish and freshwater species to more salt-tolerant plants. This shift could severely reduce waterfowl populations that depend on freshwater habitats. The timing, magnitude, and location of phytoplankton production would shift. Marine fish species could increase in abundance, while saltwater species that breed in freshwater areas would most likely decline.

- Higher temperatures in subalpine lakes could increase annual primary production (such as algae) by between 16 percent and 87 percent, which could degrade lake water quality and change the composition of fish species.

Agriculture

- The impacts of climate change on agriculture in California are uncertain. The effects of changes in temperature and precipitation alone would most likely reduce yields 4 to 40 percent, depending on the crop. However, with the combined effects of climate and higher CO_2 levels, yields for all crops, except corn, might increase.

- The potential growth in irrigation in some parts of the state may require increased extraction of groundwater due to current full use of surface water supplies. This would decrease water quality and affect water management options.

- Yields in California may be less than in most parts of the country. Crop acreage could increase because of the shifts in yields and the presence of irrigation infrastructure.

Natural Vegetation

- Drier climate conditions could reduce forest density, particularly pine and fir trees, and timber productivity.

(The full impacts on California forests was not assessed in this report).

Air Quality

- If today's emissions exist in a future warmer climate, Ozone levels in the San Francisco Bay will intensify and will change location due to higher temperatures. As a result, the number of people hours exposed to elevated ozone levels in excess of EPA standards (of 0.12 pphm) would triple.

Electricity Demand

- The annual demand for electricity in northern California could rise by 3 to 6 billion kilowatt hours (1 to 2 percent) by 2010 and by 21 to 41 BKWH (3 to 5 percent) by 2055.

- By 2010, 2-3 gigawatts (GW) would be needed to meet the increased demand. By 2055, 10-20 GW would be needed -- a 10 to 20 percent increase over baseline additions that may occur without climate change. The additional capital cost by 2055 would be $10 to 27 billion.

Policy Implications

- Water management institutions, such as the U.S. Bureau of Reclamation and the California Department of Water Resources, should analyze the potential impacts of climate change on water management in California. They should consider whether and how the Central Valley Project and State Water Project should be modified to meet diminishing supplies and increasing demands due to climate change. They may also consider whether to change water allocation rules to encourage more efficient use of water.

- State and local entities should consider the impacts of climate change on levee management in San Francisco Bay and the Delta.

- The California Air Quality Board should review the long-term implication of climate change on air quality management strategies.

- The California Energy Commission should consider the impacts of climate change on the energy supply needs for the State.

CLIMATE SENSITIVE RESOURCES OF CALIFORNIA

California's Central Valley is the most productive and diverse agricultural region of its size in the world. The Central Valley Basin, which includes the drainages of the Sacramento and San Joaquin Rivers, encompasses several large metropolitan areas, dispersed manufacturing, major port facilities, important timber reserves, heavily used recreational areas, and diverse ecosystems.

Much of the region's economic and social importance is derived from its water resources. Over 40% of California's total surface water runoff drains from the Central Valley Basin into the San Francisco Bay area (Miller and Hyslop, 1983). The basin supplies water for irrigated agricultural, municipal, and industrial uses, and for a host of other resources and activities.

The Central Valley Basin encompasses approximately 40% of California's land area (Figure 4-1). Elevations range from just below sea level on leveed islands in the Sacramento-San Joaquin River Delta to peaks of over 4,200 m (14,000 feet) in the Sierra Nevada (Figure 4-2). Mountains ring most of the basin: the Sierra Nevada along the eastern side and the Coast Ranges on the west. The only outlet to the Pacific Ocean is through the San Francisco Bay estuary (Figure 4-2).

Current Climate

California's climate is characterized by little, if any, summer precipitation, and winters are generally wet (Major, 1977). Temperature and precipitation vary with elevation and latitude in the Central Valley basin. Extremes in mean annual precipitation range from about 15 cm (6 inches) in the southern San Joaquin River Basin to about 190 cm (75 inches) in the mountains of the Sacramento River Basin. While almost all valley floor precipitation falls as rain, winter precipitation in the high mountains often falls as snow. Storage of water in the snowpack controls the seasonal timing of runoff in the Central Valley rivers and has shaped the evolution of strategies for water management and flood protection. Under current climatic conditions, peak runoff occurs between February and May for individual rivers within the Central Valley Basin (California Department of Water Resources, 1983; Gleick, 1987b).

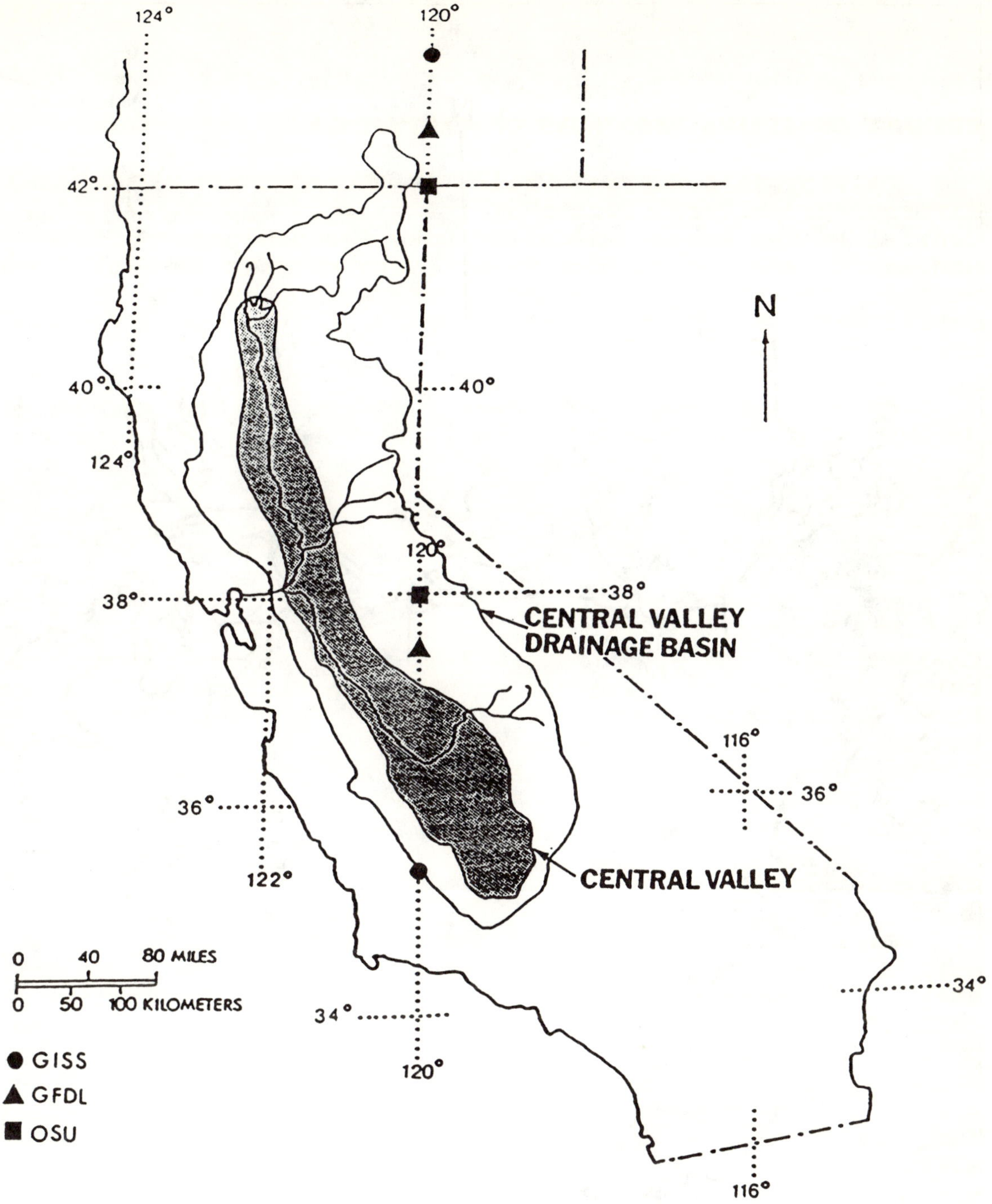

Figure 4.1. The Central Valley (shaded) and Central Valley Drainage Basin of California. Symbols refer to locations of General Circulation Model (GCM) gridpoints (see California Regional Climate Scenarios section of this chapter for details on GCMs).

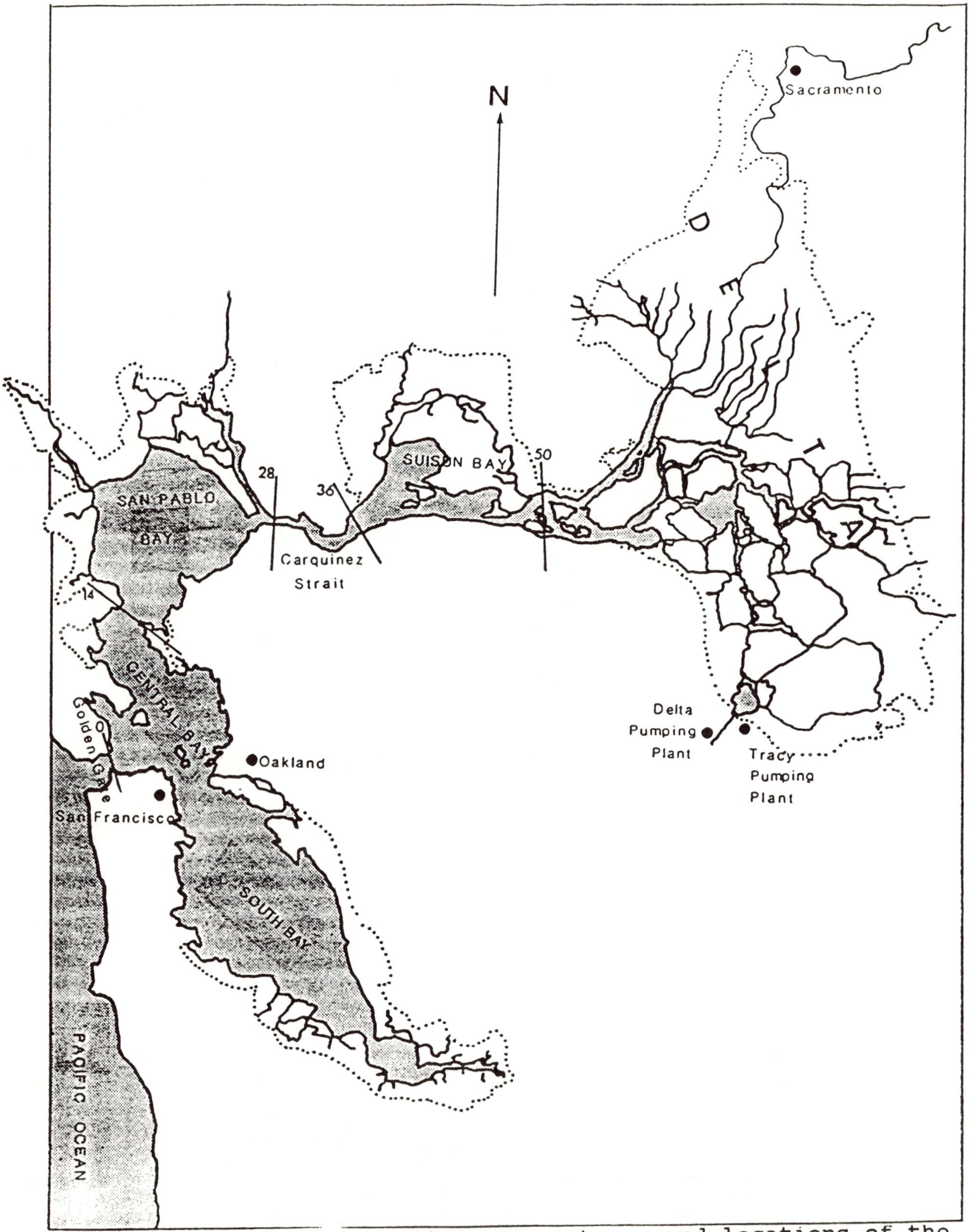

Figure 4.2. The San Francisco Bay estuary and locations of the freshwater pumping plants in the Delta. The numbered bars indicate distance (in miles) from the Golden Gate. The dotted line indicates the maximum area affected by 100-yr high tide with a 1-m (40-inch) sea level rise.

Chapter 4

Water Resources

Water Distribution

California's water resources are poorly distributed, relative to human settlement patterns in the State. Over two-thirds of the State's surface water supply originates north of Sacramento, and 70% of its population and 80% of its total demand for water lie to the south (California Department of Water Resources, 1985). In addition, about 85% of the Central Valley Basin's total annual precipitation occurs between November and April, whereas peak water use occurs during the summer.

In working to solve these water distribution problems, the U.S. Government and California have built two of the largest and most elaborate water development projects in the world: the Federal Central Valley Project (CVP) and the California State Water Project (SWP). Both are essentially designed to move water from water-rich northern California to the water-poor south, and to supply water for agriculture and municipal and industrial purposes. Currently, the CVP has a surplus of water and the SWP has a shortage, especially in relationship to users' projected requirements. Thus, the SWP is particularly susceptible to dry years.

Flood Control and Hydroelectric Power

Another objective of the CVP and SWP is flood control. By 1984, CVP facilities had prevented almost $500 million in flood damages (U.S. Bureau of Reclamation, 1985). Flood control, however, comes at the expense of water storage (and hence water deliveries), because reservoir levels must be kept low to absorb high river flows. A warmer climate could increase the severity of winter floods by reducing the proportion of total precipitation falling as snow in the Sierra Nevada during a given storm. Controlling these floods could reduce water deliveries.

Hydroelectric power generation is also an objective of the CVP and SWP, and surplus power is sold to utility companies. CVP powerplants produce an average of 5.5-6 billion kilowatt-hours (kWh) per year. In 1976 and 1977, precipitation was 35 and 55% below normal, respectively, and hydroelectric power generation fell to 50 and 40%, respectively, of target production.

Sacramento and San Joaquin River Delta

The delta at the confluence of the Sacramento and San Joaquin Rivers is the focal point of major water-related issues in

California. For example, many islands in the delta lie below sea level and are protected by levees, some of which are made of peat and are relatively fragile. These islands would be vulnerable to inundation from rising sea level associated with climatic warming. The deep peat soils on these islands support highly productive agriculture that would be eliminated under inundated areas.

In addition to agricultural importance, the delta is also the source of all CVP and SWP water exports to points farther south. The freshwater pumping plants (see Figure 4-2) in the delta are the two largest single freshwater diversions in California (Sudman, 1987). Delta outflow must be maintained at a required level to prevent saltwater intrusion into the pumping plants (see Figure 4-2). The volume of water released from upstream reservoirs to achieve this is called carriage water. Higher sea levels could exacerbate the existing problem of saltwater intrusion and require the release of greater volumes of carriage water.

Commerce

The San Francisco Bay estuary includes the largest bay on the California coast (see Figure 4-2). The bay's northern reach between the Golden Gate and the Sacramento-San Joaquin River Delta is a brackish estuary dominated by seasonally varying river inflow (Conomos et al., 1985). The southern reach between the Golden Gate and the southern terminus of the bay is a tidally oscillating lagoon-type estuary. The port facilities of the San Francisco Bay area are vital to California's internal trade, to Pacific Coast commerce, and to foreign trade, particularly with Asian countries. The ports of Oakland and San Francisco, combined, ranked fourth in the United States in tonnage of containerized cargo handled in 1983 (U.S. Maritine Administration, 1985). These facilities and operations are sensitive, in varying degrees, to both sea level change and fluctuation in freshwater runoff. For example, an increase in sea level could make navigation easier, reducing the need for dredging of navigation channels. On the other hand, a rising sea level could threaten fixed port terminals and piers.

Agriculture

California annually produces about 10% of cash farm receipts in the United States with $14.5 billion in farm income in 1986 (U.S. Department of Agriculture, 1987). Central Valley farms make up significant proportions of total U.S. production of many crops, including cotton, apricots, grapes, almonds, tomatoes, and lettuce. Since virtually all Central Valley agricultural production requires irrigation, climatic change effects on water availability and water quality could significantly affect California agriculture.

Agriculture, the primary land use and the largest consumer of water in the Central Valley Basin, accounts for 87% of total net water use in the region. Furthermore, the region accounts for 72% of total net water use for the entire State and almost 80% of net agricultural use (California Department of Water Resources, 1987a).

Forestry

Silviculture is extensively practiced in California's mountains. The nine National forests substantially within the Central Valley Basin recorded over $88.6 million in timber sales in fiscal year 1986 (U.S. Department of the Interior, 1986). Climatic change could harm timber production in the region. For example, the drought of 1976-77 contributed to significant tree mortality because of large infestations of bark beetles (California Division of Forestry and Fire Protection, 1988).

Natural Vegetation

Approximately one-fourth of all the threatened and endangered plants in the United States are found in California. About 460 species, or about 9% of the California species listed by Munz and Keck (1959), are either extinct or in danger of becoming extinct.

California contains about 5,060 native vascular plant species; of these, about 30% occur only in California (Munz and Keck, 1959; Raven, 1977). These species are more numerous than those present in the entire central and northeastern United States and adjacent Canada, a region about eight times larger than California (Fernald, 1950).

Within the Central Valley Basin, terrestrial vegetation may be grouped into the following broad classes, listed according to decreasing elevation: alpine, subalpine forest, montane forest, mixed evergreen forest, chaparral and oak woodland, and valley grassland (Barbour and Major, 1977).

Wetlands

The San Francisco Bay estuary includes approximately 90% of the salt marsh area in California (Macdonald, 1977). Nichols and

Wright (1971) documented a 60% reduction in San Francisco Bay marsh between 1850 and 1968. This reduction was largely the result of reclamation for salt ponds, agriculture, expanding urbanization, shipping facilities, and marinas. Further loss of wetlands from climate change and rising sea level could result in substantial ecological and economic losses for the region. For example, the managed wetlands north of Suisun Bay support a hunting and fishing industry valued at over $150 million annually (Meyer, 1987). Tourism, rare and endangered species, and heritage values also could be harmed.

Wildlife and Fisheries

San Francisco Bay estuary provides vital habitat for many sport and commercial fish species (California Department of Water Resources, 1983). Important sport fish include striped bass, chinook salmon, sturgeon, American shad, and steelhead rainbow trout. These species are anadromous (i.e., saltwater species that enter freshwater areas for breeding) and the delta is an important nursery for these species. Chinook salmon is also an important commercial fish species, and Central Valley rivers support about 75% of California's chinook salmon catch, valued at $13.4 million at 1981 prices. The populations of these species are affected by water quality in the estuary.

In order to protect aquatic organisms in the delta the State Water Resources Control Board (SWRCB) adopted the water right Decision 1485 in 1978 that sets water quality standards to protect the delta and Suisun Marsh. The standards vary year to year, with less stringent requirements in dry years. The standards are met by meeting minimum delta outflow requirements. If delta outflow falls below the required level, then releases from upstream state and federal reservoirs must be increased so the outflow requirement is met. The water quality standards take precedence over water export from the delta.

Recreation and Nature Preservation

Recreation and nature preservation have expanded in recent years. Important recreational areas in the Central Valley Basin include four National parks (Lassen Volcanic, Sequoia, Kings Canyon, and Yosemite) and nine National forests that lie either completely or largely within its boundaries. Two National recreation areas and 13 designated wildlife refuges and management areas also are situated in the region. Downhill skiing and other winter sports are important economically in the State. Water projects throughout the Central Valley Basin provide significant recreational opportunities.

Chapter 4

PREVIOUS CLIMATE CHANGE STUDIES

Two of the few previous studies undertaken to assess the potential effects of climatic change on the region are discussed here.

Forests

Leverenz and Lev (1987) estimated the potential range changes, caused by CO_2-induced climatic change, for six major commercial tree species in the western United States. Two of the species, ponderosa pine and Douglas-fir, have significant populations in California. Leverenz and Lev based their estimates of range changes on the species' response to increased temperature, decreased water balance, and higher CO_2 concentrations. The scenario of climate change used was based on a simulation using the Geophysical Fluid Dynamics Laboratory (GFDL) model (a different run than used for this study), with CO_2 concentrations double their present levels. Their results suggest that in California, ponderosa pine could increase in range and abundance because of its ability to withstand long summer drought. Douglas-fir could be eliminated from coastal lowlands in California but might occur in the coastal area at higher elevations.

Water Resources

Gleick (1987a,b) applied 18 General Circulation Model (GCM)-based and hypothetical scenarios of climatic change to a hydrologic model of the Sacramento River Basin and used a two-part water balance model to estimate the monthly runoff and soil changes in the basin. Application of the scenarios to his model indicates that winter runoff could increase substantially, and summer runoff might decrease under most of the scenarios. Summer soil-moisture levels might also decrease substantially. These changes are driven by higher temperatures which decrease the amount of winter precipitation falling as snow and cause an earlier and faster melting of the snowpack that does form.

CALIFORNIA STUDIES IN THIS REPORT

Seven studies were completed as part of this regional study of the possible impact of climatic warming on California (Figure 4-3). These studies were quantitatively integrated as much as possible within the overall timeframe of this Report to Congress to obtain as complete a picture of those impacts as possible.

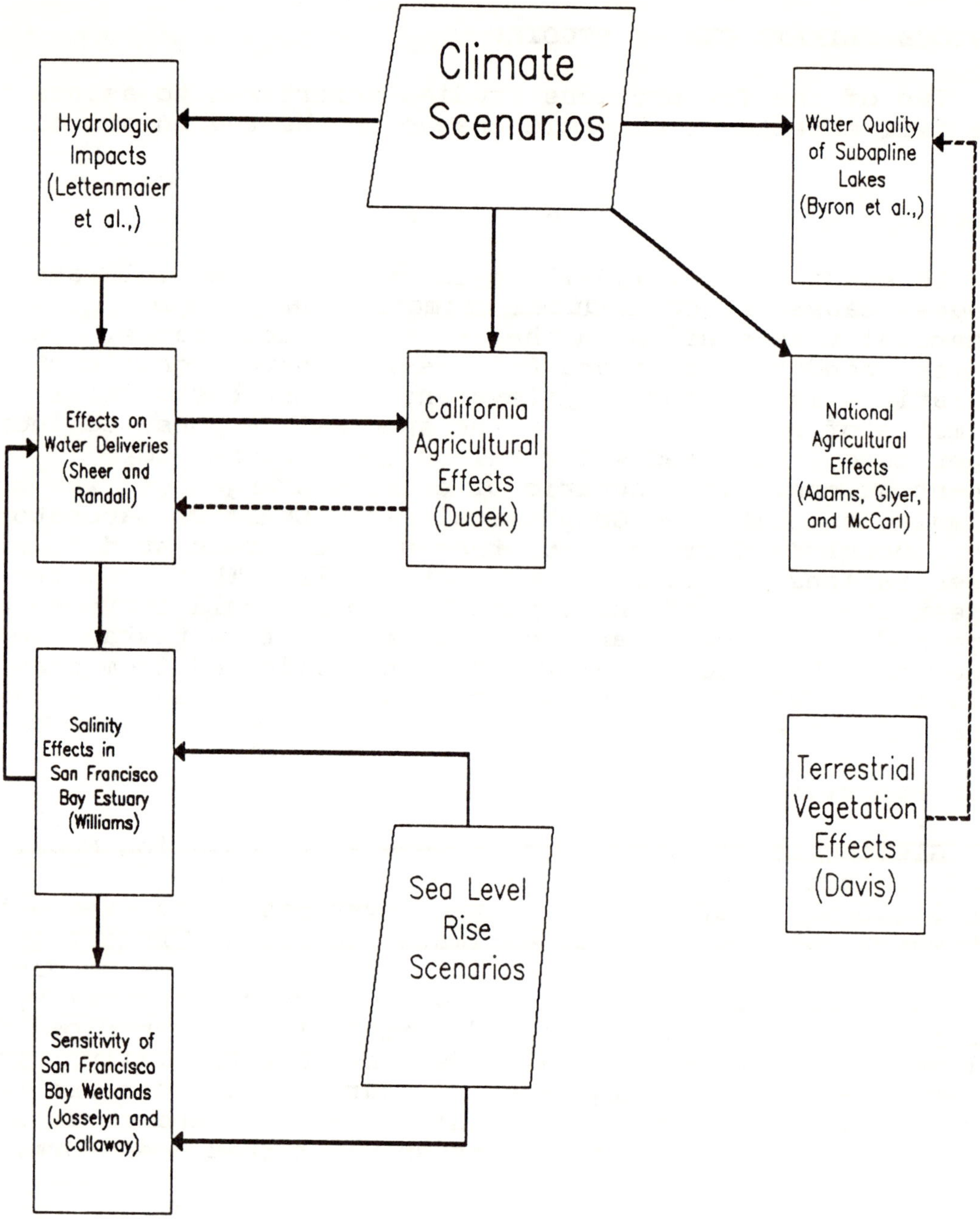

Figure 4-3. Organization of the study, showing paths of data input from scenarios and between projects (solid lines). Dashed lines indicate some important information flows that were not directly addressed in this study.

Also, several of the national studies have results pertaining to California.

Water is a key limiting resource in both managed and unmanaged ecosystems in the Central Valley Basin, and freshwater is important in estuarine ecosystems in the delta region. Consequently, the California studies were organized so that the impacts of climatic warming on the entire hydrologic system could be examined, starting at subalpine lakes in the mountains surrounding the valley and finishing at the freshwater outflow into the delta region and estuary (Figure 4-3). The individual projects examined the potential impacts of climatic change and sea level rise on particular ecosystems and water-delivery systems in the Central Valley. (See Methodology chapter of this report.) One of the major goals of this regional study was to determine the amount of runoff that would flow into the Central Valley from the surrounding mountains under different scenarios of climatic change, how much of that runoff would be available for delivery to the water users in the State, and how much would reach the delta.

Analyses Performed for this Study

o Hydrologic Effects on Central Valley Basin Rivers - Lettenmaier and Gan, University of Washington, and Dawdy, consultant.

The Lettenmaier et al. project (volume x) is the first of a series of four projects that are designed to determine the impact of climatic change on runoff and water deliveries within the Central Valley basin (Figures 4-3 and 4-4). Their project was designed to estimate changes in runoff from the mountains to the water resource system in the floor of the valley. Lettenmaier et al. used data from climate scenarios supplied to them by EPA as input to their modeling studies (see Scenario chapter and later section in this chapter on the California scenarios).

o Effects on Water Deliveries - Sheer and Randall, Water Resources Management, Inc.

Sheer and Randall (volume x) used the projected runoff from the mountains determined by Lettenmaier et al. to simulate the

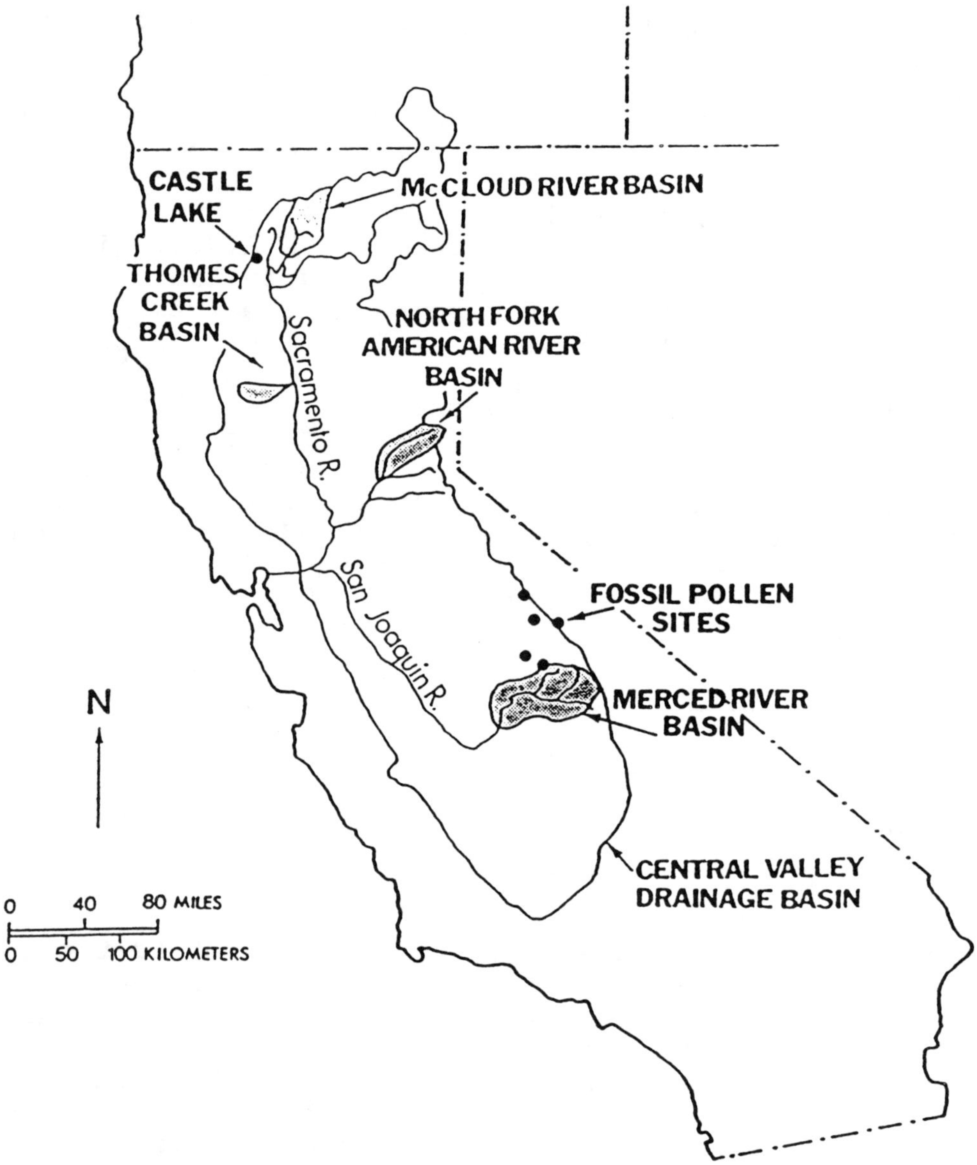

Figure 4-4. The Central Valley Drainage Basin of California. Shaded areas refer to the four study catchments used by Lettenmaier et al. The positions of the Castle Lake study site (Byron et al.) and the five fossil pollen sites (Davis) are indicated by dots.

response of the water supply and delivery systems of the Central Valley and State Water Projects to climatic change. Output from this project includes projected total water deliveries to the water users in the state.

- Salinity Effects in San Francisco Bay Estuary - Williams, Philip Williams and Associates.

Determining the impact of sea level rise and changing freshwater outflow into the Delta on salinity within the bay was the main goal of Williams' project (volume x). Williams also determined how much carriage water might be required to hold back salinity intrusions from the Delta pumping plants after sea level rise. The new carriage water requirements were than factored into Sheer and Randall's simulation of the water resource system, and represents an important feedback between the hydrologic effects of climatic change and sea level rise effects in the Delta (Figure 4-3).

- Effects on Wetlands in San Francisco Bay Estuary - Josselyn and Callaway, San Francisco State University.

Josselyn and Callaway (volume x) used results from Williams and Park (Sea Level Rise chapter) to assess the impact of changing salinity (as determined in Williams' project) and sea level rise on the wetlands within San Francisco Bay.

- Effects on California Agriculture - Dudek, Environmental Defense Fund.

Dudek (volume x) simulated the impact of changing climate on California agriculture. Besides using the climate data from the different climate scenarios, Dudek used estimates of mean annual water deliveries for irrigation under the different climate scenarios as input to his model of agricultural effects. One of the outputs of Dudek's study is an estimate of how water usage by agriculture could change under the different climate scenarios. This information was qualitatively used to compare available future water supplies and future water demand (Figure 4-3).

- Water Quality of Subalpine Lakes - Byron, Jassby, and Goldman, University of California at Davis.

Byron et al. (volume x) studied the impact of climate change on the water quality of a subalpine lake in northern California (Figure 4-4).

National Studies That Included California Results

o Economic Effects on Agriculture - Adams and Glyer, Oregon State University, and McCarl, Texas A&M University.

Adams et al. (volume x) conducted a national study of agriculture to estimate shifts in land and water use, and results pertaining to California are discussed in this chapter.

o Response of Vegetation to Past Climatic Changes - Davis, University of Arizona.

Davis (volume x) reconstructed the vegetation present in the Sierra Nevada during warm analogue periods of the Holocene to estimate the potential impact of warming on the present day vegetation in these mountains (Figure 4-4).

o Electrical Demand - Linder and Inglis, ICF, Inc.

As part of a national study, Linder and Inglis (volume x) estimated future northern Californian electrical demands in response to climatic change.

Air Pollution - Morris et al., Systems Applications, Inc.

Morris et al. (volume x) describe possible interactions of climatic change and air pollution. Results pertaining to California are discussed in this chapter.

CALIFORNIA REGIONAL CLIMATE SCENARIOS

Results from two GCM gridpoints were used to drive the effects models used in the California studies. (See the Methods chapter of this report for a discussion of how scenarios were developed and applied.) Both gridpoints lie at 120°W, with the northern gridpoint near the Oregon-California border and the southern gridpoint south of Sacramento (Figure 4-1). Average temperature and precipitation changes for both gridpoints are displayed in Figure 4-5. Generally large seasonal increase in mean temperature are projected by the models. Winter temperatures are between 1.9°C (OSU) and 4.9°C (GISS) warmer, and summer temperatures are between 3.1°C (OSU) and 4.5°C (GISS and GFDL) warmer. The OSU model generally projects less warming than the other two GCM models.

Annual precipitation increases in GISS by 0.28 mm/day, remains virtually unchanges in GFDL, and falls in OSU by 0.1 mm/day. Seasonal changes are more varied with spring rainfall in GFDL, 0.17 mm/day lower, with summer and fall rainfall in OSU being 0.32 to 0.35 mm/day lower. Rainfall in GISS increases in all seasons.

Overall, the OSU scenario represents less of a change from the present climate, and GFDL and GISS show larger temperature changes. The GISS scenario has higher precipitation than the other two scenarios. Generally, temperature increases are larger in the northern gridpoints than in the southern gridpoints. Annual precipitation is higher in the north in GISS and shows little regional difference for the other models.

RESULTS OF THE CALIFORNIA STUDIES

Hydrology of Catchments in the Central Valley Basin

Changes in mountain snowpack and runoff could have a major impact on water supply and quality in the Central Valley Basin. A hydrologic modeling approach was used by Lettenmaier et al. to simulate runoff under different climate scenarios, and these estimates then served as input to the simulation of the response of the water resource systems in the Central Valley Basin to climatic change (Sheer and Randall, Volume x).

Study Design

The approach taken was to model the hydrologic response of four representative medium-sized catchments in the Central Valley Basin and based on these four catchments, estimate streamflows for 13 larger subbasins in the Central Valley Basin. The four catchments chosen (Figure 4-4) for modeling range in size from 526 to 927 km^2 (203 to 358 mi^2). Outflows for each basin were determined using two hydrologic models that estimate snow accumulation, ablation, and daily runoff. The models were calibrated using a subset of the historic record and were verified using an independent subset of the data.

Lettenmaier et al. developed two additional climate scenarios to test the sensitivity of the results to changes in

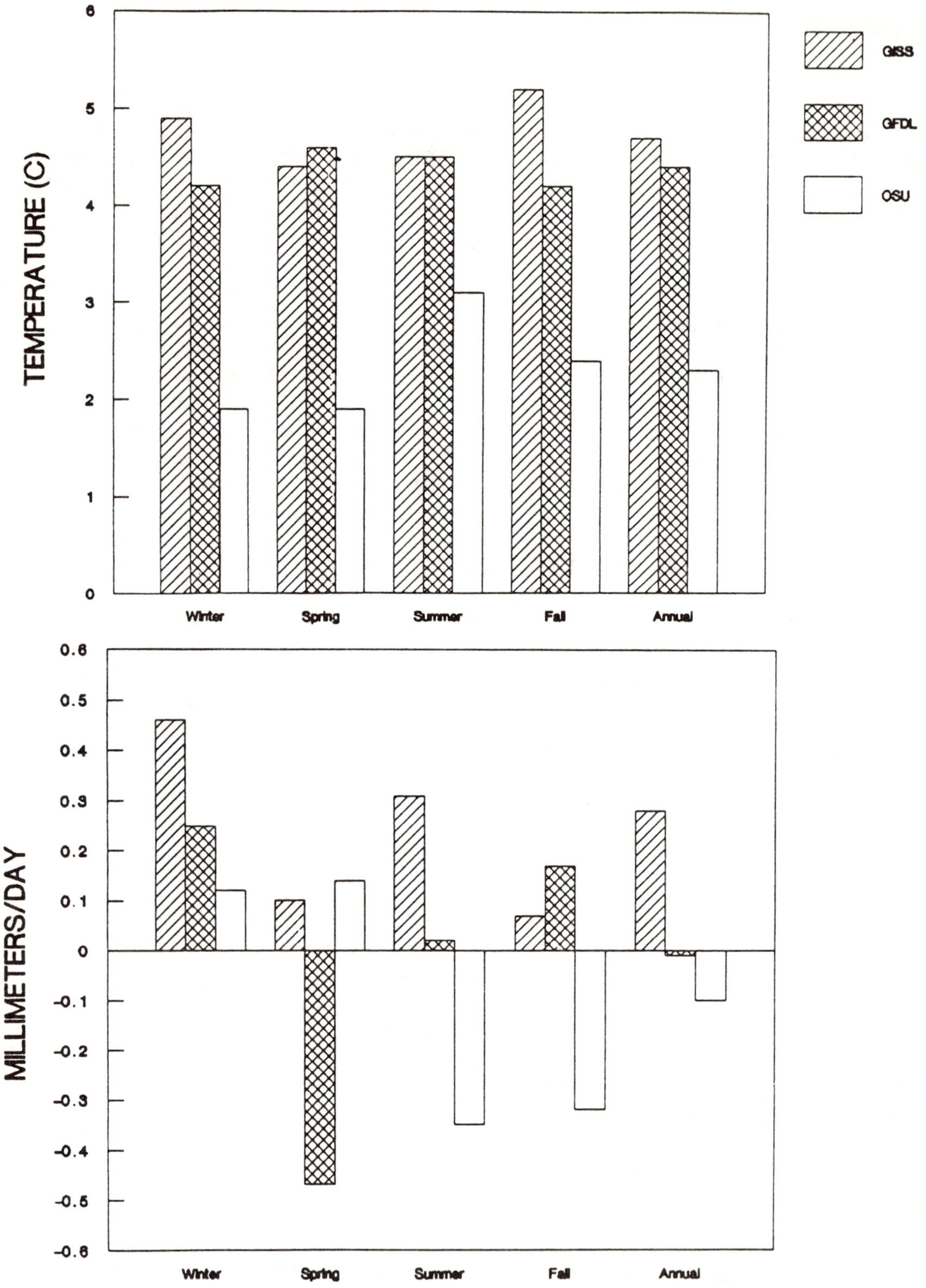

Figure 4-5. General Circulation Model (GCM) scenario results showing seasonal and annual (a) temperature changes and annual (b) precipitation ratios between GCM model runs at double current carbon between GCM model runs at double current carbon dioxide concentrations ($2xCO_2$) and at current dioxide concentrations ($1xCO_2$).

the scenarios. First, to determine the sensitivity to temperature alone, they developed a scenario with only the GISS $2xCO_2$ temperature estimates used and precipitation unchanged from the current values. Historic meteorologic data from the dry decade from 1931 to 1940 served as another scenario.

To provide input for the water resource simulation model of Sheer and Randall (Volume x), Lettenmaier et al. developed a statistical model that relates the historic flows in the four study catchments to historic flows from the 13 larger subbasins in the Central Valley basin. This statistical model was then used to estimate the flows in the 13 subbasins under the different climate scenarios.

Limitations

Results would be different if geographic and temporal variability were not held constant within each grid. Several assumptions made in this study are important considerations in terms of limitations of the results. The intensity of rainfall is the same. Fewer rainfall events of higher intensity could increase runoff relatively more than a greater number of rainfall events of lower intensity. One implicit assumption is that no long-term changes in vegetation cover and composition would occur, when in fact such changes are virtually certain (but their hydrologic manifestations are difficult to predict). If vegetation cover decreases, runoff could increase since less precipitation will be absorbed. Further, decreased cover may result in reductions in actual evapotranspiration and increased runoff.

Lettenmaier et al. assumed that the flows into the water resource system are adequately estimated from the study catchment flows using their statistical model. One limitation of this model is that the study catchments are at high elevations and their runoff is strongly affected by changes in snowfall, whereas some of the areas contributing runoff to the water resource system are at lower elevations with runoff driven primarily by rainfall under present climatic conditions. Since the principal change in the scenarios is a change in snowfall accumulation patterns, the statistical model is biased toward these effects and may somewhat overestimate the total effect of snowfall change on the water resource system. However, because basins at lower elevations have a relatively small impact on total hydrology, there is little bias in this approach.

Nevertheless, the results from this study are qualitatively robust, and any improvement in the hydrologic modeling probably would not alter the general nature of the results, although their precision probably would increase.

Results

Total annual runoff from the four subbasins remains about the same or increases slightly under the double CO_2 scenarios, but major changes occur in the seasonality of the runoff. Runoff could be higher in the winter months than it is today, because less of the precipitation would fall as snow and the snowpack could melt earlier (Figure 4-6a). As a consequence of higher early winter snowmelt, spring and summer runoff substantially decreases under these scenarios. The variability of the runoff could substantially increase in the winter months. Winter soil moisture could increase; evapotranspiration could increase in the spring; and late spring, summer, and fall soil moisture could decrease.

When only temperature changes are incorporated into the climate scenario and precipitation is held equal to the base case, total annual runoff is lower in all four catchments than in the scenario in which both temperature and precipitation were changed (Figure 4-6). However, the seasonal shift in runoff, which is the dominant effect of a general warming, is similar.

The scenario producing the most different results compared to the other scenarios is the 1930's analogue. In this case, runoff is lower in most months in the four subbasins, but the seasonal distribution of runoff is similar to the base case (Figure 4-6b). The reason for this difference is that the 1930's drought was mainly caused by a reduction in precipitation, rather than an increase in temperature, which is similar to the base case.

These results are consistent with those of Gleick (1987b), in that higher temperatures cause a major change in the seasonality of runoff. Since two different modeling approaches using many climate change scenarios produced similar results, these results can be viewed as relatively robust.

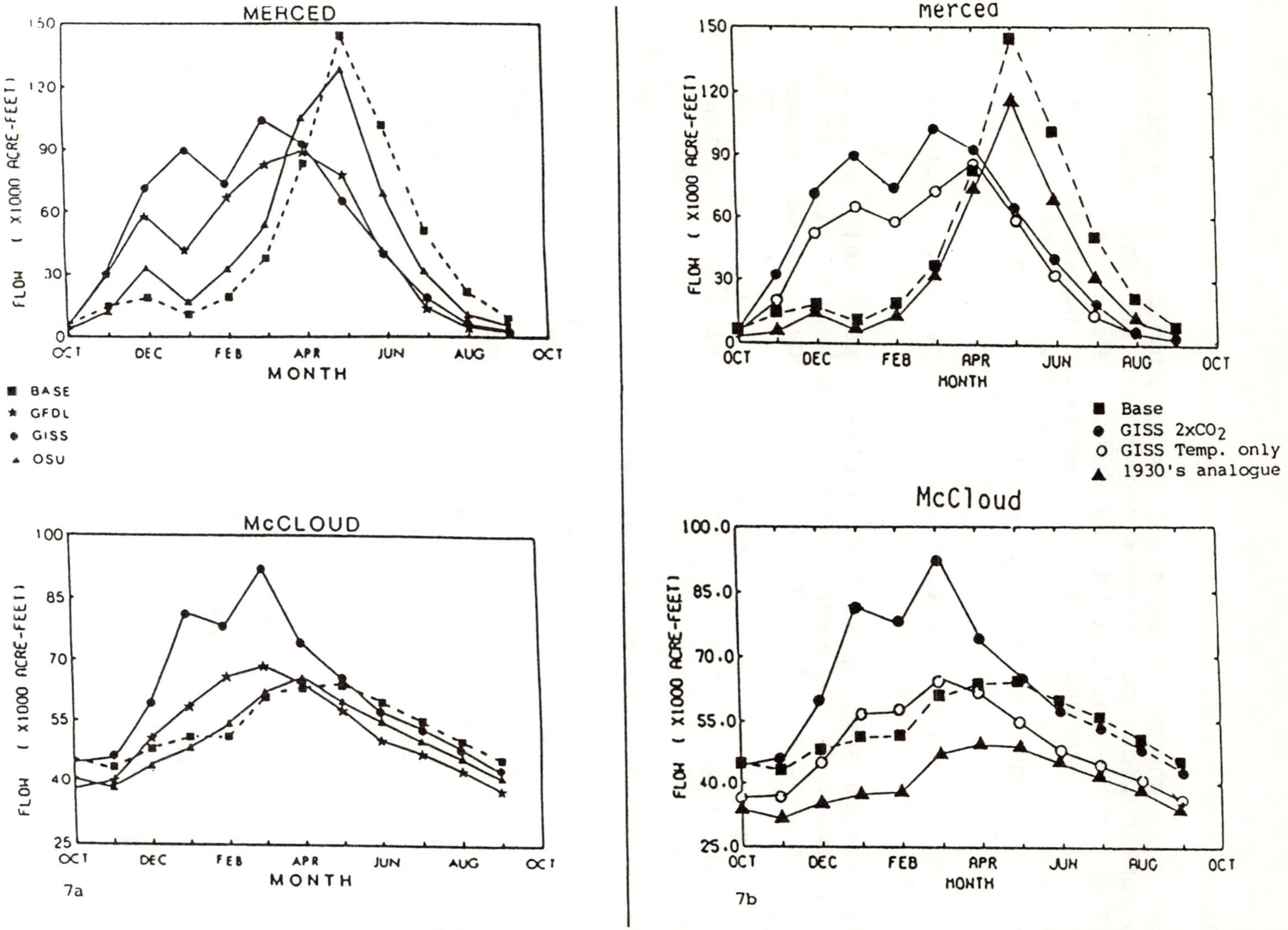

Figure 4-6. Mean monthly streamflows under different climate scenarios in two study catchments (see Figure 5 for locations of the study catchments). a. Results from the three GCM scenarios. b. Results from the scenario incorporating only the temperature change projected in the GISS model run, and from the 1930's analogue scenario.

Source: Lettermann et al., Volume __.

Implications

The potential change in seasonality of runoff could have significant implications for stream ecosystems and the water resource system in the Central Valley Basin. Reduction in streamflows in the late spring and summer could negatively affect aquatic organisms simply because of decreased water volume. Wildlife using streams for food and water also could be harmed. Water quality probably would be degraded because pollutants could become more concentrated in the streams as their flows decrease. The possible impacts on the water resource system are discussed in the next section.

The decrease in spring, summer, and fall soil moisture could have a strong impact on the vegetation in the basin, with plants adapted to drier conditions becoming more abundant at the expense of plants adapted to higher moisture conditions. These potential vegetation changes also could affect wildlife, and perhaps water quality, through changes in the nutrient composition of upland runoff and changes in erosion rates.

Water Resources in the Central Valley Basin

Changes in runoff under the different climate scenarios could have a major impact on the water resources in the Central Valley. The study by Sheer and Randall (Volume x) was designed to use estimates from Lettenmaier et al. of streamflows into the Central Valley to simulate how the water resource system would perform under these conditions. Particular importance is given to how water deliveries to users would be affected by climate change.

Study Design

Sheer and Randall used an existing model of the water resource system of California that is currently used by the Metropolitan Water District (MWD) of southern California to estimate the impact of the climate scenarios on water deliveries (Sheer and Baeck, 1987). The model emulates the State of California's Department of Water Resources Planning Simulation Model (California Department of Water Resources, 1986). The model used hydrologic inputs to project water use demands, instream and delta outflow requirements, and reservoir operating policies. Water demands were set at levels projected for 1990.

Two different sets of runs were made with the model. The first involved running the model for the different climate scenarios using current carriage water requirements. Williams (see Salinity in San Francisco Bay section of this chapter) determined that in response to rising sea level and levee failure, carriage water might have to be doubled to maintain the water quality at the delta pumping plants (see Figure 4-2). Consequently, Sheer and Randall ran the model a second time to determine the effects of doubling the carriage water requirement on water deliveries. Both simulations were run with a monthly time step, with water deliveries summarized on a yearly basis. Interannual variation is used as an indicator of reliability of delivery.

Sheer held a meeting with representatives of the State Department of Water Resources and the Bureau of Reclamation to discuss the results of his analyses and to get their response on how the water resource system would handle the changes in runoff.

Limitations

The limitations to Lettenmaier's study carry over to this one. Thus, interpretation of the results of the simulation of the water resource system's response to climatic change should focus on how the system deals with the change in seasonality of runoff, rather than the absolute values of the model output. Also, the model was run using 1990 conditions, and changes in future management practices, operating rules, physical facilities, water marketing, agriculture, and demand are not considered in the simulation.

Results

The simulation results suggest that both the amount and reliability of water deliveries could decrease after global warming. The decreases in mean annual SWP deliveries range from 7% (OSU) to 14% (GFDL) to 16% (GISS). In some years, the decreases are over 20% for all three double CO_2 scenarios. The projected decrease in water deliveries occurs despite a slight increase in precipitation over current levels in the climate scenarios and greater total outflow from the delta. Average monthly outflow from the delta increases in the late fall and winter under the climate scenarios and is lower in the spring (Figure 4-7).

The driving factor behind this decrease is the change in seasonality of runoff. Higher winter temperatures would lead to more of the winter precipitation in the mountains falling as rain rather than snow, and also to an earlier melt of the snowpack. Consequently more water would flow into the system early in the rainy season. Given current operating rules, the current reservoirs would not have the capacity to store the early runoff and at the same time retain flood control capabilities. Thus, in the simulation, much of the early runoff must be released, resulting in less water in storage at the end of the rainy season and lower water deliveries during the dry summer months. With system changes, it would be possible to store the extra runoff. The shift in the seasonality of runoff and the response of the water resource system to that shift drives the changes in monthly delta outflow (Figure 4-7).

Doubling the carriage water requirement in the model run for the GISS scenario only minimally affects SWP deliveries. This is because the baseline period (1951-80) does not include a lengthy drought period, during which the doubled carriage water requirement could have a substantial impact on deliveries.

The consensus of the meeting of the representatives from the State DWR and the Bureau of Reclamation concerning the potential changes in the seasonality of runoff is that the magnitude of this change is such that <u>operational</u> changes alone would not markedly improve the system's performance. One factor limiting the potential for adjusting the system to the projected changes is the likely need to provide for additional flood control storage during the winter months because of higher peak flows.

<u>Implications</u>

Under the three double CO_2 climate scenarios, water deliveries are less than the base case and could fall short of 1990 demand. Moreover, if carriage water requirements are doubled, shortages during a prolonged drought could become more significant. In addition, users and managers project a water demand of 3.6 million acre feet (maf) by 2010 (California Department of Water Resources, 1983). This demand could not be reliably supplied under the current climate and resource system, and the shortage might be exacerbated under the three double CO_2 scenarios. The potential decrease in water deliveries could affect urban, agricultural, and industrial water users in the state. How the potential decrease should be managed has many policy implications, which are discussed at the end of this chapter.

On the other hand, the increase in delta outflow shows that more water could flow through the Central Valley Basin under these scenarios, and water deliveries could be increased if major new storage facilities were constructed. However, this would be an environmentally and politically controversial option (see Policy section).

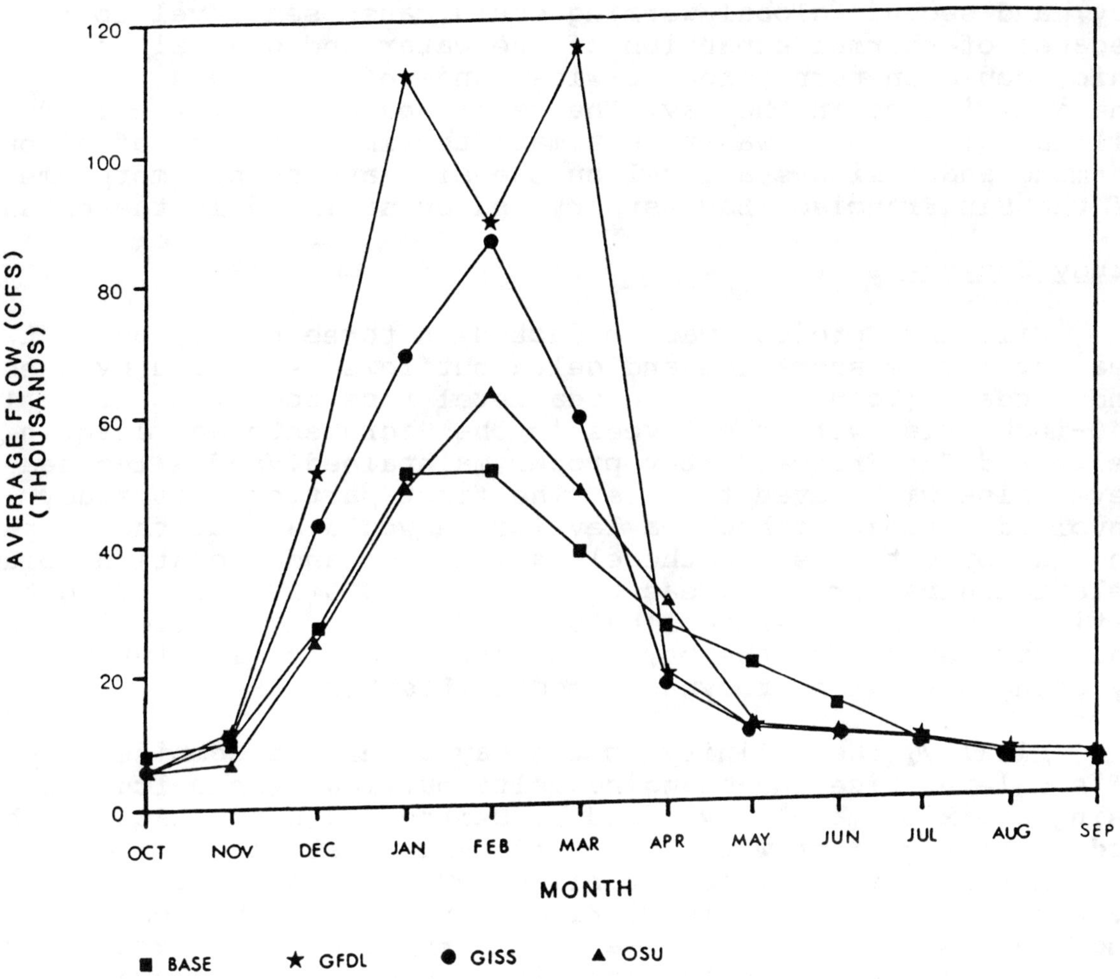

Figure 4-7. Projected monthly Delta Outflows under Different General Circulation Model Climate Scenarios.

Source: Adapted from Sheer, Volume __.

Salinity in San Francisco Bay

Climate change could affect the San Francisco Bay estuary in two ways: first, changes in precipitation and temperatures could affect the amount of freshwater runoff that will flow into the bay; and second, global warming could cause sea level to rise because of thermal expansion of the water and glacial melting, which could in turn affect a wide range of physical characteristics in the bay. The major objective of the study by Williams (Volume x) was to estimate the implications of global warming and rising sea level on the size and shape (morphometry) of the San Francisco Bay estuary and on salinity in the estuary.

Study Design

Williams' project was conducted in three parts, using two sea level rise scenarios and delta outflows estimated by Sheer and Randall (Volume x). The sea level rise scenarios are a 1-m (40-inch) rise with the levees in the Sacramento San Joaquin Delta and San Francisco Bay proper maintained, and a 1-m sea level rise with levee failure. The first part of this study involved estimating how sea level rise would affect the shape of the bay by establishing the elevation/area and elevation/volume relationships for all areas below +3m (+10.0 ft) according to National Geodetic Vertical Datum (NGVD). Then the tidal exchange characteristics for the bay were determined for its future shape by using a tidal hydrodynamic model (Fischer, 1970).

Finally, the salinity in the bay under the combined impacts of sea level rise and changing delta outflows was calculated by using a mixing model developed by Denton and Hunt (1986). This model was run first with nine different constant delta outflows (all months the same) to establish new carriage water requirements after sea level rise. Once these were established, and Sheer and Randall (Volume x) had run their simulation model with the new requirements, the mixing model was run again to determine the salinity regime in the estuary after climatic change. Included in the model output are average monthly and average annual salinities in different parts of the estuary under the different scenarios.

Limitations

Because of the short time frame, Williams used some old and inaccurate surveys in the morphometric analysis instead of making new surveys. These could produce errors of plus or minus 20% in the estimates of the estuary's volume. In addition, some levees probably will be maintained under any delta management plan, and thus the flooding of the delta islands would not be as extensive as assumed in the levee failure scenario. Williams did not consider changes in siltation and erosion of sediments that would likely occur under the different climate change scenarios. However, erosion would probably have a significant impact on water flow in the delta. For instance, deepening of the tidal channels in the delta could lead to intrusion of salinity farther upstream than projected in this study. In addition, more sophisticated models of salinity and tidal ranges and exchanges might improve the accuracy of the results. Finally, the new carriage water requirements were based on a steady-state analysis (e.g., constant delta outflows). Changes in the hydraulics of the Sacramento, San Joaquin delta and Suisun Bay with sea level rise could increase these requirements. Williams' results should be viewed as a preliminary estimate of estuarine changes, with emphasis placed on the direction of change, rather than on the absolute amount of change.

Results

The morphometric analyses suggest that given a 1-m (40-inch) sea level rise and failure of the levees, the total area of the estuary might triple, and its volume could double. If the levees are maintained, the increases in area and volume could be about 30 and 15%, respectively (see Figure 4-2). Interestingly, the amount of sea level rise is less important to the physical size of the bay than whether or not the levees are maintained.

Under the sea level rise scenarios with levees maintained, tidal ranges do not change significantly from current conditions. If levees fail, downstream constrictions at Carquinez Strait and to the east of Suisun Bay (see Figure 4-2) limit tidal transport and reduce tidal range in the delta, assuming that erosion does not alter the tidal characteristics of the delta.

The results from the initial application of the salinity model to constant delta outflows indicate that monthly carriage water requirements might have to be doubled to repel saline water from the upper part of the delta. Also, whether or not the levees are maintained has little effect on the salinity regimes in the bay according to the model's results. However, because possible scouring of tidal channels is not incorporated into the model, the predicted salinity after levee failure is probably underestimated.

Using Sheer and Randall's estimated delta outflow with double carriage water, Williams also estimated annual salinity in the bay. The results suggest that after a climatic warming, a 1-

m sea level rise, and failure of the levees, water of a given average annual salinity could migrate inland between 4 km (2.5 miles) (GFDL scenario) and 9.6 km (6 miles) (OSU scenario) (Figure 4-8).

Williams also calculated the average monthly salinity for Suisun Bay for the three climate scenarios, levee failure, and double carriage water requirements. Monthly salinities are higher for all months compared to the base case, except for winter and early spring months in the GFDL scenario. The greatly increased runoff of the GFDL scenario (Figure 4-7) during these months kept the salinity at the same level as the base case. Williams also modeled how the frequency of a given salinity value in any month would change. For example, in June salinities that were exceeded in 50% of the years in the base case might be exceeded in 80% of the years in both the GFDL and OSU scenarios, because of the lower outflows predicted under these scenarios.

Implications

Rising sea level could place the delta islands under increased risk of inundation, not only because of higher water levels but also because the larger area and volume of the San Francisco Bay estuary could result in greater wave energy and higher erosion rates of the levees. Improving the levees just to protect them against flooding at the current sea level could cost at least $4 billion (California Department of Water Resources, 1982). With higher sea levels, the cost of maintaining the levees could increase.

The large water body created if all the levees fail will have a longer water residence time. This means that any contamination (salt or other pollutant) will be more difficult to flush out of the delta region. Also, if saline water fills the islands when levees fail, significant amounts of freshwater will be needed to flush out the salt.

Increasing salinity could necessitate increases in carriage water to maintain freshwater at the export point in the delta or could require developing a different method to convey freshwater from reservoirs to users. The increase in carriage water coupled with the decrease in reservoir storage would most likely mean reduction in water deliveries to at least some of the system's users during extended droughts. With higher future water demands, shortages caused by the higher carriage water requirements may not be limited to extended droughts.

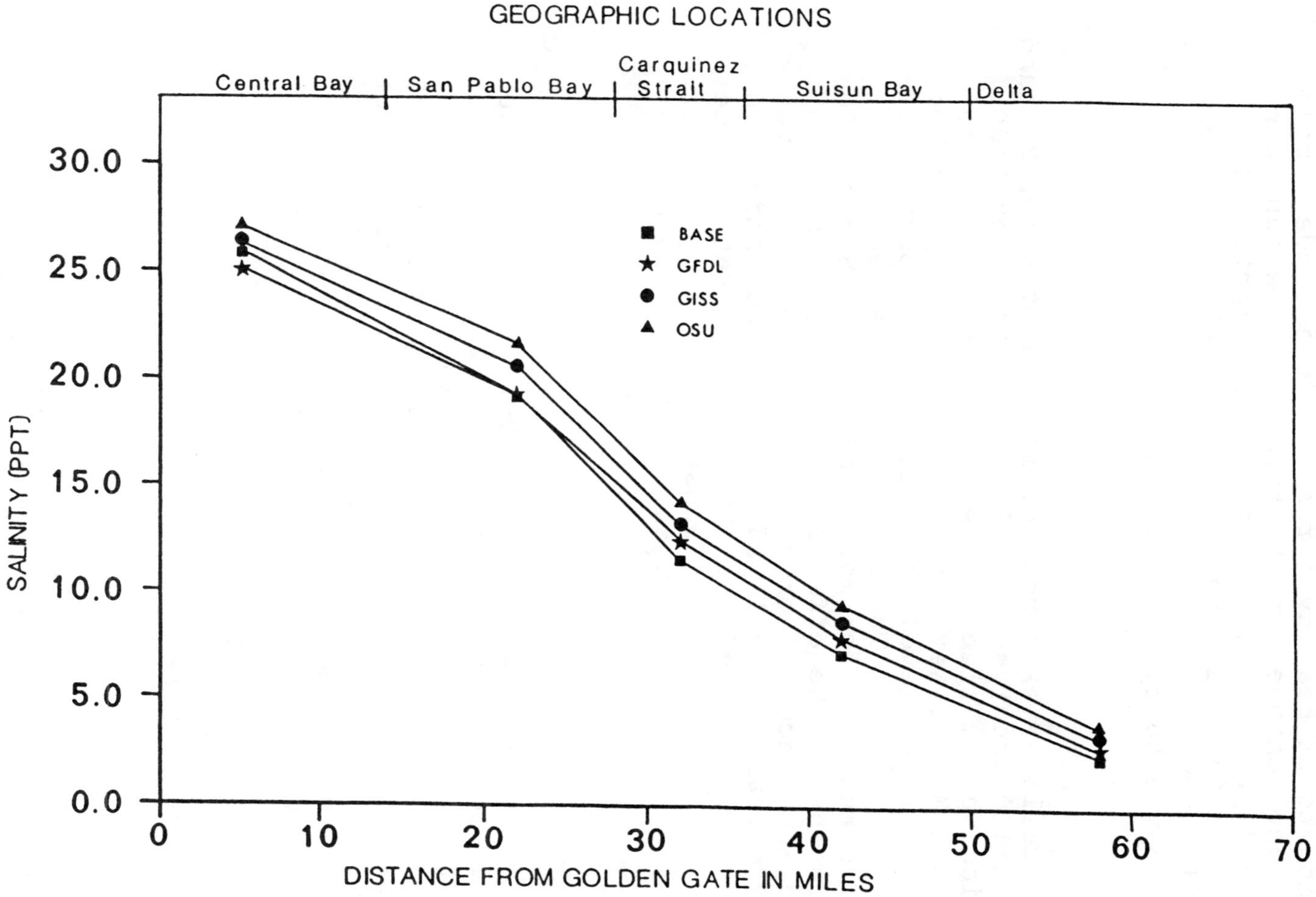

Figure 4-8. Average annual salinity in San Francisco Bay estuary under different hydrology scenarios versus distance from the Golden Gate. The corresponding geographic location of a particular distance is given at the top of the graph.

Source: Williams, Volume __.

Wetlands in the San Francisco Bay Estuary

Climatic warming could alter two important physical factors that affect wetland distribution: sea level and freshwater inflow. Major impacts of sea level rise could include erosion and marsh inundation. Changes in freshwater inflow can change the distribution and productivity of estuarine plants and animals. Josselyn and Callaway (Volume x) estimated the possible effects of climatic warming on deep-water and wetland habitats of the San Francisco Bay estuary (see Figure 4-2).

Study Design

Josselyn and Callaway examined the impacts of a 1-, 2-, and 3-m (40-, 80-, and 120-inch) sea level rise by the year 2100. Of the three scenarios, a 1-m rise by the year 2100 is regarded as the most realistic. Models were used to project nearly linear changes in the rate of sea level rise from 1990 through 2100 under these three scenarios. The relationship between sedimentation rates required for marsh maintenance and sea level rise rates was examined. The effects of salinity changes on the distributions and abundances of organisms were related to various freshwater inflow scenarios developed by Sheer and Randall (see Figure 4-7). In the absence of appropriate quantitative models, biotic changes in the estuary in response to changing salinity were qualitatively determined based on literature review and expert judgment.

Limitations

Circulation and sedimentation in the estuary will change dramatically as sea level rises and if levees fail. The specific characteristics of these biologically important changes are unknown at present and were not considered in this study. The sea level rise scenarios did not consider the possibilities of sudden changes in sea level. Increased water temperature, which may directly affect the reproduction, growth, and survival of estuarine organisms, or may have an indirect effect through changes in oxygen availability, also was not considered. Although specific impacts on plants and animals in the estuary are difficult to assess, the general impacts would most likely be similar to those reported here.

Results

Rates of sea level rise from 1990 to 2040 for the three scenarios are presented in Figure 4-9. Once the rate of sea level rise exceeds the rate of sediment accretion, tidal marsh habitats would become inundated and erosion of the marsh edge could increase. For the 1-m rise scenario, the rate of rise is not projected to exceed maximum accretion rates until about the year 2040, while the 2- and 3-m (80- and 120-inch) rise scenarios suggest rise rates in excess of accretion rates after 2010 and 2000, respectively.

Peak primary productivity, at present, occurs in early spring in San Pablo Bay and in the summer in Suisun Bay. These maximum productivity levels could be substantially reduced, particularly for brackish and freshwater plant species, under the higher salinities of the OSU scenario (see Figure 4-8), and peak spring production might shift upstream into the delta if levees fail. However, under GFDL and GISS conditions, the locations of maximum production levels might remain in their present positions if the levees are maintained. If the levees fail, primary production could increase in the extensive shallow water and mudflat areas created.

Since many areas currently protected by levees are 1 to 2 m (40 to 80 inches) or more below sea level, levee failure would cause them to become deepwater areas rather than marshes. Eventually, some of these formerly leveed areas might receive enough sediment deposition to support marsh development. Inundation of marshes and salinity impacts on freshwater and brackish-water plant species could reduce food and cover sources for waterfowl (see Figure 4-9). Loss of emergent vegetation could significantly reduce the numbers of migratory waterfowl using the managed wetlands along Suisun Bay's north shore.

If levees are maintained under conditions of sea level rise, salt may build up behind them from the evaporation of standing water. This salt will cause marsh vegetation to die back and will reduce the value of these wetlands to wildlife.

Freshwater inflows projected during springtime under the climatic change scenarios (see Figure 4-7) may be too low to support anadromous fish (saltwater fish that enter freshwater areas for spawning). Lower inflows could result in declines among these populations in this region (Kjeldson et al., 1981).

If levees fail, a large inland lake with fresh to brackish water quality could be created in the delta. Striped bass and shad spawn in essentially freshwater conditions and their spawning could be reduced under increased salinity, especially if they did not move upstream to relatively fresh water. Marine fish species could increase in abundance in the Suisun and San

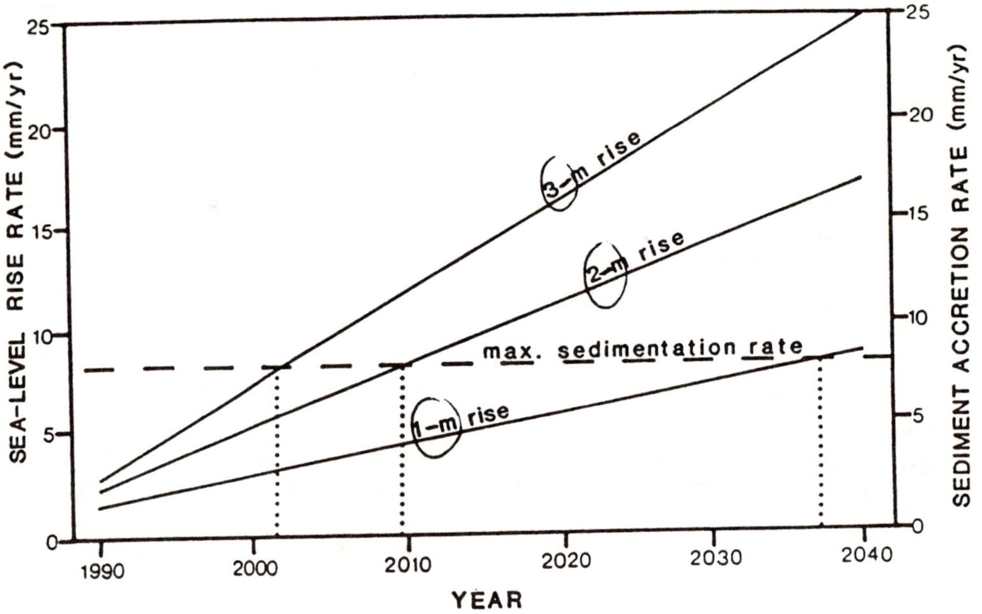

Figure 4-9. Sea level rates at San Francisco from 1990 to 2040 under 1-, 2-, and 3-m (40-, 80-, 120-inch) sea level rise scenarios. When the rate of sea level exceeds maximum recorded rates of sediment accretion, marsh habitats would most likely be converted to deeper-water habitats.

Source: Josselyn and Callaway, Volume __.

Pablo Bays in response to the higher salinities projected, and freshwater and anadromous species could decrease.

Implications

The loss of wetlands could result in substantial ecological and economic losses for the region. For example, the managed wetlands north of Suisun Bay support a hunting and fishing industry valued at over $150 million annually (Meyer, 1987). Tourism, hunting, fishing, rare and endangered species, and heritage values also could suffer.

California Agriculture

California's agricultural production is highly dependent on irrigation which accounts for approximately 80% of the states net annual water use. Dudek (volume x) used existing agro-ecological models to explore potential responses of California agriculture to climatic change.

Study Design

Climatic changes from the GISS and GFDL 2xCO_2 scenarios were linked to an agricultural productivity model adapted from Doorenbos and Kassam (1979). Growth responses to both climatic change and climatic change plus CO_2 direct effects were modeled. These productivity responses were then introduced into the California Agriculture and Resources Model (CARM; Howitt and Mean, 1985) which estimates the economic and market implications of such changes. Mean surface water supplies under the base, GISS, and GFDL scenarios were calculated from the simulations of Sheer and Randall (volume x) and were input to CARM.

Limitations

Further, the CO_2 direct effects results should be viewed as preliminary since they are based on data from growth chamber experiments that may poorly represent field conditions. This study did not consider changes in agricultural technology, crop varieties, planting dates, or changes in water costs. Also, new crop/location combinations were not considered. The interaction between climatic change and CO_2 direct effects on productivity were not examined but may significantly limit potential growth increases. No national or international context of agricultural price signals were considered. These could be a major factor in determining how California farmers respond to climatic change. Given these limitations, realistic estimates of agricultural responses to climatic change may be difficult to obtain. The results may be more valuable as indications of potential trends rather than specific impacts.

Results

Relative to the 1985 base, average yields could be significantly reduced for California crops in response to climatic changes alone (i.e., without consideration of direct effects of CO_2). Generally, the greatest impacts are projected under the hotter GISS scenario. Table 1 presents regional yield changes for sugarbeets, corn, cotton, and tomatoes. These projections were generated by the agricultural productivity model and do not consider economic adustments or water supply limitations. Tomatoes might suffer the least damage, with average yields reduced from 5% to 15%. Sugarbeets could be hardest hit, with average declines from 20% to 40%. Yield reductions in sugarbeets are projected to be greatest in the relatively hot interior southern regions. Differences in growth response between the two climatic scenarios were greatest for corn and least for tomatoes.

Without economic adjustments, corn yields declined 20% to 41% as projected by the agricultural productivity model under the GISS scenario (Table 1). With economic adustments, declines of roughly 21% are projected (Table 1); a result at the lower end of the direct productivity impacts.

When the direct effects of CO_2 on crop yields are considered, yields of sugarbeets, cotton, and tomatoes generally increased over the 1985 base (Table 1). Corn is unable to increase growth in response to increases in CO_2 concentration, although yield reductions are not as great as with climate change along (Table 1). Cotton could benefit the most from inadvertant CO_2 fertilization with yield increases ranging from 6% to 41%.

Potential increases in yields in response to CO_2 fertilization might only be achieved at a cost of increased groundwater extraction in many areas. For example, when surface water utilization is projected at 100% of capacity, as in the Northern San Joaquin region, excess demand would require increased groundwater usage (Figure 4-10). However, increased crop yields may offset increased economic costs for water.

Regionally, across all scenarios, (not considering potential changes outside of California) the largest reductions in crop acreage are projected in the Imperial Valley while the delta

Table 1. Regional and statewide percentage yield changes (relative to 1985) under different General Circulation Model climate scenarios. Regional changes are projected by the Doorenbos and Kassam agricultural productivity model while statewide production changes are projected by the California Agriculture and Resources Model (CARM). The latter estimates include economic adjustment "Net" includes the direct effects of increases in CO_2 and climatic change (CC).

		Crop							
		sugarbeets		corn		cotton		tomatoes	
Region[b]	Scenario	CC	Net	CC	Net	CC	Net	CC	Net
South Coast									
Los Angeles	GISS	-26	21	-41	-37	-22	11	-7	18
	GFDL	-20	30	-26	-22	-4	41	-5	21
North Interior									
Red Bluff	GISS	-28	12	-27	-22	-18	22	-15	11
	GFDL	-23	21	-14	-8	-10	35	-13	13
Sacramento Valley									
Sacramento	GISS	-30	17	-20	-14	-17	25	-12	16
	GFDL	-27	22	-6	0	-11	34	-8	21
Southern San Joaquin									
Fresno	GISS	-33	5	-26	-21	-16	25	-15	11
	GFDL	-39	-8	-14	-8	-10	36	-13	14
Southern Deserts									
Blythe	GISS	-40	-2	-31	-27	-28	6	-13	13
	GFDL	-39	0	-14	-8	-19	21	-12	15
CARM Statewide									
	GISS	-30	12	-21	-17	-17	23	-13	5
	GFDL	-26	15	-9	-4	-10	35	-11	16

refer to Figure 1 for city locations.
Source: Dudek, Volume __.

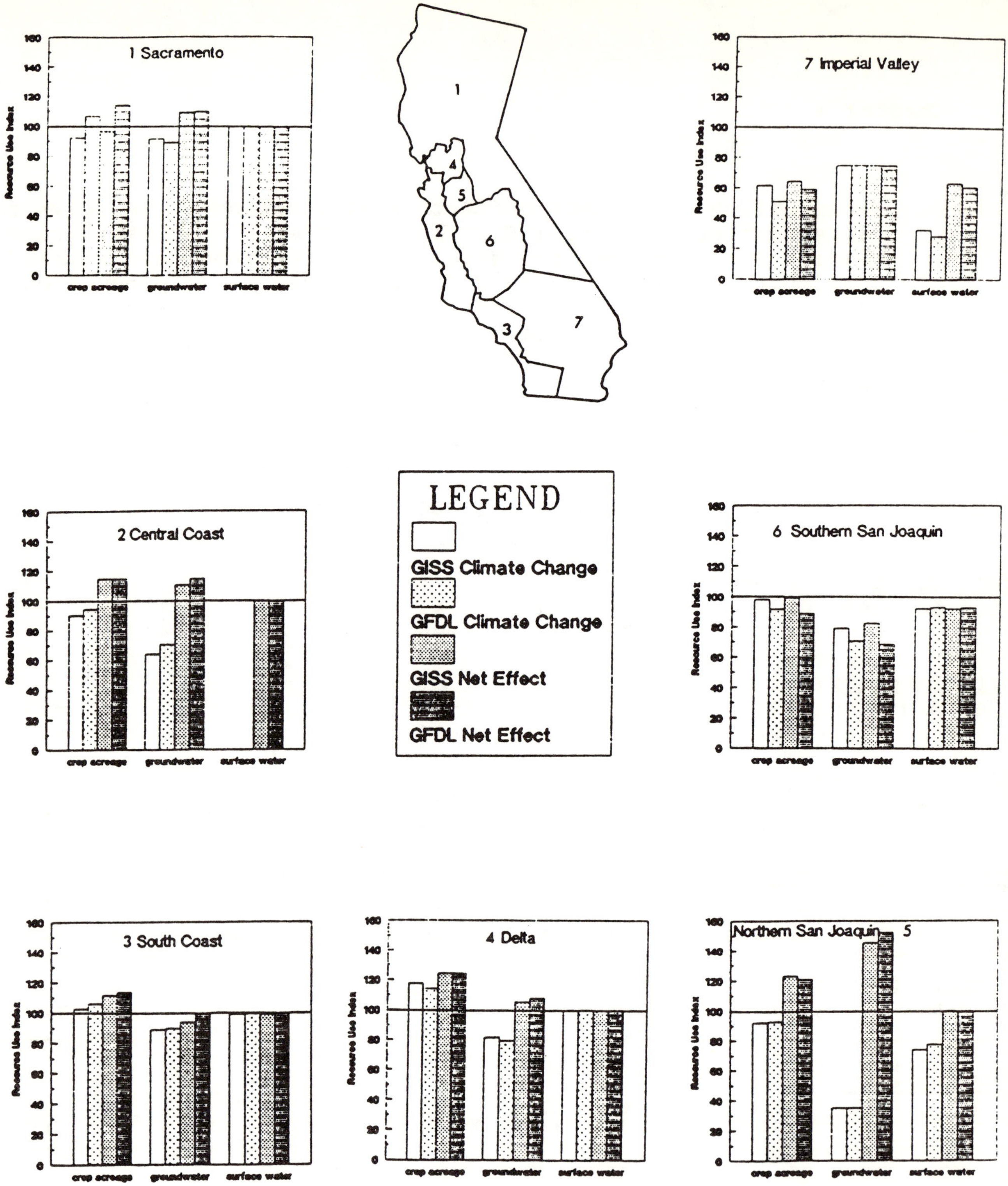

Figure 4-10. Regional crop acreage, groundwater use, and surface water use changes under different General Circulation Model climate scenarios. Net effect includes the direct effects of increases in CO_2 and climatic change.

Source: Dudek, Volume __.

region shows the largest gains in acreage (Figure 4-10). This expansion of agriculture in the delta region would depend on maintenance of levees protecting farmland. Without considering CO_2 fertilization, statewide crop acreage was reduced about 7% from the 1985 base. When CO_2 direct effects are considered, statewide crop acreage approximately equals 1985 base levels.

Implications

Regional changes in cropping locations and patterns of water use imply potential exacerbation of existing nonpoint source pollution and accelerated rates of groundwater overdraft with ensuing environmental impacts.

Changing water supply demands may result in increased conflicts between water users. In addition, shifts in the location of agricultural production could play a substantial role in the future viability of natural systems. Such shifts could also have a significant impact on the economic viability of small agricultural communities.

National Agriculture

Adams, et al. conducted a national agricultural study that includes results relevant to California (Adams et al., volume x). The results of the Adams et al. study are not directly comparable with the results from Dudeks' study (discussed above) since Adams et al. considered national agricultural impacts and aggregated California into a Pacific region with Oregon and Washington. Further, the two studies did not examine the same set of crops and modeled productivity differently (see the Agriculture chapter of this report for a description of the study's design and methodology).

Results

Nationally, Adams et al. (volume x) project 2% to 4% declines in crop acreage in response to climatic change but shows a 18% to 20% increase in acreage in the Pacific coast states. This increase in the Pacific region is attributable to the regions extensive use of irrigated agriculture. In contrast, most other regions of the U.S. predominantly use dryland farming and crop acreage might decline in response to moisture stress. The Adams et al. approach is based on maximizing farmers profits and indicates that higher yields associated with direct CO_2 effects might result in further declines in crop acreage (or, in the case of the Pacific coast states, less of an increase) since less acres might be required to produce the necessary crops.

Water Quality of Subalpine Lakes

Subalpine lakes are common in the California mountains, and many of these are the source of streams and rivers flowing down into the lowlands. Changes in the water quality of these lakes could significantly alter their species composition and nutrient dynamics and also could have an impact on downstream water quality and ecosystems. The sensitivity of California's subalpine lakes to weather variability and climate change has not been extensively studied. Consequently, Byron et al. studied how climate controls the water quality of Castle Lake, a subalpine lake in northern California.

Study Design

Goldman et al. (1989) correlated an index of water quality, primary production (i.e., the amount of biomass produced by algae in the lake), with climate variation at Castle Lake. Consequently, Byron et al. (Volume x) were able to develop empirical models relating primary production with various climate parameters.

Limitations

The model is limited to estimating annual values of primary production; seasonal variability is not calculated. The model also does not project changes in species composition and nutrient dynamics, which could have important consequences for water quality. Changes in upland vegetation and nutrient cycling, which could also affect the lake's water quality, were not part of the model.

The estimates of annual primary production produced by this model are relatively precise, although the results are general in the sense that no species-specific projections are made.

Results

The model projects that mean annual primary production could increase under all three of the double CO_2 scenarios, with increases ranging from 16% (OSU scenario) to 87% (GISS scenario) (Figure 4-11). The OSU results are within one standard error of present productivity, and thus do not represent a significant decrease in water quality. The transient scenario produced generally increasing but variable primary production estimates, with the mean production of the last decade in the transient scenario about 25% higher than the base (control) case.

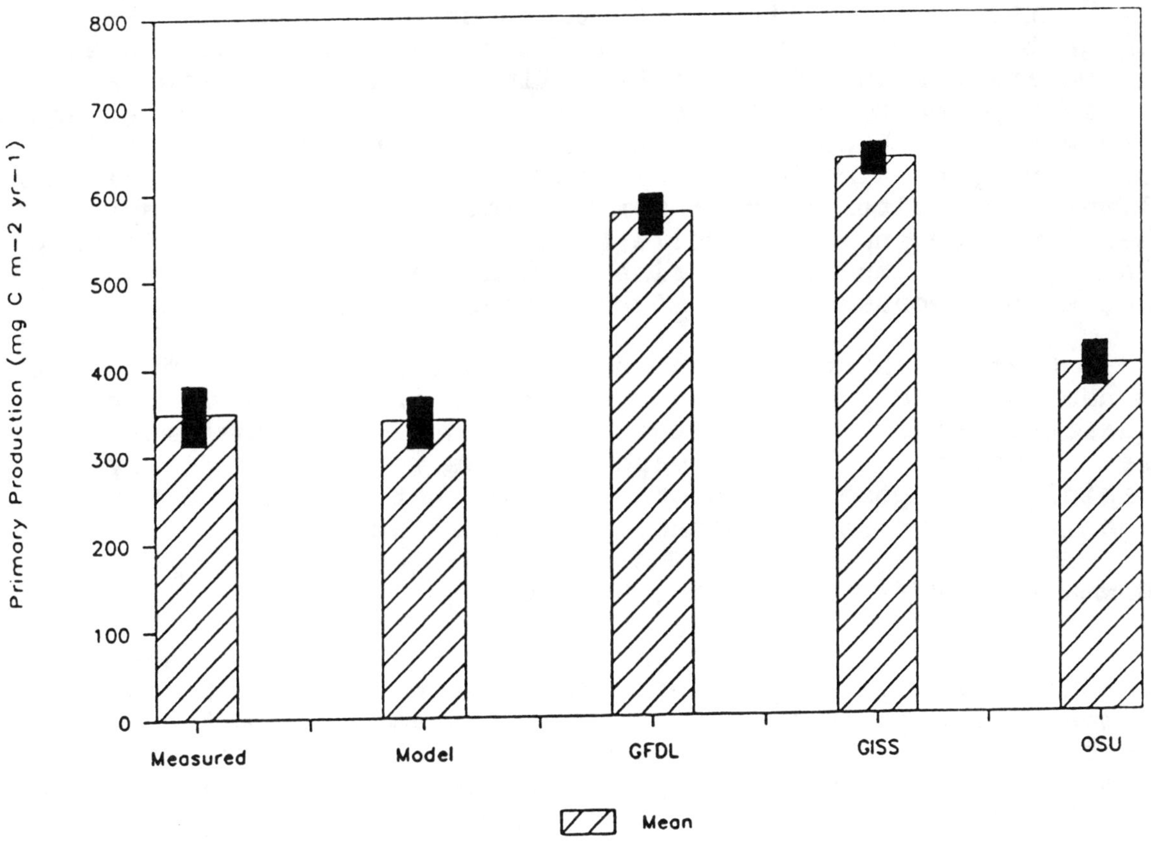

Figure 4-11. Annual primary production estimates for Castle Lake showing actual and model values for present conditions and model values for three General Circulation Model climate scenarios (see Figure 4 for location of Castle Lake). The solid bars show one standard error for each estimate.

The increase in annual primary production is attributed principally to the temperature increase projected by the scenarios. The higher temperatures resulted in less snow accumulation, which is correlated with an earlier melting of the lake ice, and a longer growing season.

Implications

Higher primary production could result in climatic effects being indirectly felt at higher points in the food web in the lake and could affect the lake's nutrient dynamics.

Extrapolating these results to other subalpine lakes suggests their water quality could decrease and their species composition might change after climatic warming. Increased primary production can provide additional food for other aquatic organisms, such as fish, but can also degrade water quality by ultimately causing a decrease in dissolved oxygen and blocking light from filtering to lower levels. Fisheries in unproductive lakes may be enhanced, while trout populations may suffer in lakes where temperatures rise past a threshold value and oxygen decreases because of the higher production.

Changes in production and concomitant changes in nutrient dynamics could affect downstream river and reservoir water quality. However, since the streams draining subalpine lakes are well oxygenated, the increased biomass would most likely be rapidly decomposed and probably would not affect the water quality of the lower reaches of streams and rivers.

Summary of Effects on Water Resources

In terms of economic and social importance, changes in water resources are among the most important possible effects of climatic change in California. A wide variety of factors related to climatic change could affect water resources, ranging from factors changing water supply to those affecting water demand. All the individual projects discussed above addressed some aspect of climatic impacts on water resources in the State. However, these studies did not consider all the major factors that will affect California water resources in the next century, mainly because of the complexity and inherent difficulties in forecasting future demand for water. This section discusses other factors that will affect future water demands that were not directly considered by the individual studies, including future changes in agriculture, population, water efficiencies, and sources of water, including groundwater.

Dudek's study used estimates of water deliveries from Sheer and Randall's study, but changes in agriculture that he determined, and hence changes in agricultural demand for water, are not factored back into the water simulation model. For instance, Dudek's results indicate that because of climatic conditions, crop acreage in the Imperial Valley decreases, freeing water used there for irrigation to be used elsewhere in the State. Also, as cropping patterns change, so would the pattern of needed water transfers via the water resource system, thus affecting water deliveries. Finally, Dudek found that groundwater usage could increase after climatic change when the direct effects of CO_2 are included in his model. Estimated groundwater usage increases when full use of surficial water sources does not meet agricultural demands estimated in the model. Thus, Dudek's results suggest that agricultural demand for water could exceed surficial supplies after climatic warming, further exacerbating water shortages.

Not considered in the overall California study, but critical to determining the magnitude of potential water shortages in the next century, are population growth and accompanying changes in water demands. Projections of population growth place the State's population at about 35 million in 2010 compared with 24 million in 1980, an increase of 45% (California Department of Water Resources, 1983). An increase of this magnitude will significantly intensify municipal water demands, exacerbating shortfalls. Meeting the needs of the growing population in face of competing demands from agriculture raises some important policy questions that are discussed in the policy section.

If water shortages become more common, it is reasonable to assume that agricultural, industrial, and residential users will change water-use efficiency. Changes in efficiency were not considered in any of these projects but could moderate possible future shortages. Any change in water pricing or water law also could affect water demand and supply, but these factors are probably impossible to project very far into the future.

Groundwater usage is discussed in Dudek, but the overall impacts of climatic change on groundwater are not addressed in this project. As demand for water increases beyond the capability of the water resource system to deliver the needed water, mining of groundwater (as Dudek shows for agriculture) is one option users could adopt to meet their demand. Using groundwater could lessen the severity of water shortages in the short term but presents environmental problems, such as land subsidence, over the long term.

In general, given the current water resource system, the projected response of that system to climatic change and qualitative considerations of future changes in water demand suggest that water shortages could be significantly greater than estimated here.

Vegetation of the Sierra Nevada

To better understand the sensitivity of natural vegetation in California, Davis studied how the terrestrial vegetation growing in the California Sierra Nevada has changed over the past 12,000 years. The climatic changes that occurred during this period suggest how the vegetation that currently exists in the mountains could respond to future climatic changes. In particular, the middle latitudes of the Northern Hemisphere are believed to have been warmest (1-3°C warmer than today) about 6,000 years ago (Budyko, 1982), and parts of western North America were apparently warmest 9,000 years ago (Ritchie et al., 1983; Davis et al., 1986). Thus, the period that existed between 6,000 and 9,000 years ago in California could present a possible analogue to a warmer future climate.

Study Design

The composition of the vegetation that has existed in the central Sierra Nevada over the last 12,000 years was determined using fossil pollen analysis. Fossil pollen samples were collected from five lakes situated along an east-west transect (see Figure 4-4) passing through the major vegetation zones of the Sierra Nevada. Dissimilarity values were calculated between modern and fossil pollen samples to determine the past vegetation at a particular site.

Limitations

The climate projected in the three double CO_2 scenarios is different from the climate that probably existed between 6,000 and 9,000 years ago in the Sierra Nevada, according to Davis. The vegetation that persisted then suggests that the climate was drier and a little cooler than it is today. This is a marked contrast to the warmer and slightly wetter climate scenarios used in this report. Consequently, the results from this study do not provide a past analogue for the climate scenarios, but nonetheless present a possible analogue for how Sierra Nevada vegetation could respond to a warmer Northern Hemisphere climate.

A relatively small set of modern pollen samples was available for comparison to the fossil samples; therefore, the precision of the vegetation reconstruction is somewhat uncertain. Also, the precision of the estimated elevational shifts in the vegetation zones is low because of the limited number of fossil sites available for the analysis. Nevertheless, this study provides a good general summary of the vegetation changes in the Sierra Nevada during the past 12,000 years.

Results

The forests existing in the western Sierra Nevada 9,000 years ago resembled those found east of the crest today (Figure 4-12), with lower forest cover and tree density. Pine and fir densities, in particular, were lower. Between 9,000 and 6,000 years ago, the vegetation gradually became similar to the modern vegetation in the same area, and by 6,000 years ago the modern vegetation zones were established on both sides of the Sierra crest. The vegetation at 6,000 years ago is subtly different from that in the area today, with less fir and more sage. The forests may have been slightly more open than today.

Implications

If climatic conditions of the Sierra Nevada in the next century are similar to those that existed 9,000 years ago, major changes could occur in the forests of the region, both in composition and density. The vegetation changes could generate significant environmental impacts, ranging from changes in evapotranspiration and related feedbacks to local hydrology, to changes in nutrient cycling and soils, which could degrade the water quality of mountain streams. Fire frequencies could increase as a function of changes in fuel loads, ignition frequencies, and vegetation. If dead wood rapidly builds up because of the decline in one or more tree species, large catastrophic fires could occur.

If future forests west of the Sierra crest become more similar to current forests east of the crest, timber production could significantly decline. Based on National Forest Inventory data, timberlands east of the crest currently support only about 60% of the wood volume of timberlands west of the crest (U.S. Forest Service, Portland, Oregon, personal communication). The different climate could also necessitate a change in timber practices (e.g., reforestation techniques).

Vegetation change in response to climatic change could produce additional stress for endangered animal species as their preferred habitats change. Populations of nonendangered wildlife also could be affected as vegetation changed, with some species

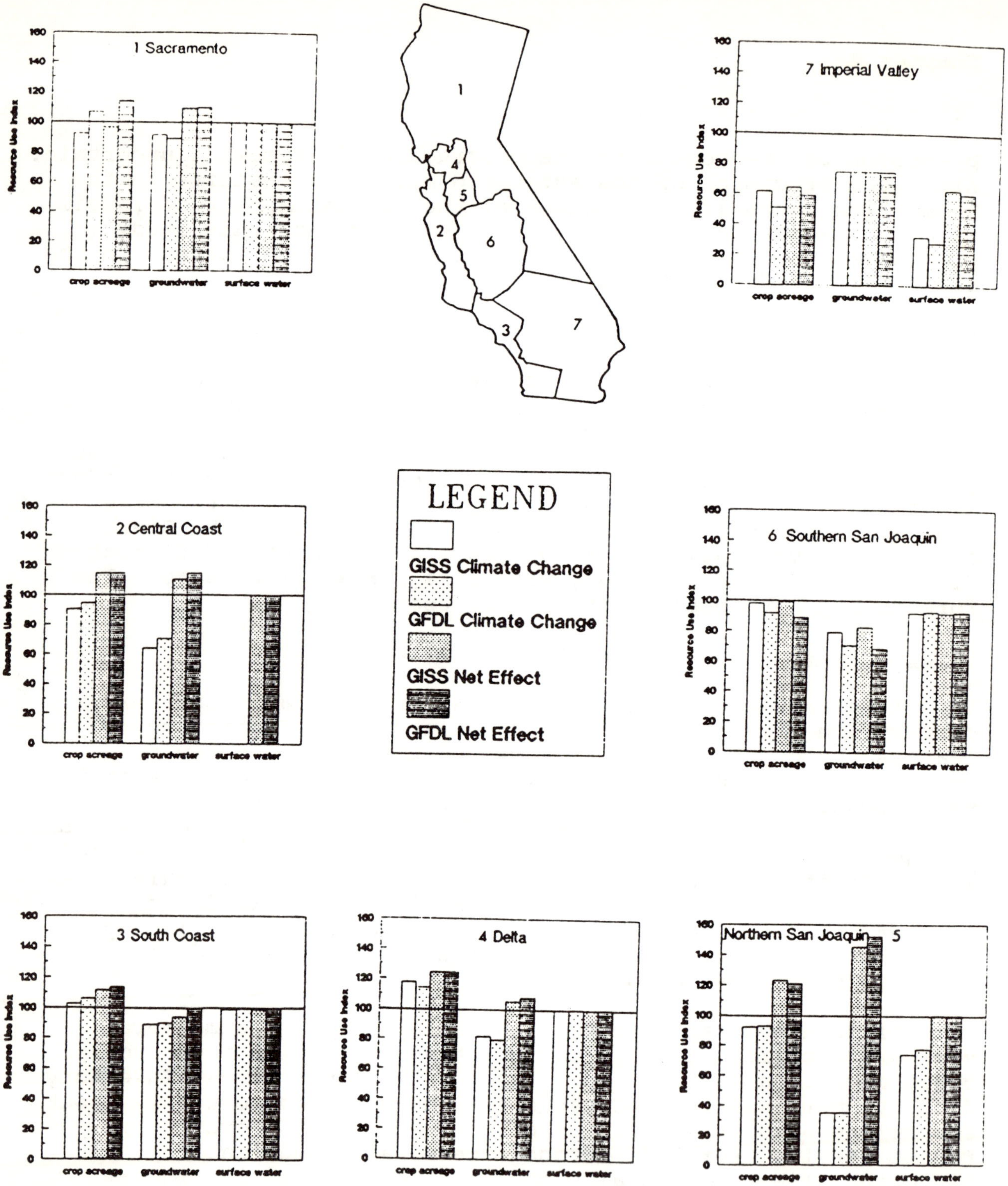

Figure 4-12. Regional crop acreage, groundwater use, and surface water use changes under different General Circulation Model climate scenarios. Net effect includes the direct effects of increases in CO_2 and climatic change.

increasing in abundance and others decreasing. Some plant species, especially alpine species, also could become endangered. Warmer temperatures could favor subalpine species over alpine species at some of the higher elevation peaks and ridges in the Sierra Nevada, leading to local extinction of the alpine species.

Electricity Demand

The electric power industry is highly sensitive to uncertainty in future energy demands associated with potential climate change. As part of a national study, Linder and Inglis estimated northern California's energy demand for the years 2010 and 2055. (See the Energy Demand chapter of this report for a description of the study's design and methodology.)

Results

In California, climatic change scenarios result in only small changes in electrical utility generation and costs by the year 2010. Annual power generation is estimated to increase by 1 to 2%, and new generation capacity requirements would be less than 1% greater than increases without climate change. By the year 2055, annual power generation is estimated to increase by 3 to 5%, and new generation capacity requirements would be 14 to 20% greater than non-climate induced needs. Then, cumulative investments in new capacity would cost $10 to $27 billion.

Implications

More powerplants may need to be built; these will need more cooling water, further depleting the water supply. Climate-induced changes in hydrology may reduce hydropower generation and increase dependence on fossil fuels and nuclear power. Increased use of fossil fuels may provide positive feedback for the greenhouse effect and may deteriorate local air quality.

Air Pollution

Morris et al. (Volume x) studied possible interactions of climatic change and air pollution in California. (The study's design and methodology are discussed in the Air Quality chapter of this report.)

Results

Morris et al. (Volume x) project an increase in ozone concentration of up to 20% during some days in August in response to a 4°C (7°F) climatic warming in central California. The

National Ambient Air Quality Standard (NAAQS) for ozone is 12 ppm. The modeling of Morris et al. estimates a 30% increase in August days that exceed this standard. Furthermore, the area exceeding the NAAQS could increase by 1,900 km^2 (730 mi^2), and the number of people exposed to these elevated ozone levels could increase by over 450,000.

Implications

The greenhouse effect climatic changes may significantly change the air's chemistry on urban and regional scales. These changes may exacerbate existing air quality problems around California metropolitan areas and agricultural areas of the Central Valley.

POLICY IMPLICATIONS

An overall question that applies to resource management in general is, "What is the most efficient way to manage natural resources?" Currently, management is based on governmental jurisdiction with, for example, forests managed at the local, State, or Federal level. Management of hydrologic systems is also based on governmental jurisdiction. An alternative would be to manage these systems using natural boundaries as the criteria for determining management jurisdiction. The pros and cons of such a management strategy deserve at least some preliminary research.

Water Supply and Flood Control

Water supply is the basis for most economic development in California. Yet, almost all the water available in the SWP is allocated for use. A major problem is to accommodate rising demand for water, interannual climate fluctuations, and the need to export water from northern California to southern California.

In addition, the results from these studies suggest that climate change over the next 100 years could cause earlier runoff, thus reducing water deliveries below their projected 1990 level. This situation would create a set of major policy problems for the water managers and land use planners in California.

Two major policy questions can be raised concerning the possible reduction in water deliveries: How can the water resource system be changed to prevent a decrease in water deliveries caused by climate change? If water deliveries fall short of demand, how should potential water shortages be allocated?

Approaches for Modifying the Water Resource System

Several possible approaches can be attempted to increase water deliveries. First, management of the system could be modified. For instance, the most recent SWP development plan suggests the possibility of State management of both SWP and CVP facilities (California Department of Water Resources, 1987a). Complete joint management could produce more than 1 million acre feet (maf) additional reliable yield in the system. Steps toward greater cooperation have been taken. The Coordinated Operating Agreement (H.R. 3113) between the SWP and the CVP, ratified in 1986, allows the SWP to purchase water from the CVP. Using conservation techniques and improving the efficiency of transfer might also increase water deliveries.

Operating rules for the reservoirs also could be modified to increase allowable reservoir storage in April, which would increase water storage at the end of the rainy season and deliverable water during the peak demand season in midsummer. However, an increase in water storage in the late winter and early spring is likely to reduce the amount of flood protection (increase the risk of flooding) in the region; this in itself could negatively affect owners of floodplain property. Floods also place the delta islands at risk because of higher water levels. The tradeoff between water supply and flood control in northern California represents a potentially serious policy conflict affecting all levels of government in the region. In fact, a meeting between representatives of the State DWR and Bureau of Reclamation, which was held to discuss Sheer and Randall's (Volume x) results, concluded that any likely changes in reservoir operation that would avoid a significant loss of flood safety would most likely bring about little improvement in the system's performance under the given climatic scenarios. Detailed study of this point is needed, however.

The second approach to maintain or increase water deliveries might be to construct new water management and storage facilities. However, trends over the past decade have shifted away from planning large storage facilities (e.g., the Auburn Dam and Delta Peripheral Canal). Building new facilities is expensive and raises serious environmental concerns such as development of wild and scenic rivers. Another option is to use smaller facilities, such as the proposed new off-stream storage facility south of the delta, and to improve the delta's pumping and conveyance facilities. With the help of these facilities, the SWP plans to achieve a 90% firm yield (the amount that can be delivered in 9 out of 10 years) of about 3.3 maf by 2010 (California Department of Water Resources, 1987a). Another relatively inexpensive option for off-line storage is artificial recharge of groundwater during wet years. The SWP is currently pursuing a proposal to deliver surplus water to groundwater recharge areas in the southern Central Valley to provide stored water for dry years.

The third approach to increase water deliveries is to turn to other sources of water. For instance, use of groundwater could be increased. However, in many metropolitan areas, groundwater bodies are currently being pumped at their sustainable yields. Any increase in pumping could result in overdraft. Furthermore, decisions to use groundwater are made by local agencies and/or individual property owners, and groundwater is not managed as part of an integrated regional water system. Whether to include it in the system is an important policy issue.

Another option is for southern California to choose to fully use its allotment of Colorado River water (which could lead to conflicts between California and other users of that water, especially Arizona). Other possibilities include desalinization plants, cloud seeding over the Sierras, and reuse of wastewater. However, desalinization plants are energy intensive and may exacerbate air quality problems. Also, cloud seeding is controversial, since downwind users may not be willing to lose some of their precipitation.

<u>Options for Allocating Water Shortages</u>

The second major policy question is how to best allocate potential water shortages. One way would be to allow greater flexibility in water marketing. The adverse effects of this policy change (e.g., perhaps water becoming too expensive for agriculture and possible speculative price increases) could be ameliorated through a variety of governmental policies. Yet, even with regulation, any changes in the current system along these lines would most likely be very controversial.

A second way to allocate the shortages is to rely on those mechanisms used in the past to deal with droughts and water shortages, specifically governmental restrictions on water use. In the past, these have included restrictions on "nonessential" uses of water (e.g., watering of lawns), increased use efficiency, and transfers of agricultural water to municipal and industrial uses. Increased efficiency of water usage through various conservation techniques could effectively increase the number of water users without actually increasing the amount of water delivered. If climate gradually changes and water shortages become more common, these restrictions could become essentially permanent.

Sacramento and San Joaquin River Delta

The delta area of the Sacramento and San Joaquin Rivers in the San Francisco Bay estuary receives great attention from governmental bodies at all levels because of its valuable agricultural land, its crucial role in the State's water resource system, and its sensitive environment. The results of the studies in this overall project suggest that this region could be significantly affected by climatic change. Major changes could occur in delta island land use and in the water quality of the San Francisco Bay estuary. The policy implications of these possible changes are discussed below.

Delta Island Land Use

A critical land use issue is whether to maintain the levees surrounding islands threatened by inundation. Much of the land present on these islands is below sea level and is usable for agriculture, recreation, and settlement only through levee protection.

The individual delta islands have a significant range of values. For example, some islands contain communities and highways, and others are strictly agricultural. The property value of the islands is about $2 billion (California Department of Water Resources, 1987b). The islands also help repel saline water from the delta pumping plants (see Figure 4-2).

The levees have been failing at an increasing rate in recent years, and further sea level rise could increase failure probability. Improving the levees to protect the islands from flooding at the existing sea level and flood probability would cost approximately $4 billion (California Department of Water Resources, 1982).

The issue of levee failure raises three important policy questions. First, will some or all of the levees be maintained? The range of options concerning the levees includes inaction, maintenance of the status quo, strategic inundation of particular islands, and construction of polder levees.

Inaction, meaning the levees are not improved with time, could eventually lead to the formation of a large brackish-water bay as all of the levees failed. Williams (Volume x) suggests that the area of the San Francisco Bay estuary could triple if all the levees failed.

Currently, the general policy is to maintain the delta's configuration. One important policy favoring the maintenance of the levees is the Delta Levee Maintenance Subventions Program, in which State financial assistance is available for maintaining and improving levees. The value of the islands for agriculture and maintenance of water quality (see below) has created additional institutional support for maintaining the levees, even though the cumulative cost may exceed the value of the land protected. Future funding decisions for this and related programs should consider the possibility of climate change. If the levees are maintained, an important policy question is who will pay for the maintenance.

Not all the islands are of equal value, especially with regard to their value in protecting the freshwater delivery system. A possible future policy response to rising sea level would be to maintain only certain levees and not reclaim other islands as they become flooded. In essence, this would be a strategic inundation policy. Some precedence exists for this policy, as Mildred Island was flooded in 1983 and not reclaimed; the high cost of reclaiming the island relative to its value was cited as a rationale.

Construction of large levees similar to the polders in Holland is an option for protecting the islands and maintaining shipping channels. However, this approach would be expensive and, though it has been discussed, has not attracted much serious attention.

Another policy question concerns failure of the levees. If all or some levees are allowed to fail, will landowners be compensated? If so, where will the money come from? The delta islands contain some of the most valuable agricultural land in the State. Loss of this land would be a severe economic hardship for the local farmers and for the associated business community. Whether these farmers should be compensated for their loss is an important public policy issue.

A final policy question concerns management of the delta islands. How will management of the delta islands be coordinated? Currently, four government bodies have jurisdiction over the islands at the local, State, and Federal levels. Coordination of these bodies will be required to reach decisions regarding the future of individual delta islands.

Chapter 4

Water Quality of San Francisco Bay Estuary

The intrusion of saline waters into the upper reaches of the San Francisco Bay estuary could be a major problem in a warmer climate. Climate change is projected to cause increases in salinity in the estuary, largely as a result of sea level rise, levee failure, and the inadequacy of freshwater inflow to offset the increase in salinity. Furthermore, land subsidence due to groundwater extraction could augment sea level rise. In some areas of the estuary, subsidence up to 1.5 m (59 inches) has occurred within the recent past (Atwater et al., 1977).

Maintenance of current levels of salinity is addressed in the Water Rights Decision 1485 (D-1485) of 1978. This decision requires that water quality standards in the delta be maintained. If they are not, additional water must be released from reservoirs to improve the delta water quality, which could reduce the amount of water available for delivery. Current policy does not explicitly take into account the potential for future climate change. Thus, D-1485 could be interpreted as requiring maintenance of delta water quality standards even if sea level rises, causing further penetration of saline water into the Delta. Delta water quality standards are currently being reviewed at the Bay-Delta Hearing in Sacramento. This hearing began in mid-1987 and is expected to continue for 3 years. The choice of future options will be greatly affected by decisions made at the hearing.

Possible methods of combating the impacts of saltwater intrusion include maintaining levees, increasing freshwater outflows, reducing withdrawals, enlarging channels, constructing a barrier in the Carquinez Strait or lower delta, and/or constructing a canal around the delta's periphery. Alternatively, the freshwater pumping plants could be moved to less vulnerable sites. Decisions regarding response options will not be easily made. Levee maintenance and construction are costly. The water delivery agencies might be reluctant to increase delta outflows or reduce withdrawals. Enlargement of delta channels, construction of saltwater barriers, and construction of a peripheral canal are extremely controversial environmental issues. Another possible response to these climatic impacts would be a gradual, planned retreat from the delta, devoting resources to options compatible with the absence of a freshwater delta. This option would also be very controversial, both politically and environmentally.

Water Quality of Freshwater Systems

Water quality of lakes, streams, and rivers could change as climate changes. Results from the Castle Lake study indicate that primary production of subalpine lakes could increase, with the potential for changes in the water quality of mountain streams (Byron et al., Volume x). Reduction in summer flows of streams and rivers in the Central Valley Basin could concentrate pollutants in these aquatic systems. The major policy question relating to these potential changes is, "How will potential reductions in water quality below levels mandated in the current Water Quality Act of 1987 (Public Law 100-4) be prevented?"

Maintaining water quality despite decreased summer flows could be difficult and expensive. Controlling nonpoint source pollution is a goal of the Water Quality Act of 1987, and meeting this goal in the future could be more difficult and expensive because of the lower summer flows. Changes in land use near water courses to prevent runoff from agricultural land from reaching the water courses may be required. Reducing herbicide and pesticide use could also be another response, but this could harm agricultural production. Another option for preventing an increase in concentrations of pollutants in river reaches below reservoirs is to increase releases of reservoirs during summer months, which would dilute the pollutants. However, this strategy would have obvious negative impacts on water deliveries.

Municipalities located along rivers in the Central Valley Basin and that release treated sewage into them also could face increased difficulties in meeting water quality standards. Options include expanding sewage treatment facilities, which is expensive; releasing water from reservoirs to dilute the pollutants, as discussed above; or controlling the production of wastewater. Any municipalities planning for new sewage treatment plants should include climatic change as one factor in the design criteria.

Reductions in summer flows could harm populations of aquatic organisms and terrestrial organisms that use riparian habitats. To the extent that these species become threatened with extinction, laws requiring preservation of endangered species (e.g., Endangered Species Act of 1973) may be invoked as a legal basis for increasing reservoir releases to preserve these species. This could place into conflict the governmental agencies and public constituencies concerned with preservation of biodiversity and those concerned with economic impacts on agriculture and industry.

Terrestrial Vegetation and Wildlife

Changing species composition and productivity might alter the character of forestry operations and the aesthetic appeal of currently popular recreational areas. Climate-induced reductions in growth and regeneration rates, and increases in losses from wildfire and insect damage and mortality, could decrease the size and value of industrial forests in the State. How these changes will be managed is a complex question involving all levels of government as well as private landowners.

One major step in response to possible future climatic change is to incorporate climate considerations into current planning processes. The Forestry chapter of this report discusses Federal planning for effects of climate change on forests. Similar planning changes could be made at other levels of government. Coordinating the actions of government agencies involved with land management to climatic change in California is another possible response.

The flora and fauna in California are highly diverse and include many rare and endangered species. Climate could change faster than some species could adapt, leading to local extinction of these species. Species conservation (as mandated by the Rare and Endangered Species Act of 1973) might require habitat reconstruction and/or transplanting in some situations. Monitoring programs may need to be instituted to track trends in populations and communities. Extensive programs have been developed for currently endangered species in the State (e.g., the California condor), and similar efforts would probably be mounted in the future for other highly valued species.

Agriculture

Changes in water availability and temperature stresses are projected to affect agricultural production. How will changes in agricultural production and crop types be managed, and how will California agriculture respond in national and international settings? (See the Agriculture chapter of this report for further discussion.)

Historically, agriculture has quickly adapted to climatic fluctuations. New technology and reallocation of resources might offset the impact of changed climatic conditions and water availability. Improved farm irrigation efficiency, such as extensive use of drip irrigation, could mitigate the impact of water delivery shortages. Water marketing may provide a cost-effective means of meeting water demands and providing market opportunities for conserving water (Howitt et al., 1980). For example, water marketing may provide rights holders with the financial ability to invest in water conservation programs to cope with the impacts of climatic warming on water availability.

Changes in cropping locations and patterns of water use could exacerbate nonpoint source pollution and accelerated rates of groundwater overdraft. Further, changing water supply demands may heighten the conflicts between water allocation strategies and ecosystem and wildlife values.

It is uncertain how agricultural effects would be manifest in California's evolving economic and policy environment. For example, increased commodity prices could mitigate financial impacts of potential reductions in crop acreage and production.

Wetland Vegetation and Fisheries

Wetland species are valuable ecologically, aesthetically, and economically (photography, hunting, fishing, etc.). With rising sea level, areas supporting shallow water vegetation might be inundated and converted to deep-water habitats supporting different species. New shallow-water sites could be created by artificially adding sediment. This option features its own environmental impacts and would most likely be expensive. However, maintaining shallow-water vegetation is important not only to the conservation of plant species but also to migratory birds, which feed on such vegetation.

Salinity impacts on phytoplankton and fisheries might be controlled via levee maintenance coupled with increases in delta outflow.

Shoreline Impacts of Sea Level Rise

The California coast includes a diverse array of shorelines ranging from cliffs to sandy beaches. Erosion along these coastlines may increase as a consequence of sea level rise. Such erosion could substantially damage shoreline structures and recreational values, and preventing the erosion would be very costly. For example, protecting the sewer culvert of the San Francisco Westside Transport Project from potential damage caused by sea level rise may cost over $70 million (Wilcoxen, 1986). Sound planning for shoreline structures should consider future erosion that may be caused by sea level rise. (See Sea Level Rise chapter for further discussion of these issues.)

The accumulation of sediment behind water project dams and the effects of diversion structures, dredging operations, and harbor developments have limited the source of sediment for beach maintenance (particularly along the southern California coast). Individual landowners and institutions constructing such structures should consider their effects on sedimentation processes. Only through artificial deposition of sand (primarily from offshore sources) have southern Californian beaches been maintained. Beaches provide recreational areas and storm buffers, and their maintenance will require a major and continued commitment.

Air Quality

Increasing temperatures could exacerbate air pollution problems in California, causing an increase in number of days in which pollutant levels are higher than National Ambient Air Quality Standards. Devising technological and regulatory approaches to meet ambient air standards is currently a major challenge in certain regions of the state and these efforts must be continued. In the future, meeting air quality standards may become even more difficult to achieve under a warmer climate. Policy makers such as EPA and the California Air Quality Board may wish to consider possible climatic changes as they formulate long-term management options for improving air quality, to ensure air quality standards are met under warmer conditions.

Energy Demand

A warmer climate could affect both energy demand and supply. For instance, higher temperatures could cause increased cooling demands, and changes in runoff could affect hydroelectric power generation. Consequently, institutions in California that are involved with energy planning, such as the State Energy Resources Conservation and Development Commission, should begin to consider climate change in their planning efforts so that future energy demands can be met in a timely and efficient fashion.

REFERENCES

Atwater, B.F., C.W. Hedel, and E.J. Helley. 1977. Late Quaternary Depositional History, Holocene Sea Level Changes, and Vertical Crustal Movement, Southern San Francisco Bay. U.S. Geological Survey Professional Paper 1014. Menlo Park, CA: U.S. Geological Survey.

Atwater, B.F., S.G. Conard, J.N. Dowden, C.W. Hedel, R.L. MacDonald, and W. Savage. 1979. History, landforms, and vegetation of the estuary's tidal marshes. In: Conomos, T.J., ed. San Francisco Bay: The Urbanized Estuary. San Francisco, CA: Pacific Division, American Association for the Advancement of Science, pp. 347-386.

Barbour, M.G., and J. Major. 1977. Terrestrial Vegetation of California. New York: John Wiley and Sons.

Budyko, Mi.I. 1982. The Earth's Climate: Past and Future. New York: Academic Press.

California Department of Water Resources. 1982. Delta Levees Investigation. Bulletin 199. Sacramento, CA: California Department of Water Resources.

California Department of Water Resources. 1983. The California Water Plan: Projected Use and Available Water Supplies to 2010. Bulletin 160-83. Sacramento, CA: California Department of Water Resources.

California Department of Water Resources. 1985. California State Water Project. Typewritten brief. Sacramento, CA: California Department of Water Resources.

California Department of Water Resources. 1986. Operations Criteria Applied in DWR Planning Simulation Model. Memorandum report. Sacramento, CA: California Department of Water Resources.

California Department of Water Resources. 1987a. California Water: Looking to the Future. Bulletin 160-87. Sacramento, CA: California Department of Water Resources.

California Department of Water Resources. 1987b. Sacramento-San Joaquin Delta Atlas. Sacramento, CA: California Department of Water Resources.

California Division of Forestry and Fire Protection. 1988. California's Forest and Rangelands: Growing Conflict Over Changing Uses. Sacramento, CA: Forest and Rangeland Resources Assessment and Policy Act Committee.

Conomos, T.J., R.S. Smith, and J.W. Gartner. 1985. Environmental Setting of San Francisco Bay. Hydrobiologia 129:1-12.

Davis, O.K., J.C. Sheppard, and S. Robertson. 1986. Contrasting climatic histories for the Snake River Plain result from multiple thermal maxima. Quaternary Research 26:321-339.

Denton, R.A., and J.R. Hunt. 1986. Currents in San Francisco Bay. Final report. Berkeley, CA: University of California.

Doorenbos, J., and A.H. Kassam. 1979. Yield Response to Water. FAO Irrigation and Drainage Paper No. 33. Rome: FAO.

Fernald, M.L. 1950. Gray's Manual of Botany, 8th ed. New York: American Book Co.

Fischer, H.B. 1970. A Method for Predicting Pollutant Transport in Tidal Waters. Contribution No. 132. Berkeley, CA: University of California Water Resources Center.

Gleick, P.H. 1987a. Regional hydrologic consequences of increases in atmospheric CO_2 and other trace gases. Climatic Change 10:137-161.

Gleick, P.H. 1987b. The development and testing of a water balance model for climate impact assessment: modeling the Sacramento Basin. Water Resources Research 23:1049-1061.

Goldman, C.R., A. Jassby, and T. Powell. 1989. Interannual fluctuations in primary production: impact of climate and weather at two subalpine lakes. Limnology and Oceanography, In press.

Howitt, R.E., D.E. Mann, and H.J. Vaux Jr. 1980. The economics of water allocation. In: Englebert, E.A., ed. Competition for California Water. Berkeley, CA: University of California Press.

Howitt, R.E., and P. Mean. 1985. Positive Quadratic Programming Models. Working Paper No. 85-10. Department of Agricultural Economics, University of California, Davis.

Kjeldson, M.A., P.F. Raquel, and F.W. Fisher. 1981. Influences of freshwater flow on chinook salmon in the Sacramento-San Joaquin Estuary. In: Cross, R.D., and Williams, D.L., eds. Proceedings of the National Symposium on Freshwater Inflow to Estuaries, Vol. 2. Washington, DC: U.S. Department of the Interior, pp. 88-108.

Leverenz, J.W., and D.J. Lev. 1987. Effects of carbon dioxide-induced climate changes on the natural ranges of six major commercial tree species in the western United States. In: Shands, W.E., and J.S. Hoffman, eds. The Greenhouse Effect, Climate Change, and U.S. Forests. Washington, DC: The Conservation Foundation, pp. 123-155.

Macdonald, K.B. 1977. Coastal salt marsh. In: Barbour, M.G., and J. Major, eds. Terrestrial Vegetation of California. New York: John Wiley & Sons, pp. 263-294.

Major, J. 1977. California climate in relation to vegetation. In: Barbour, M.J., and J. Major, eds. Terrestrial Vegetation of California. New York: John Wiley & Sons, pp. 11-74.

Meyer, P.A. 1987. The value of wildlife in San Francisco Bay. Exhibit 38, entered by the Bay Institute of San Francisco to the State Water Resources Control Board in Sacramento, CA.

Miller, C.S., and R.S. Hyslop. 1983. California: The Geography of Diversity. Palo Alto, CA: Mayfield.

Munz, P.A., and D.D. Keck. 1959. A California Flora. Berkeley, CA: University of California Press.

Nichols, D.R., and N.A. Wright. 1971. Preliminary map of historic margins of marshland, San Francisco, California. U.S. Geological Survey Open File Map, San Francisco Bay Region Environment and Resource Planning Study. Basic Data Contribution 9.

Raven, P.H. 1977. The California flora. In: Barbour, M.G., and J. Major, eds. Terrestrial Vegetation of California. New York: John Wiley & Sons, pp. 109-137.

Ritchie, J.C., L.C. Cwynar, and R.W. Spear. 1983 Evidence from north-west Canada for an early Holocene Milankovitch thermal maximum. Nature 305:126-128.

Sheer, D.P., and M.L. Baeck. 1987. Documentation of the CVP/SWP Simulation Models Developed by WRMI. Columbia, MD: Water Resources Management, Inc.

Sudman, R.S. 1987. Layperson's Guide to the Delta. Sacramento, CA: Western Water Education Foundation.

U.S. Bureau of Reclamation. 1985. Summary Statistics, 1984; Volume 1: Water, Land and Related Data. Denver, CO: U.S. Bureau of Reclamation, Division of Water and Land Technical Services.

U.S. Department of Agriculture. 1987. Agricultural Statistics 1987. Washington, DC: U.S. Government Printing Office.

U.S. Department of the Interior. 1986. National Forest Statement of Receipts, Fiscal Year 1986. Washington, DC: U.S. Government Printing Office.

U.S. Environmental Protection Agency. In prep. Ecological Effects of Global Climatic Change. Chapter 5. In: U.S. EPA Global Climatic Change Program.

U.S. Maritime Administration. 1985. Containerized cargo statistics, 1983. Washington, DC: Government Printing Office.

Wilcoxen, P.J. 1986. Coastal erosion and sea level rise: implications for ocean beach and San Francisco's Westside Transport Project. Coastal Zone Management Journal 14:173-191.

CHAPTER 5

GREAT LAKES

FINDINGS

Global climate change could affect the Great Lakes by lowering lake levels, reducing the ice cover, degrading water quality in rivers and shallow areas of the lakes. It could also expand agriculture in the north, change forest composition, decrease regional forest productivity in some areas, increase open water fish productivity, and alter energy demand and supply.

Lakes

- Average lake levels could fall by one-half to two and one-half meters because of higher temperatures under the scenarios in this report. A drop of one meter would leave average levels below historic lows. Even if rainfall increases, the levels would fall because increased temperatures reduce the snowpack and accelerate evaporation. The estimates of lake level drop are sensitive to assumptions about evaporation; under certain limited conditions, lake levels could rise.

- As a result of higher temperatures, the duration of ice cover on the lakes would be reduced by one to three months. Ice would still form in near-shore and shallow areas. Changes in wind speed and storm intensity would affect the duration of ice cover.

- Shoreline communities would have to make adjustments to lower lake levels over the next century. Hundreds of millions of dollars would have to be spent along the Illinois shoreline along, dredging ports, harbors, and channels to greater depths than without climate change. Water intake and outflow pipes may have to be relocated. On the other hand, lower levels would expose more beaches, which would enhance shoreline protection and recreation.

- Climate change could have both good and bad effects on shipping. Lower lake levels would necessitate increased dredging of ports and channels or reduced cargo loads. Without dredging, shipping costs could rise 2 to 33 percent due to a reduction in cargo capacity. However, reduced ice cover would lengthen the shipping season and would sufficiently compensate for lower lake levels up to a drop of several meters.

Water Quality and Fisheries

- Higher temperatures would change the thermal structure of the Great Lakes. The result would be a longer and greater stratification of the lakes and increased growth of algae. These two factors would combine to reduce dissolved oxygen levels in shallow areas of lakes such as Lake Erie. A study of southern Lake Michigan found that annual turnover of the lakes could be disrupted.

- Climate change could increase concentrations of pollutants in the Great Lakes Basin. Dredging of ports could suspend toxic sediments in near-shore areas. Potential reductions in river flow in the Basin would create higher concentrations of pollutants in streams. The disposal of toxic dredge spoils was not studied in this report.

- The effects on fisheries would be miked. Higher temperatures may expand fish habitats during fall, winter, and spring and accelerate the growth and productivity of open water fish such as bass, lake trout, and pike. On the other hand, fish populations could be hurt by decreased summer habitats, lower summer dissolved oxygen levels and by potential loss of wetlands due to dropping lake levels. The effects of loss of wetlands, reductions in ice cover, introduction of new exotic species, and increase in species interaction were not analyzed.

Forests

- The composition and abundance of forests in the Great Lakes region will change. Higher temperatures and lower soil moisture could reduce forest biomass in dry sites in central Michigan by 77 to 99 percent. These mixed hardwood and oak forests would become oak savannas or grasslands. In northern areas such as Minnesota, boreal and cedar bog forests could change to treeless bogs, and mixed northern hardwood and boreal forestry in upland areas would become all northern hardwoods. Productivity could decrease on dry sites and bogland sites, but increase on some well drained wet sites. Softwood species that are currently commercially important could be eliminated and replaced by hardwood, such as oak and maple, which are useful for different purposes.

- Depending on the scenario, declines in forests could be evident in 30 to 60 years. These results do not reflect additional stresses, such as acid rain and oxidant levels, nor do they reflect the possible beneficial impacts of increased CO_2 levels.

Agriculture

- Considering climate change alone, corn and soybean yields in northern areas, such as Minnesota, could increase by 50 to 100 percent and would decline in the rest of the region by up to 60 percent. The combined effects of climate and CO_2 could further increase yields in the North, and would result in net increases in the rest of the region, unless climate change is severe.

- Agricultural production in the northern part of the region could expand as a result of declines elsewhere. Acreage in the Corn Belt states may change little. However, the presence of glaciated soils in the North could limit this expansion. Wider cultivation in the north could increase erosion and runoff, and degrade surface and groundwater quality. Increased agriculture would require changes in the infrastructure base, such as transportation networks.

Electricity Demand

- There would be little net change in annual electricity demand. In northern areas, such as Michigan, reduced heating needs would exceed increased cooling requirements, while in southern areas, such as Illinois, cooling needs would be greater than heating reductions. The annual demand for electricity in the entire region could rise by 1 to 2 billion kilowatt hours by 2010 and by 8 to 17 BKWH (less than 1%) by 2055. This study did not analyze the reduced use of other fuels such as oil and gas in the winter, changes in demand due to higher prices, and the impacts on hydroelectric supplies. Previous studies have suggested that reduced lake levels and river flows would lead to reductions in these supplies.

- By 2010, approximately 3 to 8 gigawatts would be needed to meet the increased demand, and by 2055, 23 to 48 GW would be needed -- an 8 to 11 percent increase over baseline additions that may be needed without climate change. These additions would cost $23 to 35 billion by 2055.

Policy Implications

- U.S. and Canadian policy makers, though such institutions as the International Joint Commission and other institutions, should consider the implications of many issues for the region. This study raises additional issues concerning:

- -- The water regulation plans for Lake Ontario and possibly for Lake Superior lake levels.
- -- The potential increased demands for diverting Great Lakes water for uses outside of the basin. Before such a potential demand could be accommodated, additional analysis would be required. This is not currently allowed by federal statutes.
- -- Long range industrial, municipal, and agricultural water pollution control strategies will be affected by climate change. Agencies such as EPA may wish to examine the implications for long-term point and non-point water pollution control strategies.
- -- The research, planting, and land purchase decisions in northern forests supported research by federal, state, and private institutions.

CLIMATE-SENSITIVE NATURAL RESOURCES IN THE GREAT LAKES REGION

The Great Lakes region* is highly developed, largely because of its natural resources. The steel industry developed along the southern rim of the lakes, in part because iron ore from the north could be inexpensively transported over the lakes. Rich soils, moderate temperatures, and abundant rainfall have made the southern part of the region a major agricultural producer. Forests are abundant in the north and support commercial and recreational uses. The basin has become the home of over 29 million Americans and produces 37% of U.S. manufacturing output. (U.S. EPA, 1987; Ray et al.). Although these resources provide many benefits, they are also quite vulnerable to climate changes, oftentimes with negative impacts upon society.

Current Climate

The Great Lakes region has a mid-latitude continental climate. Winter is sufficiently cold to produce a stable snow cover on land and ice on the lakes. Average January temperatures over Lake Superior are -15°C (5°F), and the average July

*This chapter will cover only the U.S. side of the Great Lakes and the eight States bordering on them (see Figure 5-1).

temperatures in the southern part of the region are 22°C (72°F). The average rainfall varies from 700 to 1,000 mm (27 to 39 inches), depending on location (Cohen, in Glantz Volume _).

The Lakes

The Great Lakes are a system of five major lakes that contain approximately 18% of the world supply of freshwater and 95% of the freshwater in the United States (U.S. EPA and Environment Canada, 1987) (see Figure 5-1, Map of Great Lakes). The natural flow of the lake system begins in Lake Superior, the largest of the lakes, which drains via the St. Marys River into Lakes Michigan and Huron.* Water from Lakes Michigan and Huron flows out through the St. Clair River into Lake St. Clair. From there, the water flows through the Detroit River and into Lake Erie, the shallowest lake. The Niagara River connects Lakes Erie and Ontario, and the system ultimately empties into the Atlantic Ocean via the St. Lawrence River and Seaway.

The greatest influence on lake levels is nature. Seasonal fluctuations are on the order of 0.3-0.5 m, with the lakes peaking in late summer because of condensation over the northern lakes and reaching minimum levels in late winter. Interannual lake level changes have been much larger, approximately 2 m.

Lake Regulation

The flow between the lakes is controlled by dams at two points: (1) the St. Mary's River to control levels of Lake Superior; and (2) Iroquois, Ontario, to control Lake Ontario. The major diversion out of the lakes is the Chicago diversion, which transfers water from Lake Michigan through the Illinois River into the Mississippi River. Man's influence on lake levels is relatively small. Doubling the flow down the Chicago diversion would lower lake levels only by 2.5 inches in 15 years (Quinn, 1987).Joint control of lake supply was codified in the Boundary Waters Treaty of 1909 between Canada and the United States, which created the International Joint Commission (IJC) consisting of representatives from both countries. The IJC regulates flow through the control structures and diversions by balancing the needs of shipping, hydropower, and consumptive uses among the lakes and along the St. Lawrence River and Seaway. Two regulatory plans (Plan 1977 for Superior and Plan 1958D for Ontario) set ranges of levels between which Lakes Superior and

*Considering a single hydrologic unit because they are connected by the Straights of Mackinac.

Figure 5-1. Map of the Great Lakes studies.

Ontario must be maintained. Diversion out of the lakes is also limited by law. Flow through the Chicago diversion was limited by the Supreme Court to 90 m^3/s (3,200 ft^3/s) (Tarlock, 1988), and the 1986 Water Resources Development Act forbids diversion out of the lakes' basin without the consent of all Great Lakes governors (Ray et al.).

Climate-Sensitive Uses of the Lakes

Shipping

The U.S. Great Lakes fleet, which consists of approximately 70 ships, transported over 171 million tons of cargo in 1987 (New York Times). The tonnage of U.S. shipping consists of iron ore, coal, and limestone, all primary inputs for steel (77%); lake grain (13%); and petroleum products, potash, and cement (10%) (Nekvasil, 1988) Cargo volumes are displayed in Table 5-1. Most of the goods are shipped within the Great Lakes, with only 7% of the tonnage (mainly grains) shipped to overseas markets (Ray et al.). Although shipping activity had declined as a result of reductions in U.S. steel production, recent increases in steel output have lead to additional demand for shipping (Agar, 1987; New York Times).

Great Lakes ships last over half a century and are designed to pass within a foot of the bottom of channels and locks. Cargo capacity is quite sensitive to lake and channel depth because of this low clearance. The presence of ice usually shuts down Great Lakes shipping up to 4 months each year.

Table 5-1. Cargo for U.S. Great Lakes Shipping in 1987 (thousands of tons)

Cargo	Weight
Iron Ore	61,670
Coal	37,731
Stone	33,164
Grain	22,338
Petroleum Products	11,491
Cement	3,806
Potash	1,702
Total	171,902

Source: Nekvasil (1988).

Hydropower

The eight Great Lakes States use the connecting channels and the St. Lawrence River to obtain 35,435 gigawatt hours of hydropower each year, which is about 5% of their electricity generation. About four-fifths of the hydropower is produced in New York State, which derives over 26% of its electricity from hydropower (Edison Electric Institute, 1987).

Municipal Consumption

Most water used for the domestic and industrial consumption in the basin is taken from the lakes. Surface waters supply 95% of the basin's water needs. By the year 2000, consumption is estimated to increase by 50 to 96% (Ray et al.; IJC, 1985).

Fisheries

In 1984, the value of the harvest to the U.S. commercial fishing industry was approximately $15 million (U.S. EPA and Environment Canada, 1987; Statistical Abstract of the United States, 1987). Although most fishing in the Great Lakes is for recreation, fisheries are managed by the States, although the Great Lakes Fishery Commission coordinates activities among the States.

Tourism

Three National and 67 State parks are located along the shores of the lakes, as are numerous local parks. Over 63 million people visited these parks in 1983. (Ray et al.; Great Lakes Basin Commission, 1975). In 1984, lake-generated recreation produced $8 to 15 million in revenues. Fishing, boating and swimming are very popular.

Shoreline Development

Over 80% of the U.S. side of the Great Lakes shoreline is privately owned. One of the most developed shorelines is the 101-km Illinois shoreline, where many parks and residential buildings, including apartment houses, are built near the water's edge. Shoreline property owners have riparian rights to use adjoining waters. The shoreline property owners cannot substantially diminish the quantity or quality of surface waters (Ray et al.).

Chapter 5

Climate and Water Quality

Water quality is directly affected by climate. Lower stream runoff increases concentrations of pollutants. Every summer, the lakes stratify into a warmer upper layer and a cooler lower layer. This stratification can limit biological activity by restricting the flow of nutrients between layers. In addition, warm temperatures and an excess supply of nutrients (phosphorous and other chemicals from agricultural runoff and sewage effluent) can lead to algal blooms which decay and cause a loss of oxygen (eutrophication) and reduction in aquatic life in the lower layers of lakes such as Lake Erie. Cool weather and the formation of ice help to deepen the mixed layer and break up the stratification and thoroughly mix the lakes in the winter.

Development, industrialization, and intensive agriculture in the Great Lakes Basin has created serious pollution in the lakes, especially Lake Erie. In the early 1970s, nutrient loadings were so high that Lake Erie experienced significant eutrophication problems for several years (DiToro, 1987).

Two measures have helped improve water quality. The U.S.-Canada Great Lakes Water Quality Agreement of 1972 called for controlling nutrient inputs and eliminating the discharge of toxic chemicals, and the Clean Water Act mandated construction of sewage treatment plants and controls on industrial pollutants. The United States and Canada spent a total of $6.8 billion on sewage treatment in the Great Lakes. By 1980, nutrient loadings into Lake Erie had been cut in half (Ray et al.; DiToro, 1987), and water quality had markedly improved.

Fluctuating Lake Levels

Recent high and low lake levels have significantly affected users of the lakes. In 1964, Lake Michigan was 0.92 m below average, making some docks and harbors unusable. Shipping loads were reduced by 5 to 10% and more shipments were required, subsequently raising the cost of raw materials and supplies by 10 to 15%. In addition, many water intakes had to be extended or lowered (Changnon). Flow through the Niagara hydropower project fell by more than 20%, with electricity generation off by more than 35%. Flow through New York's St. Lawrence hydro project was more than 30% below its mean, with electricity generation decreased by 20% (Linder, 1985). Low lake levels also provide benefits: as beaches have become enlarged, their use and enjoyment have increased.

The low water levels in the 1960s exposed the supporting structures along Chicago's shoreline to air, causing dry rot. When lake levels rose to record height in the 1980s, the wood pilings and sections of the revetment collapsed. Apartment houses that were built too close to the shoreline during the low levels of the 1960s were flooded, as were roadways built close to the shore. The estimated construction cost for rebuilding the damaged shoreline protection system is $843 million (Changnon). The last 2 years have been relatively hot and dry, causing lake levels to recede to average levels. The lower levels have forced shippers to reduce tonnage just as the steel industry in the region is undergoing a resurgence.

Land Around the Lakes

The land in the Great Lakes region is extensively used for industry, agriculture, and forestry. Many of the uses are sensitive to climate.

Land Uses

Urban Development

Approximately 29 million people live in the Great Lakes Basin, mostly in the urban areas around the cities on the southern edge of the Great Lakes: Chicago, Detroit, Cleveland, Toledo, and Buffalo. Many of the residents work in manufacturing industries, which despite recent declines, still provide 23% of payroll employment (Ray et al.).

Agriculture

Agriculture is the single largest user of land: 42% of all land in the eight Great Lakes States is devoted to crops, and an additional 10% is used for pasture. The Great Lakes States encompass most of the Corn Belt. In 1983, roughly 59% of all U.S. cash receipts for corn and 40% of the receipts for soybeans came from this region. Overall, the Great Lakes States produced 26% of the total U.S. agriculture output, or $36 billion. (Federal Reserve Bank of Chicago, 1985). Most crops are grown on dryland as only about 1% of the region's croplands were irrigated in 1975 (U.S. Department of Commerce, 1987).

Livestock are also important to the agricultural economy of the region. Approximately 18% of U.S. cattle are raised in these eight States; of these, 52% are dairy cows (USDA, 1987). (The sensitivity of livestock to climate change is discussed in the Agriculture chapter of this report.)

Forests

The forests in the region have commercial, recreational, and conservation uses. The forests in the south are mainly oak and northern hardwoods, such as maple. The north has almost 21 million hectares (ha), (52 million acres), of forests consisting mostly of northern hardwoods, such as maple, birch, and beech, and boreal forests, such as spruce and fir trees. The Federal and State governments own, respectively, 11 and 13% of the forests in Michigan, Minnesota, and Wisconsin; while over half are privately owned (USDA, 1982). The pulp, construction, and furniture industries are major consumers of such species as aspen, pines, balsam fir, spruce, maples, paper birch, and oak. The forest industry is a major employer in the northern part of the region. In Wisconsin, for example, 283,000 jobs are in timber harvesting and manufacturing related to forestry (Botkin; USEPA and Environment Canada, 1987). Forestry is considered to be a growht industry in the region as Michigan has identified forest products as one of the three key industries targeted for expansion in the State (Ray et al.).

PREVIOUS CLIMATE CHANGE STUDIES

The impacts of climate change on many of the systems in the Great Lakes have been analyzed in separate and unrelated studies. Canadian researchers have examined the potential impacts of climate change on Great Lakes levels and have concluded that levels would fall. Cohen (1986) used hydrologic calculations to estimate that the lakes might fall between 0.2 and 0.8 m. More recently, Sanderson (1988) used a hydrologic model of the lakes also to estimate that the lakes would drop by an average of 0.2 to 0.6 m. Wall (1985) concluded that lower lake levels could reduce ecological diversity and dry up enclosed marshes. In another study, Cohen (1987) estimated that withdrawals of water from the lakes for municipal consumption would increase by about 2.5% on an annual basis and would only marginally affect lake levels.

Assel et al. (1985) studied the extent of ice cover during the winter of 1982-83, which had temperatures 3.3-4.4°C warmer than the 30-year mean. They found that ice cover on Lake Superior was reduced from a normal 75% coverage to 21%. On Lake Erie, ice coverage was down to 25% from the normal 90%. Meisner et al. (1987) conducted a literature review on the possible effects of global warming on Great Lakes fish. Results are discussed in the fisheries section of this chapter.

Linder (1987) used the GISS transient to estimate impacts on electricity demand and hydropower generation in 2015 in upstate New York. He found total energy demand declining by 0.21-0.27%, but peak demand increasing by 1-2%. Meanwhile, hydropower production could decline between 6 and 8.5% owing to reductions in streamflow.

Impacts on managed and un-managed vegetation have also been studied. The Land Evaluation Group examined the potential impacts of climate change on agriculture in Ontario and found that yields could decrease in southern Ontario and farming could become feasible in northern Ontario. The study also indicated that the direction of change for yields depends on whether rainfall increases or decreases (Land Evaluation Group, 1986). Solomon and West (1986) used a stand simulation model (see forest section of the chapter) to estimate impacts of doubling and quadrupling of CO_2 levels on a northwest Michigan coniferous-deciduous transitional forest. They found that doubled CO_2 would lead to an eventual disappearance of boreal forests and an increase in deciduous trees. Total biomass would decline at first and rebound in about two centuries.

Two studies by Canadian researchers examined the possible impacts of climate change on tourism and recreation in Ontario. Both studies used climate change scenarios based on the GISS and GFDL models (although these may have been earlier model runs). Crowe (1985) estimated that snowfall would decrease by 25-75%, and the ski season in southern Ontario would be cut by 75-92% (7-12 weeks) and 13-31% (2-4 weeks) in the north. Wall found similar results. He concluded that reduced snowfall could eliminate skiing in southern Ontario and would shorten the northern Ontario ski season by 30-44%. A longer summer season could increase such summer tourism activities as camping. Wall (1985) also thought that lower lake levels could decrease ecological diversity and dry up enclosed marshes.

GREAT LAKES STUDIES IN THIS REPORT

Unlike previous studies, the studies for this report used common scenarios to address some of the potential impacts of climate change on a number of natural and societal systems in the Great Lakes region. The studies address the direct effects of climate change on the resources and some of the indirect effects on infrastructure and society. They focused on the lakes themselves, examining such issues as lake levels, ice cover, thermal structure, and fisheries. They also looked at the effects of these changes on shipping and shoreline properties, and examined the sensitivities of agriculture and forest to climate change. Finally, the studies examined the implications of climate change for Great Lakes policies and institutions. Some of the studies were linked quantitatively, but most were conducted independently of each other.

The studies involved either new topics or approaches that were not used in previous studies. For example, the analysis of lake levels used a more complex hydrologic model than was used previously. The agriculture analysis complements the Land Evaluation Group's study of Ontario by using a different model to examine impacts on the U.S. side of the lakes. The potential impacts of climate change on thermal structure were examined for the first time. Also for the first time, models were used to analyze impacts on fisheries. This study complements previous studies on forests by using a combination of modeling techniques to test the similarity of results.

The following analyses were performed for this report:

Direct Effects on Lakes

- o Lake Levels - Croley and Hartmann, Great Lakes Environmental Research Laboratory.

- o Ice Cover - Assel, Great Lakes Environmental Research Laboratory.

Impacts of Lake Changes on Infrastructure

The results from the first two studies were used in the following studies:

- o Great Lakes Shipping - Keith, DeAvila, and Willis, Engineering Computer Optecnomics, Inc.

- o The Impacts of Low Lake Levels on the Illinois Shoreline - Changnon, Leffler and Shealy, Illinois State Water Survey.

Water Quality

The following studies concern water quality and the effects on aquatic life in the lakes. The first two studies examined the direct effects of climate on the thermal structure of some of the lakes.

o Thermal Structure of Southern Lake Michigan - McCormick, Great Lakes Environmental Research Laboratory.

o Eutrophication of the Central Basin of Lake Erie - Blumberg and DiToro, Hydroqual, Inc.

The results from these studies were used in the following:

o Great Lakes Fisheries - Magnuson, Regier, Sheeter, Hill, Holmes, and Meisner, Universities of Wisconsin and Toronto.

Forests

A series of studies on forests was commissioned to examine shifts in ranges, transient impacts, and the potential for migration of some Great Lakes forests. Basically, these are different analytic techniques for understanding how climate change may affect the composition and abundance of forests in the region.

o Transient Effects on Great Lakes Forests - Botkin, Nisbet, and Reynales, University of California at Santa Barbara.

o Forest Migration - Zabinski and Davis, University of Minnesota.

o Ecological Response Surfaces - Overpeck and Bartlein, Lamont-Doherty. (Regional results will be taken from this study.)

Agriculture

The potential changes in agriculture in the Great Lakes were analyzed by studying changes in crop yields in the region and integrating the results in a national analysis of production changes. That national analysis was used to determine if production in the region could increase or decrease. The results of these studies were used to examine potential farm level adjustments.

o Crop Yields in Great Lakes States - Ritchie, Baer, and Chou, Michigan State University.

o Farm Level Adjustments by Illinois Corn Producers - Easterling, Illinois State Water Survey.

This chapter will use regional results from the following:

- o Economics Effects on Agriculture - Adams, Glyer and McCarl, Oregon State University.

Energy

This project analyzed potential changes in the national demand for electricity and estimated changes in regional demands. Results for the Great Lakes region are presented in this chapter.

- o Electric Utilities - Linder and Inglis, ICF, Inc.

Policy

The potential policy implications of the changes indicated by these and previous studies for local, State, Federal, and international decisionmaking are examined. This project provided information for the background and policy implications sections. Results are not presented as part of the results section.

- o The Implications for Policies and Institutions - Ray, Brah, and Lindland, The Center for the Great Lakes.

GREAT LAKES REGIONAL SCENARIOS

All three GCMs that provide the basis for the climate change scenarios show rather large increases in temperature for the Great Lakes region under the doubled CO_2 climate. The seasonal and annual temperatures and precipitation are displayed in Figure 5-2. The OSU scenario has an annual temperature rise of 3.5°C, with no change in seasonal pattern. The GISS scenario is about a degree warmer on average and has the largest warming in the winter and fall. The GFDL scenario has the largest warming of the three models, about 6.5°C annually, with the largest warming in the summer. All three scenarios have annual increases in precipitation. OSU has an increase of approximately 0.1 mm/day, with precipitation rising in all seasons. GISS has an increase of approximately 0.2 mm/day, with precipitation declining slightly in the fall. GFDL has an annual precipitation increase of only 0.05 mm/day, but rainfall drops by 5 mm/day in the summer. The large temperature increase and small rainfall increase combine to make GFDL the most severe scenario. This is especially true in summer months, when GFDL has the largest temperature rise of any scenario and virtually reduces rainfall to zero. OSU is the mildest scenario owing to the smaller

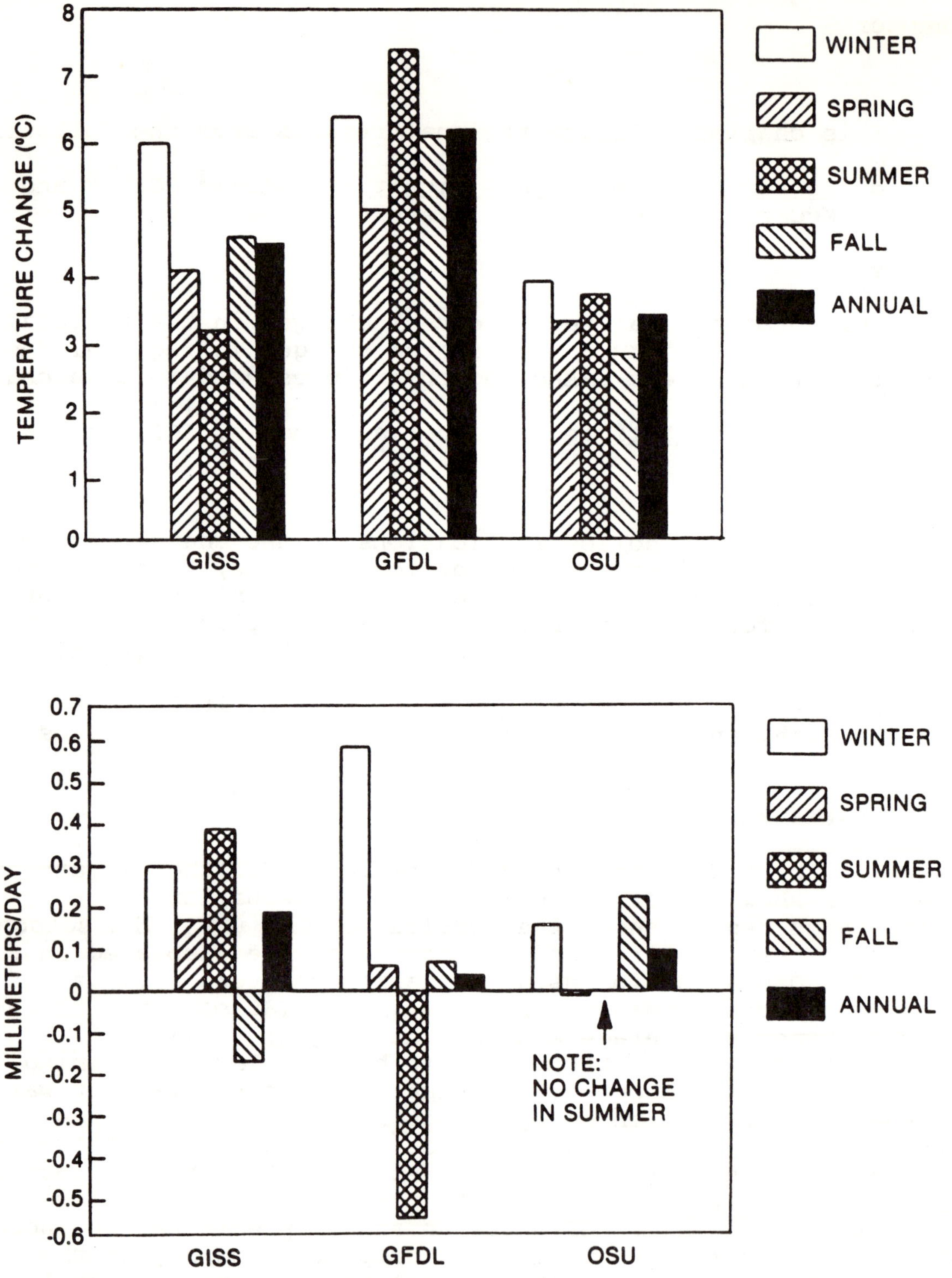

Figure 5-2. Average change in temperature (upper) precipitation (lower) over Great Lakes grid points.

temperature increase and GISS is in the middle in term of severity.

One limitation related to using the GCMs as a basis for climate change scenarios for the Great Lakes region is that the lakes are not well represented in the GCMs. The relatively large size of the GCM grid boxes results in little feedback from the lakes to the regional climate estimates from the GCMs.

RESULTS OF THE GREAT LAKES STUDIES

Lakes

Lake Levels

Geologic records indicate that Great Lakes levels have fluctuated as paleohistoric climates have been wetter and drier (Larson, 1985). Recent short-term variations have been the result of short-term changes in precipitation patterns. Croley and Hartmann examined the potential impacts of global warming on average lake levels.

Study Design

Croley and Hartmann used a water supply and lake level model of the Great Lakes Basin to estimate the potential impacts of climate change on levels of the Great Lakes (Croley, 1983a,b; Croley, 1988; Quinn, 1978). This model includes a separate model for each of the 121 watersheds in the basin. Croley and Hartmann simulated runoff in each of the sub-basins, overlake precipitation, and evaporation. Lake levels are very sensitive to evaporation; therefore, Croley and Hartmann ran each GCM scenario with different assumptions about evaporation.* Finally,they used the current plans (Plan 1977 for Superior and Plan 1958-D for Ontario) and hydraulic routing models of outlet and connecting channel flow and estimated water levels on each of the Great Lakes.

---

*In Volume _, Croley focuses on results from his latest run. This run includes assumptions that lead to relatively high amounts of evaporation and larger drops in lake levels. Earlier runs had less evaporation increases and less of a drop in lake levels. Results in this chapter include the latest run and an earlier run.

The regulation plan for Lake Superior failed under the GFDL scenario. To obtain an estimate of changes in levels for Superior-Huron, St. Clair, and Erie, Croley and Hartmann assumed that over a 30-year period, total inflows into Lake Superior (runoff + over-lake precipitation + diversions - evaporation) would equal total outflows, and Lake Superior levels would not change. Only 30-year average lake levels were calculated for the other lakes.

Limitations

The relationships in this model were developed for a cool and wet climate. The analysis did not account for changes in the consumptive uses of the lakes (due to population and economic growth or climate change), and it did not consider changes in the regulation plans, or increases or additions to diversions into or out of the lakes. The analysis also used the difference in vector winds from the GCMs as a proxy for the difference in scaler winds because GCM estimates of changes of scaler winds were not available. Thus, the wind estimates probably underestimate changes in windspeed (Rind, 1988). The uncertainty on winds is complicated by the uncertainties concerning evaporation. Different assumptions of evaporation affect the magnitude of lake level drop, but they do not affect the direction of change -- lake levels fall under all evaporation assumptions.

Results

Lake levels fall significantly under all three scenarios (see Table 5-2). The lake level changes are displayed in ranges from low to high evaporation.

Average levels for Lake Superior are about 0.4-0.5 m below average levels for the 1951-80 period under the OSU and GISS scenarios. No figure for Lake Superior is given for the GFDL scenario, since levels were held constant. These average levels are generally lower than extreme lows of recent history. Even though precipitation rises in all three scenarios, lake levels fall, primarily due to the higher temperatures. Apparently, only a large increase in rainfall or a large decrease in windspeeds can offset these changes. Lake levels are estimated to continue fluctuating on an annual basis. Specific estimates of fluctuation are not discussed here, since climate is assumed to remain constant.

Table 5-2. $2xCO_2$ Scenarios: Reduction in Great Lakes Levels from 1951-80 Averages (meters)

Scenario	Superior	Michigan/Huron	Erie	Ontario
GISS	-0.43 to -0.47	-1.25 to -1.31	-0.95 to -1.16	NA
GFDL	NA	-2.48 to -2.52	-1.65 to -1.91	NA
OSU	-0.39 to -0.47	-0.86 to- 0.99	-0.63 to- 0.80	NA
	Transient Scenario (average rate of change per decade 1980-2060)			
GISS-A	-0.006	-0.055	-0.04	NA

NA = Not applicable.

Source: Croley and Hartmann, Volume __.

Croley and Hartmann also found that the flow in the St. Mary's increases by less than 1% in the GISS high rainfall scenario and drops by 13% in the drier OSU scenario for Lake Superior. The flow in the Niagara is estimated to be reduced by 2-30%. Croley and Hartmann did not estimate the flow of these rivers for the GFDL scenario.

The lowering of lake levels appears to be correlated with increased temperatures in the scenarios. All of the doubled CO_2 scenarios lead to declines in runoff to the lakes and increases in evaporation from the lakes. The reduction in runoff is largely the result of changes in snowpack accumulation and ablation. Snowpack in the Lake Superior basin is reduced by one- to two-thirds, and in the other basins, farther to the south, the snowpack is almost entirely absent. The reduction in runoff would reduce average streamflow in the basin. These results appear to be driven mainly by the temperature increase, since precipitation rises in all scenarios.

Evaporation increases under all model runs. The increase in evaporation varied under different assumptions about the relationship of evaporation to meteorology and ranged from 20 to 48%. For a given assumption about evaporation, higher temperature scenarios generally caused more evaporation. Lake level reductions could also be higher or lower, depending on these assumptions.

All of these changes cause a reduction in net basin supply (the sum of overlake precipitations and runoff minus evaporation) by 14-68%. The exception to this is the GISS scenario for Lake Superior. In that scenario, rainfall increases by 18%, leading to a 1% increase in net basin supply.

The Ontario regulation plan failed under all scenarios, including the transient run. Under these conditions, the system would not contain enough water to keep the level of Lake Ontario and the flow in the St. Lawrence River within ranges currently specified by the plan. The Lake Superior regulation plan failed under the GFDL scenario. Although net basin supply in Lake Superior increased under GISS, the regulation plan would require increased flow through the St. Mary's River to the water-short lower lakes, resulting in a net drop in Lake Superior levels.

These results are consistent with other studies done on lake levels and climate change. Both Cohen and Sanderson agree with Croley and Hartmann that lake levels would drop under various climate change scenarios. The other two studies, however, estimated lake levels would drop less than 1 m. Croley and Hartmann may have estimated greater changes because they used a more sophisticated runoff, evaporation, and routing model and because of different assumptions made about evaporation. Croley and Hartmann also used a more integrated approach and more variables from the GCMs. The estimates for GFDL may also be higher because the GFDL scenario used in this study had a higher temperature than the GFDL scenarios used by Cohen and Sanderson.

The results of the transient run (GISS A) are expressed as the average change in lake level per decade and are not indicative of what would happen in any particular decade. Lake Superior levels drop only 0.006 m per decade, while the other lake levels fall 0.04-0.055 m per decade. An extrapolation of the transient results to the decade of the 2060s (when the GISS A transient run reaches doubled CO_2 climate conditions) results in lake level reductions less than for the doubled CO_2 GISS scenario. This is because lake levels do not respond immediately to climate change, but must catch up. By the end of the transient scenario, the 2050s, lake levels fall by more than 0.05 m per decade.

Croley and Hartmann found that enough heat would reside in Lakes Superior, Michigan, Huron, and Ontario to maintain water surface temperatures at a sufficiently high level throughout the year, so that buoyancy-driven turnovers of the water column may not occur at all. This could significantly affect lakewater quality and aquatic life (see McCormick). Croley estimated that average surface water temperatures, which would be above 0°C, would significantly reduce ice concentrations.

Implications

Higher temperatures may lead to increased withdrawals of water from lakes for municipal consumption. If more powerplants are needed (see Energy Demand), the demand for cooling water will rise. Climate change may also result in more calls for diversion of water out of the basin for use elsewhere. However, lake levels may be lowered even more as a result of higher demand for withdrawals for use in the Basin owing to population and economic growth.

Hydropower production would be reduced, as flows through the St. Mary's, the Niagara, and the St. Lawrence Rivers would fall. Losses to hydropower were not estimated for the EPA study, although Linder's earlier work on hydropower losses by 2015 in New York State showed potential loss of 1600 to 2066 gigawatt hours (Linder, 1987). Presumably, hydropower losses would be higher under the doubled CO_2 scenarios. The impacts of lower lake levels on wetlands were not estimated, and the impacts on shipping and on shoreline infrastructure are discussed later in this chapter.

Lower lake levels and reduced river flow would likely adversely affect water quality in the Basin. Less water would allow for reduced dilution of pollutants. Forty-two hot spots occupy many bays and harbors along the Great Lakes. These are contaminated with a wide variety of halogenated organics and heavy metals, as well as remobilizable nutrients. Lower lakes may cause emergence and near emergence of these toxic sediments, through erosion, leaching, oxidization, or volatilization.

Effects of Lower Lake Levels

Coastal infrastructure around the Great Lakes has generally been built assuming average lake levels would not change. A drop in levels could make much of the current infrastructure unusable and necessitate reconstruction. Changnon examined the potential impacts and adjustments to infrastructure along the 101-km Illinois shoreline. This study and shipping analysis used the lower range of the lake level drops from Table 5-2 because subsequent analyses that gave different lake levels were performed too late to be incorporated.

Study Design

Changnon et al. interviewed experts about the possible impacts and costs of adjustment along the Illinois shoreline to the lower lake level estimates described above. Results are expressed in current dollars.

Limitations

This analysis did not use economic models and did use current prices. Results are based on expert judgment. Changnon et al. also assumed that lakes would reach the levels described above by 2030. Analysis of this assumption indicates that a doubled CO_2 climate may not be reached until the latter half of the next century. This study examined only the costs of rebuilding infrastructure and did not examine ecological impacts.

Results

The largest costs appear to accrue to recreational and commercial harbors (see Table 5-3). The major expenses are associated with dredging harbors and lowering bulkheads, which could cost approximately $200-400 million. If lake levels fall enough, keeping some harbors open (e.g., Waukegan, Illinois) may not be a cost-effective choice.

Changnon et al. concluded that slips and docks would be only slightly affected. Many of these probably would have been replaced anyway and could be set at lower levels as the lakes fall. (The impacts on commercial shipping in Lakes Superior and Erie are discussed in the shipping study.)

Intake valves for municipal and industrial consumption could be exposed and may have to be lowered or moved farther offshore. Outfalls for stormwater would have to be extended. Changnon et al. estimated that extending urban water intakes and stormwater outfalls could cost $16-17 million.

Although the exposure of more land could present some erosion problems, it could also enlarge many beaches. An additional 1-2.2 km^2 of beaches would be added to the Illinois shoreline. In all, Changnon et al. estimated that the costs of adjusting to lower levels of 1.25 to 2.5 m along the Illinois shoreline, excluding normal replacement of docks and piers, would be $220-430 million.

Table 5-3. Estimated Economic Impacts of Lowerings of the Levels of Lake Michigan Over a 50-Year Period (1990-2040).

Costs in millions of 1988 dollars to address future lake levels at indicated depths below average level of Lake Michigan (1951-1980).

		1.25 meters lower	2.5 meters lower
1.	Recreational Harbors		
	Dredging	30 to 50	75 to 100
	Sheeting	15	35
	Slips/Docks	20*	40*
2.	Commercial Harbors		
	Dredging	108	212
	Sheeting/Bulkheads	38	38
	Slips/Docks	40*	90*
3.	Water Supply Sources		
	Extending urban intakes	15	15
	Wilmette Harbor Intake	1	2
4.	Beaches		
	Facililty relocations	1-2	1-2
5.	Outfalls for Stormwater		
	Extensions and modifications	2	4
	TOTALS	$270 to $292 million*	$512 to 540 million

*Some costs could be partly covered by normal replacement expenditures over the 50-year period of changing level.

Source: Changnon et al., Volume __.

Ice Cover

Warmer winters would reduce ice cover on the Great Lakes. Some analysts have speculated that ice would be completely eliminated. Assel used a model to estimate the potential extent and duration of ice cover.

Study Design

Assel developed a statistical relationship between temperature and ice cover for this study. The models were developed for the three basins of Lake Erie, for the Lake Superior Western and Eastern Basins, and Whitefish Bay in Lake Superior. Whitefish Bay was included because it has the longest period of ice cover and acts as a choke point on shipping in and out of Lake Superior. Lakes Superior and Erie represent extremes in terms of air temperature regimes, lake depth, and heat storage capacity, and bound the range of potential ice cover changes.

Limitations

Assel's study did not consider the effects of wind and other variables on ice formation. Implicitly, the analysis assumed that winds stay the same. Stronger winds would make the ice season shorter than estimated, and weaker winds (and calmer waters) would make it longer. The model was built based on the relatively cool years of the 1960s and 1970s; therefore, the doubled CO_2 scenario temperatures are outside the range of winter temperatures in those years. However, the model simulated ice duration within 3 weeks of actual ice duration for the warm winter of 1982-83.

Results

Assel found that although average ice cover might be significantly reduced, ice would still form on the lakes (Table 5-4). The central basin of Lake Erie now averages 83 days of ice cover. Ice cover could be reduced to a total of 6-19 days and ice formations would generally be limited to near-shore and shallow areas. Whitefish Bay in Lake Superior currently averages about 115 days of ice cover. Under the doubled CO_2 scenarios, ice duration would be reduced to 69-80 days. Also, the maximum percentage of Whitefish Bay covered by ice would be reduced from close to 100% to 70-20%.

Table 5-4. Reduction in Ice Cover in Lakes Erie and Superior (average annual days of cover)

Lake	Base 1951-80	GISS Transient 2000s	GISS Transient 2030s	Doubled CO_2 GISS	Doubled CO_2 GFDL	Doubled CO_2 OSU
Erie West	93	83	51	26	23	35
Erie Cent	83	68	37	8	6	19
Erie East	97	79	46	6	5	13
Supr West	112	109	86	46	24	75
Supr East	108	105	83	43	19	69
Supr WFB	115	111	91	55	26	80

Supr = Superior; WFB = Whitefish Bay.

Source: Assel, Volume __.

The temperature rise does not appear to be warm enough to eliminate ice cover on the Great Lakes, but many winters could have no ice at all. The Lake Erie Central Basin is estimated to be ice-free from 11 to 22 years out of 30 years, rather than 1 out of 30 years, as estimated for base climate conditions. This result appears to be sensitive to depth, as estimates indicate that the deeper Lake Erie East Basin would be ice-free 60-84% of the time, and the shallow West Basin would be ice-free in 7-17% of the winters. Lake Superior would have ice cover in virtually all winters under the scenarios.

Assel found that ice cover reductions during the first 30 years of the transient scenario (model years 1981-2010) are not significantly different from reductions during the modeled current conditions. The length and extent of ice cover noticeably decline, beginning in the second 30 years of the transient scenario (2011-2040). By the last decade of the transient scenario, the 2050s, the extent of ice cover is almost identical to the GISS $2xCO_2$ coverage.

Croley also found that ice cover would be reduced. His analysis shows that average surface temperatures on all the lakes in the winter is above 0°C. Even if average temperatures are that high, water temperatures in near-shore and shallow areas, the areas to which Assel said ice would be limited, would be sufficiently cold to cause ice formation.

Implications

Ice cover reductions could have positive and negative effects. On the positive side, the shipping season would be extended (see below). Water would flow more freely through rivers and connecting channels allowing for more hydropower production in the winter. On the other hand, ice protects some aquatic life, such as whitefish, and protects shorelines against the erosive impact of high-energy waves (Meisner et al., 1987).

Shipping

With lower lake levels, ships would have to lower their cargo, or ports and channels would have to be dredged. However, the reduced ice cover would allow a longer shipping season. The additional days of transport may make up for the loss of capacity on each voyage.

Study Design

Keith et al. studied the potential impacts of changes in lake levels and ice cover on shipping in six ports: Two Harbors; Duluth/Superior and Whitefish Bays in Lake Superior; and Toledo, Cleveland, and Buffalo in Lake Erie. They used the "ECO Great Lakes Shipping Model," which includes current data on major ports and commercial ships in the Great Lakes, types of cargo, costs of transport, and operating costs. Keith et al. used lake level reductions from Croley and Hartmann to study the change in cargo capacity and costs per ton and they used the change in cargo capacity to estimate how many days of shipping would be needed to transport the same amount of cargo as transported at present. The latter figure was compared to ice duration reductions estimated by Assel to determine whether the shipping season was sufficiently extended to allow for transport of the same amount of annual cargo as currently transported.

Limitations

The analysis did not consider changes in the composition of the fleet or in the mix and amount of cargo. It also assumed that demand for shipping of goods did not change, even in response to availability of shipping. The analysis did not examine whether goods would shift to or from alternate means of transportation and how changes in the costs of shipping and in the shipping season would affect users. Keith et al. also assumed that channels were not dredged to be deeper. The analysis is useful for estimating the direction and approximate magnitude of change.

Chapter 5

Results

The costs of shipping were estimated to increase due to lower lake levels. The effect on the cargo load for ships using the port of Buffalo are displayed in Figure 5-3. Under drops of 0.7-1.0 m in Lake Erie, which are the lake level reductions estimated by Croley for the OSU and GISS scenarios, cargo capacity would decrease by about 5-13%, and costs per ton would rise by the same amount. Croley's estimate from the GFDL scenario was that Lake Erie would fall 1.65 m (5.4 feet), but the shipping model does not include lake level drops of more than 5 feet. A drop of 5 feet would decrease cargo capacity per voyage by 27% and increase costs by 33%. Since lake levels in Lake Superior were not estimated to fall as much, the corresponding reduction in cargo capacity for ships on those ports would be in the range of 2 to 8%.

Sanderson estimated that lake level reduction of 0.2 to 0.6 m would increase Canadian shipping costs by 5%, assuming the current fleet and mix stayed the same. Although results are not directly comparable, since Keith et al. examined U.S. flag ships and ports, while Sanderson studied Canadian ships and ports, the estimates are of the same magnitude.

Whether the same amount of annual cargo can be transported depends mostly on how much lake levels drop. If the drop is sufficiently large, annual tonnage could be reduced. Figure 5-3 also displays the additional days needed to transport the same amount of cargo as is currently shipped through Buffalo. Under the approximate 2-3 foot drop of the wetter and relatively cooler OSU and GISS scenarios, another 15-40 days of shipping are needed. Assel estimated ice duration in eastern Lake Erie would be reduced by 84-91 days. Thus, under these scenarios, even with reduced capacity per voyage, there would be enough additional days of travel to transport even more goods. If lake levels fall 5 feet, which is less than estimated by GFDL, an additional 100 days of transport would be needed to handle the same amount of cargo. Ice duration in eastern Lake Erie is reduced by 92 days in this scenario, which would not allow enough time to transport the same amount of cargo, assuming the current fleet and demand for transport. The results appear to be more sensitive to changes in lake levels than reductions in ice cover.

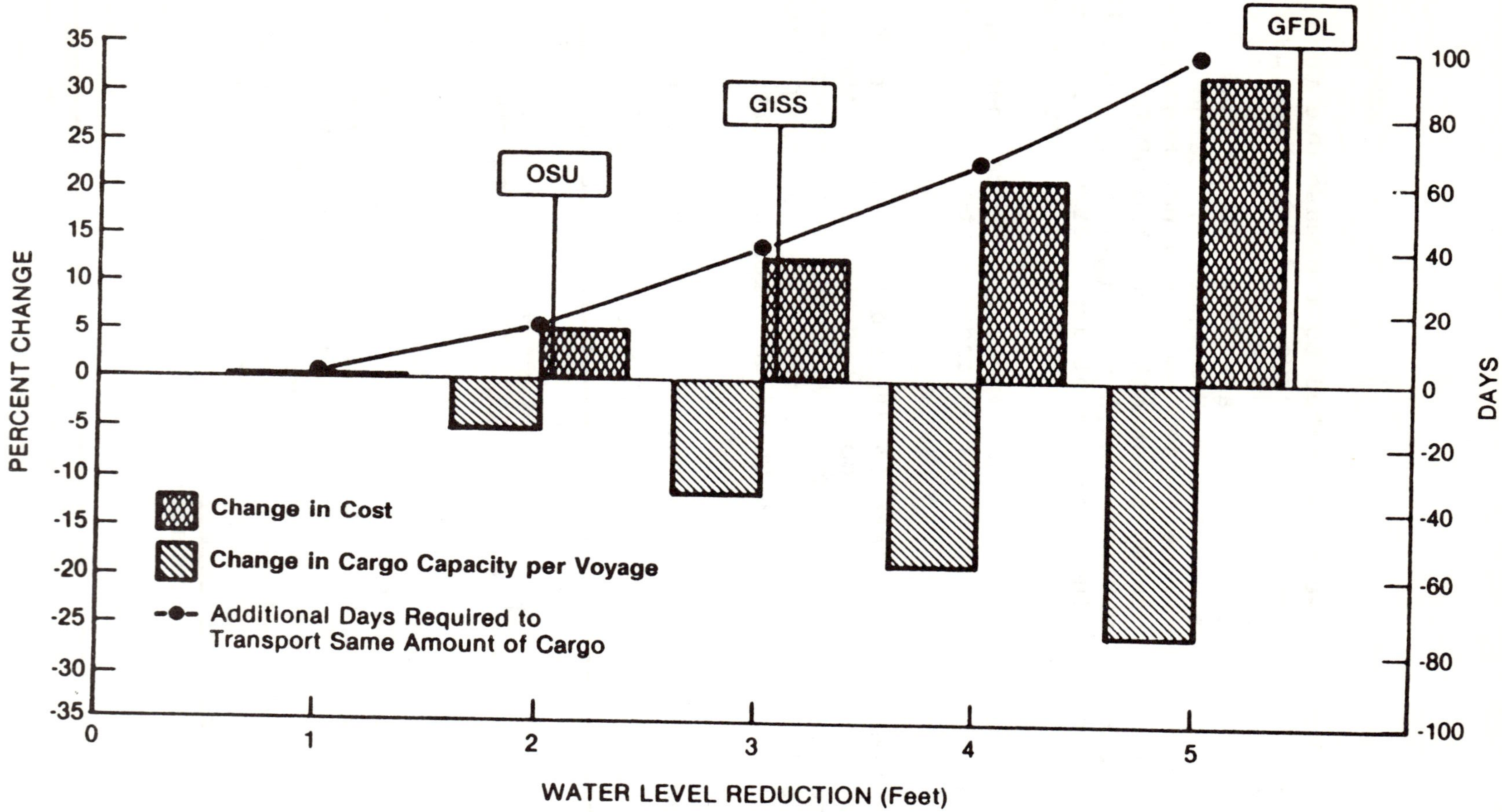

Figure 5-3. Impacts of lower lake levels and reduced ice cover on shipping, cargo capacity, costs, and days of transport for the Port of Buffalo.

Source: Keith et al., Volume __.

Keith and Willis used current dredging costs to estimate the cost of dredging the ports to restore current channel depths. The total costs of dredging the three ports in Lake Erie range from $26-82 million (1987 dollars).

Implications

Reduction in the tonnage per voyage or increased costs for dredging would raise shipping costs. However, with a longer shipping season, users of shipping such as powerplants would not have to carry large inventories to last through the winter and own enough land to store those inventories. Besides reducing costs, this could allow current lakefront storage areas to be used for other purposes. Whether these savings would offset higher shipping costs was not examined.

Dredging the ports and channels could degrade the water quality of the lakes. The sediments in many of these ports are toxic (which included many of the 42 hot spots), and disposal of the sediments could be complicated by their toxicity and by the reduced disposal areas resulting from lower lake levels.

Water Quality

Two studies estimated the temperatures and thermal structures of southern Lake Michigan and the Lake Erie Central Basin. The Lake Erie study estimates algal production and changes in dissolved oxygen levels. The Michigan and Erie analyses are used by Magnuson et al. to study changes in the thermal habitats of fish.

Thermal Structure of Southern Lake Michigan

Study Design

McCormick used a one-dimensional thermal structure model (Garwood, 1977) to estimate the heat content and structure of a site in south-central Lake Michigan. The model has been successfully applied to oceans and inland seas and was used by McCormick to analyze a site 150 m deep. GCM data for windspeed, temperature, humidity, solar radiation, and cloud cover were applied to hourly data from 1981 to 1984.

Limitations

McCormick used the years 1981-84 as his base case because hourly water temperature data are not available for 1951-80. Three years provide very limited baseline climate variability, although these years include cold and warm periods. Since the scenario may underestimate reductions in windspeed from the GCMs (see the discussion of the limitations of the lake level study), this analysis may overestimate wind-driven mixing in the upper layer and underestimate the length of time and degree of stratification. On the other hand, if the intensity of summer storm increases, then stratification may be weakened and shortened.

Results

McCormick estimated that the length of the stratified season would increase under all three scenarios. Figure 5-4 displays the mixed layer depth (the thermocline) over an average year. The higher heat content may cause the lake to begin thermally stratify, on average, about 2 months earlier than in the base case (in April as opposed to June). The stratified layers begin to deepen around late fall, as under current climate conditions.

Surface lake temperatures were estimated to be up to several degrees higher and warm more than lower levels. There appears to be a larger warming of the entire water column in the winter, about 2-3°C, than in the summer, which has a warming of about 2°C. The warmer lake temperatures are consistent with the studies of Croley and Assel, which suggest that midlake water will generally be ice-free. The earlier onset of stratification, reduced winds in the scenarios and greater temperature differences between lake layers could yield stronger density differences between upper and lower layers.

McCormick detected a significant decrease in the frequency of complete mixing of the lakes. The surface layer could be warmer and more buoyant, making it more difficult for entrainment and mixing to occur. Temperatures were too warm in the winters of some years to allow the lake to become isothermal, (the mixed layer would stay above the botton of the lake akk year) leading to a year-long stratification. This result is consistent with Croley's analysis.

Implications

Reduced turnover of the lakes could have serious implications for aquatic species in the lakes. Mixing of oxygen and nutrients would be disrupted, possibly affecting the abundance of life in the lower and upper layers of the lakes.

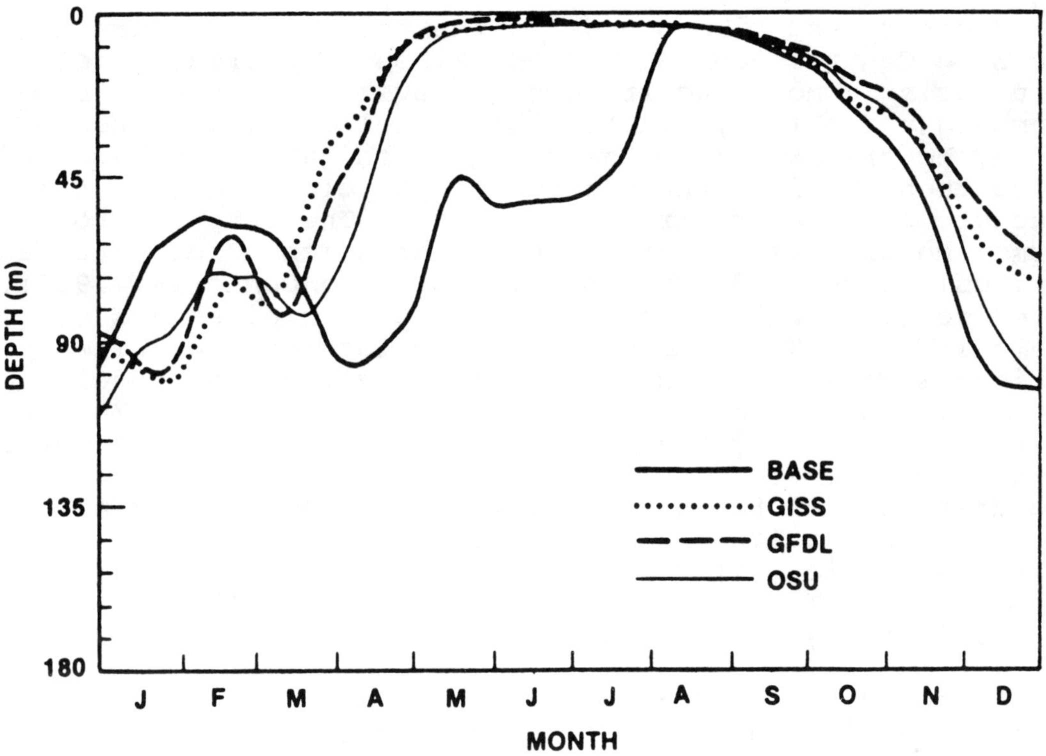

Figure 5-4. Lengthening of thermal stratification of southern Lake Michigan.

Source: McCormick, Volume __.

Eutrophication of the Lake Erie Central Basin

Nutrient loadings have made many areas of the shallow Lake Erie eutrophic at times. The shallow western and central basins of the lake are particularly vulnerable to eutrophication. Installation of pollution controls in recent years has improved water quality. Blumberg and DiToro analyzed whether climate change would have an effect on eutrophication in the Lake Erie Central Basin.

Study Design

Blumberg and DiToro modeled the thermal structure of the Lake Erie Central Basin. They developed a thermal model for the basin, using a modeling framework previously designed by Blumberg (Blumberg and Mellor, 1983). This model is similar to the one used by McCormick for southern Lake Michigan. Blumberg and DiToro then examined the direct effects of changes in the thermal structure on aquatic life in the basin. The outputs from the thermal model were fed into a eutrophication model that had been previously developed by DiToro (DiToro and Connolly, 1980). The latter model estimates what would happen to dissolved oxygen levels in the lakes by simulating the interactions between nutrient availability and plankton activity. The models were only run using two base years, 1970 and 1975. In 1970, the thermocline (density gradient between the upper and lower layers) was deep, and over 60% of the hypolimnion (lower level) in the Lake Erie Central Basin was anoxic (depleted of oxygen). In 1975, the thermocline was shallow, and less than 10% of the lower layer was anoxic (DiToro, 1987).

Limitations

Although the two base years encompass a wide range of baseline anoxic conditions, they do not represent a full range of climate variability. The analysis did not incorporate the actual reduction in nutrient loadings from the base years, or the estimated drop in lake levels from Croley's work. Lower lake levels would reduce the volume of the lower layer in Lake Erie, possibly increasing eutrophication. The models were not run for the winter, but Blumberg and DiToro tested the sensitivity of results to higher water column temperatures (due to warmer winter air temperatures) in the spring and found no significant difference in results. Blumberg and DiToro used the vector wind estimates from the GCMs, which may overestimate mixing in the upper layer.

Chapter 5

Results

Blumberg and DiToro estimated that the Lake Erie Central Basin could remain stratified about 2 to 4 months longer than under current conditions, with the stratified season starting 2 to 6 weeks sooner and ending 2 to 7 weeks later. The temperature difference between the upper and lower layers of the basin are estimated to be greater under all scenarios, leading to less exchange of nutrients across the thermocline. The depth of the thermocline appears to be most sensitive to estimated changes in windspeeds. In two scenarios, GISS and GFDL, windspeeds are generally lower, and the thermocline is about 2 m higher than current depths. In the OSU scenario, windspeeds are estimated to increase and the thermocline becomes approximately 1 m deeper than current levels. A lowering of the thermocline depth by 2 m in the 25-m-deep Lake Erie Central Basin can reduce the volume of the lower layer by 20%, limiting total oxygen availability.

All three scenarios generally lead to decreases in dissolved oxygen levels compared to base case conditions despite differences in thermocline depth. The increase in area of the Lake Erie Central Basin that is estimated to become anoxic is shown in Figure 5-5. Dissolved oxygen levels increase only in the July 1970 case, and this occurs because the levels were near zero to begin with. Blumberg and DiToro concluded that the difference in oxygen content is caused by warmer lake temperatures, which raise bacterial activity enough to increase oxygen demand. The enhanced biological activity is combined with a more intense and longer stratified season to further lower dissolved oxygen levels. Lower thermocline depths, such as in the OSU scenario, result in even greater decreases in dissolved oxygen levels.

The estimated changes in the thermal structure of Lake Erie are comparable to McCormick's results for southern Lake Michigan. Both estimated that average temperatures in the water column would rise, that there would be greater differences in temperature between the epilimnion and hypolimnion, and that stratification would last longer. One major difference in the results is that stratification begins earlier and lasts longer in Lake Erie and begins earlier and breaks up at the same time as the present stratification in Lake Michigan. It is not clear whether this difference is attributable to different lake depths, surface meteorology used to force the models, or to surface boundary conditions in the calculations.

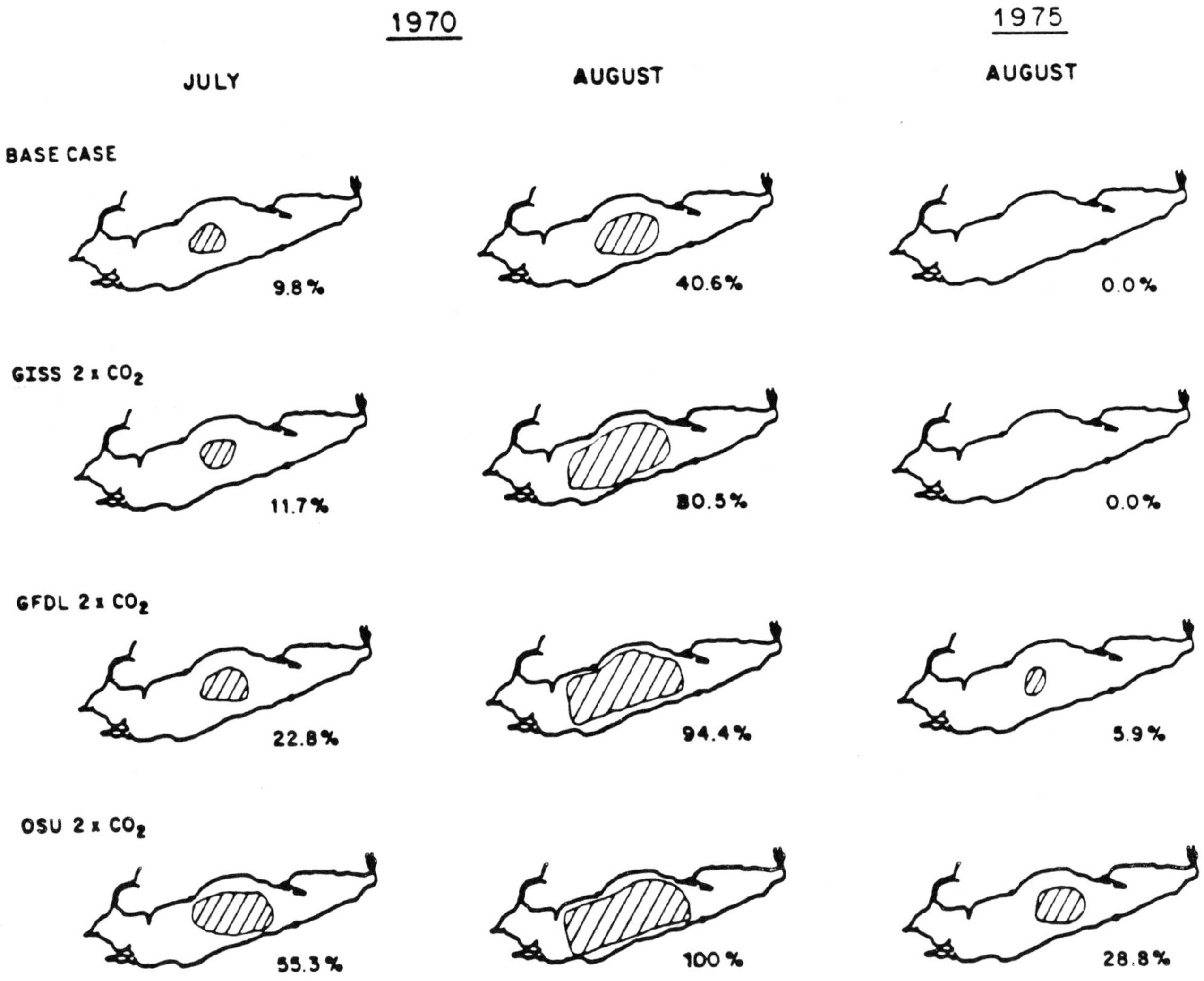

Figure 5-5. Area of central basin of Lake Erie that becomes anoxic.

Source: Blumberg and DiToro, Volume __.

Implications

Increased eutrophication could make the Lake Erie Central Basin uninhabitable for finfish and shellfish during the summer. This could reduce recreational uses of the lake such as swimming, fishing, and boating. It also could put more pressure on reducing sources of pollutants, especially such nutrients as phosphorous, from point and nonpoint sources.

Fisheries

The Blumberg and McCormick studies show that climate change would probably raise lake temperatures and reduce oxygen levels in certain areas. To get an initial sense of what these changes might mean for Great Lakes fish, Magnuson et al. examined the potential ecosystem, organism, and population responses to warmer temperatures.

Study Design

Magnuson et al. estimated changes in fish habitat, growth, prey consumption, and population for sites in Lakes Erie, Michigan, and Superior. The work used several approaches and models to examine the following:

- Changes in ecosystem activity, such as changes in phytoplankton populations, were estimated by using a community "Q_{10}" rule (Ruttner, 1931), which assumes that higher biological activity is associated with higher temperatures.

- Magnuson et al. used the Blumberg and McCormick thermal structure studies to estimate the potential effects on thermal habitats -- the niche in which temperatures are tolerable for species of fish. To estimate changes in habitats, the study used laboratory estimates of the temperature regimes preferred by fish (Magnuson et al. 1979; Crowder and Magnuson, 1983) and assumed that the lower layer of the Lake Erie Central Basin is uninhabitable. This study assumed that fish migrate to habitable sites when inshore temperatures are too warm. In addition, using a thermal model for streams (Delay and Seaders, 1966), the study calculated the change in habitat for brook trout in a southern Ontario river.

- Magnuson et al. used a food consumption and conversion model (Kitchell et al., 1977) to estimate the changes in growth and prey consumption at three near-shore sites in

Lakes Superior, Michigan, and Erie. This analysis assumed there were no limits on forage availability.

Limitations

The study did not consider changes in species interactions and prey availability. If prey is not sufficiently available, fish populations could decline. It did not examine the combined effects of reduced habitat and greater need for forage in the summer, which would combine to intensify species interactions. The analysis did not incorporate impacts resulting from lower lake levels, such as possible loss of wetlands, and it did not analyze the aquatic effects of the potential reduction in the frequency of lake turnover or the impacts of a reduction in ice cover. The introduction of new species, which could have negative impacts on existing fish, was not examined.

Any uncertainties associated with the McCormick and Blumberg studies would be carried over into the analysis on habitat. These changes in the lakes and littoral systems may have negative impacts on Great Lakes fish. These uncertainties could reverse the direction of results and lead to more declines in fish populations than indicated here.

Results

Phytoplankton production, zooplankton biomass, and maximum fishery yields are estimated to increase 1.25-3 fold with the biggest increase in phytoplankton production (2 to 3 fold) (Figure 5-6). The larger increases in biological activity are generally associated with larger temperature increases. The increase in phytoplankton provides more forage for zooplankton, which in turn, provides more forage for fish. The increase in phytoplankton can also enhance eutrophication, as was estimated by Blumberg and DiToro.

Magnuson and Regier found that the average annual thermal habitat for all fishes would increase. This was especially apparent for cold coldwater fishes such as lake trout, which could have more than a doubling of habitat (see Figure 5-7). The major reason for the increase in habitat is that more habitable waters would be found in the fall, winter, and spring. On the other hand, hotter temperatures in the summer could decrease habitats by 2-47%, depending on the temperature rise and species. The length of stream suitable for brook trout in the summer could be reduced by 25-33% due to higher temperatures.

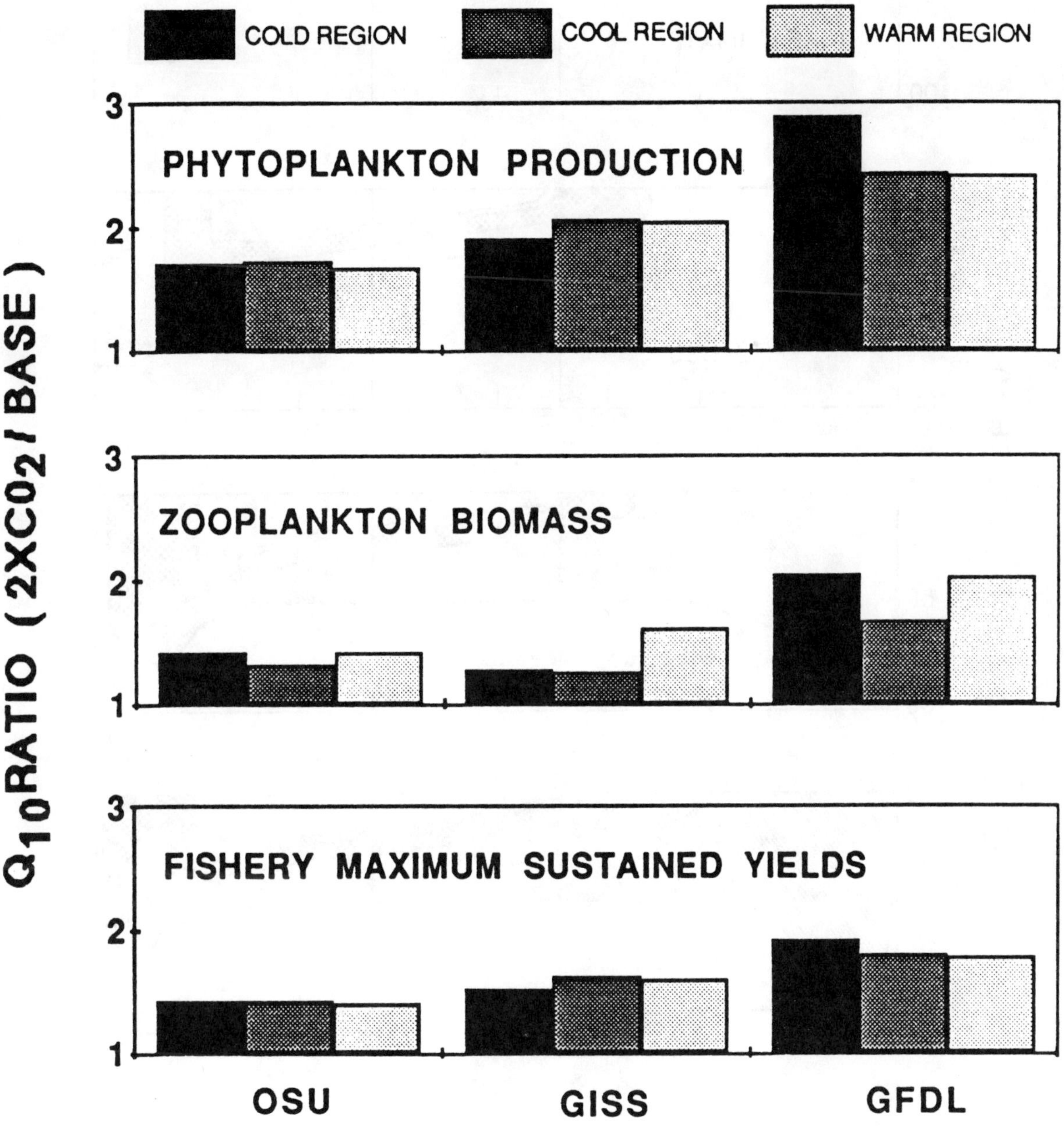

Figure 5-6. Increases in Great Lakes aquatic productivity.

Source: Magnuson et al., Volume __.

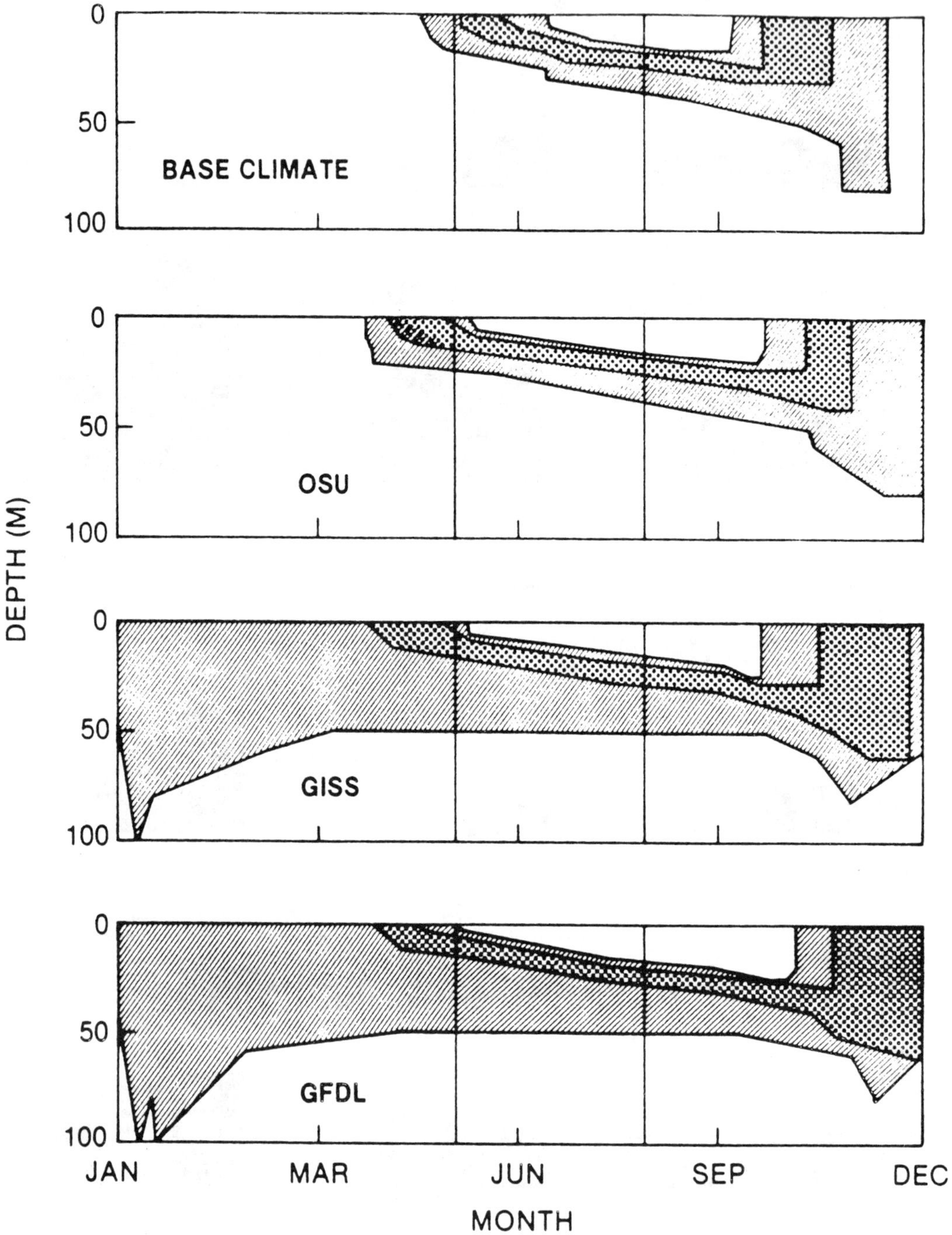

Habitat: ± 2°C of optimum temperature

± 5°C of optimum temperature

Figure 5-7. Increase in Great Lakes cold coldwater habitat

Source: Magnuson et al., Volume __.

Fishes are generally estimated to have increased body size under the scenarios. Cool and cold coldwater fishes could have 20-70% more growth, and warmwater fishes in warm areas could have 3.2 to 5.7 times greater growth. This assumes that prey availability increases. If prey availability does not increase, fish growth would also decrease owing to an inability to compensate for the increased metabolic costs of living in higher temperatures. Furthermore, the increased demand for forage may intensify species interactions and alter the food web structure.

The effects of reduced ice cover and possible reduction in wetlands on Great lakes fishes was not investigated, although Meisner et al. concluded that loss of wetlands due to lower lake levels could reduce spawning, nursery, and feeding grounds for fish in shallow areas, reducing fish populations (Meisner, et al., 1987).

Implications

If fish populations increase, there could be beneficial implications for commercial and recreational fishing. Presumably, commercial yields and sport fishing would increase, although certain species, such as brook trout in streams, may be reduced. A net increase in fisheries would lead to more employment in commercial fishing and tourism industries but, increase the need for maintaining water quality in the lakes. Increased demand on the forage base by predators and the introduction of new species could have negative effects, but these cannot be predicted and must be considered as surprises of unknown probability.

Forests

Climate change could affect the distribution and abundance of forests in the Great Lakes region. Overpeck and Bartlein examined the equilibrium range shift of forests, Botkin et al. studied transitional impacts on composition and abundance, and Zabinski and Davis analyzed the ability of trees to migrate along with a rapidly changing climate.

Potential Range Shifts

Study Design

Overpeck and Bartlein studied the potential shifts in ranges of forest types over eastern North America. This analysis suggests where trees are likely to grow in equilibrium $2xCO_2$

climate conditions. It only indicates the approximate abundance of different species within a range, what the transitional effects of climate on forests might be, or how fast trees will be able to migrate to the new ranges. (See the Forestry chapter of this report for a discussion of the study's methodology and limitations.)

Results

Under all three 2xCO_2 scenarios, the range of spruce, a major component of the boreal forests, would shift almost entirely out of the region entirely. Northern hardwoods, such as birch and northern pine species, would shift to the north, but would still be in the region. Oak trees, which are mostly found in the southern part of the region, would be found all over the region in the warmer conditions. The abundance of prairie forbs (shrubs) would increase in the region and southern pines could migrate to the southern part of the region.

Transitional Effects

In contrast to Overpeck and Bartlein, Botkin et al. examined the transitional effect of climate change on forests and equilibrium conditions.

Study Design

Botkin et al. used a model of forest species growth and competition to estimate the effects of climate change on Great Lakes forests (Botkin et al., 1972, 1973). This model, which is known as a stand simulation model, can be used to estimate the transitional changes in composition and abundance of forests species in response to environmental changes such as higher temperature and precipitation.

Botkin et al. studied two diverse sites in the Great Lakes region. The first is in Mt. Pleasant, Michigan, a heavily settled area dominated by northern hardwoods and oaks, where commercial forests are an important resource. The other site is in Virginia, Minnesota, an undeveloped area dominated by boreal forests that have commercial and recreational uses.

Limitations

The model includes all dominant tree species in the northern United States and assumes that seeds from all these trees are universally available throughout the region. Species with predominantly southern distributions are not included; therefore, the model does not estimate whether they could grow in the region under the warmer climate. (Overpeck found that southern pines may migrate into the southern part of the region.) Thus, the stand simulation model does not accurately estimate migration of trees, either within the region or from other areas. In addition, the model does not account for fertilization effects of CO_2, although CO_2 may not have positive effects in the competitive environment of unmanaged ecosystems (see Botkin). Botkin's analysis did not account for introduction of new pests into the region for the possibility of increased frequency of fires, or the combined impact of changes in tropospheric gas levels and UV-B radiation.

Results. Botkin et al. found that the abundance of species could significantly change in three to six decades. Figure 5-8 displays results from the transient scenarios for balsam fir and sugar maple at the Minnesota site. The basal area of balsam fir starts to decline in three to six decades. Potential declines in several decades are also seen in simulations of white cedar and white birch in the Minnesota site. Sugar maple, which has negligible basal area in the current climate, starts to exhibit significant growth within three decades in both transient scenarios.

Botkin et al. estimated the doubled CO climate would cause major changes in forest composition throughout the region. Results from the Mt. Pleasant site indicate that tree biomass at dry sites, which now have oak and sugar maple, could be reduced by 73 to 99% and converted to oak savannas or even prairies. Relatively wet soil sites might be converted from sugar maple to mostly oak woodlands with some red maple. Biomass at these sites could be reduced by 37 to 77%.

In the Minnesota site, the boreal forests could be replaced by northern hardwood forests, now characteristic of areas to the south (see Figure 5-9). Relatively dry areas, such as the Boundary Waters Canoe Area where balsam fir dominates, and upland areas where white birch and quaking aspen dominate, could be replaced by forests consisting mainly of sugar maples. Where currently saturated soils in these upland areas become drier and better sites for tree growth, wood production may increase. However, bogs that now contain white cedar could become treeless. This is because no species which could tolerate warmer bog conditions are currently in the region. More southerly species could be transplanted.

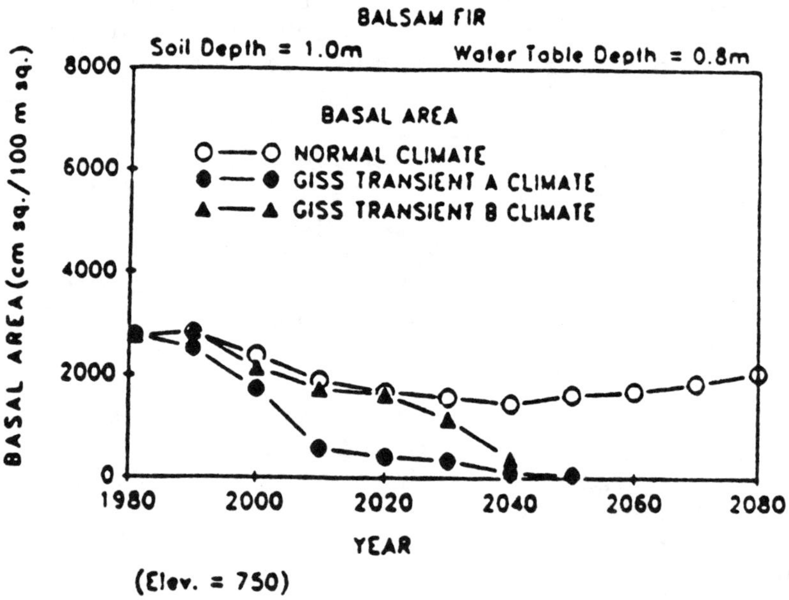

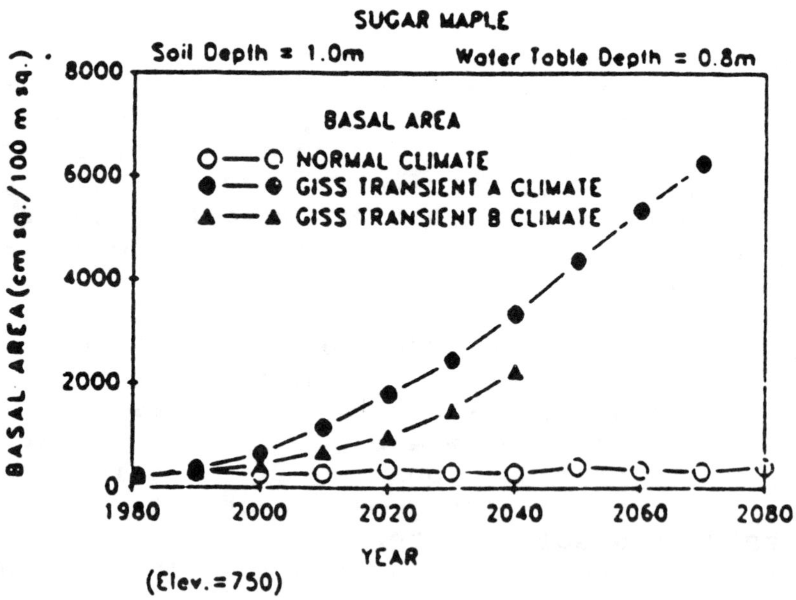

Figure 5-8. Change in forest composition during the next century for a deep, wet, sandy soil in the Boundary Waters canoe area.

Source: Botkin et al., Volume __.

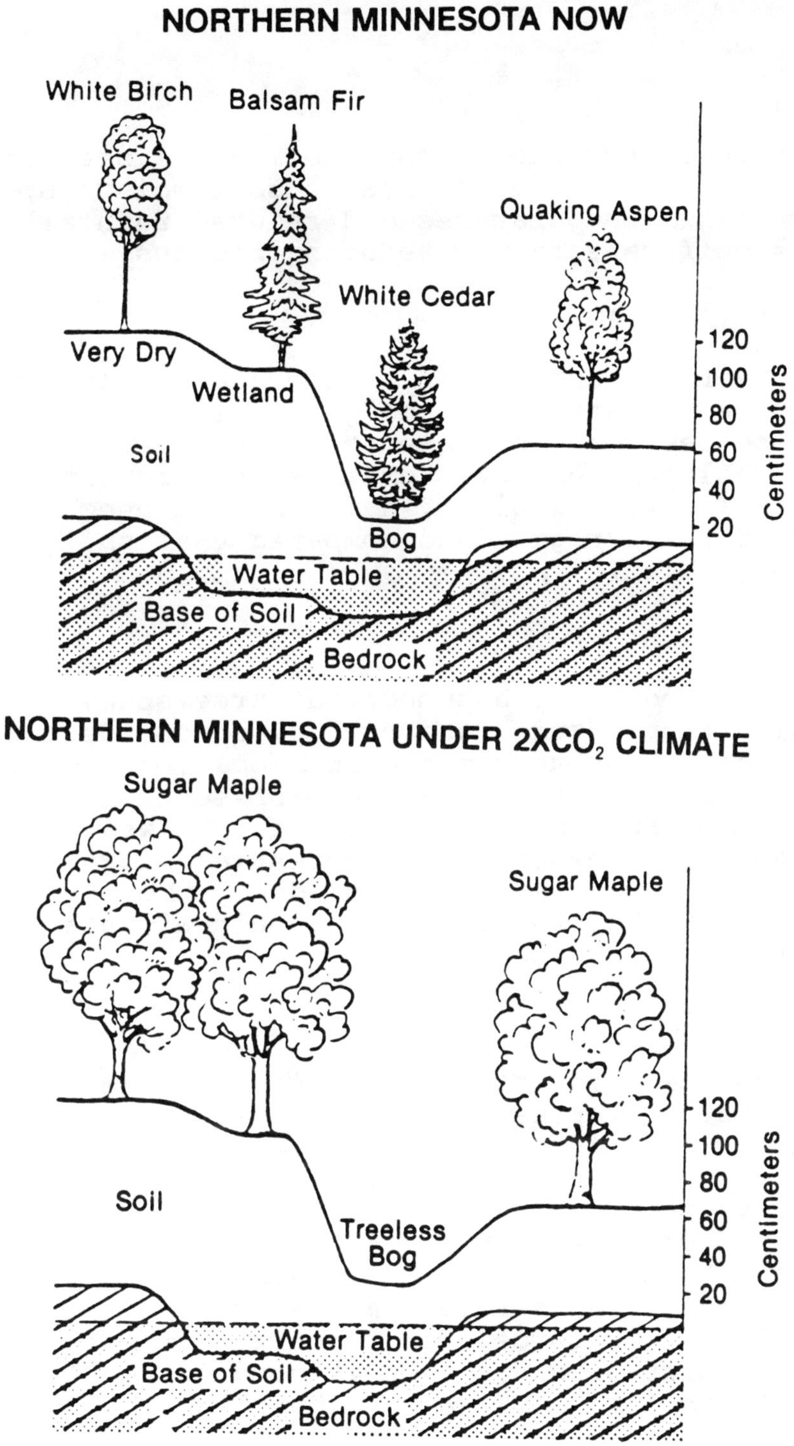

Figure 5-9. Effects on northern Minnesota forests (Virginia, Minnesota).

Source: Botkin et al., Volume __.

In both sites, the biggest decline is seen in the hotter and drier GFDL scenario. Decreased soil moisture, which is a result of higher temperatures and reduced rainfall, appears to be the most significant factor reducing biomass.

Forest Migration

Both Overpeck and Botkin assume that trees would be able to migrate to new locations (although Botkin does not assume southern species would be able to migrate into the Great Lakes region). Zabinski and Davis examined the potential range shifts of sugar maple, yellow birch, hemlock, and beech currently found in the Great Lakes region and compared that shift with potential rates of migration.

Study Design

Zabinski and Davis assumed that tree species only grow in climate with temperatures and precipitation identical to their current range. They determined the location of potential species ranges under the GISS and GFDL scenarios. The climate values were determined by extrapolating between gridpoints. Zabinski and Davis examined the potential migration of the species by assuming that the doubled CO_2 climate would not occur until 2090, and that these species could migrate into new regions at the rate of 100 km per century.

Limitations

The analysis did not consider competition among species or whether migratory routes would be blocked. It also did not analyze soil conditions, nutrient availiability, sunlight, and other relevant factors about northern areas to determine if trees would survive there. The assumptions about when doubled CO_2 climate would occur and the rate of migration are both optimistic. Doubled CO_2 climate conditions could occur sooner than 1990. The rate of forest migration used is double the maximum rate ever recorded for temperate trees. A faster warming and slower migration would make it more difficult for forests to keep up with shifts in range attributable to climate change.

Results

Under the wetter GISS scenario, the potential ranges of sugar maple, yellow birch, hemlock, and beech move markedly northward to central Canada. The results for hemlock and sugar maple are displayed in Figure 5-10. The stippled area shows the potential range, and the black area shows how far the trees could

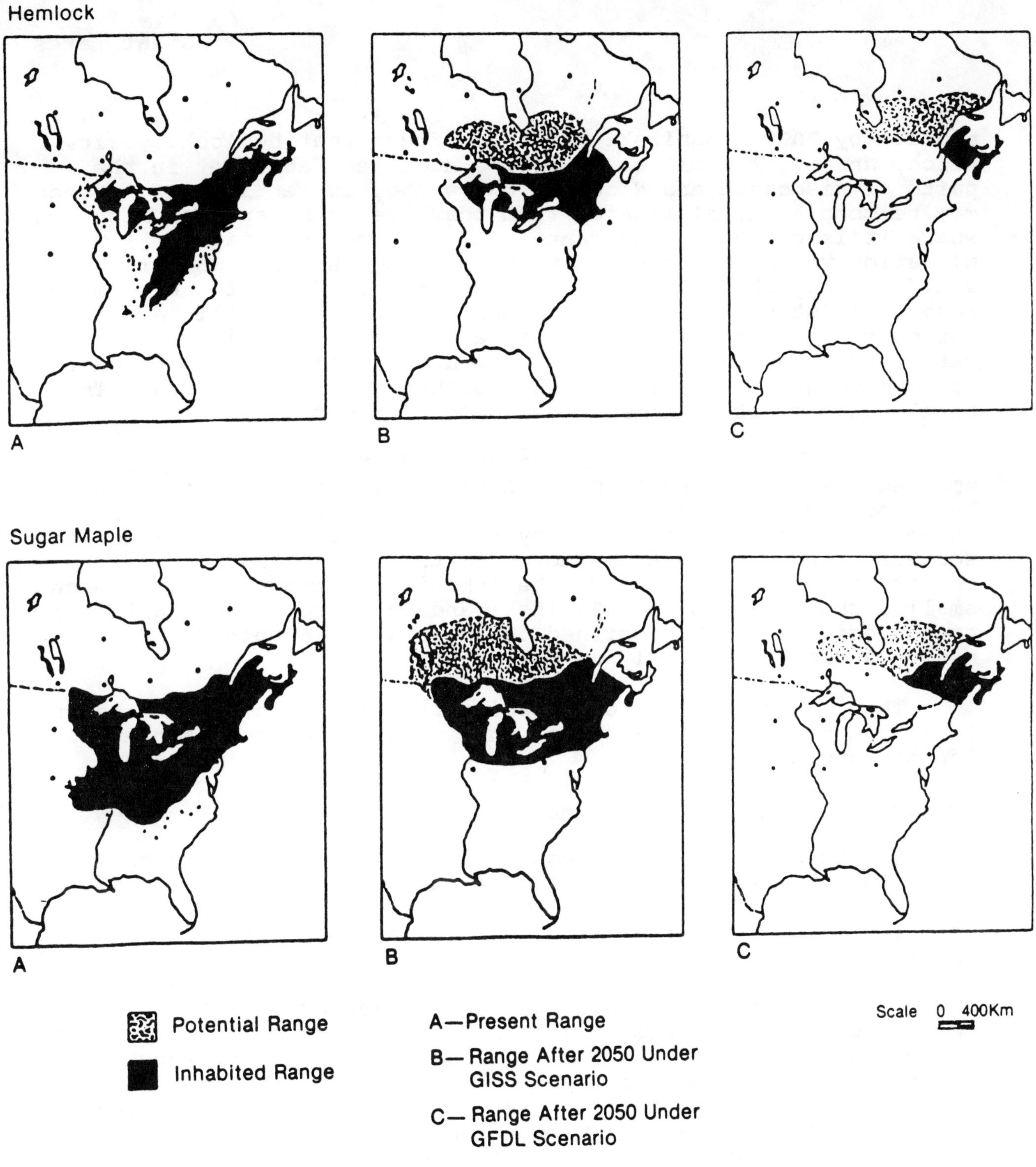

Figure 5-10. Shifts in range of hemlock and sugar maple.

Source: Zabinski and Davis, Volume __.

migrate by 2090. Zabinski and Davis found that hemlock, yellow birch, and sugar maple could become much less abundant in the parts of Wisconsin and Michigan where they currently grow. Beech may be completely eliminated from the lower peninsula of Michigan where it is presently abundant. In addition, the rate of migration would be slower than the climate change. The trees would not migrate as far as the northern boundary of the climate range (the stippled area). The southern boundary will be driven northward by climate change. Since the shift in climate zones is faster than the assured rate of migration, the southern boundary would move north faster than the northern migration rates. The total range of all four species would be reduced.

Under the GFDL scenario, which is hotter and drier, all four species are eliminated from the Great Lakes region. Northern hardwood tree species might be replaced by trees characteristic of more southern latitudes or by prairie or scrubland. Since the southern range of the trees moves farther north than in GISS but the trees migrate no faster, the inhabitated range would be much smaller than under GISS. Zabinski and Davis found that all four tree species would be confined to an area in eastern Canada having a diameter of only several hundred km.

The ability of the four species to survive in more northern latitudes may depend on whether they could adapt to different day lengths and soils.

Implications of Forest Studies

All three studies, through different analytic approaches, agree that the scenarios of climate change would produce major shifts in forest composition and abundance. Boreal forests would most likely no longer exist in the region. Northern hardwood forests might still be present, especially in the north. Much of the southern part of the region could become an oak savannah or prairie. Uncertainty exists concerning whether forests will die back or whether new species will be able to migrate and flourish. The rapid rate of climate change, coupled with the presence of urban areas and extensive farmland in the southern Great Lakes States, may impede migration of southern species into the region. Such a shift could result in increased soil erosion and decreased water quality. In addition, higher tree mortality and drier soils could increase fire frequency. Increased environmental stress might increase pathogen-related mortality in plants. Shifts in forest composition and abundance may have implications for wildlife in the region.

This shift in species also could have significant impacts on the commercial forest industry in the region. The industry currently harvests softwoods for production of pulp, paper, and construction materials. These species would decline and be replaced by oaks and maples, which are useful for furniture but take longer to become fully grown. Red maple, which may be more abundant in the southern area, is not currently used commercially. Changes in forest abundance may also affect tourism.

Agriculture

The agriculture studies combine analyses of impacts on the region and across the country. Ritchie et al. studied the potential impacts of climate change on crop yields in the region. Adams et al. then used the results from this study and other regional crop yield analyses to estimate economic adjustments by farmers. Easterling studied how a typical Illinois corn farmer would try to adapt to climate change.

<u>Crop Yields</u>

<u>Study Design</u>

Ritchie et al. used crop growth models to estimate the impacts of climate change on yields for corn and soybeans in the Great Lakes States (Jones and Kiniry, 1986). The two physiological models examine the direct effects of temperature and precipitation on crop yields. Ritchie et al. also used simple estimates of increased photosynthesis and decreased transpiration to conduct a sensitivity analysis of the combined impacts of change in weather and CO_2 fertilization on crop yields. In addition, he studied whether crops currently in southern areas may mitigate climate effects.

<u>Limitations</u>

The analysis of combined effects is new research and will need further development and refinement. The model runs use simple parameters for CO_2 effects, assume higher atmospheric concentrations of CO_2 than are predicted, and may be overestimates of crop yields. (See the Agriculture chapter of this report.)

All the scenarios assumed that soils were relatively favorable for crops, by having low salinity and no compaction, and they assumed no limits on the supply of all nutrients. In addition, the analysis assumed farmers would make no technological adjustments to improve crop yields. Possible negative impacts due to changes in storm frequency, droughts, and pests and pathogens were not factored into this study. The percentage changes for Duluth are very large because current yields are very low relative to other sites.

Results

Ritchie et al. found that temperature increases alone could reduce crop yields everywhere in the region except the northernmost latitudes, such as Duluth, where yields could increase depending on rainfall availability. Corn yields could decrease from 3 to 60% depending on climate and water regime (dryland or irrigated). However, Duluth, the most northern site, could see increases of 49-86%. Current dryland and irrigated corn yields are lower in Duluth than in the more southern sites. Dryland yields in Duluth under climate change would be equal to other sites, and irrigated yields would exceed the other locations.

Dryland soybean yields are estimated to drop by 3-65% in the region, except in the north. There, dryland yields decrease by 6% under GFDL, but increase by 109% under the wetter GISS. Under irrigated scenarios, soybean yields in the north increase by 96 to 153%. Even with the increase in output, the soybean yields in Duluth are still lower than in areas to the south.

Table 5-5. Effects of Climate Change Alone on Corn and Soybean Yields in Great Lakes States (ranges are GISS-GFDL and are % change from base)

Site	Corn		Soybeans	
	Dryland	Irrigated	Dryland	Irrigated
Duluth, MN	+49 to -30	+86 to +36	+109 to -6	+153 to +96
Green Bay, WI	-7 to -60	-3 to -44	-3 to -65	+3 to -26
Flint, MI	-17 to -48	-14 to -38	-6 to -51	+6 to -11
Buffalo, NY	-26 to -47	-18 to -38	-21 to -53	+6 to -6
Fort Wayne, IN	-11 to -51	-15 to -48	-2 to -58	0 to -19
Cleveland, OH	-26 to -50	-19 to -43	-16 to -59	-1 to -14
Pittsburgh, PA	-22 to -55	-19 to -45	-13 to -59	0 to -13

Source: Ritchie et al., Volume __.

The reduction in yields in the south is due mainly to the shorter growing period resulting from extreme summer heat. Production in the north is currently limited by the long winter, so a longer frost-free season results in increased yields.

Ritchie found that the demand for irrigation would rise between 20 and 173% under the GFDL scenario and up to 82% under GISS, although some sites under GISS were estimated to have up to 21% reductions in demand of up to 21%.

The combined effects of higher concentrations of CO_2 and climate change could increase yields if sufficient rainfall is available. If it is not, yields could rise or fall. Dryland corn and soybean yields rise up to 135% under the GISS scenario and up to 390% in Duluth. In the GFDL scenario, however, yields fall up to 30% or rise 17%, again, except for Duluth, which has an increase of 66-163%. Irrigated yields for corn rise and fall under both scenarios, but irrigated soybean yields rise 43-72% in the south and up to 465% in Duluth. The combined effects lead to an estimated reduction in demand for irrigation for corn of 26 to 100% under both scenarios, whereas irrigation needs for soybeans under GFDL rise by 65-207% and range in GISS from a reduction of 10% to an increase of 32%.

Ritchie found that use of a longer season corn variety could reduce the negative effects of climate alone, under the GFDL scenario, but still result in net losses.

It is not clear whether crop yields would rise or fall in the region. Among other factors, this will depend upon how CO_2 and climate change combine to affect crop growth and on how hot and dry the climate becomes. Yields and the potential demand for irrigation appear to be quite sensitive to rainfall, being higher under relatively drier scenarios. If climate change is severe enough, as under the GFDL scenario, yields could fall. In general, irrigation demand would rise, but some significant exceptions exist.

Implications. The potential shifts of agriculture northward are discussed below. Since the demand for irrigation is generally higher, it could become a more attractive option for farmers in the region. Whether more irrigation is actually used will depend on its costs and changes in revenue results.

Regional Shifts

Ritchie et al.'s analysis only estimates changes in potential yields for the Great Lakes region. How much farmers actually grow will depend in part on what happens elsewhere. If the relative productivity of agriculture rises, farmers will probably increase acreage. If productivity falls, they would most likely cut back. Adams et al. examined how different regions of the United States may react to potential productivity changes. Results are presented here for the Great Lakes region only.

Adams et al. modeled potential nationwide shifts in crops using the Great Lakes analysis and analyses of shifts in other regional crop yields. He did the analysis for yields attributable to climate change alone, and for the combined effects of climate and enhanced CO_2 concentrations. Adams et al's analysis did not account for the effects of climate on agriculture in other countries. How U.S. and regional agriculture respond to climate change may be strongly influenced by changes in relative global productivity and demand. (See the Agriculture chapter of this report for a discussion of the study's design and limitations.)

Results. Adams et al. estimates of acreage changes for the Great Lakes States are shown in Table 5-6. It appears that land devoted to agriculture in the Great Lakes region would not change significantly in response to climate change. The results indicate a slight tendency to increase acreage in the northern Great Lakes States, although only by small amounts. Land use in the Corn Belt would most likely be unchanged.

The relative increase in acreage in the north is consistent with Ritchie et al.'s results. Despite the yield reductions estimated by Ritchie et al. for the southern Great Lakes States, acreage would not be significantly reduced.

Table 5-6. Percentage Change in Acreage for Great Lakes States of Iowa and Missouri

Area	Climate Change Alone GISS (%)	Climate Change Alone GFDL (%)	Climate and CO_2 GISS (%)	Climate and CO_2 GFDL (%)
Lake States	+3	0	%1	+10
Corn Belt	+2	-6	-1	-6

Implications

The results of Adams et al. and Ritchie et al. suggest that northern regions will become more attractive for agriculture, although more extensive analysis is needed to confirm this result. The presence of thin, glaciated soils may limit this shift. If it occurs, such a shift could have significant implications for development of the north. Additional acreage could be converted from current uses, such as forests, to agriculture. Increased erosion and runoff from this additional acreage would pollute groundwater and streams and lakes in relatively pristine areas. Enhanced agriculture may increase the need for more shipping as lower lake levels raise shipping costs.

Adjustments by Illinois Corn Producers

Farmers may make many adjustments to climate change such as planting different crop varieties, planting earlier in the season, irrigating, and using different fertilizers. Easterling examined how a typical corn farmer in Illinois would react to climate change.

Study Design

Easterling presented several professional crop consultants with the GISS and GFDL climate change scenarios and with estimates of corn yields and prices for climate effects alone from the Ritchie et al. and Adams et al. studies. Based on the interviews, a set of decision rules were established to estimate how a typical Illinois corn farmer would alter farming practices in response to the climate and agriculture scenarios.

Limitations

The climate change scenarios involve climate conditions not experienced by the experts. Their estimates of how farmers would respond are not based on experience with similar conditions but on speculation. The results of the combined climate and CO_2 sensitivity analyses were not presented to the experts. The analysis is specifically for Illinois corn farmers and cannot be extrapolated to other areas or crops.

Results

Easterling found that the degree of adjustment depends on how much climate changes. Under the wetter GISS scenario, farmers could make adjustments to help mitigate the impacts of higher temperatures. Such adjustments could include planting earlier in the spring to avoid low soil moisture levels in the summer, using full-season corn varieties for earlier planting, and changing tillage practices and lowering planting densities to better conserve soil moisture. Under the hotter and drier GFDL scenario, corn production might not be feasible. Farmers would likely install irrigation systems; switch to short-season corn, soybeans, and grain sorghum; and perhaps remove marginal lands from production. This last conclusion is consistent with the Adams et al. study.

Implications

Although farmers have a variety of adjustment options to help cope with climate change, they may have great difficulty coping with extreme changes such as the dry climate implied by the GFDL scenario. Use of more irrigation would have negative implications for water quality, although this would be partly counterbalanced by any retirement of marginal lands.

Electricity Demand

Study Design

Linder and Inglis estimated the national changes in demand for electricity for the years 2010 and 2055. They first estimated the change in electricity demand due to gross national product (GNP) and population growth, and then factored in demand changes based on change in climate. The results for the Great Lakes States are displayed here. The Great Lakes analysis did not consider any reductions in hydropower production resulting from drops in lake levels. (See the Energy chapter of this report for a description of the study's design and limitations.)

Results

Estimates of changes in annual demand induced by climate change are displayed in Table 5-7. The results for 2010 are a range based on GISS transient scenarios A and B, and the results for 2055 are just for GISS A. A latitudinal difference exists within the Great Lakes region. In the northern States of Minnesota, Wisconsin, Michigan, northern Ohio, and upstate New York, annual demand falls. The reduced demand for winter heating is apparently greater than the increased demand for summer cooling. This is true in 2010 and 2055 when scenario temperatures are, respectively, 1 and 4° higher than the base case. Annual demand in the southern part of the region (in Illinois, Indiana, southern Ohio, and Pennsylvania) rises because increased cooling needs are apparently greater than reductions in heating.

Although annual demand could fall in some areas, new generation capacity requirements for all utilities in the region would be higher now because of increased summer cooling needs. New generation capacity requirements needs are estimated to rise by 3 to 8% in 2010 and by 8 to 11% in 2055. Whether costs would rise in the next two decades is not clear. Linder estimated that under the gradual warming of GISS B, cumulative capital costs in the region would be reduced by $1.3 billion, while under the more rapid warming of GISS A, costs would increase by $300 million. By 2055, costs would rise to $23 to 35 billion under GISS A. However, the cost to build additional capacity to meet GNP and population growth without climate change would be $488 to 715 billion.

Table 5-7. Estimated Change in Electricity Demand Induced by Climate Change for Great Lakes Utilities (%)

Utility	Annual (2010)	Annual (2055)
Minnesota	-0.2 to -0.3	-1.2
Wisconsin	0.4 to -0.5	-2.3
Michigan	-0.2 to -0.3	-1.2
Upstate New York	-0.2 to -0.5	-1.3
Ohio, north	-0.2 to -0.3	-1.3
Ohio, south	0.4 to -0.5	2.1
Pennsylvania	0.4 to -0.5	2.2
Illinois	0.5	2.0
Indiana	0.4	1.9
Total	Negligible	<1

Source: Linder and Inglis, Volume __.

<u>Implications</u>

Increased capacity requirements could place additional stress on the region. Powerplants may need cooling water and may seek to obtain it from rivers or the Great Lakes. Fossil fuel plants could add more pollutants to the air. The lake level analysis indicates that hydropower production from the lakes would be reduced, further increasing the demand for energy from other sources.

POLICY IMPLICATIONS

Climate change could raise many issues for policymakers in the region to address. Fundamentally, decisionmakers may have to cope with water use, water quality, and land management issues. They will probably have to respond to a decline in water availability, increased demand for water, poorer water quality, and shifts in land use, including the possibility of expanded agriculture in the north.

Most likely, many of the decisions in response to climate change, especially issues concerning water management, necessarily will be made on an international basis. Both Canada and the United States oversee the regulation of the lakes, water quality, and diversions of water out of the basin.

Water Supply Issues

Lake Regulation

One important issue to be faced by both countries may be regulation of the lakes. Lower lake levels may require altering plans for Lakes Superior and Ontario. This would involve tradeoffs among the needs of shippers, hydropower, shoreline property owners, and infrastructure, and downstream needs, in deciding how high to keep the lakes and rivers. For example, maintaining high water levels in the lakes to support shipping, hydropower, consumption, and improved water quality would be at the expense of shipping, hydropower, consumption, and water quality in the St. Lawrence River. Additional structures to control the flow on the lakes may be an option. The International Joint Commission should begin to consider the potential impacts of climate change on lake regulations in its long term planning.

Withdrawals

Even without climate change, population growth would increase demand for water for municipal and industrial consumption and power generation. Climate change would most likely intensify the demand for withdrawals from the lakes for even more uses within and outside the basin. Municipal consumption would rise (Cohen, 1987), additional powerplants may use water from the lakes for cooling purposes, and farmers in the region may need more water for irrigation.

Others outside the Great Lakes may demand diversion of water from the basin. The 1986 Water Resources Development Act prohibit such diversion without the agreement of all Great Lakes governors and prohibits the Federal Government from studying this issue. Increased diversion through the Chicago Ship Canal was requested this summer to raise water levels on the drought-starved Mississippi River. The U.S. Army Corps of Engineers rejected the request.

Shipping

Any response to the potential impacts on the shipping industry may be costly. Possibilities include dredging of both ports and connecting channels. Dredging could cost tens, if not hundreds, of millions of dollars. In addition to the high capital costs of dredging, substantial environmental costs could be incurred in disposing of dredge spoils contaminated with toxic chemicals. If dredging were not undertaken, cargo loads would be lower and would possibly impair Great Lakes commerce.

Pollution Control

Climate change will probably lead to stricter pollution control to maintain water quality. Reduced riverflow, lower lake levels, changed thermal structure, and potentially reduced groundwater supplies may necessitate stricter standards and additional controls on sources of pollution. A need may exist for better management of nutrient runoff from farms into shallow areas, such as the Lake Erie Western and Central Basins. Many pollution control institutions, such as EPA and State and local water quality agencies, would have the authority to impose appropriate controls on polluters.

The water quality problems directly caused by climate change could be exacerbated by other responses to climate change. Intensified agriculture in the region could increase runoff, necessitating more control of nonpoint sources of pollution. If agriculture in northern areas expands, surface and groundwater quality in relatively pristine areas may be degraded. Additional powerplants could add to pollution problems through the use of more cooling water. Pollution control authorities such as the U.S. EPA may need to impose more comprehensive controls for those areas and should consider this in their long-term planning.

Fisheries

Although the analysis on fisheries indicates that fish populations in the Great Lakes would generally increase, maintaining fisheries may require intensive management. In productive areas, the possibility of introduction of new species could mean major changes in aquatic ecosystems. Fisheries management may be needed to maintain commercially and recreationally valuable species. The Great Lakes Fishery Commission may wish to consider the possible implications of climate change on valuable fisheries and management strategies to handle these possible changes. Additional pollution controls may be needed to help maintain fisheries in such areas as western and central Lake Erie.

Land Use

Shorelines

The potential changes in land availability and uses present opportunities and challenges. Lower lake levels would open up new beaches and potential areas for recreation and development, although high capital costs may be associated with developing them. These lands could be kept undeveloped to serve as recreational areas and as protection against fluctuating lake levels and erosion. Conversely, they could be developed to provide more housing and commercial uses. Building structures closer to the shorelines would make them more vulnerable to short-term rises in lake levels.

How these lands will be used will be decided by local and State governments as well as private shoreline property owners. Under the Coastal Zone Management Act, States may identify coastal zone boundaries and define permissible land (and water) uses (Baldwin, 1984). Thus, the act could be used to help manage the use of exposed shorelines.

Lower lake levels and less ice cover may also increase shoreline erosion, decreasing the value of shorelines and degrading water quality. The Great Lakes Basin is not included in the U.S. coastal barrier system, a program that denies Federal funds for development of designated erosion or flood-prone coastal barriers (Ray et al.).

Forestry

The potential decline in forests and northward shift in Great Lakes agriculture raise many land use issues. One important issue may be how to manage potentially large and rapid shifts in forest composition. To speed northward colonization, plantings of the species might be recommended along the advancing front of suitable climate. However, unsuitable soils and day lengths shorter than the species can tolerate might limit the success of such plantings. The forest industry may consider growing different types of species and producing wood for different uses, such as for furniture rather than for pulp and paper. Potential uses of new prairies or savannas could present challenges.

Agriculture

Although forests may decline, demand for more land for agriculture in northern areas may grow; however, Adams indicated this demand may be small and will depend on market forces and policies. Federal and State land managers as well as local zoning laws may need to consider that the demand for land use may change. Rules on these lands could have a major influence on how, if at all, the north is developed.

Demographic Shifts

This report does not study the demographics associated with climate change and cannot say whether people will migrate north along with warmer climates. A workshop on climate change and the Great Lakes region, conducted by Ray et al. and attended by Government representatives, academics, and citizens group representatives who have studied climate-related Great Lakes resources, concluded that populations from other regions of the United States could migrate to the Great Lakes. The region could have a more favorable climate than more southern areas. Although lake levels may fall, the lakes will still contain a large amount of freshwater while other areas have more severe water availability problems. Consequently, the Great Lakes may be relatively more attractive than other regions.

Like lower lake levels, an in-migration could present opportunities and challenges. Such a migration could revitalize the region, reversing population and economic losses of recent decades. However, it also could exacerbate some of the problems associated with climate change. More people and industries would require more water and add more pollution, further stressing water supplies and quality. Population growth could increase pressure to develop exposed shorelines along the lakes.

CONCLUSION

Climate change could significantly alter the character of the region. It will be costly to respond to and will require new rules and laws. Some changes, such as loss of forests and dependent wildlife, may be irreversible. The lakes are a precious resource and may become even more valuable when climate changes. Maintaining their quality may require careful management of supplies and strengthened efforts at limiting pollution. Preserving forests and wildlife in a rapidly changing climate also may be difficult for planners. The challenge for the Great Lakes region will be to maintain the quality of the lakes and the surrounding land in response to climate change and the pressures it brings.

Chapter 5

REFERENCES

Assel, Raymond A. C.R. Snider, and R. Lawrence. 1985. Comparison of 1983 Great Lakes winter weather and ice conditions with previous years. Monthly Weather Review Vol. 113.

Baldwin, M.F. 1984. The Federal Government's role in the management of private rural land. Unpublished paper. p. 22.

Blumberg, A.F., and G.L. Mellor. 1983. Diagnostic and prognotic numerical circulation studies of the South Atlantic Bight. Journal of Geophysical Research 88:4579-4592.

Botkin, D.B., J.F. Janak, and J.R. Wallis. 1972. Rationale, limitations and assumptions of a northeast forest growth simulator. IBM Journal of Research and Development 16:101-116.

Botkin, D.B., J.F. Janak, and J.R. Wallis. 1973. Estimating the effects of carbon fertilization on forest composition by ecosystem simulation. In: Woodwell, G.M., and E.V. Pecan, eds. Carbon and the Biosphere. Brookhaven National Laboratory Symposium No. 24. Oak Ridge, Tennessee: USAEC, pp. 328-344.

Cobb, C.E. Jr. 1987. The Great Lakes' troubled waters. National Geographic 172(1):2-31.

Cohen, S.J. 1986. The effects of climate change on the Great Lakes. In: Titus, J.G., ed. Effects of Changes in Stratospheric Ozone and Global Climate, Vol. 3. United Nations Environment Programme, U.S. EPA. pp. 163-183.

Cohen, S.J. 1987. Projected increase in municipal water use in the Great Lakes due to CO_2 induced climate change. Water Resources Bulletin 23(1). pp. 91-101.

Croley, T.E. II. 1983a. Great Lakes basins (U.S.A. toCanada) runoff modelling. Journal of Hydrology 64:135-158.

Croley, T.E. II. 1983b. Lake Ontario Basin (U.S.A. toCanada) runoff modeling. Journal of Hydrology 66:101-121.

Croley, T.E. II. 1988. Lumped Modeling of Laurentian Great Lakes Evaporation, Heat Storage and Energy Fluxes for Forecasting and Simulation. NOAA Technical Memo, ERL GLERL-XX, Environmental Research Laboratories, Boulder, Colorado. (In press.)

Crowder, L.B., and J.J. Magnuson. 1983. Cost-benefit analysis of temperature and food resource use: a synthesis with examples from fishes. In: Aspey, W.P., and S.I. Lustick, eds. Behavioral Energetics. Ohio State University Press. Columbus, Ohio. pp. 189-221.

Crowe, R.B. 1985. Effect of Carbon Dioxide Warming Scenarios on Total Winter Snowfall and Length of Winter Snow Season in Southern Ontario. Atmospheric Environment Service - Canada. (Unpublished manuscript.)

Delay, W.H., and J. Seaders. 1966. Predicting temperatures in rivers and reservoirs. Journal of Sanitary Engineering 92:115-134.

DiToro, D.M., and J.P. Connolly. 1980. Mathematical Models of Water Quality in Large Lakes. Part 2: Lake Erie. Duluth, Minnesota: U.S. EPA. U.S.EPA -600/3-80-065.

DiToro, D.M., N.A. Thomas, C.S. Herdendorf, R.P. Winfield, and J.P. Connolly. 1987. A post audit of a Lake Erie eutrophication model. Journal of Great Lakes Research 134(4):801-825.

Edison Electric Institute. 1987. Statistical Yearbook of the Electric Utility Industry/1986. No. 54.

Eliot, J.L. 1985. A North Woods Park Primeval: Isle Royal. National Geographic. Vol. 167; No. 4 pp.534-550.

Environment Canada. 1987. Implications of Climate Change for Navigation and Power Generation in the Lakes. Climate Change Digest CCD 87-03.

Federal Reserve Bank of Chicago & The Great Lakes Commission. 1985. The Great Lakes Economy: A Resource and Industry Profile of the Great Lakes States. Boyne City, MI: Harbor House Publishers, Inc.

Garwood, R.W. Jr. 1977. An oceanic mixed layer capable of simulating cyclic states. Journal of Physical Oceanography 7:455-468.

Great Lakes Basin Commission. 1975. Great Lakes Basin Framework Study: Appendixes 5, 11, 12, 13, and 19. Ann Arbor, MI: Public Information Office. Great Lakes Basin Commission.

Chapter 5

Horvath, F.J., M.R. Jannereth, and C.A. Shafer. 1988. Impacts of Water Level Fluctuations. Water Science and Technology Board Colloquium Series, Great Lake Water Levels: Shoreline Dilemmas. Chicago, Illinois.

International Joint Commission. 1985. Great Lakes Diversions and Consumptive Uses. Report by International Great Lakes Diversion and Consumptive Uses Study Board.

Jones, C.A. and J.R. Kiniry, eds. 1986. CERES-Maize: A simulation of maize growth and development. College Station, Texas: Texas A&M Press.

Kitchell, J.F., D.J. Stewart, and D. Weininger. 1977. "Applications of a bioenergetics model to yellow perch (Perca flavescens) and walleye (Stizostedion Vitreum vitreum). Journal of Fish Research 34:1922-1935.

Land Evaluation Group. 1986. Implications of Climatic Change and Variability for Ontario's Agri-food Sector. Prepared for Atmospheric Environment Service, Environment Canada. LEG-26.

Larsen, C.E. 1985. Lake Level, Uplift, and Outlet Incision, the Nipissing and Algoma Great Lakes. Quaternary Evolution of the Great Lakes: Geological Association of Canada Special Paper 30.
Linder, K.P., M.J. Gibbs and M.R. Inglis. 1987. Potential Impacts of Climate Change on Electric Utilities. New York State Energy Research and Development Authority, Albany, NY. Report 88-2.

Magnuson, J.J., L.B. Crowder, and P.A. Medvicle. 1979. "Temperature as an ecological resource." American Zoology 19:331-343.

Meisner, D.J. et al. 1987. An Assessment of the Effects of Climate Warming on Great Lakes Basin Fishes. J. Great Lakes Research 13(3):340-352.

Nekvasil, G.G. 1988. Lake Carriers Association, Personal Communication; June 16.

The New York Times. 1988. p. 16. June 15.

Quinn, F.H. 1987. Hydrologic response model of the North American Great Lakes. Journal of Hydrology 37:295-307.

Quinn, F.H. 1988. Fluctuations of Great Lakes Water Levels. Water Science and Technology Board Colloquium Series, Great Lakes Water Levels: Shoreline Dilemmas. Chicago, Illinois.

Conversation with Frank Quinn, Great Lakes Environmental Research Lab (October, 1987).

Conversation with David Rind, Goddard Institute for Space Studies (January, 1988).

Ruttner, F. 1931. Hydrographische und hydrochemische Beobachtungen auf Java, Sumatra und Vali. Archiv for Hydrobiologic Supplement. 8:197-454.

Solomon, A.M., and D. West. 1986. Atmospheric Carbon Dioxide Change: Agent of Future Forest Growth or Decline?" In: Titus, J., ed. Effects of Changes in Stratospheric Ozone and Global Climate. United Nations Environment Program/U.S. Environmental Protection Agency. Washington, D.C. pp. 23-38.

Tarlock, D.A. 1988. Multi-Jurisdictional Issues" Presented to National Research Council, Water Science Technology Board Colloquium Series, Great Lakes Water Levels: Shoreline Dilemma. Chicago, Illinois. March, 1980.

U.S. Bureau of the Census. 1987. Statistical Abstract of the United States: 1987. (107th edition) Washington, D.C.
U.S. Department of Commerce, Bureau of the Census. 1983. 1982 Census of Agriculture: Geographic Area Series.

U.S. Department of Agriculture, 1987. Agricultural Statistics. Washington, D.C.: U.S. Government Printing Office.
USDA Forest Service. 1982. An analysis of the timber situation in the U.S. 1952-2030. Forest Resource Report No. 23.

U.S. EPA and Environment Canada. 1987. The Great Lakes: An Environmental Atlas and Resource Book.

Wall, G. 1985. Climate Change and its Impact on Ontario Tourism and Recreation. Submitted to Environment Canada.

CHAPTER 6

SOUTHEAST

FINDINGS

Global climate change could affect the Southeast by causing forests to shift to grasslands, reducing agricultural productivity and increasing the abandonment of farms, diminishing fish and shellfish populations, and increasing electricity demand higher than the national average. Approximately 90 percent of the national coastal wetland loss and two-thirds of the national shoreline protection costs due to climate change could occur in the Southeast. The impacts on rivers and water supplies are uncertain.

Agriculture

- Southeastern agriculture is generally more vulnerable to heat stress than to freezing, so the adverse impacts of more hot days would offset a longer growing season.

- Due to climate change alone, yields of soybeans and corn would vary from no change in the cooler regions to up to a decrease of 91 percent in warmer areas, even if rainfall increases. When climate and the direct effects of CO_2 are considered, yields still drop in states such as Alabama and Mississippi due to extreme temperatures. Elsewhere, if rainfall rises, yields may increase. Increased CO_2 could also affect weeds, but these impacts were not analyzed.

- If rainfall decreases, irrigation would become necessary for farming to remain viable in much of the region.

- The range of such agricultural pests as potato leafhoppers, sunflower moths, and black cutworms could move north by a few hundred kilometers. This would most likely result in changes in the uses of pesticides.

- Considering various scenarios of climate change and CO_2, the productivity of southeastern agriculture could decline relative to northern areas, and 10 to 50 percent of the region's farmland could be withdrawn from cultivation. The decline in cultivated acreage may tend to be concentrated in areas where farming is currently marginally profitable.

Forests

- o There may be a significant dieback in southern forests. Higher temperatures and drier soils may make it impossible for most species to regenerate naturally and may cause forests to convert to grassland. The decline in the forests could be noticeable in 30 to 80 years depending on the site and scenario. Southern non-coastal areas, such as Atlanta and Vicksburg, may have particularly large reductions. Relatively cool northern forests, such as those in Tennessee, may survive, although with some losses.

- o The forest industry, which has infrastructure in main coastal areas and which is structured around currently valuable tree species, would have to either move or invest in new commercial tree species.

- o Historically, abandoned farms have generally converted to forests. If large portions of the Southeast can no longer support the natural generation of forests, the character of the Southeast's landscape could gradually resemble that of the Great Plains.

Water Supplies

- o Whether river flows and water supplies in the Southeast would increase or decrease cannot be determined because of uncertainties in precipitation and minimal snow pack. Water supply is very important for agricultural, recreational, and industrial uses. Any decrease in rainfall could disrupt recreation, hydropower, power plant cooling, and dilution of effluent, while more rainfall could increase the risk of flooding.

- o For the scenarios used in this report, changes in operating rules for managed water systems would allow current water demands to be met in some instances.

Sea Level Rise

- o A one-meter rise in sea level by the year 2100 would inundate 30 to 90 percent of the region's coastal wetlands and flood 2,600 to 4,600 square miles of dry land, depending on the extent to which people erect levees to protect dry land from inundation. Louisiana alone would account for 40 percent of national wetland loss, assuming current river management practices continue.

- Holding back the sea by pumping sand or other measures to raise barrier islands and protecting mainland areas with bulkheads and levees, for a one-meter rise would cost approximately $42 to $75 billion by the year 2100.

Marine Fisheries

- Gulf Coast fisheries could be negatively affected by climate change. A loss of coastal wetlands due to sea level rise could eliminate critical habitats for shrimp, crab, and other commercially important species. Warmer temperatures in the Gulf Coast estuaries will be at or above the thermal tolerances for commercially important finfish and shellfish, such as shrimp, flounder, and oysters. Oysters and other species could be threatened by the increased salinity that will accompany sea level rise.

Electricity Demand

- The annual demand for electricity in the Southeast could rise by 14 to 22 billion kilowatt hours (2 to 3 percent) by 2010 and by 100 to 197 BKWH (7 to 11 percent) by 2055 because of increased temperature.

- By 2010, approximately 7 to 16 gigawatts would be needed to meet the increased demand, and by 2055, 56 to 115 GW would be needed -- a 24 to 34 percent increase over baseline additions that may be needed without climate change -- the cumulation costs would be $77 to $110 billion by 2055.

Policy Implications

- Federal laws constrain the Corps of Engineers and other water resource managers from rigorously considering trade-offs between many nonstatutory objectives of federal dams in the Southeast, including recreation, water supply, and environmental quality. Increased flexibility would improve the ability of these agencies to respond to and prepare for climate change.

- Strategies now being evaluated by the Corps of Engineers to protect coastal wetlands in Louisiana may need to consider a sea level rise of 0.5 to 2.0 m in order to prevent wetlands from being lost. Measures that would enable this ecosystem to survive would require major changes in federal navigation and river flow policies.

o Given the potentially important impacts on forests, agencies such as the U.S. Forest Service and state agencies, and private companies may wish to assess the potential for large losses of southern forests and the implications for research and management strategies.

CLIMATE AND THE SOUTHEAST

The climate and the coastal zone of the Southeast are among the chief factors that distinguish the southeastern United States from the rest of the nation.* The warm temperatures, abundant rainfall, and generally flat terrain gave rise in the 17th century to a strong agricultural economy with a distinctive regional culture. The combination of a benign climate and 60% of the nation's ocean beaches continues to attract both tourists and new residents to the southeastern coastal plain. Florida, for example, is the nation's fastest growing state and will be the third largest by the year 2000. (Meo, Volume _).

The projected change in climate, however, could severely disrupt life in this region. Cities that already are unpleasantly hot during the summertime would become hotter, with corresponding increases in air-conditioning demands and electric bills. Fisheries would be threatened by the hotter estuarine temperatures and the loss of coastal wetlands. Rising sea level would erode beaches, inundate wetlands, exacerbate coastal flooding, and increase the salinity of estuaries, aquifers, and wetlands.

CLIMATE SENSITIVE RESOURCES OF THE SOUTHEAST

Water Resources

When statewide averages are considered, each of the seven states in the Southeast receives more rainfall than any other state in the continental United States. Moreover, the rivers of the Southeast drain over 62% of the nation's lands, with the Mississippi River draining 38% of the nation (Geraghty et al., 1973).

---

*Except for the discussion of the economic implications for agriculture, the term "Southeast" refers to the study area shown in Figure 6-1: North Carolina, South Carolina, Georgia, Florida, Alabama, Mississippi, Tennessee, and the coastal zones of Louisiana and Texas.

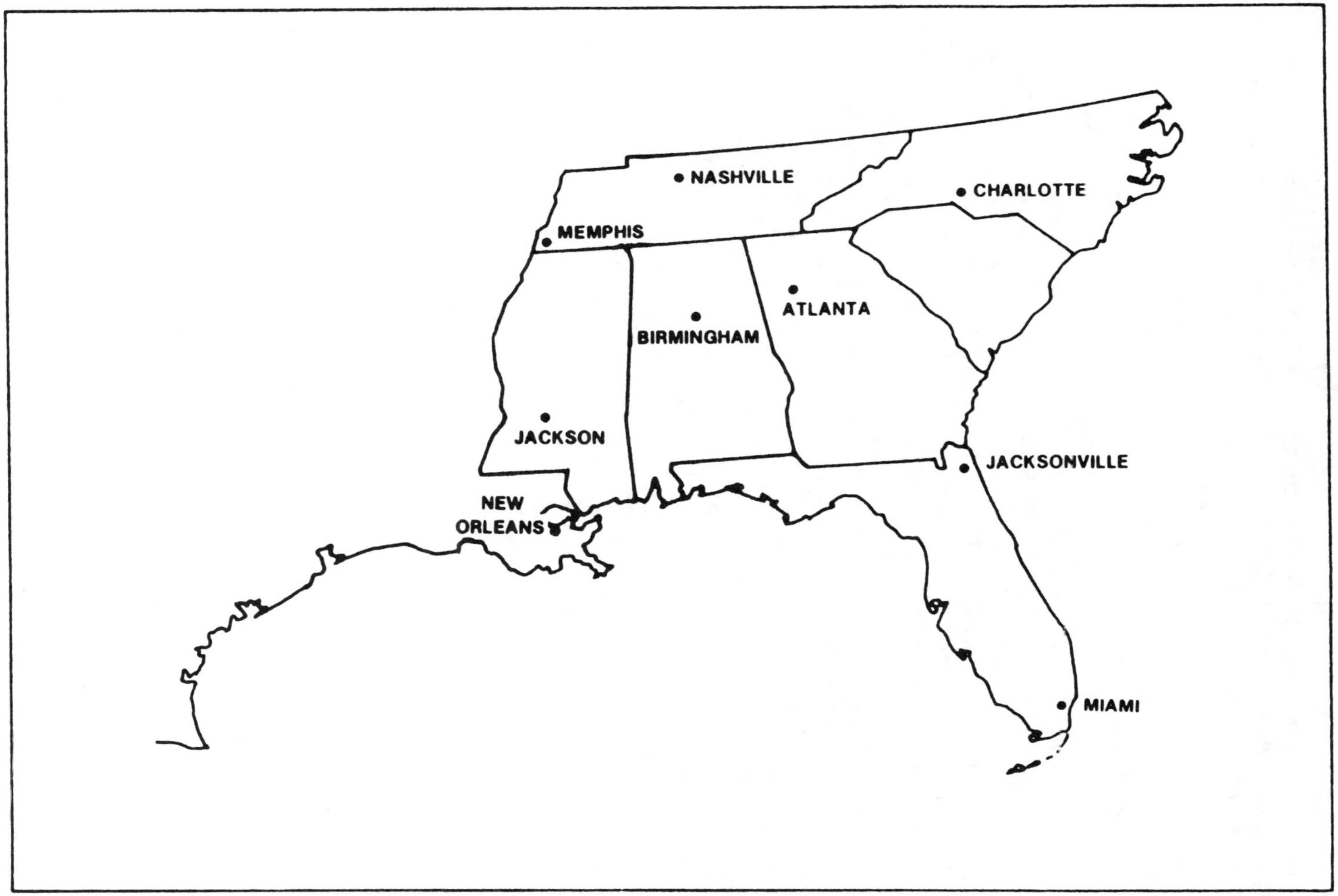

Figure 6-1. Southeast region

The Southeast supports 50,000 square miles (mi^2) of bottomland hardwood forests (Mitch and Gosselink, 1986),** which are periodically flooded areas that offer winter habitat formigratory birds such as ducks, geese, and songbirds. Bass, catfish, and panfish are found in the slow-moving rivers, and trout inhabit the fast-moving mountain streams.

Dams have been constructed along most of the region's major rivers. Although private parties have built a few dams, most of the major projects were built by the U.S. Army Corps of Engineers, the Tennessee Valley Authority, and other federal agencies. In general, these reservoirs have been designed for the following statutory purposes: to ensure a sufficient flow of water during droughts; to prevent floods; and to the extent that this purpose does not conflict with the previous objectives, to generate hydropower. The nonstatutory objectives of environmental quality, recreation, and water supply also are considered in the operation of dams. Dam construction has created large lakes along which people have built houses, hotels, and marinas. These dams generate 22.2 billion kilowatt hours (kWh) per year, approximately 7% of the region's power requirements (Edison Electric Institute, 1985). In general, the reservoirs have sufficient capacity to retain flood surges and to maintain navigation flows during the dry season. The one notable exception is the Mississippi River: levees and land-use regulations are the main toos for preventing flood damages; although the Mississippi's base flow usually is sufficient to support navigation, the navigation on navy stretches of the river was impeded during the drought of 1988.

In Florida, which accounts for 45% of southeastern water consumption, groundwater supplies about half the water used by farms and 85% of the water use for residential and industrial purposes. For the rest of the Southeast, groundwater supplies most water for agriculture and rural uses but only 30% for public supplies.

Atlanta and some other metropolitan areas derive their water supplies from Federal reservoirs. Many cities that do not obtain water from Federal reservoirs still benefit from Federal and Federal/State water management. For example, New Orleans obtains its water from the Mississippi River. Without the Old River Control Structure in Simmesport, Louisiana, which prevents the river from changing its course to the Atchafalaya River, the New Orleans water supply would be salty during droughts. Although Miami obtains its water from the Biscayne aquifer, some coastal wells would be salty without the efforts of the U.S. Army Corps

---

**This measure includes MS, AK, LA, TX, and VA.

of Engineers and the South Florida Water Management District to recharge the aquifer with supplemental freshwater from canals and Lake Okeechobee.

The various uses of water often conflict with each other. Hydroelectric power generators and lakefront property and boat owners benefit when water levels are maintained as high as possible. However, high water levels make flood control more difficult, and summer releases for navigation, hydropower, and environmental quality require that lake levels be lowered during the dry season.

Estuaries

Over 43% of the fish and 70% of the shellfish harvested in U.S. waters are caught in the Southeast (Fishery Statistics Division, NOAA, 1987). Commercially important fishes are abundant largely because over 85% of the nation's coastal wetlands are found in this region; over 40% are found in Louisiana alone.

Most of the wetlands in the Southeast are less than 1 meter (m) above sea level. The wetlands in Louisiana are already being lost to the sea at a rate of 50 mi^2 per year due to the interaction of human activities and current rates of relative sea level rise resulting from the delta's tendency to subside 1 cm per year.

Summer temperatures in many of the gulf coast estuaries are almost as warm as crabs, shrimp, oysters, and other commercially important fishes can tolerate (Livingston, Volume__). Winter temperatures along the Gulf coast are almost warm enough to support mangrove swamps, which generally replace marshes once they are established; mangroves already dominate the Florida coast south of Fort Lauderdale.

Beach Erosion and Coastal Flooding

The Southeast has 1,100 miles of sandy ocean beaches, many of which are found on low and narrow barrier islands. The Atlantic coast is heavily developed; while much of the gulf coast is only now being developed. In part because of their vulnerability to hurricanes, none of Mississippi's barrier islands has been developed, only one of Louisiana's barrier islands has been developed. Because much of Florida's gulf coast is marsh, it is still largely undeveloped.

All eight coastal states are experiencing coastal erosion. Along developed coasts, recreational beaches have narrowed, increasing the vulnerability of shorefront structures to storms. In Louisiana, some undeveloped barrier islands are eroding and breaking up. Elsewhere, narrow barrier islands are keeping pace with sea level rise by "overwashing" (i.e., rolling over like a rug) in a landward direction, while wide islands and mainland coasts have simply eroded. The coastal states of the Southeast are responding by holding back the sea in some areas and by adapting to erosion in others.

The two greatest natural disasters in U.S. history resulted from floods associated with hurricanes in Galveston, Texas, and Lake Okeechobee, Florida, in which over 8,000 people drowned. After the Mississippi River overflowed its banks and inundated most of coastal Louisiana in the 1930s, Congress directed the Corps of Engineers to initiate a major Federal program of flood control centered around the Southeast. Nevertheless, flood waters often remain over some low areas in Louisiana and Florida for several days after a major rainstorm.

Hurricanes continue to destroy recreational development in at least a few ocean beach communities almost every year in the Southeast. The region presently experiences the majority of U.S. coastal flooding and probably would sustain the worst increases in flooding as a result of global warming. Unlike the Northeast and Pacific coasts, this region has wide low-lying coastal plains and experiences several hurricanes annually. Each year, storms destroy recreational development in at least a few ocean beach communities. Florida, Texas, and Louisiana account for 62% of the $144 billion of private property insured by the Federal Flood Insurance Program.

Agriculture

In the last few years, droughts and heat waves have caused crop failures in many parts of the Southeast. Unlike many of the colder regions, cold weather is generally not a major constraint to agricultural production, except for Florida's citrus industry.

Although cotton and tobacco were once the mainstays of the Southeast's economy, agriculture now accounts for only 1% of the region's income (Bureau of Economic Analysis, 1986). Since World War II, substantial amounts of farmland have been withdrawn from agriculture, and much of this land has been converted to forest. The cotton crop has largely been lost to the irrigated Southwest, and although tobacco remains profitable, it is grown on only 1/2 million acres. However, in the last few decades, southeastern farmers have found soybean growth to be profitable; soybeans now account for 45% of all cultivated land in the Southeast. Corn

Table 6-1. Annual Revenues by State for Various Crops.

Crop	Value ($1,000)
Corn for Grain	
Alabama	56,550
Florida	31,493
Georgia	203,931
Mississippi	22,600
North Carolina	324,789
South Carolina	104,333
Tennessee	193,687
Cotton	
Alabama	145,540
Florida	8,112
Georgia	97,325
Mississippi	449,630
North Carolina	30,944
Tennessee	109,610
Sugarcane for sugar and seed	
Florida	369,899
Tobacco	
Florida	NA
Georgia	NA
North Carolina	NA
South Carolina	NA
Tennessee	NA
Peanuts for Nuts	
Alabama	133,930
Florida	48,600
Georgia	472,645
North Carolina	122,941
South Carolina	5,882
Soybeans	
Alabama	140,719
Florida	31,036
Georgia	179,676
Mississippi	365,018
North Carolina	196,673
South Carolina	125,214
Tennessee	230,373

Source: U.S. Department of Agriculture, Agricultural Statistics: 1987, pp. 32, 34, 62, 63, 80, 93, 121, 126, 127.

continues to account for 5% of southeastern agriculture (Bureau of the Census, 1982). Table 6-1 compares annual revenues by state for various crops.

Forests

The commercial viability of southeastern forests has increased greatly since World War II, primarily due to the increased use of softwoods, such as pines and firs for plywood and uses that once required hardwood. Because this transition coincided with lower farm prices and declining soils in the piedmont foothills of the Southeast, many mountain farms have been converted to forests. Conversely, in the last 10 years, 7 million acres of coastal plain forests have been converted to agriculture (Healy, 1985).

Approximately 45% of the nation's softwood (mostly loblolly pine) and 50% of its hardwood are grown in the region. Forests cover 60% of the Southeast, and 90% of forests are logged. Oak-hickory covers 35%, and pine covers another 33% of commercial forests. Only 9% of the southeastern forests are owned by the Federal and state governments, and 18% are owned by the forest industry. In contrast, 73% of the forests are owned by farmers and other private parties (Healy, 1985).

Indoor and Outdoor Comfort

The Southeast is one of the few areas that spends as much money on air-conditioning as on heating. Figure 6-2 shows temperatures throughout the Southeast for the months of January, April, July, and October. Even in January, about half the region experiences average temperatures above 50°F, and almost the entire region has a typical daily high above 50°F. Thus, with the possible exception of the mountains of Tennessee and North Carolina, a global warming would increase the number of days during which outdoor temperatures would be unpleasantly warm.

PREVIOUS STUDIES OF THE IMPACTS OF CLIMATE CHANGE ON THE SOUTHEAST

Most studies examining the impact of global warming on the Southeast have focused on sea level rise. Recent efforts have addressed other topics.

Flooding

Leatherman (1984) and Kana (1984) applied flood-forecasting models to assess the potential increases in flooding in Galveston, Texas, and Charleston, South Carolina. For the

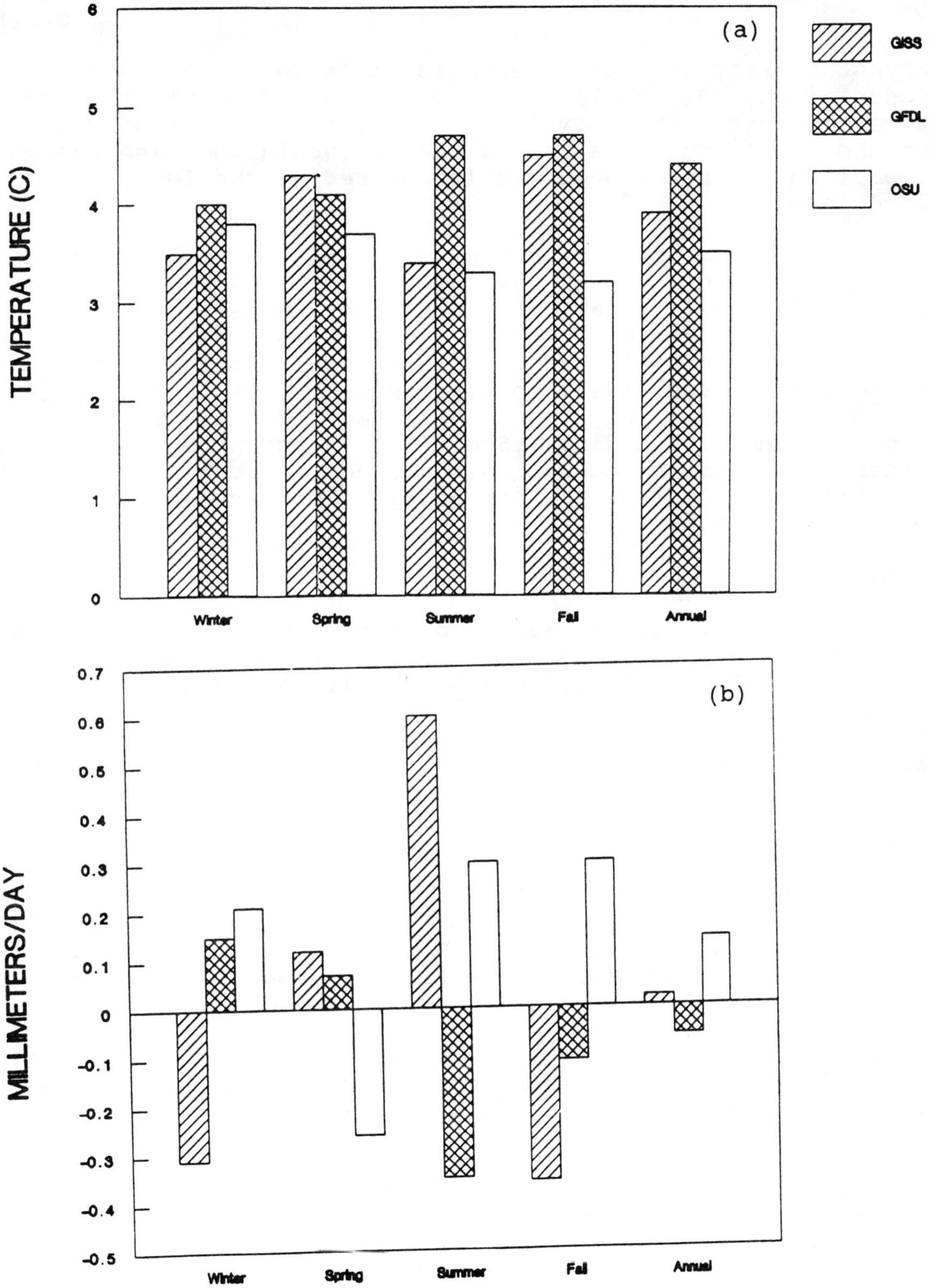

Figure 6-2. Climate scenarios for the Southeast, (a) temperature and (b) precipitation.

Galveston area, a 90-cm rise would increase the 100-year floodplain by 50%, while a 160-cm rise would enable the 100-year strom to overtop the seawall erected after the disaster of 1900. For the Charleston area, a 160-cm rise would increase the 10-year floodplain to the area currently covered by the 100-year floodplain.

Gibbs (1984) estimated that the economic impact of a 90-cm rise by 2075 could be as great as $500 million for Galveston and over $1 billion for Charleston. However, he also estimated that the adverse impacts of flooding and land loss could be cut in half if the communities adopted measures in anticipation of sea level rise. Titus (1984) focused on decisions facing Sullivans Island, South Carolina, in the aftermath of a storm. He concluded that rebuilding $15 million in oceanfront houses after a storm would not be economically sound if future sea level rise is anticipated, unless the community is prepared to continuously nourish its beaches.

Wetlands

Kana et al. (1986) surveyed marsh transects and estimated that 90- and 160-cm (3.0- and 5.2-foot) rises in sea level would drown 50 and 90%, respectively of the marsh around Charleston, South Carolina. Armentano et al. (1988) estimated the Southeast would lose 35 and 70% of its coastal wetlands for respective rises of 1.4 and 2.1 m, assuming that developed areas are not protected.

Infrastructure

The Louisiana Wetland Protection Panel (1987) concluded that a rise in sea level might necessitate substantial changes in the ports and shipping lanes of the Mississippi River to prevent the loss of several thousand square miles of coastal wetlands. Titus et al. (1987) showed that a reconstructed coastal drainage system in Charleston should be designed for a 1-foot rise in sea level if the probability of such a rise is greater than 30%. Linder et al. (1988) found that warmer temperatures would require an electric utility company to substantially increase its generating capacity.

Several dozen researchers presented papers on other global warming impacts on the Southeast at a 1987 EPA conference held in New Orleans (Meo, 1987). The proceedings include papers on forests, agriculture, coastal erosion, and water resources.

Chapter 6

CLIMATE CHANGE STUDIES IN THIS REPORT

Table 6-2 and Figure 6-3 illustrate the studies undertaken as part of this effort. Models of coastal erosion, coastal wetland loss, agricultural yields, and electricity consumption were used to derive regionwide estimates. Efforts to estimate impacts on water resources, infrastructure, forests, and the ecology were limited to conducting illustrative case studies.

SCENARIOS OF FUTURE CLIMATE CHANGE

Figure 6-2 illustrates the scenarios of future climate change from general circulation models. Table 6-3 shows the more detailed seasonal changes.

Table 6-3 illustrates how the frequency of mild days during the winter and the frequency of very hot days during the summer might change under the Goddard Institute for Space Studies (GISS) doubled CO_2 scenario. These estimates used average monthly changes in temperature and assumed no change in the frequency or duration of temperature extremes. Under this scenario, the number of days per year in which the mercury would fall below freezing would decrease from 34 to 6 in Jackson, Mississippi; from 39 to 20 in Atlanta; and from 41 to 8 in Memphis. The number of winter days above 70°F would increase from 15 to 44 in Jackson, from 4 to 14 in Atlanta, and from 5 to 24 in Memphis.

Of the nine cities shown, only Nashville has summer temperatures that currently do not regularly exceed 80°F. However, the number of days with highs below 80°F would decline from 60 to 34. Elsewhere, the heat would be worse. The number of days per year above 90°F would increase from 30 to 84 in Miami, from 17 to 53 in Atlanta, and from 55 to 85 in New Orleans. Memphis, Jackson, New Orleans, and Jacksonville, which currently experience 0 to 3 days per year above 100°F, would have 13-20 such days (Kalkstein, Vol. _).

Coastal Impacts

A number of national studies for the report presented results for the effects of climate change on the southeastern coast. Leatherman estimated the cost of maintaining recreational beaches. Park et al. and Weggel et al. examined the impacts onwetland loss and shoreline defense, and used their results to estimate the regionwide cost of raising barrier islands.

Table 6-2. Studies of the Southeast

Regional Studies

- Impacts on Runoff in the Upper Chattaboochee River Basin -- Hains, C.F. Haines, Hydrologist, Inc.

- Projected Changes in Estuarine Conditions Based on Models of Long-Term Atmospheric Alteration -- Livingston, Florida State University.

- Policy Implications of Global Climatic Change Impacts upon the Tennessee Valley Authority Reservoir System, Apalachicola River, Estuary, and Bay and South Florida -- Meo, Ballard, Deyle, James, Malysa, and Wilson, University of Oklahoma.

- Potential Impacts on Climatic Change on the Tennessee Valley Authority Reservoir System -- Miller and Brock, Tennessee Valley Authority.

- Impact of Climate Change on Crop Yield in the Southeastern U.S.A. -- Peart, Jones, and Curry, University of Florida.

- Methods for Evaluating the Potential Impacts of Global Climate Change -- Sheer and Randall, Water Resources Management, Inc.

- Forest Response to Climate Change: A Simulation Study for Southeastern Forests -- Urban and Shugart, University of Virginia.

National Studies that Included Southeast Results

- The Economic Effects of Climate Change on U.S. Agriculture: A Preliminary Assessment. -- Adams, Glyer, and McCarl, Oregon State University.

- National Assessment of Beach Nourishment Requirements Associated with Accelerated Sea Level Rise -- Leatherman, University of Maryland.

- The Potential Impacts of Climate Change on Electric Utilities: Regional and National Estimates -- Linder and Inglis, ICF Inc.

Chapter 6

- The Effects of Sea Level Rise on U.S. Coastal Wetlands -- Park and Trehau, Butler University.

- Potential Effects of Climatic Change on Plant-Pest Interactions -- Stinner, Rodenhouse, Taylor, Hammond, Purrington, McCartney, and Barrett, Ohio Agricultural Research and Development Center.

- An Overview of the Nationwide Impacts of Rising Sea Level -- Titus and Greene, U.S. Environmental Protection Agency.

- The Cost of Defending Developed Shorelines Along Sheltered Waters of the United States from a Two Meter Rise in Mean Sea Level -- Weggel, Brown, Escajadillo, Breen, and Doheny, Drexel University.

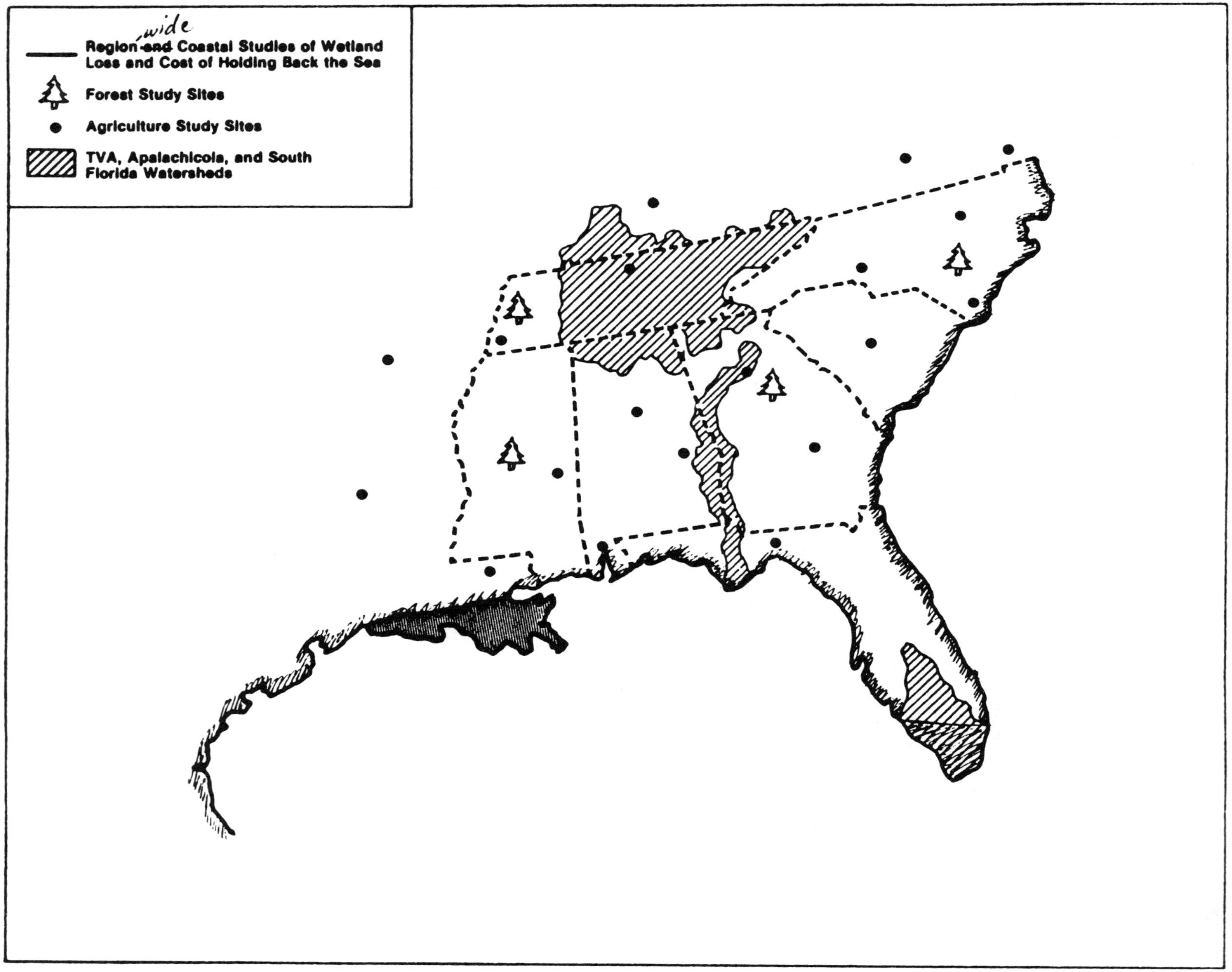

Figure 6-3. Overview of Southeastern Studies

Table 6-3. The GISS Doubled CO_2 Scenario: Frequency of Hot and Cold Days (^{o}F)

	Number of Winter days with...				Number of Summer days with....					
	Daily Low <32		Daily High >= 70		Daily High <80		Daily High >90		Daily High >= 100	
YEAR	HIST	2xCO2	HIST	2xCO2	HIST	2xCO2	HIST	2xCO2	HIST	2xCO2
Atlanta, GA	38.3	20.5	4.2	13.6	10.0	2.2	17.1	53.3	0.6	4.2
Birmingham, AL	35.5	8.1	7.1	30.7	4.5	0.4	34.1	72.5	1.5	10.7
Charlotte, NC	42.1	23.8	3.4	9.9	11.9	3.7	23.1	56.5	0.1	5.9
Jackson, MS	33.5	5.9	15.3	43.5	0.8	0.2	55.1	83.1	2.0	19.5
Jacksonville, FL	9.3	1.7	34.6	49.6	2.3	0.3	46.4	81.3	0.6	14.1
Memphis, TN	41.2	8.1	5.2	23.6	4.9	0.7	50.5	74.8	2.6	19.1
Miami, FL	0.2	0.0	72.9	82.7	0.6	0.0	29.8	83.5	0.0	2.5
Nashville, TN	42.5	15.4	0.3	8.6	60.4	33.7	10.5	20.2	0.3	3.5
New Orleans, LA	14.9	3.5	24.9	39.5	0.9	0.1	55.4	84.9	0.3	13.5

The projected rise in sea level would cause shorelines to retreat, exacerbate coastal flooding, and increase the salinity of estuaries, wetlands, and aquifers. The nationwide study of sea level rise focused on retreating shorelines and the costs of holding back the sea. (See Chapter 9 for a discussion of the rationale, methods, and nationwide results of these studies.)

Coastal Wetlands

Park (Volume __) examined 29 southeastern sites to estimate the likely regionwide loss of coastal wetlands for a variety of scenarios of future sea level rise. His analyses included such societal responses as providing structural protection for all shorelines (total protection), protecting areas that are densely developed today (standard protection), and allowing shorelines to adjust naturally without coastal protection (no protection).

Figure 6-4 illustrates Parks's projections for the year 2100 for the various scenarios of sea level rise and coastal defense. Even if current sea level trends continue, 25% of the Southeast's coastal wetlands will be lost, mostly in Louisiana.

- current trends imply a loss of 15%;
- a 50-cm rise would result in a loss of 35 to 50%; depending on how shorelines are managed;
- a 100-cm rise results in losses of 45 to 68%; and
- a 200-cm rise implies losses of 63 to 80%.

Park et al. estimate losses of 50, 75, and 98% for Louisiana under the three scenarios. However, Park did not consider the potential for mitigating the loss by restoring the flow of river water into these wetlands. Titus and Greene estimated statistical confidence intervals illustrated in Table 6-4.

Total Coastal Land Loss

Park also estimated total land loss, including both wetlands and dry land. Most of the land loss from a rise in sea level would occur in Louisiana, as shown in Table 6-4. A 50-cm (20-inch) sea level rise would result in the loss of 1900-5900 mi^2 of land, while a 200-cm rise would inundate 10000-11000 mi^2.

Cost of Protecting Recreational Beaches

In Volume___, Leatherman notes that the projected rise in sea level would threaten all developed recreational beaches. Even a 1-foot sea level rise would erode shorelines over 100 feet

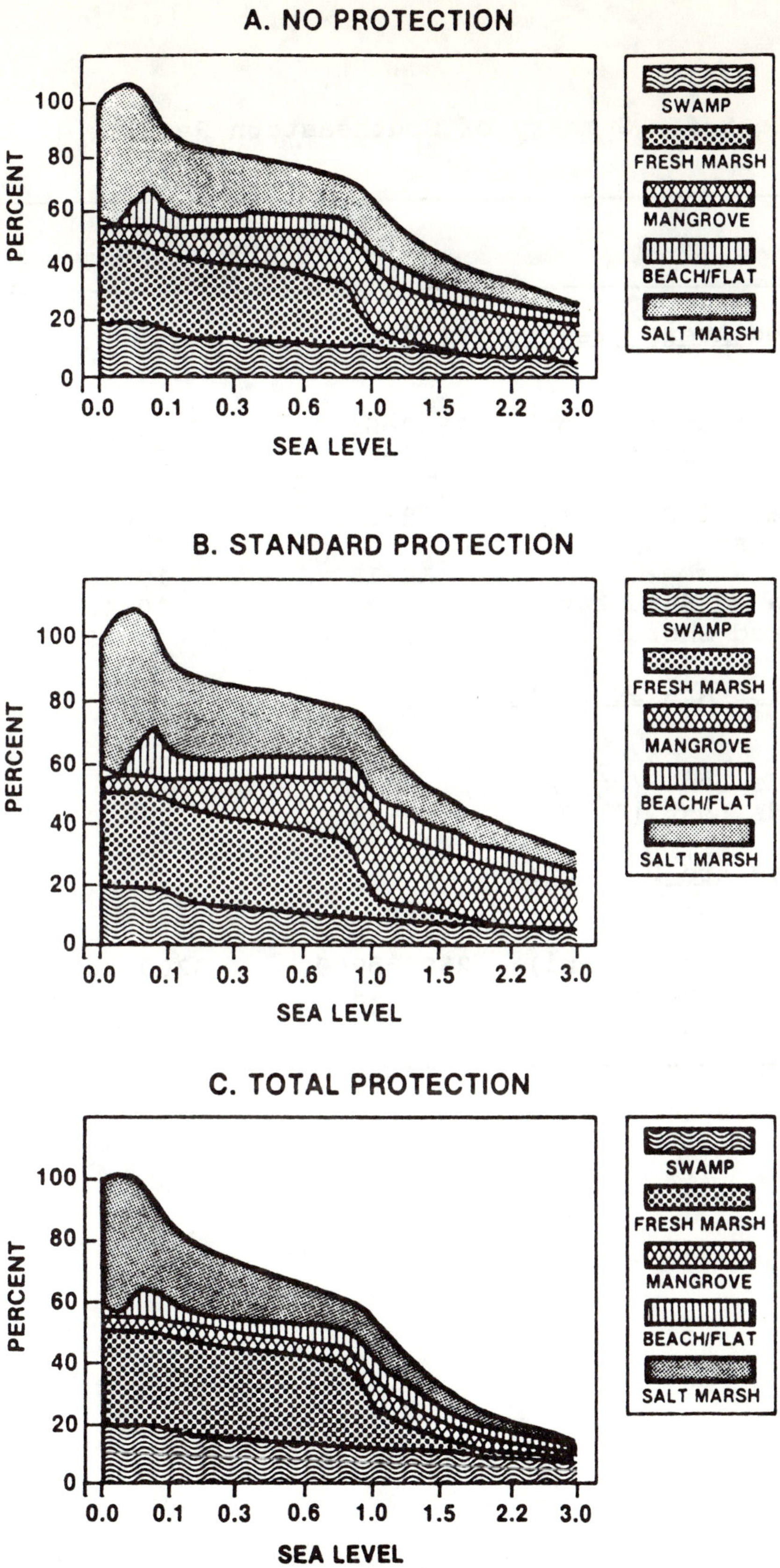

Figure 6-4. Southeastern Wetland Loss for Three Shoreline Protection Options

Source: Park et al., Volume _.

Table 6-4. Summary of Southeastern Results ($ billions)

	50	100	200
If Developed Areas Are Protected			
Land Lost			
Dryland Lost (sq. mi)	1900-5500	2600-6900	4200-10100
Wetlands Lost (%)	24-50	34-77	40-90
Cost of Coastal Defense	19-28	42-75	127-174
Open Coast			
Sand	10-15	19-30	44-74
Elevated Structures	5-9	10-40	60-75
Sheltered Shores	2-5	5-13	9-41
If All Shores Are Protected			
Land Lost			
Drylands Lost (sq. mi)	0	0	0
Wetlands Lost (%)	38-61	47-90	68-93
If No Shores Are Protected			
Land Lost			
Dryland Lost (sq. mi)	2300-5900	3200-7600	4800-10800
Wetlands Lost (%)	22-48	30-75	37-88

Source: Titus and Greene

throughout the Southeast. Along the coasts of North Carolina and Louisiana, the erosion would be considerably greater. (See Chapter 9 for a discussion of methods for estimating coastal erosion.) Because the distance from the high tide line to the first building is rarely more than 100 feet, most recreational beaches would be lost, unless either the buildings were removed or engineering measures were undertaken.

Table 6-4 illustrates Leatherman's estimates of the cost of protecting recreational beaches by pumping sand from offshore locations (see Chapter 9, Sea Level Rise for state by state results). A 1-m rise in sea level would imply almost $20 billion in dredging costs, with Texas spending $8.5 billion and Florida and Louisiana each spending over $3 billion.

Using constant unit costs (except for Florida), Leatherman estimated that a 2-m rise would only double the total cost to $43 billion. Titus and Greene estimated that if the unit costs of sand increased, 1- and 2-m rises would cost $30 and 74 billion, respectively. They also estimated that the cost of rebuilding roads and utilities on barrier islands would be 5-9, 10-40, and 60-75 billion for the three scenarios.

Cost of Protecting Calm-Water Shorelines

While Leatherman focused only on the open ocean coast, Weggel et al. estimated the regionwide costs of holding back the sea in developed sheltered and calm-water areas. Table 6-4 illustrates the estimated costs for the Southeast. Weggel et al. estimate that about $2 billion would be spent to raise roads and to move structures, and $23 billion would be spent to erect the necessary levees and bulkheads. The combined cost is $68-83 billion. These estimates do not include the costs of preventing flooding or of protecting water supplies.

Tennessee Valley Authority Studies

The Tennessee Valley Authority (TVA) was created in 1933 to spur economic growth in an area previously considered to be one of the nation's poorest. Geographically isolated by the Appalachian Mountains, the region lacked electricity and roads, and the Tennessee River could not provide reliable transportation because it flooded in the spring and dried to a trickle during the summer. By creating the TVA, Congress sought to remedy this situation by harnessing the river to provide electricity, to prevent the flooding that had plagued Chattanooga, and to ensure sufficiently stable riverflows that would permit maintenance of a 9-foot-deep navigation channel.

The region administered by the TVA covers 40,000 mi^2 and includes parts of seven states. In the last half century, the TVA has coordinated the construction of 43 major dams along the river and its tributaries, many of which are shown in Figure 6-5. The system provides power to over 7 million people, and contains 675 miles of navigable waterways with annual commercial freight of 28 million tons. The lakes created by the dams have over 10,000 miles of shoreline, which generate 75 million visits each year and along which people have invested $630 million. The lakes and streams generate 17 million visits per year, enough to indirectly boost the region's annual economy by $400 million (Miller and Brock).

To assess the potential impacts of climate change, Miller and Brock conducted a modeling study of the water resource implications, and Meo et al. examined the policy implications for the TVA.

TVA Modeling Study

Methods

Miller and Brock used the TVA's "Weekly Scheduling Model," on which the Agency currently bases its operations, to assess the impacts of climate change. This is a linear programming model that selects a weekly schedule for managing each reservoir in the TVA system by sequentially satisfying the objectives of flood control, navigation, water supply, power generation, water quality, and recreation. Miller and Brock used this model to simulate reservoir levels, riverflows, and hydropower generation for wet and dry scenarios, based on the runoff estimates from the GISS doubled CO_2 model run.*

Miller and Brock assessed the potential impacts of climate change on flood levels in Chattanooga, Tennessee, using a model that had been developed to estimate the constraints on weekly tributary releases. They also estimated the potential implications for water quality in the Upper Holston Basin of the valley, using a reservoir water quality model, a riverflow model, and a water quality model that TVA has used in the past to determine the environmental constraints affecting riverflow.

---

*TVA used runoff from the GCMs, rather than hydrologic analysis, to estimate streamflow. The GISS scenario estimates high runoff. Since the Geophysical Fluid Dynamics Laboratory (GFDCL) and Oregon State University (OSU) scenarios estimated that no runoff the $1xCO_2$ and $2xCO_2$ runs, Miller and Brock used the inverse of GISS as a dry scenario (Rind, 1988)).

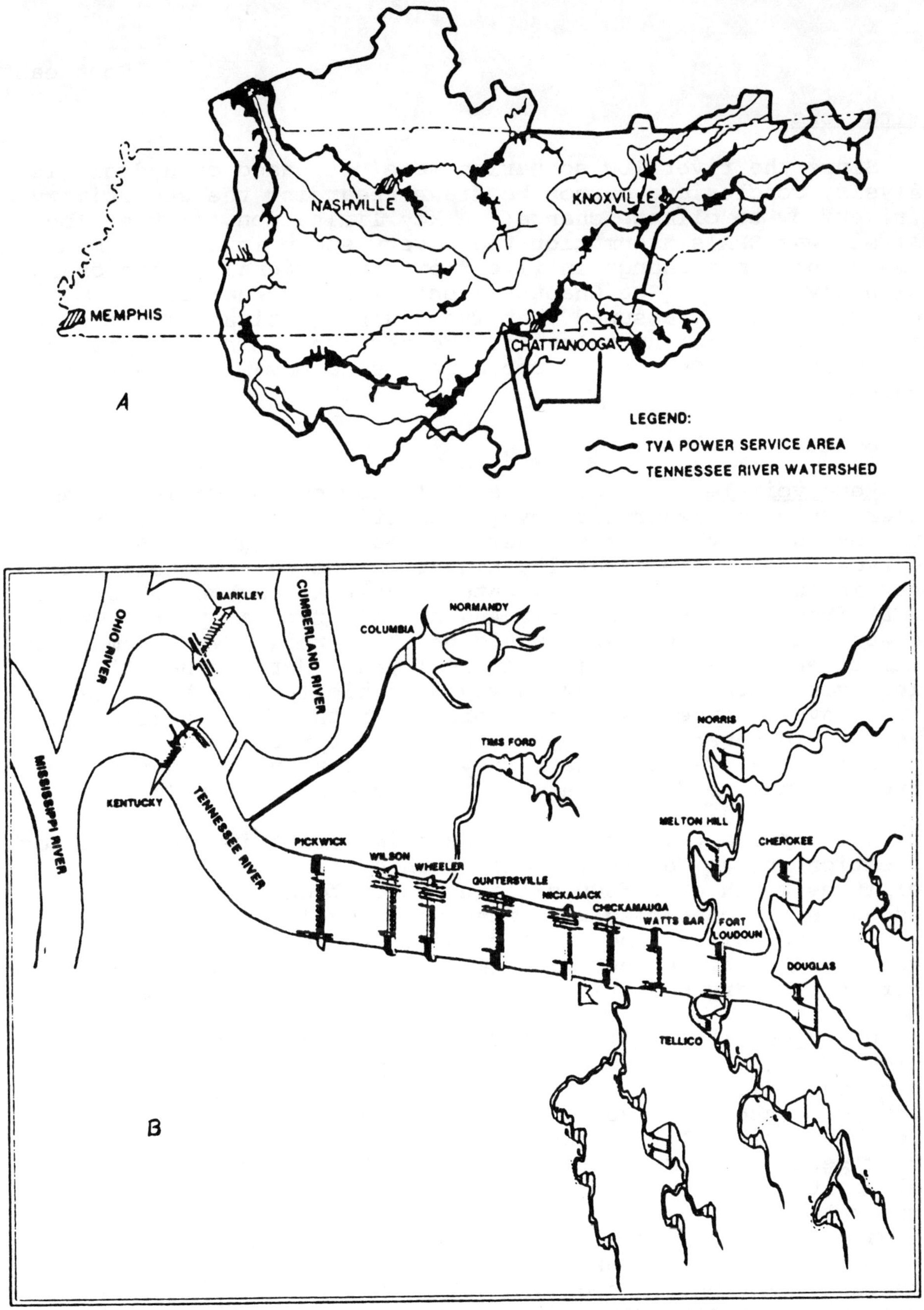

Figure 6-5. (a) Map of the TVA region and (b) schematic of the TVA reservoir system (b).

Source: Tennessee Valley Authority

Limitations

Since the riverflow scenarios were not based on hydrologic analysis, conclusions cannot be drawn regarding the sensitivity of riverflow to climate change. A key limitation for the flood analysis was EPA's assumption that every storm in a given month would result in a change in riverflow proportional to the change in monthly runoff. (See Chapter 3 on climate scenarios.) Finally, the study assumed that TVA did not mitigate impacts by changing its operating rules for the reservoirs in response to climate change. Changes in operating rules may mitigate some impacts.

Results

Reservoir levels. Assuming TVA's current operating rules, Miller estimated reservoir levels resulting from the scenarios, with levels for the Norris reservoir shown in Figure 6-6. Reservoir levels are generally higher in the wet scenario and lower in the dry. Currently, water levels are above 1,010 feet (NLVD) from early May to early August in a typical year. Under the wet scenario, the water would be above this level from early April to early September in a typical year; during the (1%) driest years, the water levels would be similar to the current normal level between May and October. In the dry scenario, water levels would never exceed 1,005 feet in a typical year and even during the (1%) wettest years they would barely exceed the current normal condition between April and September.

These projected lake levels have important implications for recreation in the Tennessee Valley, which is supported by facilities worth over $600 million. Even today, recreation proponents are concerned with reservoir levels dropping during some summers. Miller and Brock found that the wet scenario would largely eliminate current problems with low lake levels; in contrast, the dry scenario would make these problems the norm.

Water Quality. Miller and Brock found that a dryer climate could also create environmental problems. Lower flows would reduce the dilution of municipal and industrial effluents discharged into the river and its tributaries. Moreover, because water would generally remain at the bottom of reservoirs for a longer period of time, the amount of dissolved oxygen would decline; this would directly harm fish and reduce the ability of streams to assimilate wastes. Miller and Brock concluded that the water supplies from TVA probably would be sufficient, but the TVA could experience operational difficulties and customer dissatisfaction due to degraded water quality. During extended low-flow conditions, wastes would have increased opportunities to backflow upstream to water supply intakes.

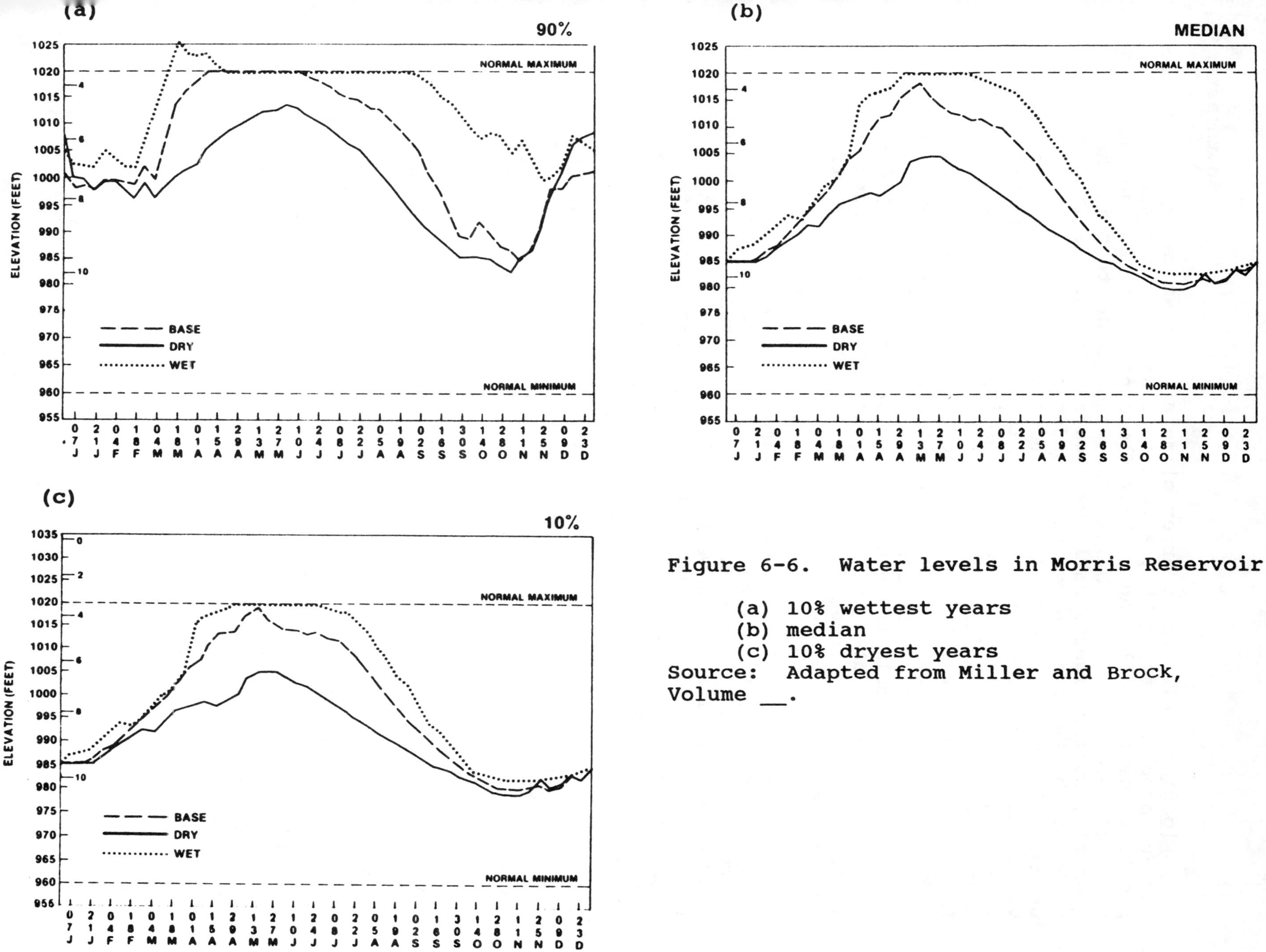

Figure 6-6. Water levels in Morris Reservoir

(a) 10% wettest years
(b) median
(c) 10% dryest years

Source: Adapted from Miller and Brock, Volume __.

Flooding. Although a dryer climate would exacerbate many current problems facing TVA, a wetter climate could create problems, particularly the risk of flooding, in areas that are currently under control. Miller and Brock found that in the wet scenario, during exceptionally wet years, storage was inadequate at the tributary reservoirs; this condition could result in uncontrolled spillage over dams. A high probability of flooding would also exist at Chattanooga. Miller and Brock took the levels of the five worst floods of the last 50 years at Chattanooga that did not overflow the banks of the Tennessee River and flood the city, and they compared these floods with floods that would result under the wet GISS scenario.Under the GISS scenario, two of the five floods would overtop the banks. The worst flood would reach a level of 56.3 feet and would cause over $1 billion in damages. (See Figure 6-7.) The second worst flood would reach a level of 46 feet, causing over $200 million in damages.

Flooding could be reduced if climate change were incorporated into operating rules that, kept water levels lower in reservoirs on tributaries (although this would diminish the hydropower benefits from a wetter climate). However, changes in operating rules would not be sufficient to protect Chattanooga from a repeat of the worst storm in which rainfall was largely concentrated over the "mainstem" reservoirs, which do not have substantial flood- control storage.

Power Generation. Miller and Brock calculated that the wet and dry scenarios imply, respectively, an annual increase of 3.2 megawatt-hours (16%, $54 million/year) and a decrease of 4.6 megawatt-hours (24%, $87 million/year), given current capacity and operating rules.

Climate change could also have an impact on fossil-fuel powerplants. If river temperatures become warmer, additional dilution water would be required for powerplants. The water may be available if the climate becomes wetter, but under the dry scenario, meeting minimum flow requirements would be more difficult. Miller suggested that the most feasible operational change would be to cut back power generation at fossil/fuel powerplants during periods of low flow. However, hydropower production would also be reduced during periods of low flow, so cutting back production might not be acceptable. One alternative would be for the plant to construct cooling towers, which would eliminate discharges of hot water, at a capital cost of approximately $75 million.

Figure 6-7. Chattanooga was vulnerable to flooding until the TVA system of dams was constructed. The upper photo shows the 1867 Flood, with levels similar to what Miller and Bruck project could happen under the wet scenario.

Source: Tennessee Valley Authority

Tennessee Valley Policy Study

Meo et al. analyzed the history, statutory authority, and institutional structure of the TVA to assess the ability of the organization to respond to climate change. Their analysis relied both on the available literature and on interviews with a few dozen officials of TVA and states within the region.

Meo et al. divided the possible responses of TVA into two broad categories: (1) continuing the current policy of maximizing the value of hydroelectric power, subject to the constraints of flood control and navigation; and (2) modifying priorities so that power generation would be subordinated to other objectives if doing so would yield a greater benefit to the region.Meo concluded that if the climate becomes wetter, current policies would probably be adequate to address climate change because the only adverse effect would be the risk of additional flooding, which is already a top priority of the system.

If climate becomes drier, on the other hand, existing policies might be inadequate, because they require power generation to take precedence over many of the resources that would be hardest hit. Although they expect that the TVA will be more successful at addressing the current (1988) drought, Meo et al. found that during the 1985-86 drought, falling lake levels impaired recreation and reduced hydropower generation, forcing the region to import power while five powerplants sat idle.

Meo et al. point out that groundwater tables are falling in parts of the region, in part because numerous tributaries recharge the aquifers whenever water is flowing but are allowed to run dry when water is not being released for hydropower. Meo et al. suggest that the deteriorating groundwater quality and availability are likely to lead a number of communities to shift to surface water supplies in the coming decades, adding another use that must compete for the water that is left over when the demands for power have been met. Even with current climate these contend, the TVA should assess whether other uses of the region's water resources would benefit the economy more. If climate became drier, the need for such a reevaluation would be even more necessary.

Studies of Effects on Lake Lanier and Apalachicola Bay

Figure 6-8 shows the boundaries of the Chattahoochee-Flint-Apalachicola River Basin, which encompasses 19,800 mi^2. The Corps of Engineers and others who manage the Chattahoochee River as it passes through Lake Lanier on its way to the Apalachicola estuary and the Gulf of Mexico face many of the same issues as those faced by the TVA. However, they also are managing the water supply of Atlanta, the 2nd largest city in the Southeast, and the flow of water into an estuary that supports the most productive fishery in Florida. (Statistical Abstract of U.S.: 1988, Department of Commerce.)

A number of researchers were involved in EPA's assessment of the potential implications of climate change for this watershed. A study of Lake Lanier and a study of the implications for the fish in Apalachicola Bay are discussed in the following sections of this chapter.

Lake Lanier

Lake Lanier, located 30 miles northeast of Atlanta, is a source of water for the city and nearby jurisdictions. Federal statutes require the U.S. Army Corps of Engineers to manage Lake Lanier to provide flood control, navigation, and hydropower. The lake is also managed to mind nonstatutory objections such as recreation, minimum flows for environmental dilution, and water supply.

Since Lake Lanier was dammed in 1957, the statutory objectives of flooding and navigation have been met; historical annual hydropower generation has been 199, 134 MWH[1], equal to 2% of today's power requirements for Atlanta[2]; and the releases of water have fulfilled the additional minimum flow needed to dilute the effluents from sewage treatment plants.

During the last two decades, the lake's shoreline has been substantially developed with marinas, houses, and hotels. To a large degree, the residents have become accustomed to the higher water levels that have prevailed from the 1970s through 1984.

Droughts from 1985 to the present, however, have lowered lake levels, disrupting recreation. In the summer of 1986, navigation for recreational boats located downstream of the lake was curtailed due to minimal releases from the lake. Recently, Atlanta imposed water-use restrictions, with the objective of cutting consumption 10-20%. A bill has been introduced to add recreation to the list of statutory purposes (HR-4257). [Rep. Jenkins, Rep. Barnard, Rep. Darden. HR-4254 "Georgia Reservoir Management Improvement Act of 1988," 100th Congress, 20 Session].

Runoff in the Chattahoochee River Basin

To assess the potential implications of climate change for Lake Lanier, Haines estimated the flows of water into the lake.

Study Design. Haines calibrated the Sacramento hydrology model developed by the National Weather Service (Burnash, 1973) to the conditions found in the watershed of the upper Chattahoochee River. He then generated scenarios of riverflow for the baseline climate and the three doubled CO_2 scenarios.

Limitations. The Sacramento model was designed primarily for flood forecasting, not base flow. In addition, the model was calibrated using the data on evaporation of water from pans, which is not perfectly correlated with evapotranspiration.

Results. As with the Tennessee River, the major climate models disagree on whether the Chattahoochee watershed would become wetter or dryer with an effective doubling of greenhouse gases. Haines estimated that the wetter GISS model would increase the average annual riverflow of the Chattahoochee River by 13%, while the drier OSU and GFDL models imply declines of 19 and 27%, respectively, as shown in Figure 6-9. The GISS scenario implies slight decreases in winter flow and increases the rest of the year. The GFDL scenario shows substantial decreases throughout the year, with almost no flow in late summer. The OSU scenario also shows reductions, but the reduction is greatest during the flood season (February to May) and negligible during the dry season (late summer/early fall).

Management of Lake Lanier

Sheer used the calculation of riverflow from Haines to estimate the management implications of climate change.

Study Design. Sheer modified a monthly water balance model/operations model previously applied in Southern California for the lake, based on current operating rules for the reservoir. For the first set of runs, the model assumes that (1) minimum flows are maintained for navigation and environmental dilution at all times, (2) lake levels are kept low enough to prevent flooding, (3) historic rates of consumption continue, and (4) peak hydropower generation is maximized. To ensure that the assumptions adequately reflect the actual decision rules used by water managers, Sheer reviewed the rules with local officials from the Corps of Engineers, the Atlanta Regional Council, and others responsible for managing the water supply. In a second set of runs, Sheer examined the impacts of climate change under alternative operating rules that assume recreation is also a statutory objective.

Limitations. The Sheer model does not consider changes in demand for water due to climate change. In addition, model results were not compared to actual historic lake levels.

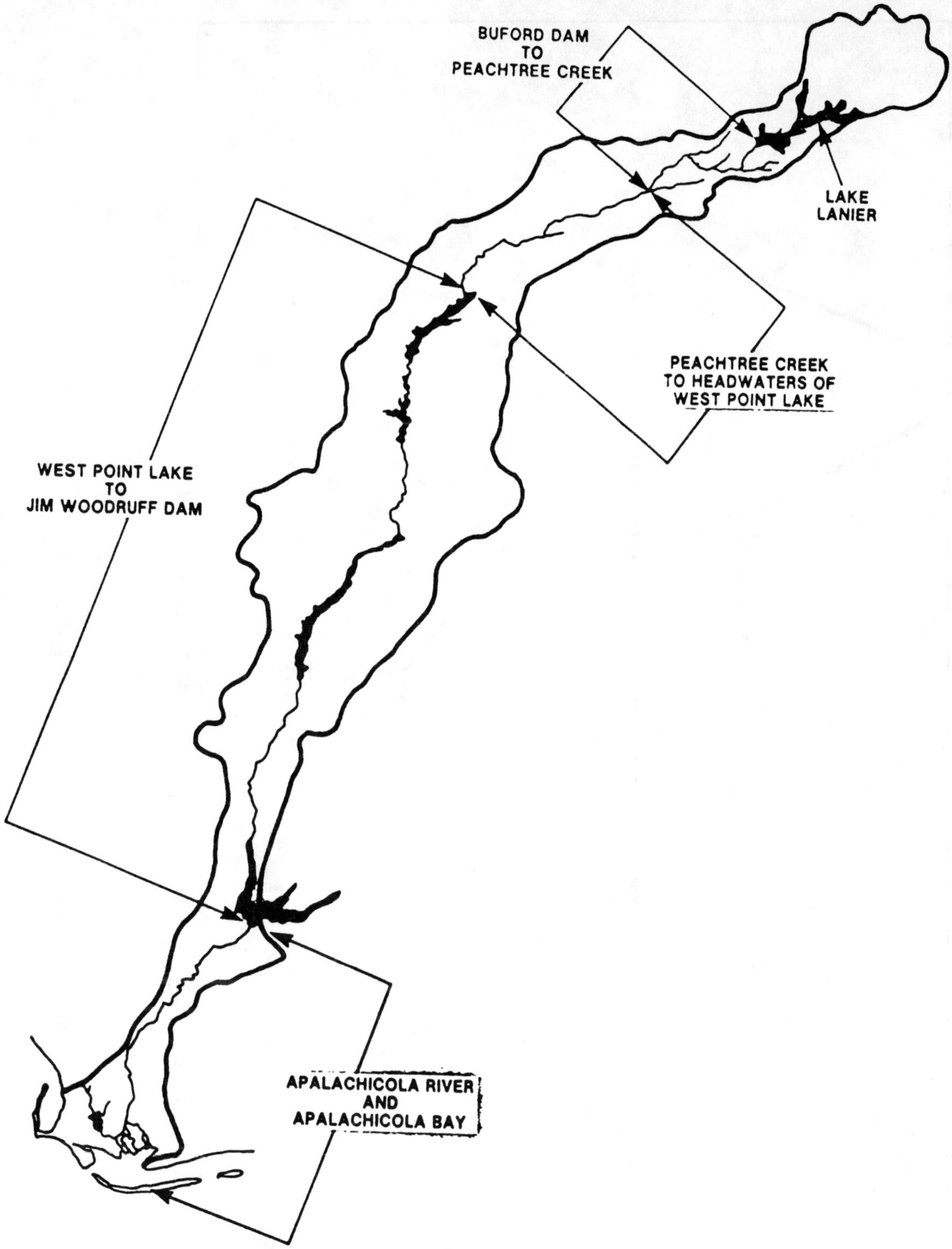

Figure 6-8. Drainage area of the Apalachicola-Chattahoochee-Flint River system.

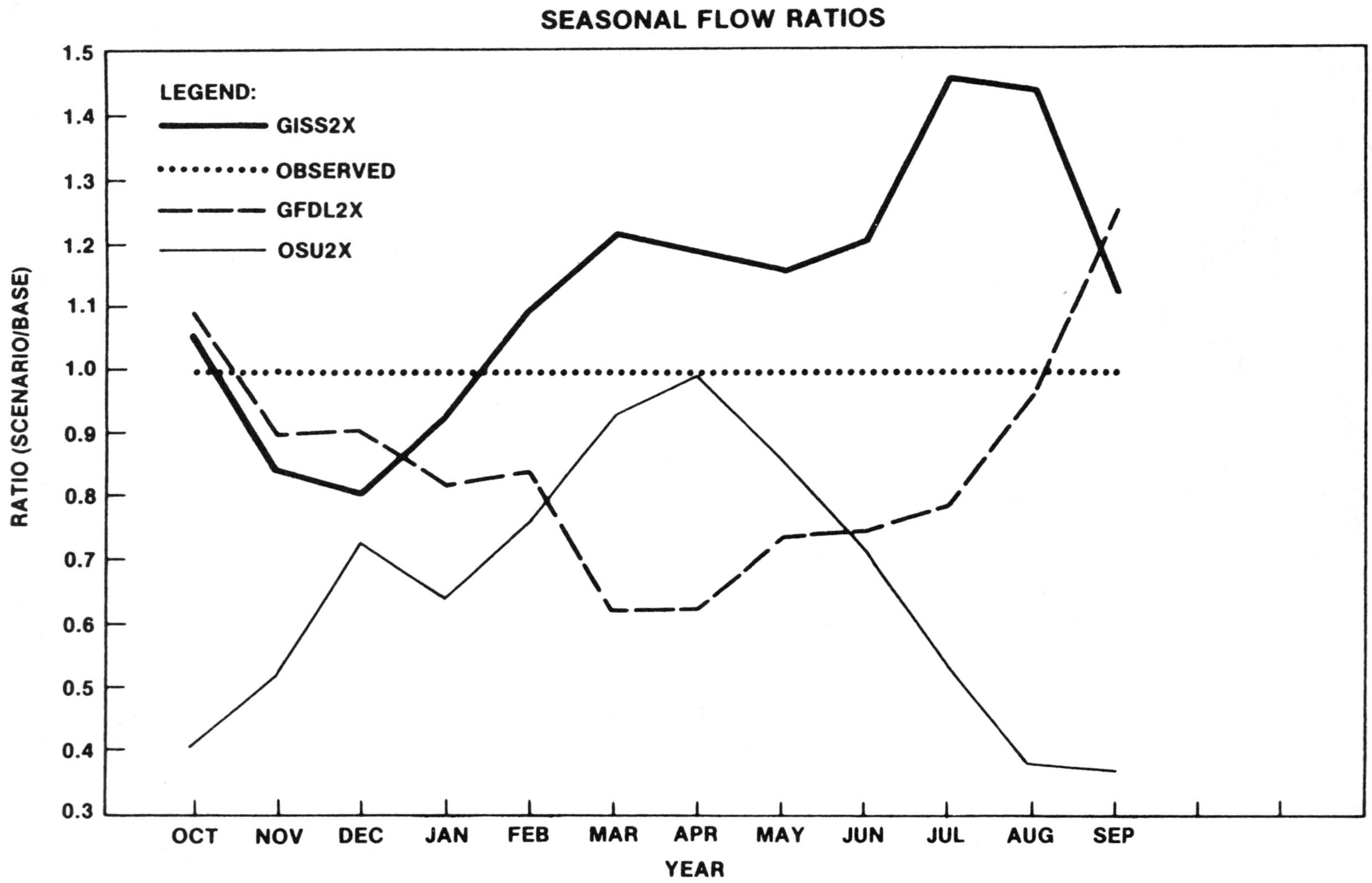

Figure 6-9. Upper Chattahoochee - $2XCO_2$

Source: Hains, Volume _.

6-32

Results. Figures 6-10 shows Sheer's estimates of lake levels. Figure 6-11 shows quarterly hydropower production. Under the wet GISS scenario, annual power production would increase by 9%. The higher streamflows in this scenario would still be well below those that occasionally occurred before Lake Lanier was closed; hence, no significant threat of flooding would exist for a repeat of the climate of 1951-80. Under the dry GFDL scenario, however, power production would drop 47%, and lake levels would be likely to drop enough to substantially disrupt recreation. This scenario assumes that Atlanta would continue to take as much water as it does currently (a conservative assumption given the area's rapid growth).

Sheer also examined the implications of making recreation a statutory objective. Although, it would be possible to maintain lake levels, even in the dry GFDL scenario. Altanta's water supply would be threatened. With the current climate, strict enforcement of such a policy would result in Lake Lanker. Although the wet GISS scenario would reduce this to 1 month every30 years, under the dry GFDL scenario, Atlanta would have to use an alternative source of water one to three months each summer.

Implications. Climate change combined with population growth may require water managers to reexamine the tradeoffs between the various uses of the Chattahoochee River and Lake Lanier. A number of local water officials who met with Sheer suggested that an appropriate response to changing water availability might be to relax minimum flow requirements for navigation and environmental quality. They reasoned that minimum flows for environmental purposes are based on conditions that prevail only when sewage treatment plants are at their maximum discharge and temperatures are high, and that little is accomplished by maintaining minimum flows for navigation because ship traffic is light in the lower Chattahoochee. Others argued, however, that it would be unwise to assume that minimum flows could be decreased because future growth may increase the need for dilution of effluents, and warmer temperatures would speed biological activity. The likely impacts of climate change on Apalachicola Bay may also increase the need to maintain minimum flows.

Apalachicola Bay, Florida

Apalachicola Bay supports hundreds of commercial fishermen; over 80% of Franklin County earns a livelihood from the bay (Meo, Vol. _). The contribution of fishing to the area was estimated at $20 million for 1980, representing 90% of Florida's oyster harvest and 10% of its shrimp harvest. This figure is projected to grow to $30-60 million by 2000.

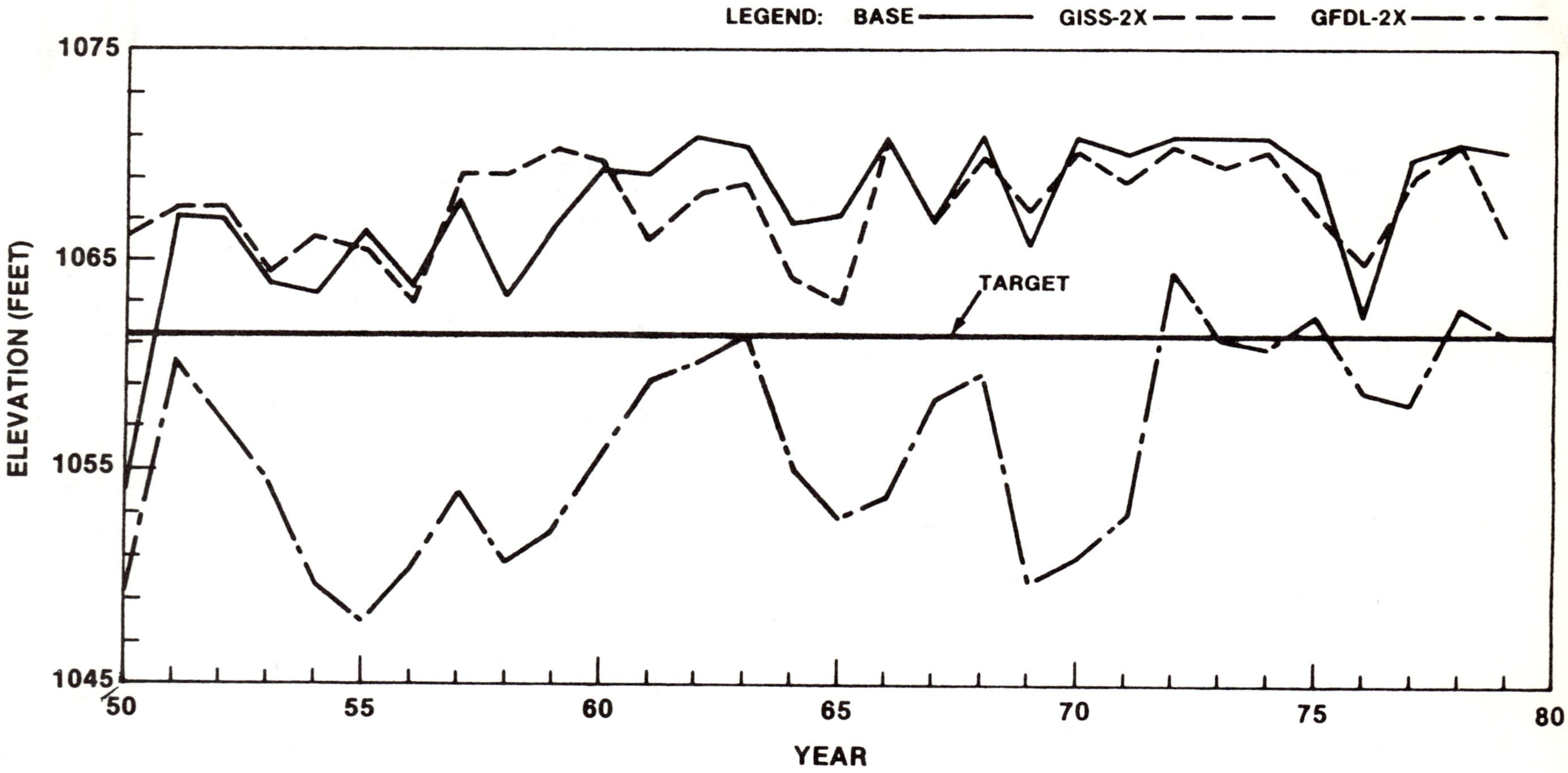

Figure 6-10. Lake Lanier elevation (September).

Source: Sheer and Randall, Volume __.

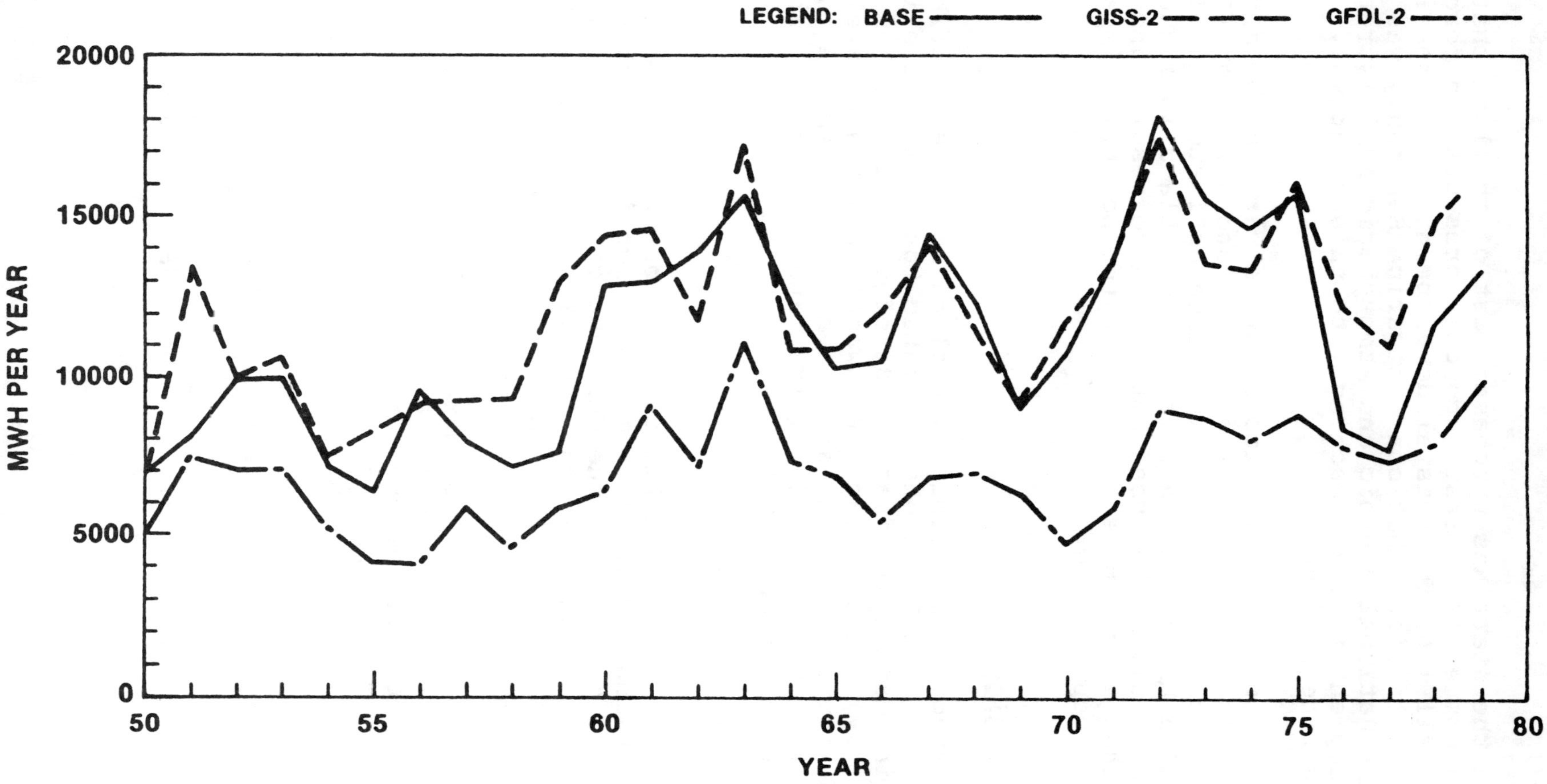

Figure 6-11. Lake Lanier power generation.

Source: Sheer and Randall, Volume __.

Although the state has purchased most of the land that is not part of a commercial forest, economic pressures on forestry companies to sell land for coastal development are increasing. In 1979, the Apalachicola National Estuarine Sanctuary was created by the National Oceanic and Atmospheric Administration to prevent development from encroaching into this relatively pristine estuarine environment.

The biology of the Apalachicola Bay estuary may be affected by higher temperatures, higher sea levels, and different flows of water into the Apalachicola River. Haines estimated the flow of the Apalachicola River, and Park estimated wetland loss due to sea level rise. Livingston used both of these results and the temperature change scenarios to evaluate the potential impacts on the bay's fish populations.

Sea Level Rise

Park's methods for estimating wetland loss are described in chapter 9: Sea Level Rise. He estimated that a 1-m rise in sea level would inundate approximately 60% of the salt marshes in Apalachicola Bay, and that mangrove swamps, which are rarely found outside southern Florida at present, would replace the remaining salt marsh. Table 6-6 illustrates Park's estimates of wetland shifts due to sea level rise.

Apalachicola River Flow

Study Design. Haines estimated the impact of climate change on riverflow, using a regression model, which is simpler than the Sacramento model he used for the Chattahoochee River analysis. The regression expressed the logarithm of riverflow as a function of the logarithms of precipitation and evapotranspiration for a few weather stations located in the basin.

Limitations. Haines' procedure greatly oversimplified the relationships between the causal variables and riverflow, ignoring the impacts of reservoir releases and the failure of the relationships to fit the simple log-linear form. These results should be interpreted as an indication of the potential direction of change.

Results. Figure 6-12 illustrates Haines' estimates of average monthly flows for the Apalachicola estuary. Annual riverflow would decrease under all scenarios, although it would increase in the summer and fall for the GISS and OSU scenarios, respectively.

Table 6-6. Loss of Coastal Wetlands in Apalochicola Bay (hectares)

	1987	Baseline	50	100	200
Swamps	9.46	6.71	6.26	5.47	4.16
Fresh Marsh	1.46	1.27	1.17	1.00	0.25
High Marsh	1.19	0.37	0.035	0.035	0.020
Low Marsh	3.42	2.33	0.388	0.062	0.032
Mangrove	0	0	3.057	2.13	1.80
Wetlands					

Source: Park

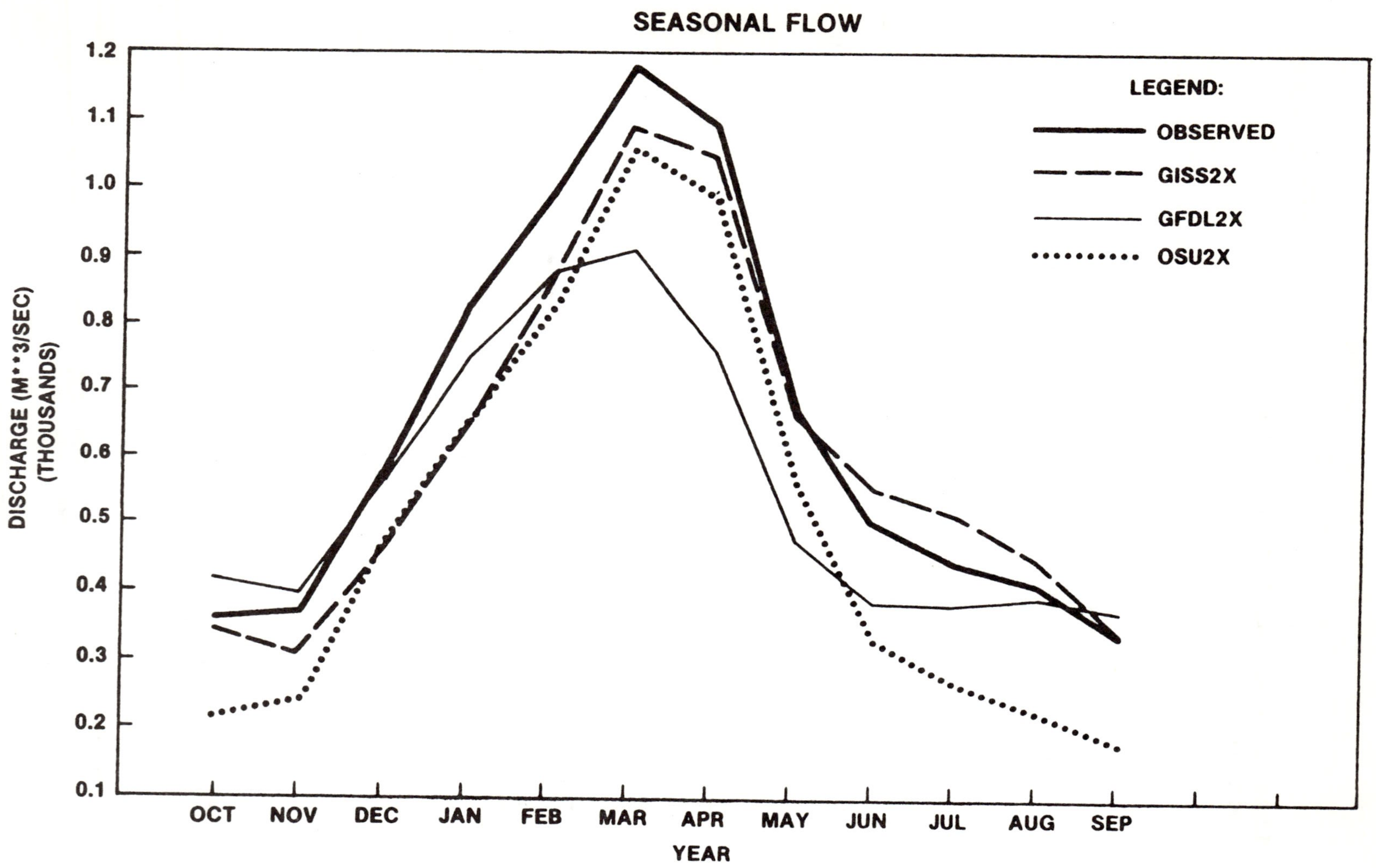

Figure 6-12. 2XCO2 Flow into Apalachicola Bay

Source: Hains, Volume __.

6-38

Chapter 6

Fish Populations in Apalachicola Bay

Study Design. Using data from the literature on the tolerance of various species to warmer temperatures, Livingston estimated the number of months in a typical 30-year period during which the estuary would be too hot for these species and extrapolated this information to estimate reductions in populations.

Hydrologic modeling was not used to estimate the combined impacts of sea level rise and changing riverflow on salinity. Since Haines estimated that riverflow would be lower under all three scenarios, salinity in Apalachicola Bay would increase. Recent droughts have decreased riverflow by approximately the same magnitude as would be suggested by the dry OSU and GFDL scenarios. Livingston used historic data to estimate regression equations relating riverflow to salinity and salinity to populations of some commercially important seafood species.

Limitations. There is no historical record by which to estimate the impact of warmer temperatures on the Apalachicola (or any other) estuary. Livingston did not model the relationships between various aquatic species or how they would change. He did not consider how finfish and shellfish might adapt to climate change, and he was unable to estimate the impact of wetland loss on populations of finfish and shellfish.

The limitations in Haines' estimates of riverflow do not significantly affect the results of Livingston's study because riverflow was one of several variables to be considered. The uncertainties surrounding changes in rainfall probably dwarf any errors due to Haines' simplified hydrology, and higher temperatures and sea level rise appeared to be more important.

Results. The results of this study suggest a dramatic transformation of the estuary from subtropical to tropical conditions.

Warmer temperatures. Livingston concluded that warmer temperatures would have a profound effect on seafood species in the estuary because they cannot tolerate temperatures much above those that currently prevail. Figure 6-15 compares the number of months in a six-year period (based on 1971-76) in which temperatures exceed a particular level for the current climate and the GISS, GFDL, and OSU scenarios, with known thresholds for major commercial species.

Livingston concluded that crabs, shrimp, oysters, and flounder could not survive in the estuary with the warming projected by the GISS and GFDL scenarios, which imply close to 100% mortality for blue crab larvae and juveniles. The GFDL scenario would also imply over 90% mortality for spotted sea trout, oyster larvae, panfish, and flounder. The mortality under the milder GISS scenario would be only 60%.

Although Livingston concludes that the oysters would probably be eliminated, he cautions that shrimp and other mobile species might adapt by fleeing the estuary for cooler gulf waters during the summer. However, such a flight would leave them vulnerable to predators.

Increased Salinity. Although it appears that sea level rise and warmer temperatures are likely to substantially reduce the productivity of the estuary, the probable impact of precipitation changes is less clear. However, lower flow in the Chattahoochee would combine with sea level rise to increase salinity concentrations in the estuary. Livingston concluded that oysters appear to be the most vulnerable to increases in salinity because oyster drill and other predators, as well as the disease MSX, generally require high salinities. Livingston estimated losses of 10 to 35% for oysters, blue crabs, finfish, and white shrimp under the GFDL scenario due to salinity increases alone.

Sea Level Rise. Livingston also concludes that the loss of wetland acreage would have important impacts on the estuary. Table 6-7 also includes Livingston's estimates of losses in particulate organic carbon, the basic source of food for fish in the estuary. Sea level rise between 50 and 200 cm would reduce available food by 42 to 78%. A proportionate loss in seafood populations would not necessarily occur, since organic carbon food supplies are not currently the constraining factor for estuarine populations. However, wetlands also are important to larvae and small shrimp, crabs, and other species, serving as a refuge from predators. A rise in sea level of a meter or more could lead to a major loss of fisheries.

Despite the adverse impacts on shellfish and flounder, a number of species might benefit from global warming. For example, Livingston concludes that pink shrimp could become more prevalent. Moreover, some finfish spend their winters in Apalachicola Bay and occasionally find the estuary too cold. Other species, such as rock lobster, that generally find the waters too cold at present also may be found in the estuary in the future.

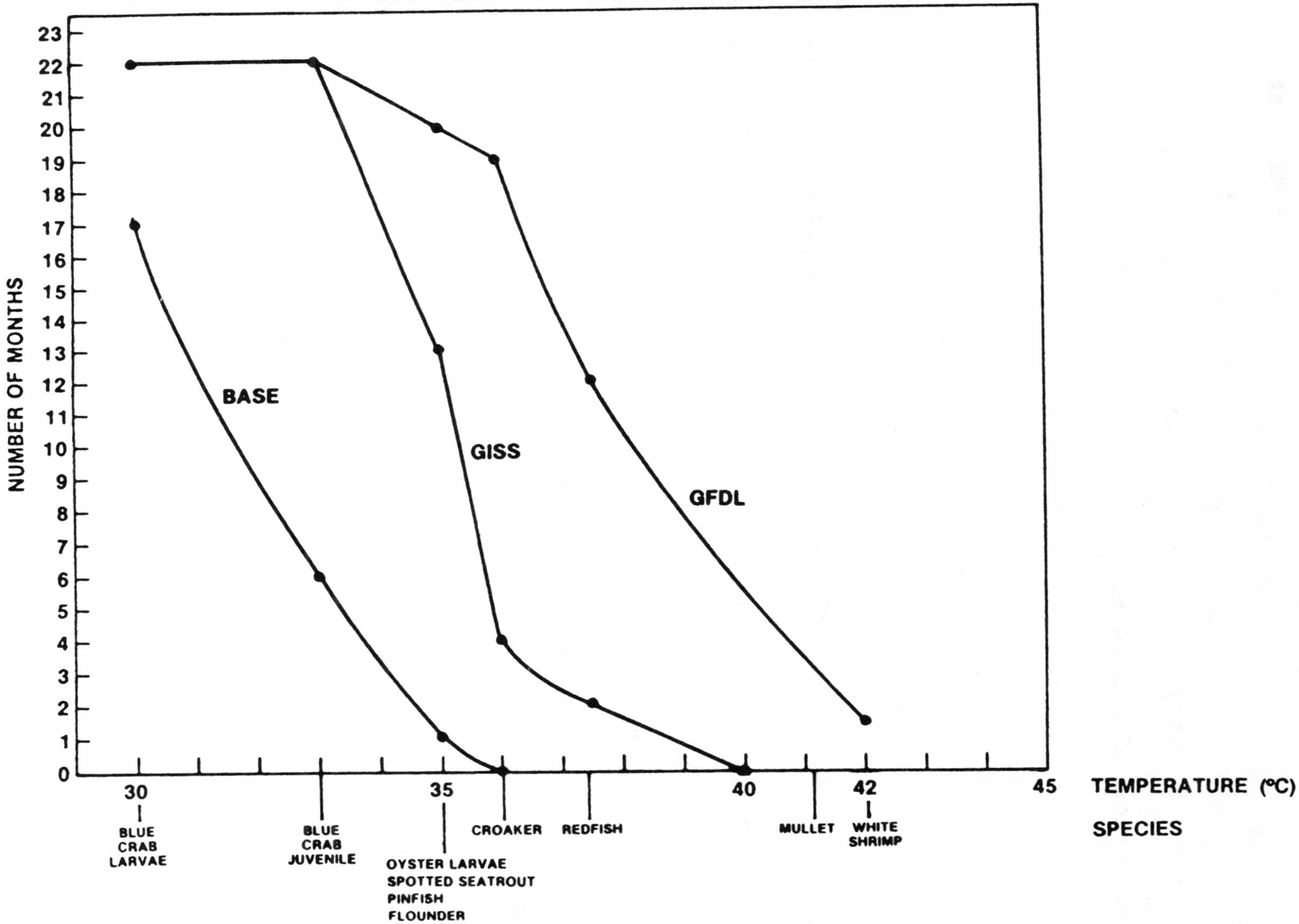

Figure 6-15. Months in a six-year period during which temperatures would be too high for selected species.

Source: Adapted from Livingston, Volume _.

Table 6-7. Projected Changes of the Net Input of Organic Carbon (mtons/year) to the Apalachicola Bay System for Various Scenarios of Sea Level Rise

Factor	Fresh Wetlands	Sea-Grass	Salt Marshes	Phyto-Plankton	Total
Current Scenario for 2100	30,000	27,200	46,905	233,280	337,385
Baseline Sea Level Rise	26,100	28,700	23,500	144,640	222,940
0.5 M Rise	24,000	28,800	4,690	71,450	128,940
1.0 M Rise	21,300	30,100	940	58,790	111,130
2.0 M Rise	4,980	31,035	780	15,160	51,955

Source: Livingston, Volume __.

Implications. Based on Livingston's projections, Meo et al. (Volume __) used current retail prices of fish to estimate that the annual net economic loss to Franklin County would be $5-15 million under the GFDL scenario, $1-4 million under GISS, and $4-12 million under the OSU scenario.

Livingston's results should not be interpreted to mean that fishing will be eliminated from Apalachicola Bay. The extent to which commercially viable tropical species could replace the species that are lost simply cannot be estimated at this time.

Agriculture

Agriculture in the Southeast will be directly affected by changes in climate and indirectly affected by changes in economic conditions and pests. This section presents results from a crop modeling study of yield changes by Peart et al., and regional results from national studies of agricultural production shifts by Adams et al. and of impacts of changes in pest populations by Stinner et al.

Crop Modeling Study

Study Design

Peart et al. used the crop models CERES-Maize (Jones and Kiniry, 1986) and SOYGRO (Wilkerson et al., 1985) to estimate the impacts of climate change on yields of corn and soybeans for 19 sites throughout the Southeast and adjacent states. These models have been used for several years by agricultural scientists to project the impacts of short-term climatic variations. They incorporate the responses of crops to solar radiation, temperature, precipitation, and soil type, and they have been validated over a large range of climate and soil conditions in the United States and other countries.

The major variable not considered by these and other existing agricultural models is the direct "fertilization effect" of increased levels of atmospheric carbon dioxide. Peart et al., therefore, modified their models to consider both the increased rate of photosynthesis and the increased water-use efficiency that corn and soybeans have exhibited in field experiments.

Limitations

The analysis of combined effects is new research and will need further development and refinement. The model runs use simple parameters for CO_2 effects, assume higher atmospheric concentration CO_2

than are predicted, and may be overestimates of crop yields (see Chapter 10 of this report).

The scenarios examining the effects of climate alone on crops assumed that soils were relatively favorable for crops, with low salinity or compaction, and assumed no limits on the supply of all nutrients, except nitrogen. In addition, the analysis assumes no technological adjustments are made by farmers to improve crop yields. Possible negative impacts due to changes in storm frequency, droughts, and pests and pathogens are not factored into this study.

Results

Soybean Yields. Table 6-8 illustrates the results of the soybean model for 13 nonirrigated sites in the study area, as well as Lynchburg, Virginia, a colder site included for comparison purposes.

The results differ between the wet GISS and dry GFDL scenarios. In the GISS scenario, the cooler sites in Georgia and the Carolinas mostly show declines in soybeans yields of 3 to 25%, and the other sites show declines of 20 to 40%, ignoring CO_2 fertilization. When the latter effect is included, the Atlantic states experience gains of 11 to 39%, and the other states vary from a 13% drop in Memphis to a 15% gain in Tallahassee. (Tennessee fares worse than the North Carolina sites at similar latitudes because its grid cell does not receive as favorable an increase in water availability.)

By contrast, the dry GFDL scenario results in very large drops in soybean productivity, with all but one site experiencing declines greater than 50% and eight sites losing over 75%, considering only the impact of climate change. Even when CO_2 fertilization is considered, all but four sites experience losses greater than 50%.

Corn Yields. Table 6-9 shows that the two scenarios differ in a similar fashion for nonirrigated corn. However, in the case of irrigated corn, where the analysis primarily reflects the impact of temperature increases, the two scenarios show more agreement. Ignoring CO_2 fertilization, the GISS model shows drops of 13 to 20%, and GFDL shows drops of 20 to 35%. When CO_2 fertilization is included, the GISS scenario implies declines of less than 8% for all sites, and the GFDL model shows similar declines for two sites and respective declines of 17 and 27% for Charlotte, North Carolina, and Macon, Georgia.

Irrigation. The two scenarios show more agreement for agricultural fields that are already irrigated. Since the changes in water availability are irrelevant here, the impacts are dominated by the increased frequency of very hot days.

The results are mixed on whether currently dry land areas would be shifted to irrigation. Table 6-10 shows the percentage increases in

Table 6-8. Impacts of 2xCO_2 Climate Change on Soybean Yields for Selected Southeastern Sites (yield as a percentage of current yield)

State	Climate Change and CO_2 Fertilization		Climate Change Only	
	GISS	GFDL	GISS	GFDL
Memphis, TN	87	30	62	18
Nashville, TN	104	79	70	48
Charlotte, NC	132	12	93	8
Raleigh, NC	139	24	97	13
Columbia, SC	118	38	80	22
Wilmington, NC	125	59	89	38
Atlanta, GA	127	33	89	22
Macon, GA	111	18	75	9
Tallahassee, FL	115	83	80	49
Birmingham, AL	100	71	69	46
Mobile, AL	92	error	66	57
Montgomery, AL	90	32	61	16
Meridian, MS	91	36	63	22
Lynchburg, VA[a]	146	45	101	26

[a] Peart et al. investigate a number of sites in states adjacent to the Southeast. Lynchburg is included to permit comparison of Southeast results with a colder site.

Source: Peart et al, Volume __.

Table 6-9. Impacts of $2xCO_2$ Climate Change on Corn Yields

Scenario	Charlotte, NC	Macon, GA	Meridian, MS	Memphis, TN
Climate Change Only				
Irrigated				
GISS	87	81	80	82
GFDL	71	65	80	80
Rainfed				
GISS	100	98	84	93
GFDL	24	18	30	37
Climate Change and CO2 Fertilization				
Irrigated				
GISS	101	92	93	95
GFDL	83	73	93	94
Rainfed				
GISS	116	114	98	113
GFDL	28	20	36	37
CO_2 Fertilization Only				
Irrigated	117	113	117	118
Rainfed	116	116	116	116

Source: Peart et al, Volume __.

Table 6-10. Increases in Corn Yields from a Shift to Irrigation (percent, assuming no CO_2 fertilization)

State	No Climate Change	GISS	GFDL
Memphis, TN	70	50	370
Nashville, TN	65	49	205
Charlotte, NC	64	43	486
Raleigh, NC	51	28	444
Columbia, SC	58	47	386
Wilmington, NC	16	8	50
Atlanta, GA	15	7	79
Macon, GA	61	33	589
Birmingham, AL	6	9	61
Mobile, AL	36	41	91
Montgomery, AL	72	39	593
Meridian, MS	62	53	423
Lynchburg, VA*	56	37	361

Source: Derived from Tables 6-8 and 6-9 (see text)

yields that would result from adding irrigation for no climate change and the two climate change scenarios. All but four sites could increase yields today by 50-75% by irrigating. Under the wetter GISS scenario, irrigation only would increase yields 25-50%. However, under the dry GFDL scenario, irrigation would increase yields by 200-600% -- that is, it would mean the difference between crop failure and a harvest slightly above today's levels in most years. Even without CO_2 fertilization, 75% of the nonirrigated southeastern sites could gain more from irrigation than they would lose from the change in climate resulting from the GFDL scenario.

A farmer's decision to irrigate, shift to other crops, or remove land from production would depend to a large degree on what happens to future prices of both crops and water. Even though water is plentiful today, the capital costs of irrigation prevent most farmers in the Southeast from taking advantage of the potential 50% increases in yields. But if crop failures due to drought became as commonplace as Peart et al. project for the dry GFDL scenario, a major increase in irrigation probably would be inevitable.

Shifts in Production

Adams et al. (Volume __) examined the impacts of changes in crop yields on farm profitability and cultivated acreage in various regions of the United States. (The methods for this study are discussed in Chapter 10: Agriculture) Their results suggest that the impact of climate change on southeastern agriculture would not be directly proportional to the impact on crop yields (Table 6-11).

Considering only the impact of climate change, Adams et al. found that the GISS and GFDL scenarios would reduce crop acreage by 10 and 16%, respectively. This is a small difference, given Peart's projection that all but one site would lose more than half of their production under the GFDL scenario, while none of the sites would lose that much under the GISS scenario. When CO_2 fertilization is considered, however, Adams et al. project respective declines in farm acreage of 57 and 33% for the GISS and GFDL scenarios.

The conclusion that the greatest decline in farming could occur under the scenario in which yields increase is not unreasonable in light of the substantial decline in farming that took place during the period of rapid technical innovation after World War II. As yields increase, prices decline. Adams et al. project that most areas of the nation would lose farm acreage. However, Adams et al. estimate that the Southeast would experience the worst losses: while the Southeast has only 13% of

Table 6-11. Impact of Climate Change on Cultivated Land in the Southeast

Region	Baseline	With Direct CO_2 GISS	With Direct CO_2 GFDL	Without Direct CO_2 GISS	Without Direct CO_2 GFDL
Acreage (millions)					
SE coast	12.5	8.7	7.8	11.5	11.2
Appalachia	15.5	2.8	7.4	14.1	12.9
Delta	19.9	9.3	16.7	17.7	16.2
Total	47.9	20.8	31.9	43.3	40.3
Percent loss					
SE coast		30	38	8	10
Appalachia		82	52	9	17
Delta		53	16	11	19
Total		57	33	10	16

Note: SE coast includes Florida, South Carolina, Georgia, Alabama.
Appalachia includes North Carolina, Tennessee, Virginia, West Virginia, Kentucky.
Delta includes Mississippi, Louisiana, Arkansas.

Source: Adams et al, Volume __.

the cultivated acreage, it would account for 60-70% of the nationwide decline in farm acreage.

When the CO_2 fertilization effect is ignored, the reductions in acreage would be much smaller, although the Southeast would still account for 40-75%. The general decline in yields would boost prices, which would make it economical for many farmers to irrigate and thereby avoid the large losses associated with a warmer and possibly dryer climate.

Agricultural Pests

The modeling and economic studies of agriculture do not consider the impact of pests on crop yields. However, Stinner et al. (Volume ___) suggest that global warming would increase the range of several agricultural pests that plague southeastern agriculture. (See Chapter 10: Agriculture for details on the methods of this nationwide study.)

Stinner et al. point out that the northern ranges of potato leafhoppers, sunflower moths, black cutworms, and several other southeastern pests are limited by their inability to survive a cold winter. Thus, milder winters would enable them to move farther north, as illustrated in Figure 6-16. Stinner also points out that more droughts could also increase the frequency of pest infestations.

<u>Implication of Agriculture Studies</u>

Agriculture appears to be at least as vulnerable to a potential change in climate in the Southeast as in any other section of the country. Unlike many of the colder regions, the benefits of a longer growing season would not appreciably offset the adverse impacts of warmer temperatures in the Southeast, where cold weather generally is not a major constraint to. agricultural production.

Florida may present an important exception to the generally unfavorable implications of climate change for crop yields. Although Florida is the warmest state in the Southeast, its agriculture appears to be harmed by cold temperatures more than the agriculture of other states in the region. In recent years, hard freezes have destroyed a large fraction of the citrus harvest several times. As a result, the industry is moving south into areas near the Everglades, and sugarcane, which also thrives in warm temperatures, is expanding into the Everglades themselves. Global warming could enable the citrus and sugarcane areas to include most of the state. Warmer temperatures also would help coffee and other tropical crops that are beginning to gain a foothold in the state.

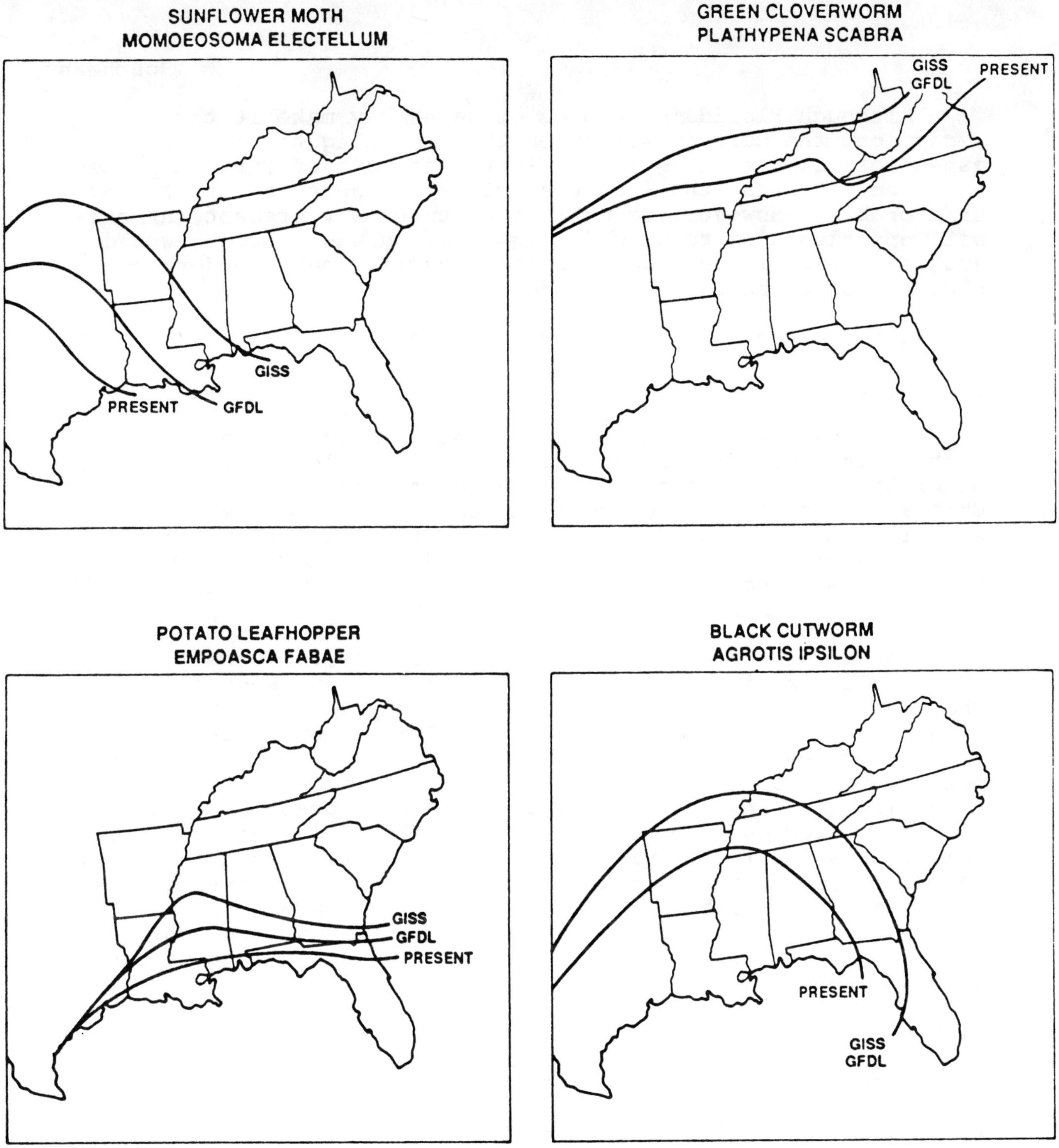

Figure 6-16. Present and predicted northern ranges of various agricultural pests.

Source: Stinner et al, Volume __.

Although Florida's surplus of water may make it the exception, the current situation there highlights an important aspect of climate change: Within the context of current prices and crop patterns, the impact of climate change appears to be unfavorable. However, warmer temperatures may present farmers with opportunities to grow different crops whose prices would justify irrigation or whose seasonal cycles would conform more closely to future rainfall patterns.

Simulation Study of Southeastern Forests

Study Design

To assess the potential impact of climate change on southeastern forests, Urban and Shuggart applied a forest simulation model to a bottomland hardwood forest along the Chattahoochee River in Georgia and to upland sites near Knoxville, Tennessee; Macon, Georgia; Florence, South Carolina; and Vicksburg, Mississippi. Their study considered the OSU, GFDL, and GISS scenarios for doubled CO_2, as well as the GISS transient A scenario through the year 2060.

The model these researchers used was derived from FORET, the "gap" model originally developed by Shuggart and West (1977). The model simulates forest dynamics by modeling the growth of each tree in a representative plot of forest land. It keeps track of forest dynamics by assigning each of 45 tree species optimal rates of growth, seeding rates, and survival probabilities, and by subsequently adjusting these measures downward to account for less than optimal light availability, temperature, soil moisture, and soil fertility. In the case of the bottomland hardwood site, the model also considered changes in river flooding, based on the flows in the lower Chattahoochee calculated in the Lake Lanier study. The researchers applied the model to both mature forests and the formation of a new forest from bare ground.

Limitations

Urban and Shuggart caution against taking their results too literally owing to a number of simplifying assumptions they had to make. First, they had to assume that certain major species, such as loblolly pine, could not tolerate more than 6,000 (cooling) degree-days per year. These species are not currently found in warmer areas, but the southern limits of their range are also limited by factors other than temperature: the Gulf of Mexico and the dry climate of Texas and Mexico. Although the 6,000 degree-day line coincides with their southern boundary across Florida, the peculiar environmental conditions of that state make it impossible to confidently attribute an estimate of thermal tolerance to that observation alone. Fortunately, this caveat does not apply to most of the oaks, hickories, and other species found in the cooler areas of the Southeast.

Another important caveat is that the model does not consider the potentially beneficial impact of CO_2 fertilization on photosynthesis, changes in water-use efficiency, or leaf area.

Results

The simulations by Urban and Shugart call into question the ability of southeastern forests to be generated from bare ground, particularly if the climate becomes drier as well as warmer. For the Knoxville site, the dry GFDL scenario implies that a forest could not be started from bare ground, while the GISS and OSU $2xCO_2$ scenarios imply reductions in biomass of 10-25%. For the South Carolina site, only the GISS climate would support a forest, albeit at less than 50% of today's productivity. The Georgia and Mississippi sites could not generate a forest from bare ground for any of the scenarios.

The transient analyses suggest that mature forests could also be lost -- not merely converted to a different type -- if climate changes. Figure 6-17 shows that none of the forests would decline significantly before the year 2030; however, all would decline substantially before 2060. The Mississippi forest would mostly die by 2040, and the South Carolina and Georgia sites by 2060. Only the relatively cool Tennessee site would remain somewhat healthy, although biomass would decline 35%.

Although the simulation results suggest that southeastern forests are unlikely to benefit from the global warming, the impact on forests may not be as bad as the model suggests. Nevertheless, major shifts in forest types are almost certain to occur from the warmer temperatures alone.

Electric Utilities

Linder (Volume __) examined the impact of global warming on the demand for electricity throughout the Southeast for the two GISS transient scenarios. (See Chapter 16: Electricity Demand for additional details on the methods and limitations of this study.) Because Linder's study was limited to electricity, it did not consider the reduced consumption of oil and gas for space heating that would result from warmer temperatures.

Table 6-12 shows the percentage changes in electric power requirements for various areas in the Southeast. Along the Gulf coast, annual power requirements would increase 3-4% by 2010 and

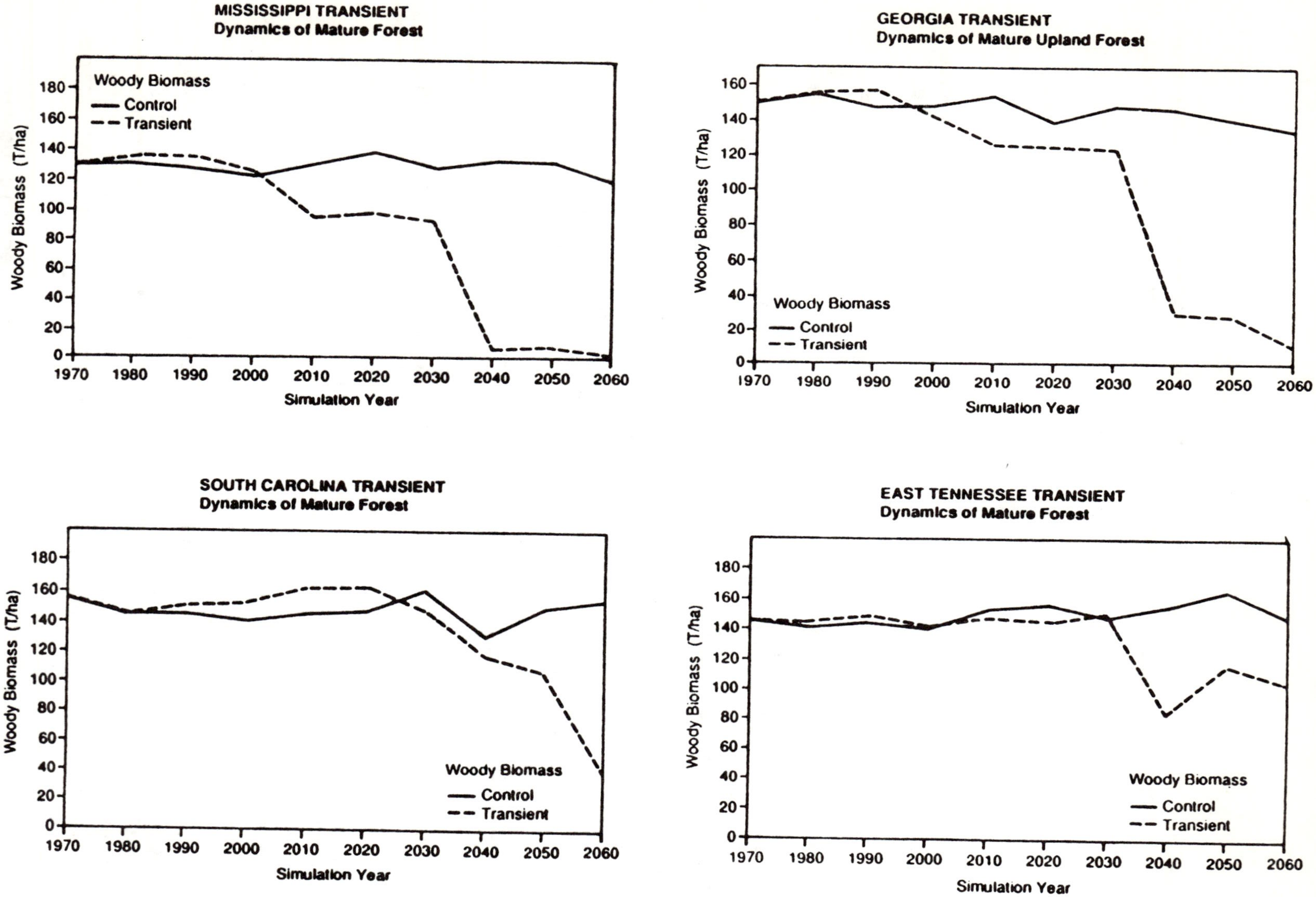

Figure 6-17. Southeastern response of forests to GISS transient scenarios or climate change.

Source: Urban and Shugart, Volume __.

Table 6-12. Percentage Increases in Peak and Annual Demand for Electricity by 2010 and 2055 Due to Climate Change

Area	GISS A (2010) Annual	GISS A (2010) Peak	GISS B (2010) Annual	GISS B (2010) Peak	GISS A (2055) Annual	GISS A (2055) Peak
North Carolina, South Carolina, Georgia	1.6	7.3	1.3	2.4	5.9	24.4
Florida	2.7	4.9	2.7	3.6	9.3	20.0
Eastern Tennessee	1.6	3.7	1.3	1.2	5.9	12.2
Alabama, Western Tennessee	1.9	3.8	2.2	5.7	6.8	13.5
Mississippi	3.8	7.6	4.4	11.4	13.6	26.9
Louisiana	2.9	7.6	2.7	6.6	10.2	23.4
East Texas	3.1	7.9	2.8	6.6	11.3	25.3

Source: Linder and Inglis, Volume __.

10-14% by 2055; elsewhere, the increases would be somewhat less. Because peak demand for electricity generally occurs during extremely hot weather, peak demand would rise more than annual demand. Global warming could increase new generating capacity requirement 10-24% by 2010 and 24-34% by 2055.

Linder compared increases in electric capacity required by climate change with those necessitated by economic growth. He estimated that through 2010, climate change could increase the expected capital costs of $137 billion 6 to 9%; through 2055, it could increase expected requirements of $350-500 billion by as much as 20%.

OTHER STUDIES

Coastal Louisiana

An estuary accounting for half of the Southeast's wetlands is also the most vulnerable: the delta at the mouth of the Mississippi River. Because previous studies assessed the vulnerability of Louisiana in detail, the current effort supported no additional research into this issue.

Over the last few thousand years, the sediment washing down the Mississippi River has formed the nation's largest delta at the river's mouth, almost all of which is in Louisiana. Composed mostly of marsh, cypress swamps, and small distributary channels that carry water, sediment, and nutrients from the river to these marshes and swamps, Louisiana's wetlands support half of the nation's shellfish, one-fourth of its fishing industry, and a large trapping industry. They also provide flood protection for metropolitan New Orleans and critical habitats for bald eagles and other migratory birds.

Water management and other human activities of the last 50 years are now causing this delta to disintegrate at a rate of over 100 km^2 (about 50 km^2) per year. The delta has always subsided as the sediment compacted, but the rate of sinking was more than balanced by new sediment reaching the wetlands through the distributaries and the annual river flooding. But that sediment now largely washes into the deep waters of the Gulf because flood-control and navigation guide levees confine the flow of the river. Thus, the delta is gradually being submerged, and cypress swamps are converting to open-water lakes as saltwater penetrates inland. If current trends continue, almost all of the wetlands will be lost in the next century.

A rise in sea level would further accelerate the rate of land loss in coastal Louisiana. As shown in Figure 6-18, even a

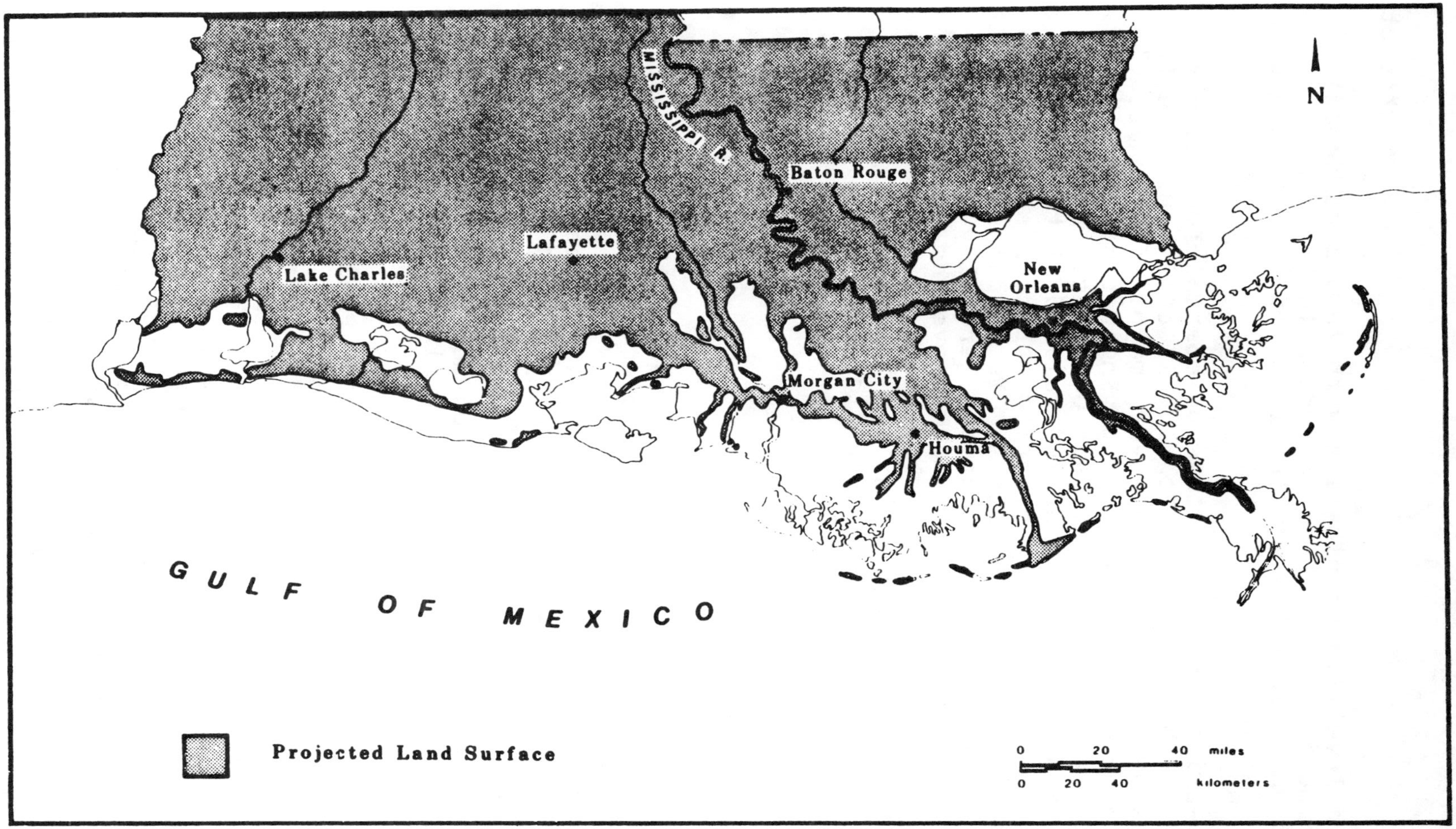

Figure 6-18. Projected Future Coastline of Louisiana for the Year 2033 A.D. Given a Rise in Sea Level of 55 cm as Predicted in the High Scenario

Source: Louisiana Wetland Protection Panel, 1987.

50-cm rise in sea level (in combination with land subsidence) would inundate almost all of the delta and would leave New Orleans, most of which is below sea level and only protected with earthen levees, vulnerable to a hurricane.

Strictly speaking, the entire loss of coastal Louisiana's estuaries should not be attributed to global warming because the ecosystem is already being lost. However, major efforts are being initiated by the U.S. Army Corps of Engineers, the U.S. Fish and Wildlife Service, the Louisiana Geological Survey, several local governments, and other Federal and state agencies to curtail the loss, generally by erecting structures to provide freshwater and sediment to the wetlands. Technical staff responsible for developing these solutions generally fear, however, that a 1-m rise in sea level could overwhelm current efforts, and that if such a rise is ultimately going to take place, they already should be planning and implementing a much broader effort (Louisiana Wetland Protection Panel, 1987).

IMPLICATIONS AND POLICY

Land Use

The impact of climate change on land use in the Southeast must be examined in the context of the projected abandonment of 10-50% of the farmland in the Southeast and declines in forests. A major issue is how this land will be used. In the past, forests have been cleared for agriculture, and when abandoned, they have been converted to forest again. But the forest models suggest that the impact of climate change on the generation of new forests from bare ground would be even more adverse than the impact on existing forests. If the forest simulations are correct, the abandoned fields would become grasslands or overgrown with weeds, and the Southeast could gradually come to resemble the scenery found today in the Great Plains.

Water Resources

The water resource "problems" faced by the Southeast are the type that would make western water managers envious. Rainfall and runoff would increase in the GISS scenario. Although most other assessments suggest that runoff would decline, the magnitude of the decline does not appear to threaten the availability of water for municipal, industrial, or residential use. However, the nonconsumptive uses for hydropower, navigation, environmental quality, and recreation could be threatened. Although sufficient time exists to develop rational strategies to implement the necessary tradeoffs, current Federal statutes in a rigorous fashion discourage water managers from formally considering them.

Chapter 6

Impacts of Wetter Climate

Although most water resource problems have been associated with too little water, it does not necessarily follow that a wetter climate would be generally beneficial. The designs of water management infrastructure and the location of development along lakes and rivers have been based on current climate. Hence, shifts in either direction would create problems.

The chief problem from a wetter climate would be more flooding, particularly in southern Florida and coastal Louisiana, where water often lingers for days and even weeks after severe rainstorms and river surges. Inland communities, such as Chattanooga, also might face flooding if wetter periods exceed the ability of dams to prevent flooding.

Impacts of Drier Climate

A drier climate, on the other hand, would exacerbate current conflicts over water use during dry periods. Hydropower would decline, increasing the need to use fossil or nuclear power, both of which would require more water for cooling. Conflicts between municipal water users and recreational interests also would intensify. Lake levels could drop more during the summer, even if municipal use of water did not grow. However, warmer temperatures probably would increase municipal water demand for cooling buildings and watering lawns.

These conflicts could be further exacerbated if farmers increase the use of irrigation. Groundwater is available in reasonably shallow aquifers that drain into rivers. Any consumptive use of water from these aquifers would reduce, and in some cases reverse, the base flow of water from aquifers into these rivers. Water also could be drawn directly from rivers for irrigation in some areas.

A decline in riverflows could be important for both navigation and environmental quality. For the Tennessee, as well as the Chattahoochee and other small rivers, adequate reservoir capacity exists to maintain flows for navigation, if this use continues to take precedence over water supply and recreation. However, the 1988 drought has graphically demonstrated that there are not enough dams to guarantee navigation in the Mississippi. If this situation became more commonplace, the economic impact on New Orleans could be severe. On the other hand, traffic on the Tennessee and Ohio Rivers might use the Tennessee-Tombigbee Canal as an alternative, which would benefit the port of Mobile.

Lower flows also would reduce the dilution of municipal and industrial effluents discharged into rivers and would decrease the level of dissolved oxygen. This would harm fish populations directly, and indirectly by reducing the abilities of streams to assimilate wastes. Reduced flows also would threaten bottomland hardwood and estuarine ecosystems. To prevent these problems, factories and powerplants might have to more frequently erect cooling towers or curtail their operations.

Is Current Legislation Adequate?

The same issues that face the TVA and Lake Lanier would likely face decisionmakers in other areas. Federal laws discourage water managers in the Southeast from rigorously evaluating the tradeoffs between the various uses of water. Most dams are more than sufficient to meet the statutory requirements for navigation and flood safety and to continue generating substantial hydropower on demand. Consequently, there has been little need to analyze the tradeoffs between these factors. For example, a literal application of the law would not allow the Corps of Engineers to cut hydropower production or navigation releases to ensure a supply of water for Atlanta. Therefore, agencies have not analyzed the allocation of water that best serves the public for various levels of water availability (although the TVA is beginning to do so).

At a practical level, Federal water managers have shown flexibility, as in the case of cutting navigation along the Chattahoochee instead of further cutting Atlanta's water supply. If climate changes and more than a modest level of flexibility is necessary, water resource laws could be changed; the physical infrastructure is largely in place to address water problems of the Southeast. But until the laws are changed, the Federal agencies in the Southeast often would be forced to allocate water inefficiently. Moreover, people making decisions concerning siting of recreational and industrial development, long- term water supply sources, powerplant construction, and other activities sensitive to the availability of water would risk basing their decisions on incorrect assumptions regarding the future allocation of water.

Estuaries

Coastal plants and animals across the Southeast may have difficulty surviving warmer temperatures. For example, along the northern coast of the Gulf of Mexico, several types of fish spend at least part of their lifetimes in estuaries that are already as hot as they can tolerate. If climate becomes warmer, however, migrating north would not be feasible. While these species could escape the summer heat by fleeing to the cooler waters of the gulf, such a flight would make them vulnerable to larger fish.

Although the gulf coast estuaries are the warmest, estuaries as far north as North Carolina reach temperatures only about 3° cooler than those that prevail in the Apalachicola estuary. Thus, all southeastern estuaries are likely to lose at least as great a proportion of their oysters, shrimp, and crabs to high temperatures as the Apalachicola loses today, and most would lose more.

In addition to the direct effect of climate change on estuaries, human responses to climate change and sea level rise also could hurt coastal estuaries. The flood control works erected in response to disasters in Florida and along the Mississippi River now appear to be destroying the Everglades and coastal wetlands of Louisiana. Additional flood control levees and bulkheads to stop erosion could result in additional wetland loss, as discussed in Chapter 9: Sea Level Rise. Increased reservoir construction would decrease the amount of sediment flowing down the river and nourishing the wetlands. If the climate becomes drier, irrigation could further reduce freshwater flow into estuaries.

To a large extent, the policy implications for wetland loss in the Southeast are similar to those facing the rest of the U.S coastal zone. Previous studies have identified several measures to reduce the loss of coastal wetlands in response to sea level rise (e.g., Office of Wetland Protection, 1988; Titus, 1988). These measures include the following:

- o increase the ability of wetlands to keep pace with sea level,
- o remove impediments to landward creation of new wetlands, and
- o dike the wetlands and artificially maintain water levels.

All of these measures are being employed or actively considered.

Congress has authorized a number of freshwater and sediment diversion structures to assist the ability of Louisiana's wetlands to keep up with relative sea level rise. These structures are engineered breaches in river levees that act as spillways into the wetlands when water levels in the river are high. Although decisions on where to build diversion structures are being based on current climate and sea level, consideration of global warming would substantially change the assumptions on which current analyses are being based and the relative merits of alternative options. More frequent or higher surges in the Mississippi River would increase the amount of water delivered to the wetlands. And if climate change resulted in more soil erosion, more sediment might also reach the wetlands; lower flows could have the opposite effect. Sea level rise might shorten the useful lifetimes of these projects, but because the flood-protection benefits of protecting coastal wetlands would be greater with a higher sea level, the economic justification of these projects would not necessarily be undermined by the prospect of sea level rise. Nevertheless, they would change the relative merits of alternative options (Louisiana Wetland Protection Panel, 1987).

Artificially managing water levels also has been proposed for Louisiana, particularly by Terrebonne Parish, whose eastern wetlands are far removed from a potential source of sediment. Such an approach also might be possible for parts of Florida, where wetlands already are confined by a system of dikes and canals, and water levels already are managed. Although no one has yet devised a practical means by which shrimp and other fish could migrate between ocean and estuary, other species spend their entire lifetimes within the estuary, and freshwater species could remain in artificially maintained freshwater wetlands.

A final response would be to accept the loss of existing wetlands, but to take measures to prevent development from blocking the landward creation of new wetlands. This approach has been enacted by the state of Maine (1987) and would be consistent with the proposals to discourage bulkheads that have been widely discussed by coastal zone managers and enacted by the state of South Carolina. Titus Greene estimates that 1800 mi^2 of wetlands in the Southeast could be created if developed areas were not protected. Although this area represents a small fraction of the potential loss, it would increase the remaining areas of wetlands by 30-90%, and it would maintain and perhaps increase the proportion of shorelines on which at least some wetlands could be found.

Beach Erosion

The implications of sea level rise for recreational beaches in the Southeast are similar to the implications for the mid-Atlantic and the Northeast. If shore-protection measures are not taken, the majority of resorts will have no beach at high tide by 2025 under the midrange scenario of future sea level rise. The cost of undertaking the necessary measures through 2025 would probably be economically justified for most resorts (see Chapter 9: Sea Level Rise). However, the cost of protecting all recreational beaches through 2100 would be $100 to $150 billion, which would probably lead some of the more vulnerable areas to accept a landward migration much as areas on North Carolina's Outer Banks are facing today.

The potential responses to global warming should be viewed within the context of current responses to erosion flooding. Florida has a trust fund to nourish its beaches and has received Federal assistance for pumping sand onto the shores of Miami Beach. Although Mississippi has no reason to combat the landward migration of its barrier islands, it has nourished the beaches of Biloxi, Gulfport, and other resort communities that lie on the mainland along the protected waters behind the barriers. Louisiana is rebuilding its undeveloped barrier islands because they protect the mainland from storms. Most states have employed a variety of structural solutions, such as groins (jetties) and seawalls. However, they are moving toward "soft engineering" solutions, such as beach nourishment, because of doubts about the effectiveness of hard structures in universal erosion and their interference with recreational uses of the beach.

Land-use measures also have been employed to adapt to erosion. Because of unusually high erosion rates on the Outer Banks, houses along the coast are regularly moved landward. North Carolina requires houses, hotels, and condominiums to be set back from the shore by the distance of a 100-year storm plus 30 years' worth of erosion on the assumption that after 30 years, the house could be moved back. To maintain the integrity of its publicly owned recreational beaches and to discourage shorefront property owners from expecting state efforts to protect these beaches, Texas requires that any house left standing in front of the vegetation line after the shore erodes must be torn down.

If a global warming increases the frequency of hurricanes, a number of southeastern communities will be devastated. There is no way to know which communities might experience a hurricane resulting from the greenhouse effect. However, the overall impact of increased hurricane frequency would be small compared to the impact of sea level rise. While a doubling of hurricanes would convert 100-year floodplains to 50-year floodplains throughout much of the Southeast, a 1-m rise would convert them to 15-year floodplains.

Because the open-coast areas most vulnerable to sea level rise are generally recreational beach resorts, the costs of erosion and flooding should be viewed within the larger context of why people go to the beach. People from the north visit southeastern beaches to escape winter, and residents of the region go to escape the summer heat. As temperatures become warmer, Georgia and the Carolinas will be able to compete with Florida for northerners. Hotter temperatures also may increase the desire of the region's residents to visit the beach.

Thus, it is possible that the cooler communities will reap benefits from a longer and stronger tourist season that are greater than the increased costs for erosion control. Areas that already have a year-round season are less likely to benefit, and in a few areas like Miami Beach, the off-season may be extended.

REFERENCES

Armentano, T.V., R.A. Park, and C.L. Cloonan. 1988. Impacts on Coastal Wetlands Throughout the United States." In Titus (ed) 1988. op. cit.

Barth, M.C. and J.G. Titus. 1984. Greenhouse Effect and Sea Level Rise: A Challenge for This Generation. New York: Van Nostrand Reinhold Company.

Bureau of Economic Analysis. Department of Commerce. 1986.

Bureau of the Census. Department of Commerce. 1982 Census of Agriculture. Vol. 1. Geographic Area Series.

Bureau of the Census. Department of Commerce. 1988. Statistical Abstract of the United States: 1988.

Burnash Robert J.C., R. Larry Ferral, and Richard A. Mcguire. 1973. "A Generalized Streamflow Simulation System, Conceptual Modeling for Digital Computers." National Weather Service and California Department of Water Resources. Sacramento, California.

Edison Electric Institute. 1985. Statistical Yearbook of the Electric Utility Industry.

Energy Information Administration. May 1988. Electric Power Monthly.

Fisheries Statistics Division, National Marine Fisheries Service, NOAA. 1987 Preliminary Statistics for United States Domestic Catch.

Geraghty, J., D. Miller, F. Van Der Leeden, and F. Troise. 1973. Water Atlas of the United States. Water Information Center, Port Washington, N.Y.

Gibbs, M. 1984. Economic analysis of sea level rise: Methods and results. In Barth and Titus 1984.

Healy, Robert G. 1985. Competition for Land in the American South. Washington, D.C.: The Conservation Foundation.

Jenkins, Representative, Rep. Barnard, and Rep. Darden. 1988. HR-4254: Georgia Resevoir Management Act of 1988. 100th Congress. 2nd session.

Jones, C.A. abd J.R. Kiniry (eds.). 1986. CERES-Maize: A simulation model of maize growth and development. College Station, Texas: Texas A&M Press.

Kana, T.W., J. Michel, M.O. Hayes, and J.R. Jensen. 1984. The physical impact of sea level rise in the area of Charleston, South Carolina. In Barth and Titus (eds) 1984.

Leatherman, S.P. 1984. Coastal geomorphic responses to sea level rise: Galveston Bay, Texas. In Barth and Titus (eds) 1984.

Linder, K.P., M.J. Gibbs, and M.R. Inglis. 1988. Potential Impacts of Climate Change on Electric Utilities. New York, N.Y.: New York State Energy Research and Development Authority.

Louisiana Wetland Protection Panel. 1987. Saving Louisiana's Coastal Wetlands: The Need for a long-term Plan of Action. Washington, DC: U.S. Environmental Protection Agency/Louisiana Geological Survey. EPA-230-02-87-026.

Meo, M. 1987. Proceedings of the Symposium on Climate Change in the Southern United States. Norman, OK: University of Oklahoma.

Mitch, W. and J. Gosselink. 1986. Wetlands. New York: Van Nostrand Reinhold Company, Inc.

Shugart, H.H. and D.C. West. 1977. Development of an Appalachian deciduous forest succession model and its application to assessment of the impact of the chestnut blight. Journal of Environmental Management. 5:161-179.

Titus, J.G., C.Y. Kuo, M.J. Gibbs, T.B. LaRoche, M.K. Welts, and J.O. Waddell. 1987. Greenhouse effect, sea level rise, and coastal drainage systems. J. Water Res. Planning and Management ASCE 113(2):216-227.

Titus, J.G. 1987. "The Greenhouse Effects, Rising Sea Level, and Society's Response." In: R.J.N. Devoy. 1987. Sea Surfact Studies Croom Helm: New York.

Titus, J.G. 1988. Greenhouse Effect, Sea Level Rise, and Coastal Wetlands. Washington, D.C.: U.S. Environmental Protection Agency.

Titus, J.G. 1988. "Sea Level Rise and Wetland Loss: An Overview." In Titus (ed). op. cit.

Titus, J.G. 1984. Planning for Sea Level Rise Before and After a Coastal Disaster. In: Barth, M.C. and Titus, J.G. (eds).

Greenhouse Effect and Sea Level Rise: A Challenge for This Generation. New York: Van Nostrand Reinhold Company.

Wilkerson, G.G., J.W. Jones, K.J. Boote, and J.W. Mishoe. 1985. SOYGRO V5.0: Soybean Crop Growth and Yield Model. Technical Documentation. Gainesville, FL: University of Florida.

Table 6-11. Impact of Climate Change on Cultivated Land in the Southeast

Region	Baseline	With Direct CO_2 GISS	With Direct CO_2 GFDL	Without Direct CO_2 GISS	Without Direct CO_2 GFDL
Acreage (millions)					
Southeast coast	12.5	8.7	7.8	11.5	11.2
Appalachia	15.5	2.8	7.4	14.1	12.9
Delta	19.9	9.3	16.7	17.7	16.2
Total	47.9	20.8	31.9	43.3	40.3
Percent loss					
Southeast coast		30	38	8	10
Appalachia		82	52	9	17
Delta		53	16	11	19
Total		57	33	10	16

Note: SE coast includes Florida, South Carolina, Georgia, Alabama.
Appalachia includes North Carolina, Tennessee, Virginia, West Virginia, Kentucky.
Delta includes Mississippi, Louisiana, Arkansas.

CHAPTER 7

GREAT PLAINS

FINDINGS

Global warming in the Great Plains may reduce agricultural output, increase irrigation demand, change water quality, and increase electricity needs.

Agriculture

- Generally, the effects of a warmer climate alone would reduce wheat and corn yields. Yield changes range from +15 to -90%. The direct effects of CO_2 on crop photosynthesis and water use may mitigate these effects, unless climate change is severe.

- Crop yields in southern areas, such as Texas and Oklahoma may decline relative to northern areas of the United States. This change in productivity could lead to a 4 to 22 percent reduction in cultivated acreage in the Great Plains.

- Because of the increased reliability of yields from irrigated lands relative to dryland yields, and because of higher crop prices, demand for irrigation water on remaining farms will probably increase. The number of acres irrigated may increase by 5 to 30 percent.

Water Quality

- Water quality in the southern Great Plains would be altered due to changing pesticide loadings in surface runoff and erosion. Quality of groundwater may be less at risk than surface water quality, because of increased evaporation and less leaching. These results are very sensitive to changes in the amounts and frequency of rainfall.

Electricity Demand

- The annual demand for electricity in Kansas, Nebraska, Oklahoma, and West Texas could rise by 5 to 10 billion kilowatt hours (2 to 4 percent) by 2010 and by 41 to 81 BKWH (10 to 14 percent) by 2055.

- By 2010, approximately 3 to 6 gigawatts would be needed to meet the increased demand and by 2055, 22 to 45 GW would be needed -- a 27 to 39 percent increase over baseline additions that may be needed without climate change -- the cumulative cost of these additions by 2055 would be $24 to $60 billion.

- o Agencies with responsibility for agricultural land use, such as the USDA Agricultural Stabilization and Conservation Service and the Soil Conservation Service, could begin to analyze how their missions may be affected by climate change and to consider development of flexible strategies to deal with potential impacts. Water resource managers, such as those on the River Basin cimmissions and in State natural resource agencies, may wish to factor the potential effects of cllimate change into planning of long-term irrigation, drainage, and water-transfer systems.

CLIMATE-SENSITIVE ASPECTS OF THE GREAT PLAINS

Agriculture

The Great Plains consist of a treeless region of relatively flat topography between the Rocky Mountains and the Mississippi lowlands of central North America. Since agricultural settlement began in the late 1800s, the area has developed into a key small-grains agricultural production region. The southern Great Plains developed rapidly as a wheat production area and became an extensive dryland agricultural region dominated by the cultivation of wheat (Figure 7-1). In the last two decades, exploitation of water from the Ogallala Aquifer has led to the development of significant irrigated agricultural production. In many areas, irrigated farming of corn, rice, and cotton has replaced dryland wheat production, especially in western Kansas and the Texas Panhandle (Figure 7-1).

The Great Plains portions of Kansas, Nebraska, Oklahoma, and Texas constitute a vital part of the United States' agricultural base. Nearly 100,000 farms encompassing over 111 million acres produce an important array of dryland and irrigated crops. Major dry farming crops include winter wheat and grain sorghum, and irrigated grains include corn and rice. In all, the four States produce over 80, 40, and 15%, respectively, of the nation's grain sorghum, wheat, and corn (Table 7-1).

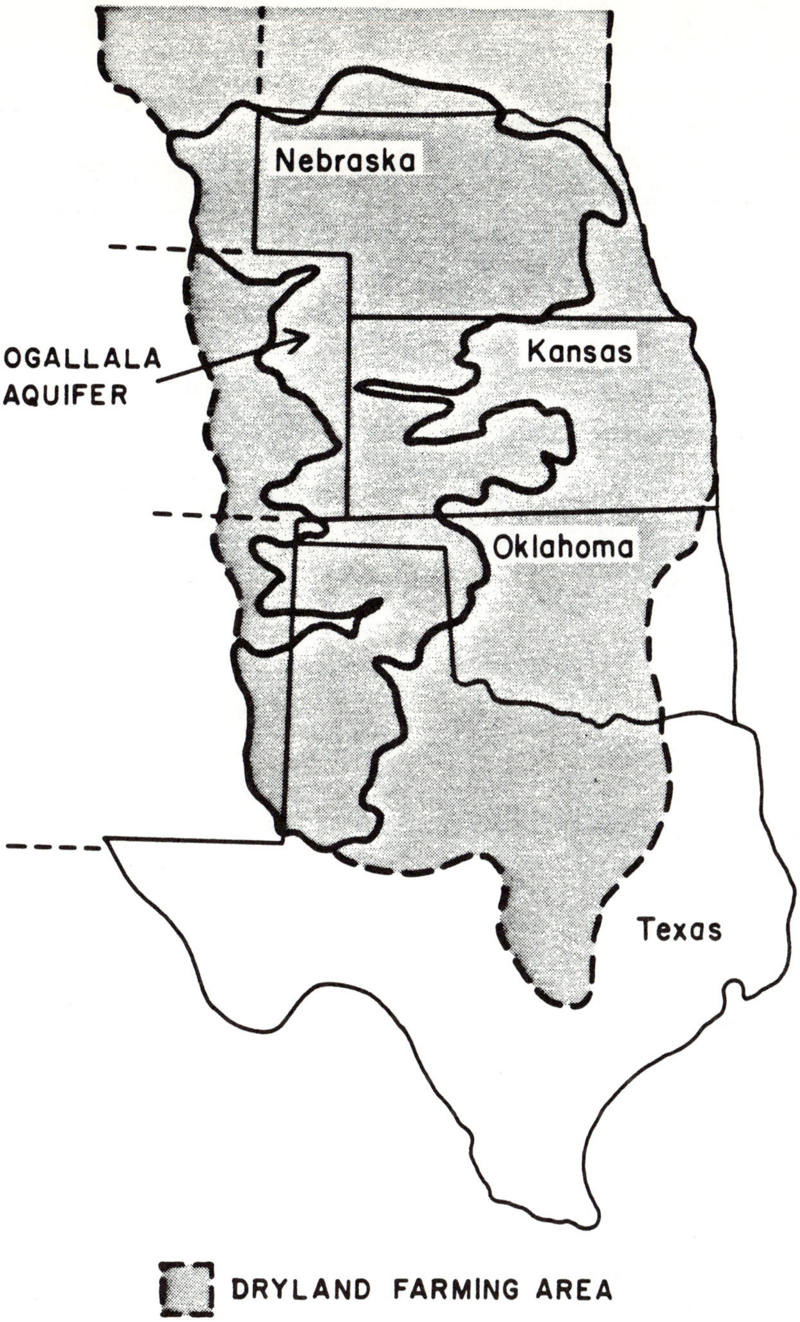

Figure 7-1. Dryland wheat production in Great Plains.

Source: Science of Food and Agriculture (1987, 1988).

TABLE 7-1. U.S. Agricultural Ranking and Percent of U.S. Total for Selected Products, 1982.

	Kansas	Nebraska	Oklahoma	Texas	% U.S. Total
Sorghum harvested	2	3	5	1	80.5
Cattle fattened on grain and concentrates sold	2	3	9	1	46.7
Value of cattle and calves sold	2	3	7	1	40.7
Wheat harvested	1	9	3	6	31.8
Cotton harvested			9	2	25.8
Hay harvested	9	2	16	7	15.9
Market value of all agricultural products	6	5	20	3	18.5

Source: 1982 Census of Agriculture. U.S. Bureau of the Census. 1983. U.S. Government Printing Office. Washington, D.C.

Livestock constitute another important agricultural commodity in the region. Almost 50% of all cattle fattened in the country are raised in the four States, accounting for 40% of the total U.S. value of marketed livestock.

In addition to contributing substantially to national production, the four States are also major exporters of agricultural products. Foreign exports of grain and animal products are especially notable (Table 7-2). In total, these four States provide approximately one-fifth of the dollar value of all U.S. agricultural exports. Yet, dependence on foreign markets puts Great Plains farmers at high risk. While large historical fluctuations in grain and livestock production levels are partly related to climatic variability, changing international markets have also played an important role in the region's failure to achieve long-term economic and social stability.

Farmers in the Great Plains have been particularly vulnerable to climate variability. The southern and central Great Plains States of Nebraska, Kansas, Oklahoma, and Texas, the focus of this regional case study, were among the hardest hit during the Dust Bowl of the 1930s (Worster, 1979; Hurt, 1981). Yields of wheat and corn dropped as much as 50% below normal, causing the failure of about 200,000 farms and migration of more than 300,000 people from the region.

The Dust Bowl, other droughts, and the desire for continued expansion and intensification of dryland farming have led to numerous technological and social adjustments to climate and market fluctuations. Especially critical, from a dryland farming perspective, has been the improvement of conservation tillage practices (Warrick and Bowden, 1981; Riebsame, 1983). These practices, now prescribed for the area, aim to conserve moisture, reduce energy input, and minimize erosion, thus increasing yields and profits.

Besides the developments in cropping systems, policies have also been devised to absorb or mitigate the impacts of climate stresses. These include Federal programs for crop insurance, disaster grants and low-interest loans to farmers, and Government-sponsored drought research (Warrick, 1980). Such programs can be costly. For example, the Drought Relief Bill proposed for the current drought of 1988 is projected to cost about $3.9 billion nationally (New York Times, 1988).

Table 7-2. Agricultural Exports from Selected Plains States, Fiscal Year 1984 (Millions of Dollars).

	U.S.	Kansas	Nebraska	Oklahoma	Texas	% U.S. Total
Feed grains and products	7,585	372	903	-	385	22
Wheat and products	4,526	797	150	353	276	35
Live animal and meats	1,161	130	134	18	161	38
All agricultural products	31,187	1,719	1,762	71	2,031	19

Source: USDA Economic Research Service. 1985. Foreign Agricultural Trade of the United States. Washington, D.C.

Chapter 7

The Great Plains is one of the most marginal agricultural regions in the United States. Some observers feel that the southern Plains are simply too sensitive to climatic swings and that intensive dryland farming should be abandoned (Worster, 1979). Yet in a wet year, the Plains produce bumper crops of small grains that add significantly to the nation's export trade balance.

Despite the adoption of conservation tillage techniques, drought-resistant cultivars, and risk management programs, some analysts argue that the region remains particularly vulnerable to climate-induced reductions in crop yields and might manifest the first impacts of climate change (e.g., Lockeretz, 1978). Rapid acreage increases in the 1970s, destruction of windbreaks for larger fields to accommodate bigger machinery, and speculative farm expansion all raise the specter of renewed land degradation similar to that of the Dust Bowl period, if climate change creates an increased frequency of heat waves and droughts in the region.

The Ogallala Aquifer

A major response to the semiarid and highly variable climate of the Great Plains has been exploitation of surface and groundwater resources for irrigation. In 1982, 75,000 km^2, or 12% of all Great Plains cropland, were irrigated. Groundwater provides most of the water for irrigation: 61-86% of the water used in Nebraska, Oklahoma, and Kansas compared with only 20% nationally.

The improvement and application of well drilling and pumping technology after World War II permitted the use of water from the immense Ogallala Aquifer (Figure 7-1). Today, the aquifer supplies irrigation for approximately 14 million acres in the Great Plains States of Colorado, Nebraska, Kansas, Oklahoma, New Mexico, and Texas (High Plains Associates, 1982). Use of the aquifer allows the irrigation of upland terrain previously too far from surface supplies. The aquifer also provides water for municipal and industrial purposes.

Nebraska recently started to use the aquifer to irrigate corn, which is grown mostly for livestock feed. Corn, wheat, and some sugar beets are irrigated further south, while in Texas the Ogallala is tapped chiefly for cotton. The aquifer varies in its depth from the surface of the land, its rate of natural discharge, and its saturated thickness in different regions. Nebraska has a higher recharge rate (i.e., the rate at which the aquifer is replenished) than the other Great Plains States and has not yet suffered significant drawdown problems. In Texas, high withdrawal and low recharge rates of the aquifer have already resulted in "mining" of the resource (i.e., the rate of water withdrawal is greater than rate of recharge) and in the abandonment of thousands of irrigated acres (see Wilhite, Volume _).

PREVIOUS CLIMATE CHANGE STUDIES

Several studies of climate change impacts on agriculture in the Great Plains have been performed, using a variety of approaches and models. Warrick (1984) considered the vulnerability of the region to a possible recurrence of the 1930s drought. Using a crop yield model tuned to 1975 technology, Warrick found that a recurrence of 1934 and 1936 conditions on the Plains would result in wheat yield reductions of over 50%.

Terjung et al. (1984) investigated water regimes for corn production under differing climate change scenarios, using a crop water-use and yield model. They found that the middle latitudes of the Great Plains were the most sensitive to the climate change scenarios both in evapotranspiration and total water applied for irrigation. This modeling work was continued by Liverman et al. (1986), who found that the lowest irrigated yields occurred under cloudy, hot, and very dry climate change scenarios. Under dryland conditions, minimum yields occurred under sunny-hot and sunny-warm scenarios with very dry conditions.

Using an agroclimatic approach, Rosenzweig (1985) found that lack of cold winter temperatures in the southern Great Plains may necessitate a change from winter to spring wheat cultivars. Decreased water availability may also increase demand for irrigation. In a later study, Rosenzweig (1987) showed that although the combined effects of climate change and increased CO_2 compensated for the negative effects of climate change in years with adequate rainfall, many simulated crop failures occurred in years with poor weather.

Robertson et al. (1987) estimated the combined impact of climate change and increased CO_2 on corn and wheat yields and erosion using the Erosion Productivity Impact Calculator (EPIC). Modeled wheat yields in Texas decreased and modeled corn yields increased only marginally due to moisture stress.

Glantz and Ausubel (1984) suggested that the Great Plains' mining of the Ogallala Aquifer and its susceptibility to future incidence of drought projected by global climate change models be combined in analyses of the region, since both are critical to the habitability of the area.

GREAT PLAINS STUDIES IN THIS REPORT

The EPA studies for this report examine the implications of climate change for agricultural production and economics, demand for irrigation water, and water quality. Climate change impact research on energy use and resource management policy relevant to the Great Plains are also described.

The studies performed for this report are listed in Table 7-3.

SUMMARY OF CLIMATE CHANGE SCENARIOS FOR THE GREAT PLAINS

The predicted changes in seasonal and annual temperatures and precipitation for all three scenarios are shown in Figure 7-2. All three scenarios show large increases in temperature for the Great Plains States under a doubled CO_2 climate. The GISS scenario has an average warming of 4.5°C, the GFDL scenario has a warming of 4.9°C and OSU has an average warming of 3.3°C. In general, winter temperatures increase more than summer temperatures in the GISS model, summer temperature changes are greater than winter temperature changes in the GFDL and OSU scenario. The studies used only the GISS and GFDL climate change scenarios.

Annual precipitation decreases by 0.26 mm/day in the GISS scenario, and 0.12 mm/day in the GFDL while OSU has a slight increase. However, these annual values mask a pronounced reduction in rainfall in Nebraska and Kansas in the GFDL scenario (see Figure 7-3). The large temperature increase and pronounced summer drying in these states combine to make the GFDL scenario more severe than the GISS scenario for the Great Plains.

The magnitudes of climate changes from the GFDL scenario and the climate of the 1930s drought in Nebraska and Kansas are compared in Figure 7-3. While the predicted decreases in precipitation are about the same as those during the most severe drought years (1934 and 1936) in the area, CO_2-induced temperatures are predicted to be about 3°C higher than the Dust Bowl temperatures.

Table 7-3. Great Plains Studies for EPA Report to Congress on the Effects of Global Climate Change

---

Analyses Performed for This Case Study

- Potential Effects of Climate Change on Agricultural Production in the Great Plains: A Simulation Study

  Cynthia Rosenzweig
  Department of Geography, Columbia University
  NASA/Goddard Institute for Space Studies

- Effects of Projected CO_2-Induced Climatic Changes on Irrigation Water Requirements in the Great Plains States

  Richard G. Allen
  Utah State University

National Studies That Included Great Plains Results

- Economic Effects of Climate Change on U.S. Agriculture: A Preliminary Assessment

  Richard M. Adams
  Oregon State University

- Impacts of Climate Change on the Movement of Agricultural Chemicals Across the U.S. Great Plains and Central Prairie

  Howard L. Johnson
  Oklahoma Climatological Survey
  University of Oklahoma

- The Potential Impacts of Climate Change on Electric Utilities: Regional and National Estimates

  Kenneth P. Linder
  ICF Incorporated

- Adjusting Resource Management Policy to Climate Change

  William E. Riebsame
  University of Colorado

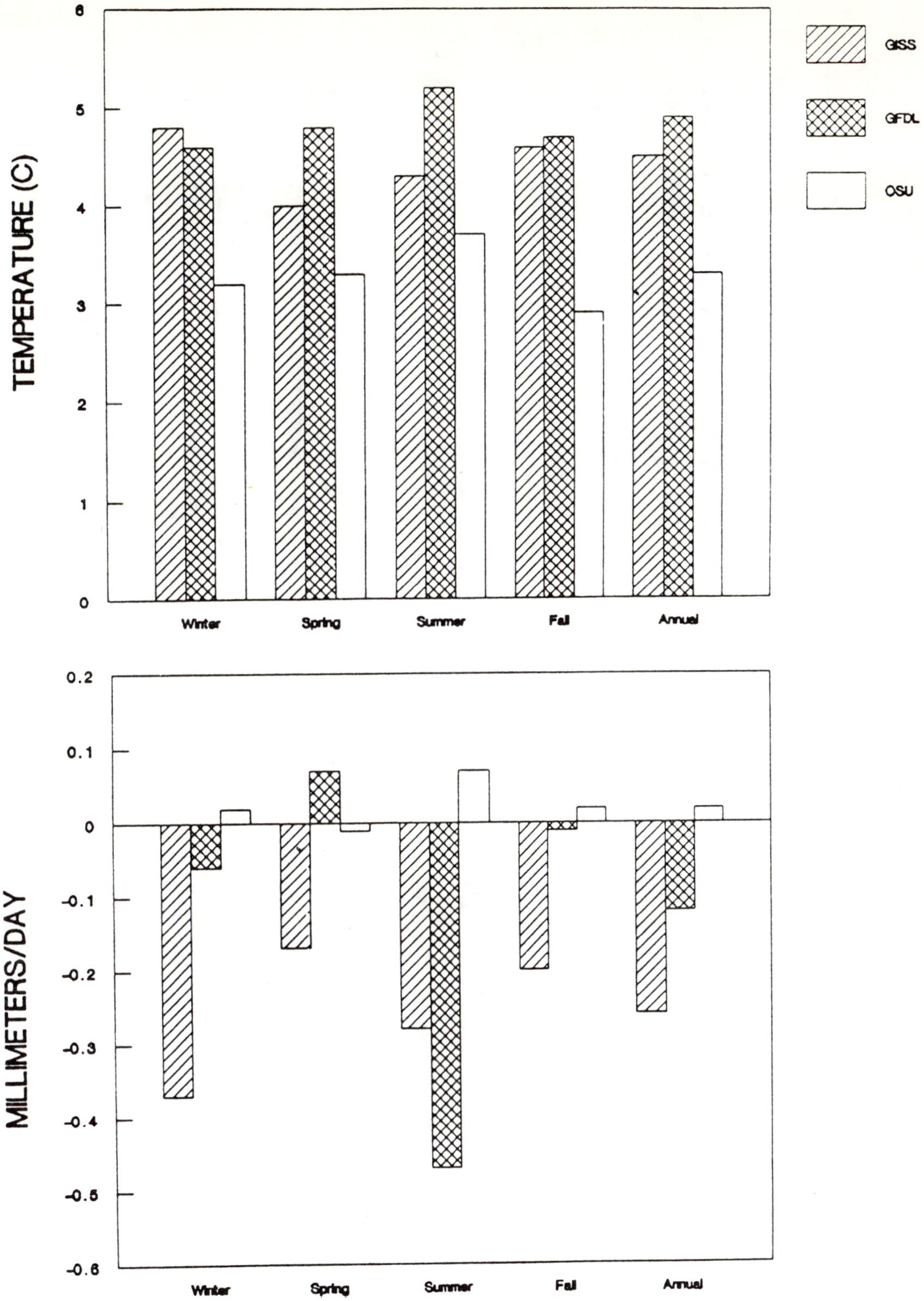

Figure 7-2. Average change in (a) temperature and (b) precipitation over Great Plains gridpoints ($2XCO_2$ less $1XCO_2$).

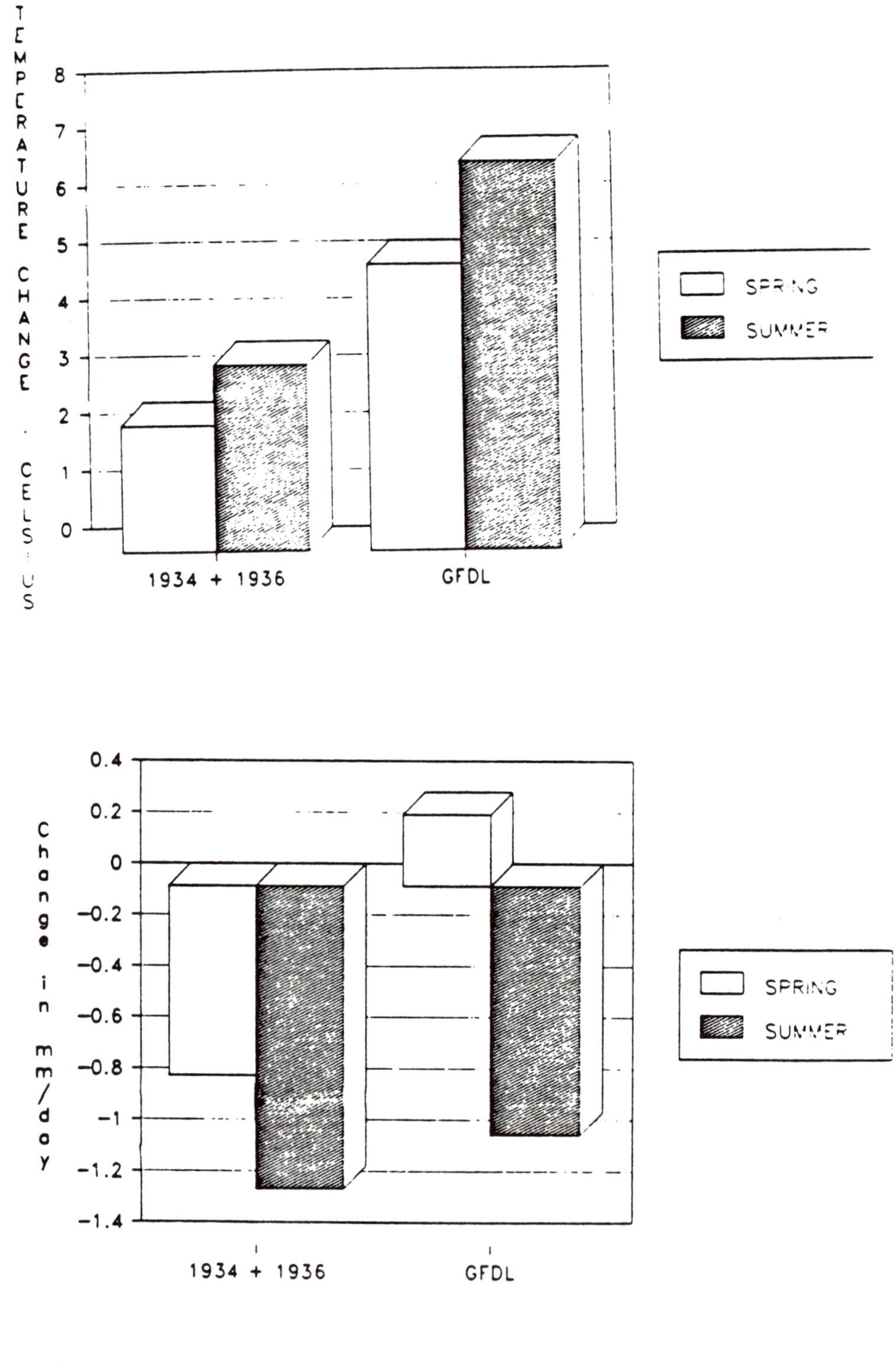

Figure 7-3. Comparison of observed drought and GFDL climate change for (a) temperature and (b) precipitation.

Chapter 7

RESULTS OF THE GREAT PLAINS STUDIES

Crop Production

To better understand the potential physical impact of climate change on crops, Rosenzweig modeled changes in corn and wheat yields in the Great Plains using crop growth models.

Study Design

The crop growth models, CERES-Wheat (Ritchie and Otter, 1985) and CERES-Maize (Jones and Kiniry, 1986), were used to test the sensitivity of crop yields to the GISS and GFDL climate change scenarios. These models are designed for large-area yield prediction and for farm decisionmaking.

They respond to a wide range of soil, climatic, and planting conditions. The models have a whole-plant framework.

At each of 14 locations, the crop models were run on three soils present in the region representing low, medium, and high productive capacity. Model results were generated for change of yield, water used for irrigation (if crop is irrigated), crop evapotranspiration, and planting and maturity dates for both dryland and irrigated cases.

The direct effects of CO_2, i.e., increased photosynthesis and decreased transpiration per unit leaf area, were simulated with the climate change scenarios in another set of runs. A method for approximating the direct effects was developed by computing ratios of daily photosynthesis and evapotranspiration rates of a canopy exposed to elevated CO_2 to the corresponding rates of the same canopy if exposed to ambient CO_2 conditions (see Peart et al., Volume _).

Limitations

This work does not consider changes in frequencies of extreme events, even though extremes of climatic variables, particularly runs of extremes, are critical to crop productivity (see Mearns, Volume _). Development of the CERES models was based on current climate; the relationships in the models may or may not hold under differing climatic conditions, particularly the high temperatures predicted for the greenhouse climate.

The direct effects of CO_2 are only approximated in the crop modeling study, because the models do not include a detailed simulation of photosynthesis. Also, experimental results from controlled environments may show more positive effects of CO_2 than will actually occur in variable, windy, and pest-infected (e.g., weeds, insects, and diseases) field conditions.

Results

Climate change scenarios cause simulated wheat and corn yields to decrease in the southern and central Great Plains. Decreases in modeled yields appear to be caused primarily by increases in temperature, which shorten the duration of crop life cycle (the period during which crops grow). This results in reduced yields. Wheat yields decrease more at lower latitudes. The changes in corn yields are displayed in Figure 7-4.

When the direct effects of CO_2 on crop photosynthesis and transpiration are included in the climate change simulations, modeled crop yields overcome the negative effects of climate change in some cases, but not in others (Figure 7-5). The more severe the climate change scenario, the less compensation provided by direct effects of CO_2.

The amount of water needed for irrigation increases in the areas where precipitation decreases. Irrigation appears to reduce interannual variability in yields.

Adjusting the planting date of wheat to later in the fall does not significantly ameliorate the effects of the GISS climate change scenario on CERES-Wheat yields. Changing to varieties with lower vernalization requirements (period of cold weather necessary for reproduction) and lower photoperiod sensitivity (sensitivity to daylength), in addition to delaying planting dates, overcomes yield decreases at some sites but not at others.

Implications

With high temperatures, increased irrigation may be needed to produce acceptable and stable yield levels. This could place increased demand on the already depleted Ogallala Aquifer and other water resources in the region.

Regional Agricultural Shifts

Many economic consequences are likely to result from the physical changes in crop yields and water availability caused by climate change. Adams et al. introduced yield changes from the Great Plains and other regional crop modeling studies and changes in water demand and availability from the GISS and GFDL scenarios into an economic model to translate the physical effects of

a)

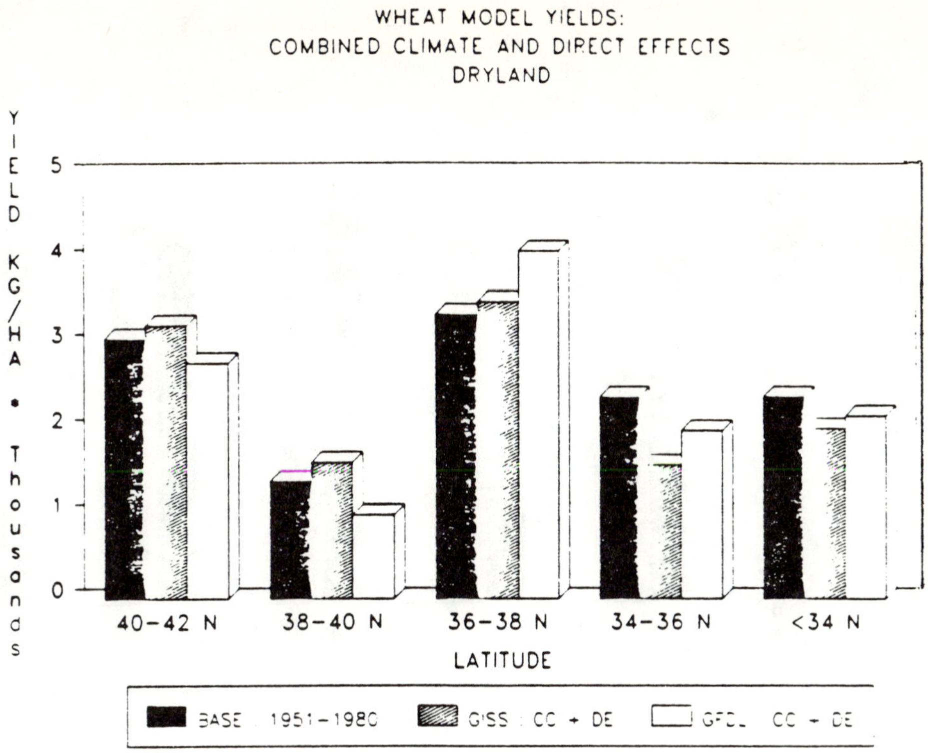

b)

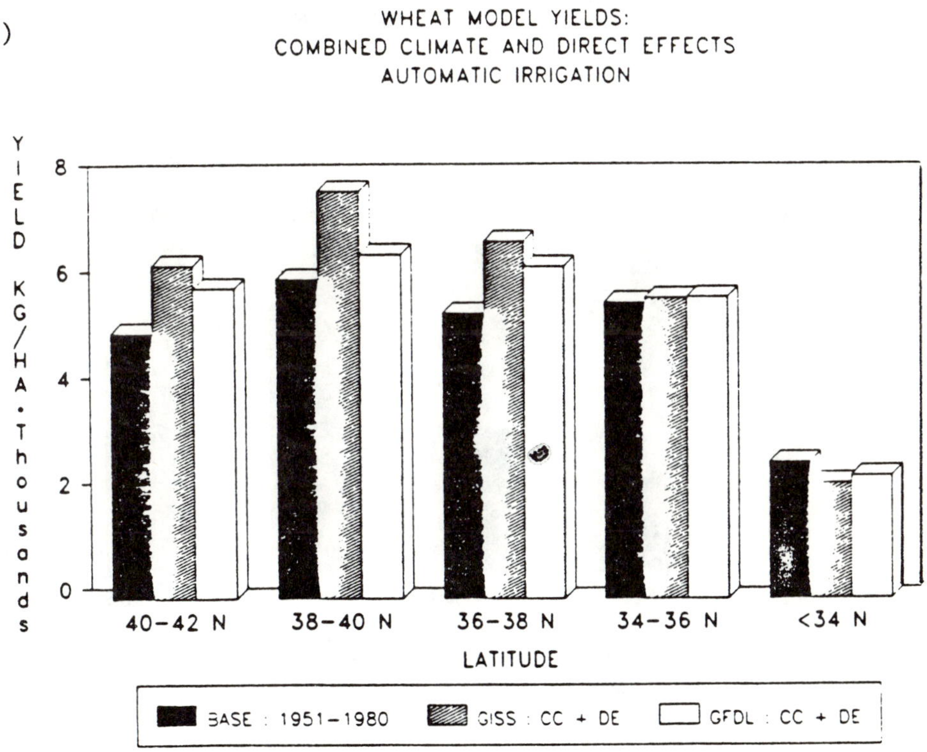

Figure 7-4. CERES-Wheat yields with climate change scenarios alone: (a) dryland (b) irrigated

Source: Rosenzweig, Volume __.

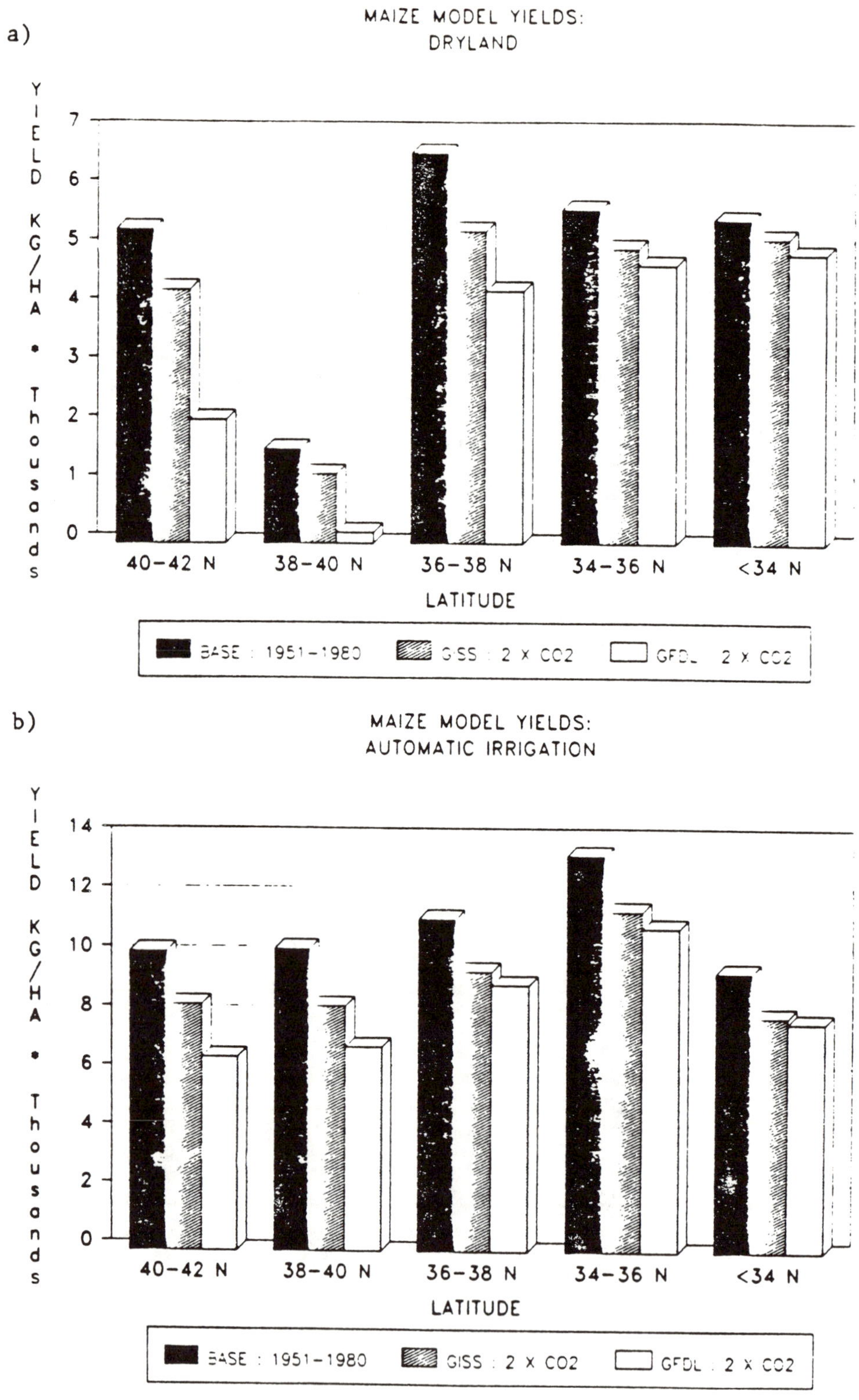

Figure 7-5. CERES-Maize yields with climate change and direct CO_2 effect: (a) dryland (b) irrigated

Source: Rosenzweig, Volume __.

climate change into economic consequences (see Chapter 10, Agriculture, for study design and limitations). Analyses were done both for climate change alone and for the combined effects of climate change and enhanced CO_2 concentrations. This analysis did not address the issues of whether the physical and institutional changes required to accommodate increased demand for irrigated acreage are feasible.

Regional Results

The estimates of Adams et al. (see Volume _) for total agricultural and irrigated acreage changes in the southern Great Plains States (Oklahoma and Texas only) are shown in Table 7-4. Agricultural land decreases in the southern Great Plains in all scenarios, both GISS and GFDL with and without the direct effects of CO_2. Decreases range from -22 to -4%. Irrigated acreage on the other hand increases in all scenarios, from 9 to 30%. This is because of increased stability in irrigated yields relative to dryland yields and a rise in commodity prices that makes expansion of irrigation production economically feasible. The higher limits of these ranges indicate the possibility of significant changes in the agriculture of the region.

Regional Implications

The results of the agricultural economics study imply that wheat and corn production may shift away from the southern Great Plains, causing dislocations of rural populations. For many rural communities in the region, this may further weaken an economic base already under pressure from long-term structural changes in U.S. agriculture.

If irrigated acreage does increase in the area, groundwater overdrafts also would be likely, along with associated increases in surface and groundwater pollution and other forms of environmental degradation.

Demand for Water for Irrigation

The projected higher air temperatures due to the growth of trace gases in the atmosphere may increase evaporative demands. Crop water use and irrigation water requirements are governed largely by the evaporation process.

Study Design

Allen and Gichuki (see Volume _) evaluated the effects of climate change and variations in canopy resistance on crop irrigation water requirements in the Great Plains. Allen and

Table 7-4. Changes in Land Usage in the Southern Plains (Millions of acres)

		GISS			GFDL		
	Base Acreage	Acreage	Change	%Change	Acreage	Change	%Change
NET LAND USE							
Without direct effects	54.7	42.6	-12.1	-22.1	52.0	-2.7	-4.9
With direct effects	54.7	48.8	-10.9	-19.9	52.7	-2.0	-3.8
IRRIGATED ACREAGE							
Without direct effects	5.3	6.9	1.6	+29.6	5.6	0.3	+4.9
With direct effects	5.3	5.8	0.5	+9.4	6.1	0.8	+15.3

Source: Adams et al., Volume __.

Gichuki used a soil water balance-evapotranspiration model to calculate daily soil moisture balances, evapotranspiration, and irrigation water requirements. The model employed the Penman-Monteith combination method to estimate crop evapotranspiration for corn, wheat, and alfalfa (Monteith, 1965). The model was run for four levels of potential direct effects of CO_2 on transpiration.

Limitations

Air temperature data at airport locations were adjusted to reflect temperature profiles to be expected over an irrigated crop field. This study assumed that alfalfa, corn, and wheat all would respond similarly to increased CO_2 (which may reduce transpiration), although published reports of experimental results show different responses among crops (see Rose, Volume _).

Results

The results indicate that irrigation requirements may increase or decrease, depending on the type of crop, changes in climatic factors and variations in canopy resistance. The perennial crop alfalfa showed persistent increases in seasonal net irrigation water requirements (see Figure 7-6). These increases were driven primarily by higher temperatures, with less influence from stronger winds, greater solar radiation, and a longer growing season.

On the other hand, decreases in seasonal net irrigation requirements were predicted for winter wheat and corn in most regions, depending on the projected direct effects of CO_2 on transpiration. These modeled decreases were generally due to shorter seasons caused by higher temperatures, which accelerates crop life cycles. When crop varieties appropriate to the longer growing season were modeled, irrigation requirements for winter wheat increased. Irrigation water requirements during peak periods increased in almost all cases (Figure 7-7).

Plant canopy (leaf) temperatures were predicted to increase above current baseline values for all crops and sites studied. Projected increases in leaf temperatures may reduce photosynthetic activity and crop yields. (See discussion on direct effects of CO_2 in Chapter 10, Agriculture.)

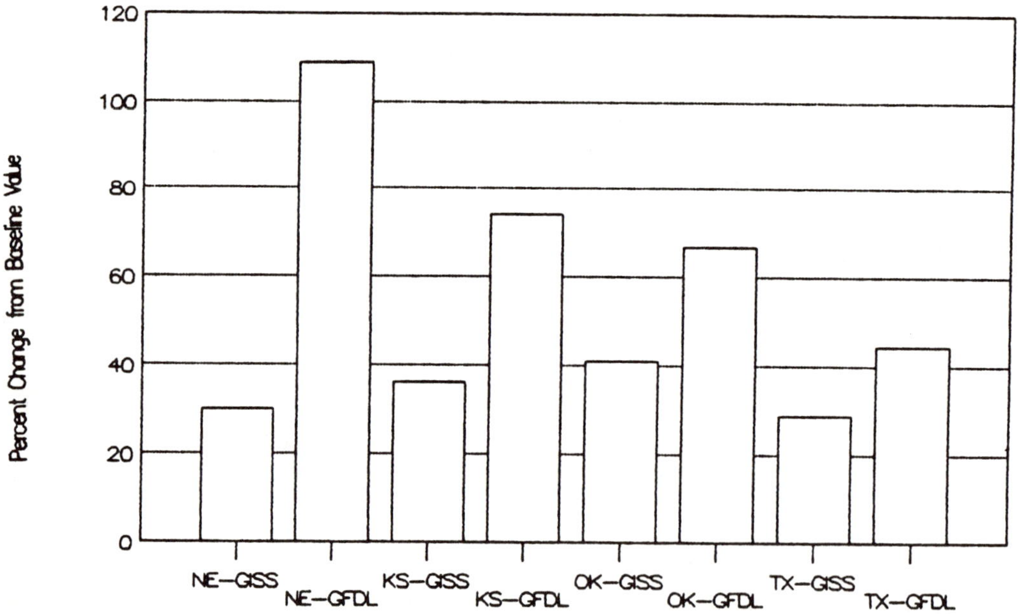

Figure 7-6. Seasonal irrigational requirement for 20% increase in bulk stomatal resistance of alfalfa.

Source: Allen and Gichuki, Volume __.

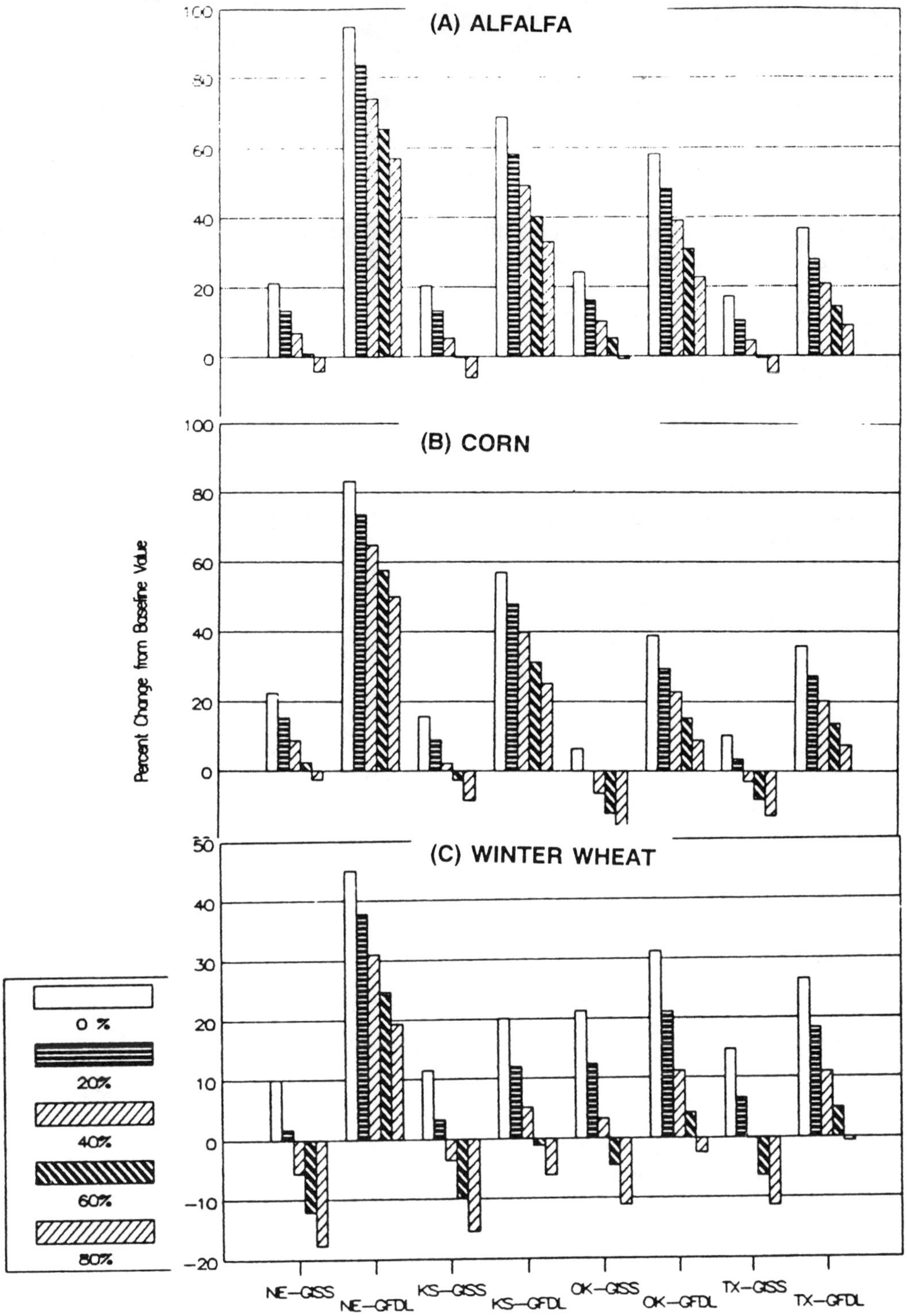

Figure 7-7. Percent change in net peak monthly irrigation requirement from baseline values for alfalfa, corn and winter wheat vs. postulated increases in bulk stomatal diffusion resistance.

Source: Allen and Gichuki, Volume __.

Implications

Any reduction in irrigation requirements for corn and winter wheat would be beneficial because less water and energy would be required to produce the crops. However, the shortened crop durations might allow for longer-season varieties or double-cropping (planting two crops in one season), increasing total irrigation requirements.

Expanded farm irrigation systems may require increased capital investments and larger peak drafts on groundwater systems and on energy supplies. Increased groundwater extraction would pose serious environmental and economic problems, especially where "water mining" is currently a major problem. Any increases in irrigation efficiency by irrigators as an attempt to cope with projected water shortages may lead to increased salinity problems if leaching requirements are not met.

Water Quality

Agricultural pesticides are a high-priority pollution problem in at least half of the States within the U.S. Great Plains and Central Prairie. How the movements of agricultural chemicals may be affected by climate change is essential to a thorough evaluation of potential changes in drinking water quality.

Study Design

Johnson et al. used the Pesticide Root Zone Model (PRZM) to (Carsel et al., 1984) to simulate the partitioning of pesticides between plant uptake, chemical degradation, surface runoff, surface erosion, and soil leaching in the Great Plains under baseline climate and climate change scenarios. The locations modeled were representative of cropping practices for winter wheat and cotton in the region. (See the national agriculture chapter of this report for further discussion of the study's design and limitations.)

Results

As Figure 7-8 shows, in the winter wheat regions of the central Great Plains, surface runoff and surface erosion increase under the GISS scenario, and results are mixed in the GFDL scenario. In the southern Great Plains cotton simulations, the GISS and GFDL scenarios produced increases in surface pesticide losses with runoff and eroded soils.

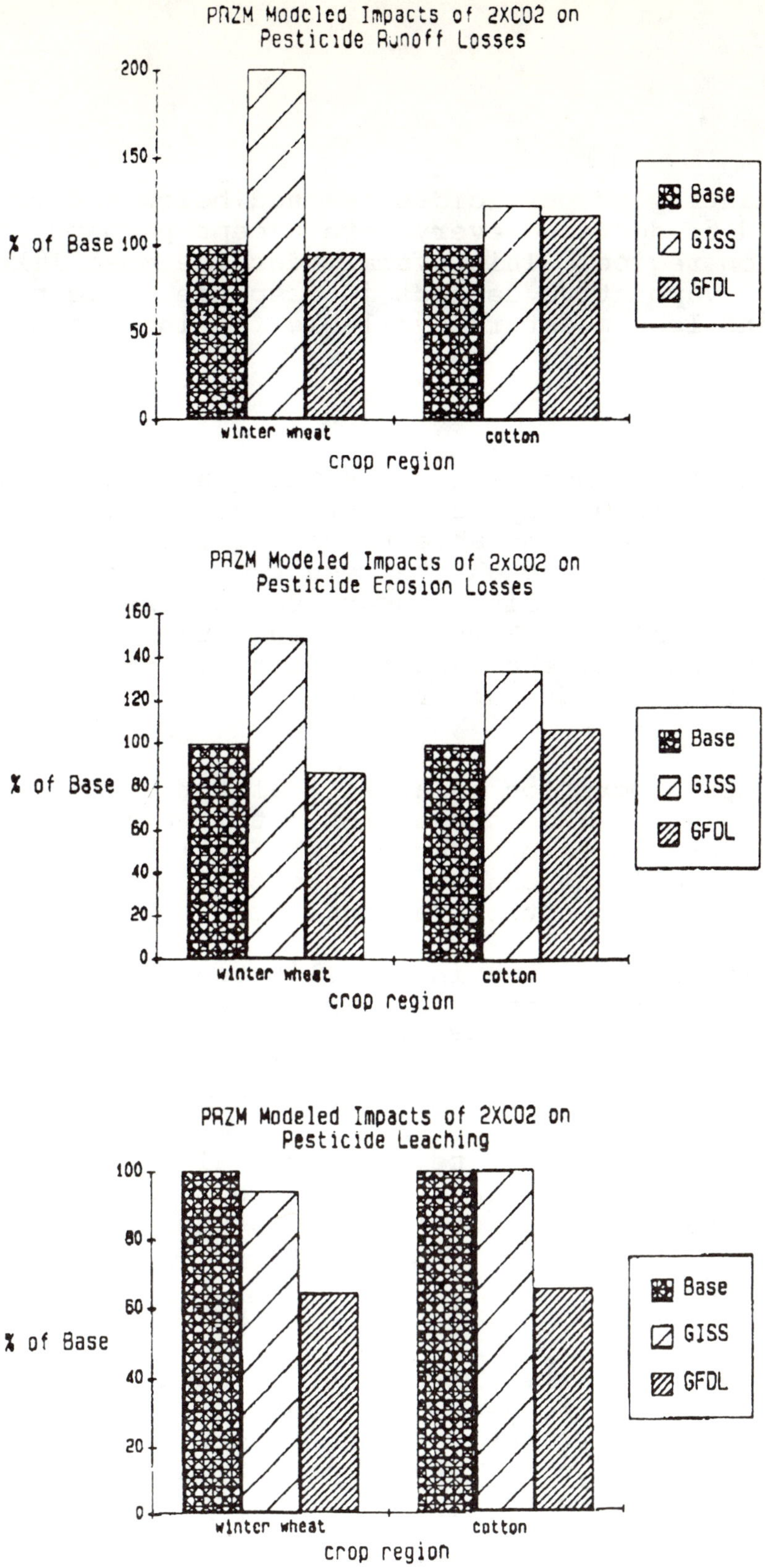

Figure 7-8. Regional summary of surface and subsurface pesticide loss as a percentage of the base climate scenario losses.

Source: Johnson et al., Volume __.

The quantity of pesticides leached below the crop root zone is estimated to decrease everywhere except perhaps on silty soils in the cotton region. This overall decline most likely results from higher evaporative demands in response to temperature increases and less moisture available for infiltration and deep percolation.

Implications

Results of the modeling imply that water quality in the Southern Great Plains may be affected by climate change. Surface water appears to be vulnerable to deterioration under climate change conditions, although the result does not hold for all cases. Groundwater quality appears to be less at risk than surface water quality. However, because these results are highly dependent on the frequency and intensity of precipitation events, directions of change are uncertain.

Livestock

Livestock production is a critical agricultural activity in the Great Plains. Since the approximately 8 million head of beef cattle in Texas represent nearly 40% of all beef cattle in the United States, climate change factors that affect Texas cattle will affect beef production nationally.

The warming predicted in the climate change scenarios may alleviate cold stress conditions in the winter but will most likely exacerbate heat stress in the summer. Warmer summers may necessitate more hours of indoor cooling. Reproductive capabilities may also decline because of higher temperatures. (See the Agriculture chapter in this report for a discussion of livestock issues.) Higher temperatures also may enable tropical diseases to extend their ranges into the Great Plains. High temperatures may also reduce insect pest activities in some locations and increase them in others.

Energy

Linder and Inglis estimated the national changes in demand for electricity for the years 2010 and 2055. (See the Energy chapter of this report for a further description of study's design and methodology.) They first estimated the change in electricity demand due to projected growth in GNP and population, and then factored in changes in demand based on the GISS transient climate change scenarios A and B. The results for the southern and central Great Plains are discussed here.

Table 7-5. Estimated Change in Peak Demand and Annual Energy Requirements Induced by Climate Change (%)

	2010				2055	
	GISS A		GISS B		GISS A	
Utility Area	Ann.	Peak	Ann.	Peak	Ann.	Peak
Kansas/Nebraska	1.7	6.8	1.3	5.2	5.7	22.1
Oklahoma	3.0	7.9	2.8	6.6	11.3	25.3
Texas, east	3.0	7.9	2.8	6.6	11.3	25.3
Texas, south	3.3	10.0	1.7	5.1	10.6	24.6
Texas, west	3.1	8.6	2.4	6.1	11.1	25.1

Source: Linder and Inglis, Volume __.

Results

Estimates of changes in peak demand, capacity requirements, and cumulative and annual costs projected for the climate change scenarios are shown in Table 7-5. New capacity requirements are estimated to increase by 15-28% by 2010 for the climate change scenarios compared to the base case. By 2010, cumulative capital costs induced by climate change may grow by $3.7-6.7 billion, and annual costs may rise by 3-6%.

In 2055, new capacity generating requirements are estimated to increase by 22 to 45 gigawatts or 27-39%. Power generation in the region increased by about 10-14% under the climate change scenarios.

Linder calculated that cumulative capital costs for energy in the region would increase by $20-53 billion with climate change. The estimated changes in annual costs induced by climate change range from $5 to 10 billion -- nearly the same as the $11 billion difference in high and low GNP estimate.

Implications

Increased energy capacity requirements and the need to maintain the reliability of utility systems could place additional stress on the Great Plains. Power plants may take the cooling water they need from rivers or from the already overused Ogallala Aquifer. Fossil fuel plants could add more pollutants to the air.

POLICY IMPLICATIONS

Policymakers may have to respond to a decline of agricultural production in the area, increased demand for water, and poorer water quality. The major issues for policymakers to address include land-use management, water resource management, water quality regulations, and risk management (see Riebsame, Volume _). Regional utility planners and policymakers should also begin to consider climate change as a factor -- along with other uncertainties -- affecting their resource availability analyses and planning decisions.

Land-Use Management

State land managers should analyze how their missions and holdings may be affected by climate change and should develop flexible strategies to deal with potential impacts. Federal agencies, such as the Department of Agriculture, the Forest

Service, the Fish and Wildlife Service, and the Department of Interior, should work with State agriculture, forest, and park agencies on such plans.

Agriculture and other uses may shift across several States. This adjustment process can be made more efficient and less disruptive if individual jurisdictions, such as municipalities, States, and federal regions respond in a coordinated manner. Decisions made by managers of agriculture will affect decisions on forests, wildlife, water resources, and other areas. Decisionmakers should work together on developing sound and flexible anticipatory strategies, and new institutional arrangements may be needed.

Some programs already in place can help to lessen the negative effects of climate change on the Great Plains. Federal legislation such as the "Sod-Buster Bill" and the related "Swamp-Buster Bill," and the Conservation Reserve Program, are examples of new policies designed to reduce the use of marginal lands for agriculture. These laws have the basic goals of protecting the most erodible farmlands by removing them from crop production and of using conservation as a tool for reducing overproduction. These policies are prudent now for reducing runoff and erosion, and may become even more important for protecting water quality in light of climate change. Protection of marginal lands may have to be weighed against the need for greater crop production if climate change lowers yields.

Water Resource Management

With climate change, parts of the Great Plains are likely to suffer increasing aridity. Farmers may demand more water for irrigation, although groundwater sources are already taxed. Competition for water resources between agricultural and nonagricultural demands may be exacerbated. Managers need to factor these potential effects of climate change into their decisions on irrigation, drainage, and water transfer systems and as they formulate supply allocation rules, reservoir operating criteria, safety protocols, and plans for long-term water development. Water conservation techniques, water reallocation between competing uses, water transfers and markets, and land-use adjustments must be evaluated for their ability to absorb the effects of future climate change. Primary consideration should be given to environmentally sound programs for managing water resources.

Water Quality Regulations

Decisionmakers should consider the potential effects of climate change on water quality and the use of pesticides. They should examine alternative pest control strategies, such as Integrated Pest Management, which uses such techniques as biological control, genetic resistance, and innovative cropping systems to reduce pesticide applications.

Risk Management

Several government, semi-private and private institutions have a large financial stake in Great Plains agriculture through land credit, commodity and equipment loans, and insurance. Additionally, the federal government provides disaster relief for climate extremes affecting regional agriculture. Climate warming poses a potential long-term risk to the financial institutions supporting agriculture, the resources available for emergency relief, and to individual farmers. This consideration should be carefully assessed and the risk monitored as climate changes.

REFERENCES

Carsel, R.F., C.N. Smith, L.A. Mulkey, J.D. Dean, and P. Jowise. 1984. Users Manual for Pesticide Root Zone Model: PRZM. USEPA/ERL Report EPA-600-3-84-109. Washington, DC: U.S. Government Printing Office.

Glantz, M.H., and Ausubel, J.H. 1984. The Ogallala Aquifer and carbon dioxide: Comparison and convergence. Environ. Conser. 11(2):123-31.

High Plains Associates. 1982. Six-State High Plains-Ogallala Aquifer Regional Resources Study. Austin, Texas.

Hurt, R.D. 1981. The Dust Bowl. Chicago: Nelson-Hall.

Jones, C.A., and J.R. Kiniry. 1986. CERES-Maize: A Simulation Model of Maize Growth and Development. College Station, TX: Texas A&M Press.

Liverman, D.M., W.H. Terjung, J.T. Hayes, and L.O. Mearns. 1986. Climatic change and grain corn yields in the North American Great Plains. Climatic Change 9:327-347.

Lockeretz, W. 1978. The lessons of the Dust Bowl. American Scientist 66:560-569.

Monteith, J.L. 1965. Radiation and crops. Experimental Agriculture Review 1(4):241-251.

New York Times. August 12, 1988. Drought cutting U.S. grain crop 31% this year. By K. Schneider. p. A1.

Riebsame, W.E. 1983. Managing agricultural drought: The Great Plains experience. In: Platt, R., and G. Macinko, eds. Beyond the Urban Fringe: Land Use Issues in Non-Metropolitan America. Minneapolis: University of Minnesota Press, pp. 257-270.

Riebsame, W.E. 1987. Human Transformation of the United States Great Plains: Patterns and Causes. Proc. Symp on The Earth as Transformed by Human Action, Clark University.

Ritchie, J.T., and S. Otter. 1985. Description and performance of CERES-Wheat: A user-oriented wheat yield model. In: Willis, W.O., ed. ARS Wheat Yield Project. USDA-ARS. ARS - 38. pp. 159-175.

Robertson, T., V.W. Benson, J.R. Williams, C.A. Jones, and J.R. Kiniry. 1987. Impacts of climate change on yields and erosion for selected crops in the southern United States. In: Meo, M., ed. Proc. Symp. on Climate Change in the Southern U.S.: Impacts and Present Policy Issues, Science and Public Prog., Univ. of Oklahoma, Norman, OK.

Rosenzweig, C. 1985. Potential CO_2-induced climate effects on North American wheat-producing regions. Climatic Change 7:367-389.

Rosenzweig, C. 1987. Climate change impact on wheat: The case study of the High Plains. In: Meo, M., ed., Proc. Symp. on Climate Change in the Southern U.S.: Impacts and Present Policy Issues, Science and Public Prog., Univ. of Oklahoma, Norman, OK.

Terjung, W.H., D.M. Liverman, J.T. Hayes et al. 1984. Climatic change and water requirements for grain corn in the North American Great Plains. Climatic Change 6:193-220.

Warrick, R.A. 1984. The possible impacts on wheat production of a recurrence of the 1930s drought in the U.S. Great Plains. Climatic Change 6:5-26.

Warrick, R.A., and M.J. Bowden. 1981. The changing impacts of drought in the Great Plains. In: Lawson, M.P., and M.E. Baker, eds. The Great Plains: Perspectives and Prospects. Lincoln: University of Nebraska Press, pp. 111-137.

Worster, D. 1979. Dust Bowl: The Southern Great Plains in the 1930s. New York: Oxford University Press.

CHAPTER 8

WATER RESOURCES

FINDINGS

Many regions of the country are likely to experience changes in water availability and water quality. The frequency of droughts and floods may change. Under some scenarios, summer soil moisture and water availability in middle latitudes may be reduced due to temperature increase, a northward shift of the rainbelt, and an earlier onset of winter snowbelt and spring runoff. Even under warm-wet scenarios, in many regions the positive effect on supply of increased precipitation may be more than offset by temperature increases that reduce supply by evaporation and increase water demand.

- A warmer and drier climate in some regions would decrease water supplies and increase water demand, especially demand for irrigation and electric power production.

- Lower river flows resulting from drier conditions could negatively affect hydropower production, navigation, aquatic ecosystems, wildlife habitat, and recreation.

- Lower stream flow and lower lake levels could cause power plants to locate in coastal areas to obtain a water source that is reliable and that may be used without violation of thermal restrictions. This would have important implications for land use, transmission lines, and the costs of power.

- Municipal and industrial water users may have greater incentives to purchase water rights in the West to ensure the reliability and quality of supply.

- Because of earlier runoff and potentially higher evaporation, increased rainfall may not increase water availability in all regions. However, flooding may increase. Flood-prone regions may require structural modifications, new capital improvements, and modifications in operating rules for reservoirs.

Policy Implications

- Changes in water availability and water quality and conflicts among water users are the major impacts of climate change on water resources, although the magnitude and distribution of hydrologic and water resource effects are still uncertain. Responses could include the following:

-- build new storage capacity, provided that the structures show positive net benefits under a variety of possible climatic conditions;

-- modify water system operations to improve performance under extreme conditions, to enhance recovery from extreme conditions, and to accept greater risk to low-valued uses in order to protect high-valued uses; and

-- encourage a reduction in water demand and an increase in water use efficiency through conservation, water markets, water quality control, contingency planning, and coordinated uses of regional and interstate water resources, provided that such measures do not reduce the performance and recovery capabilities of supply systems.

CLIMATE, HYDROLOGY, AND WATER RESOURCES IN THE UNITED STATES

The United States is endowed with a bountiful supply of water, but the water is not always in the right place at the right time or of the right quality.

On the average, 4,200 billion gallons per day (bgd) of precipitation fall on the lower 48 States. However, a large portion of this water evaporates, leaving only 1,435 bgd to run off into surface water and groundwater.

Surface and groundwaters are managed by controlling and diverting flows through impoundments and aqueducts; by withdrawing water for such "off-stream" uses as irrigation and municipal use; by regulating flows to maintain "in-stream" water quality and such uses as navigation, hydropower, and recreation, and by controlling flows under flood conditions to avoid loss of life, damage to property, or inconvenience to the public. Water may be "withdrawn" and returned to the source more than once, or "consumed" and not returned to the source.

Control and use of water are costly. Water supply for municipal and industrial use represented a $108 billion national investment in infrastructure in 1984 (National Council on Public Works Improvement, 1988). Government agencies and industries spent $336 billion (in constant 1982 dollars) from 1972 to 1985 (Farber and Rutledge, 1987) on water pollution abatement and control activities. In other areas, excess water periodically floods agricultural and urban areas, causing an average of $3 billion (in constant 1984 dollars) (National Council on Public Works Improvement. 1988) per year in damages during the past decade.

Most water resource decisions in the past have been based on the assumption that the climate of a region varies around a stationary mean. Even without climate change, the ability to use water resources is greatly influenced by weather variability. As shown in Figure 8-1, weather controls hydrologic conditions through precipitation (mean and frequency), runoff, snowmelt, transpiration and evaporation, soil moisture, and the variability of storms and drought.

Water Availability

On a national scale, water supplies are adequate. Largely due to weather variability, only 675 bgd of the 1,435 bgd of runoff water in the conterminous United States is considered available in 95 out of 100 years for irrigation, municipal, industrial, and other off-stream uses. In 1985, freshwater withdrawals for off-stream uses were 338 bgd. Ninety-two bgd of the withdrawals were consumed, mostly for irrigation.

The supply of water from surface and groundwater sources for soil moisture, in-stream uses, withdrawal, and consumption is vulnerable to decreased precipitation. Warmer temperatures would increase the proportion of rain versus snow, would cause earlier runoff of winter snowpacks, and would increase evaporation. These effects would diminish water availability during the summer and reduce water levels and flows.

Although water availability exceeds withdrawals and consumption, the water is not always in the right place at the right time. In some regions, the gap between demand for water and available supply is narrow. For example, average surface water withdrawal exceeds average streamflow in the Great Basin, Rio Grande, and Colorado River Basins. In these water-short basins, off-stream water uses often conflict with in-stream uses, such as recreation and maintenance of environmental quality. Degraded water quality further limits water availability in many regions.

Water Uses

The uses primarily affected by water availability constraints are irrigation, thermoelectric power, and industrial and domestic water use. Withdrawals and consumption of freshwater by major offstream uses in 1985 are summarized in

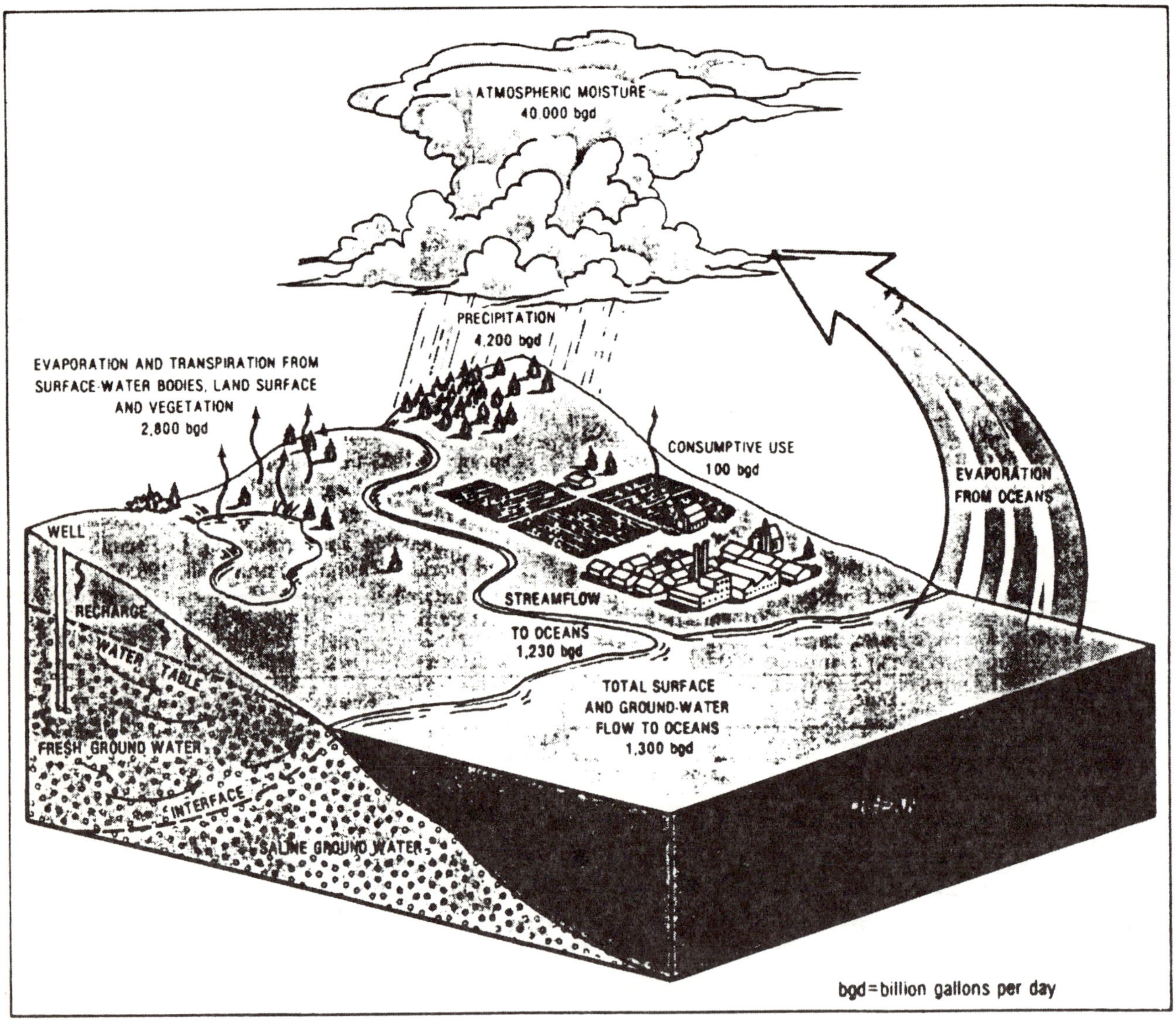

Figure 8-1. Hydrologic cycle showing the gross water budget of the conterminous United States.

Source: Langbein et al. (1949) and Solley et al. (1983).

As the largest and relatively least valuable use of freshwater, irrigation accounts for 42% of freshwater withdrawals and 82% of fresh water consumption in the United States. Although irrigated land comprises about 10% of harvested cropland acreage nationwide, it contributes 30% of the value of cropland production. Many of these crops are fruits, vegetables, and specialty crops (U.S. Water Resources Council, 1978; Bajwa et al., 1987). The 17 Western states account for 85% of the irrigated lands in the country (Bajwa et al., 1987).

In contrast, thermoelectric powerplants withdraw almost as much freshwater as irrigation but consume much less than irrigation. Although 131 bgd of freshwater are withdrawn to produce 60% of the nation's electricity (Water Resources Council, 1978), only 4.35 bgd are actually consumed (Solley et al., 1988).

Domestic uses account for 10% of total water withdrawn and 11% of consumption. Over the past 20 years, domestic water use has increased from 16 to 25 bgd owing to growth in the number of households, with little change in usage per household (Solley et al., 1988).

Water Quality

Degraded water quality reduces the amount of water for human use. The Nation has made significant progress in cleaning up water pollution since the enactment of the 1972 Clean Water Act. However, some of the States reported that some persistent pollution problems remain, especially contamination by toxic pollutants and nonpoint source pollution (USEPA, 1987b). Water quality in the nation's rivers, lakes, and streams depends in part on water quantity. Water supply is needed for dilution of wastewaters that flow into surface and groundwater sources. Freshwater inflows are needed to repel saline waters in estuaries and to regulate water temperatures in order to forestall changes in the thermal stratification, water quality, aquatic biota, and ecosystems of lakes, streams, and rivers.

Flood Control

Major water management systems and structures, which are designed to reduce hazards and damages from major floods, may have the flexibility to adapt to modified flood conditions. The major concern with existing dams and levees is the consequence of failure under extreme conditions.

For instance, an increased probability of great floods, whether due to urbanization of upstream watersheds or climate

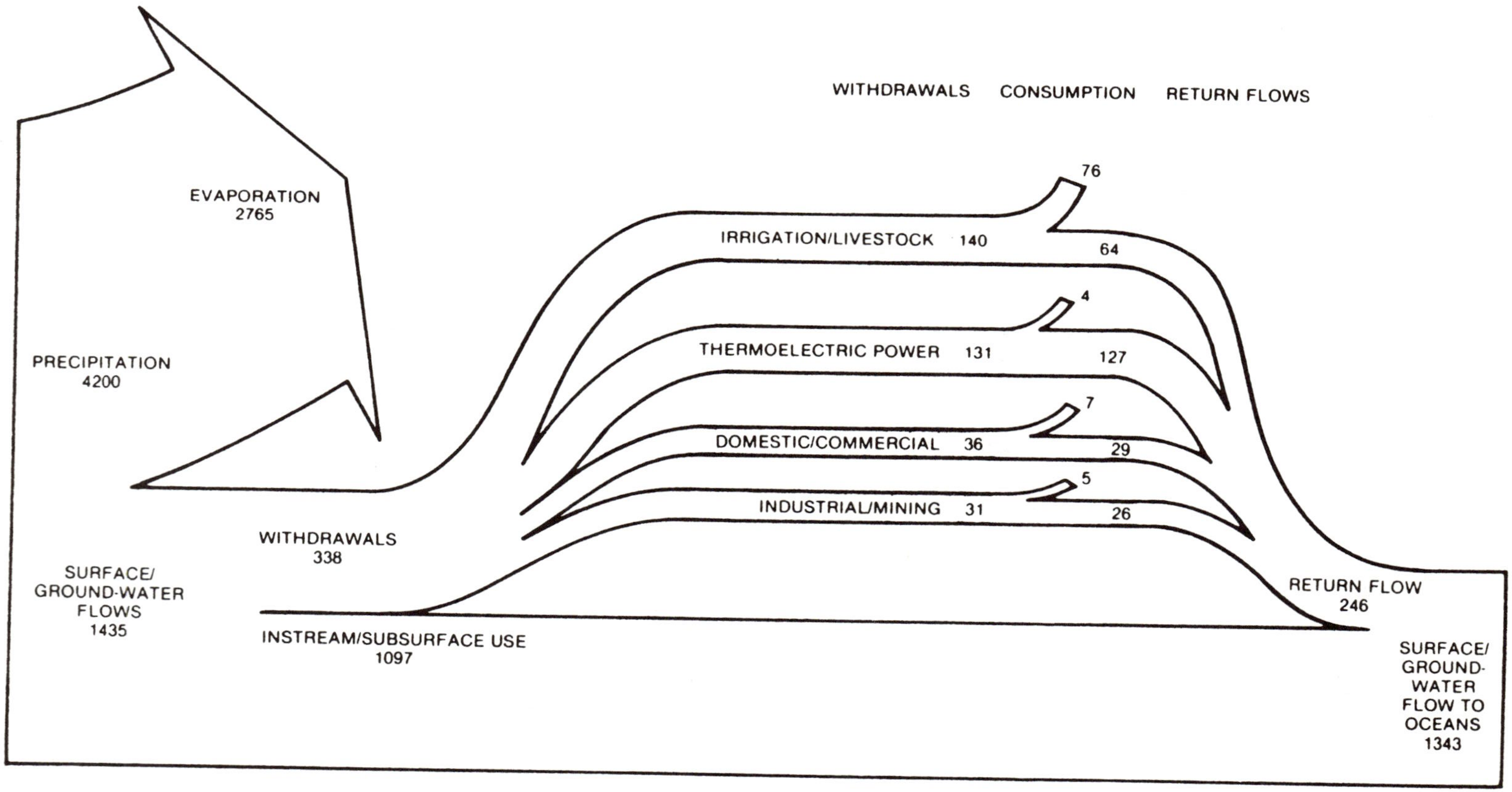

Figure 8-2. Water withdrawals and consumption by offstream uses, coterminous United States, 1985. Source: Solley et al. (1988).

change, would cause dams with inadequate spillways to fail. Some major structures may be currently vulnerable to a "probable maximum flood" (National Research Council, 1986). Smaller structures, such as urban drainage culverts and sewers and local flood protection projects, are currently more susceptible to floods and are in poorer condition than large structures (National Council on Public Works Improvement, 1988).

REGIONAL IMPACTS OF CLIMATE CHANGE

The EPA regional studies reported in this document use scenarios generated from up to four GCMs as their starting points (see Chapter 3) and match them with regional water resource models. This section combines the results from this report's EPA studies on California, the Great Plains, the Great Lakes, and the Southeast with previous studies of the impacts of climate change on these and other regions. The EPA studies are listed in Table 8-1.

Climate impacts may be greatest on regions already stressed: where water demand is great and supply is short, where water quality has deteriorated, or where periodic excess supply floods agricultural and urban lands. Table 8-2 summarizes the current status of water supply by major river basin region. The regions are delineated in Figure 8-3.

West

The arid and semiarid river basins west of the Mississippi show up as problem areas in all four of the U.S. Water Resources Council's water quantity and quality maps. Total water use exceeds average streamflow in 24 of 53 western water resource regions (U.S. Water Resources Council, 1978). The majority of the West's water withdrawals go to irrigation. Surface and groundwater quality in the West have deteriorated due to low flow, salts concentrated by irrigation, and pesticide use. The West also depends upon nonrenewable groundwater supplies for irrigation (Solley et al., 1988).

Climate change may exacerbate water shortage and quality problems in the West. Some GCM scenarios predict midsummer drought and heat, less groundwater recharge, and less groundwater and surface water available for irrigation in the middle latitudes of the country. The sensitivity analyses conducted by Stockton and Boggess (1979) indicated that the warmer and drier

Table 8-1. Water Resource Studies

Region	Study
California	
	o <u>Interpretation of Hydrologic Effects of Climate Change in the Sacramento-San Joaquin River Basin, California</u> - Lettenmaier, University of Washington.
	o <u>Methods for Evaluating the Potential Impact of Global Climate Change</u> - Sheer and Randall, Water Resources Management, Inc.
	o <u>The Impacts of Climate Change on the Salinity of San Francisco Bay</u> - Williams, Philip Williams & Associates
Great Lakes	
	o <u>Effects of Climate Changes on the Laurentian Great Lakes Levels</u> - Croley, Great Lakes Environment Research Lab.
	o <u>Impact of Global Warming on Great Lakes Ice Cycles</u> - Assel, Great Lakes Environment Research Lab.
	o <u>The Effects of Climate Warming on Lake Erie Water Quality</u> - Blumberg and DiToro, HydroQual, Inc.
	o <u>Potential Climatic Changes to the Lake Michigan Thermal Structure</u> - McCormick, Great Lakes Environment Research Lab.
Great Plains	
	o <u>Effects of Projected CO_2 - Induced Climate Changes on Irrigation Water Requirements in the Great Plain States</u> - Allen and Gichuki, Utah State University.
Southeast	
	o <u>Potential Impacts of Climatic Change on the Tennessee Valley Authority Reservoir System</u> - Miller and Brock, Tennessee Valley Authority.

Chapter 8

- o <u>Impacts on Runoff in the Upper Chattahoochee River Basin</u> - Hains, C.F. Hydrologist, Inc.
- o <u>Methods for Evaluating the Potential Impact of Global Climate Change</u> - Sheer and Randall, Water Resources Management, Inc.

Table 8-2. Current Status of Water Supply

Water-Resources Region	Average Renewable Supply in billion gallons per day	1985 Withdrawals as a percentage of Average Renewable Supply	1985 Consumption as a percentage of Average Renewable Supply	Reservoir Storage as a percentage of Average Renewable Supply	Streamflow Exceeded 95% of time as a percentage of Average Streamflow	Percentage of Groundwater Overdraft
New England	78.4	11.7	9	15	62.4	0
Mid-Atlantic	80.7	29.5	21	11	62.2	1.2
South Atlantic-Gulf	233.5	13.5	21	15	55.5	6.2
Great Lakes	74.3	42.9	29	8	62.6	2.2
Ohio (exclusive of Tennessee Region)	139.5	22.3	15	13	59.0	0
Tennessee	41.2	22.3	8	24	77.0	0
Upper Mississippi (exclusive of Missouri Region)	77.2	21.9	23	14	54.0	0
Mississippi (entire basin)	464.3	3.7	12	32	46.7	0.5
Souris-Red Rainy	6.5	4.3	19	110	32.1	0
Missouri	62.5	55.2	203	120	40.7	24.6
Arkansas-White-Red	68.6	22.3	117	41	36.5	61.7
Texas-Gulf	33.1	41.4	182	67	27.5	77.2
Rio Grande	5.1	109.8	437	182	33.3	28.1
Upper Colorado	14.7	51.4	163	229	39.0	0
Colorado (entire basin)	15.6	47.4	272	403	75.0	48.2
Great Basin	9.9	81.8	364	30	50.0	41.5
Pacific Northwest	276.2	12.9	45	20	70.7	8.5
California	70.2	53.6	299	49	44.0	11.5
Alaska	975.5	0.4	0	0.1	78.5	0
Hawaii	7.4	17.2	18	0.1	60.3	0
Caribbean	5.1	11.9	32	5	35.6	5.1

Average renewable supply is defined as the average flow potentially or theoretically available for use in the region.
Data from U.S. Water Resources Council, 1978; U.S. Geological Survey, 1983; Solley et al., 1988.

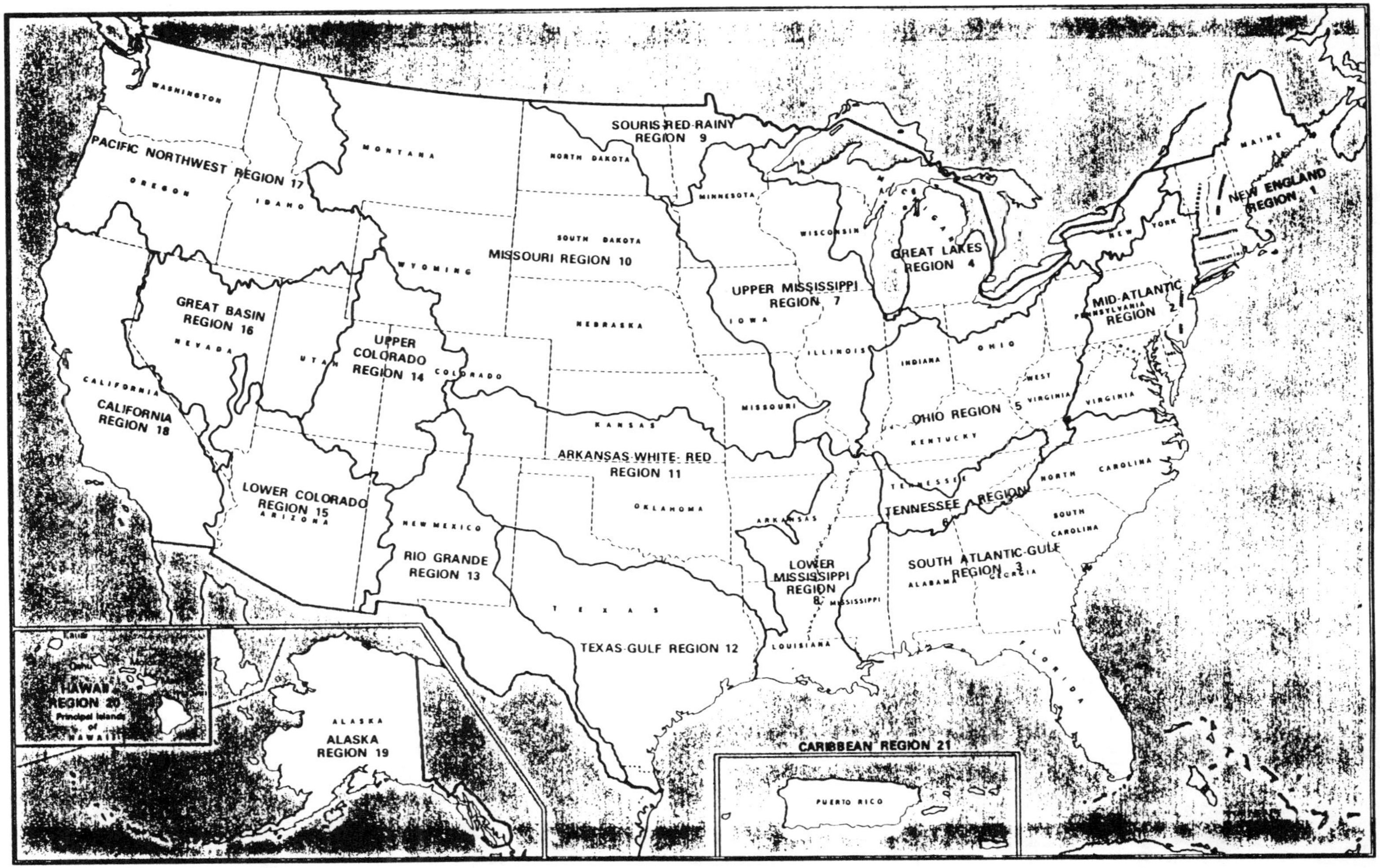

Figure 8-3. Water Resources Regions climate
Source: USGS, 1985

scenarios may have the greatest impacts in arid western river basins (Rio Grande, Colorado, Missouri, California) by increasing water shortages.

Revelle and Waggoner (1983) showed that warmer air temperatures and slight decreases in precipitation could severely reduce both the quantity and quality of western water resources. Water shortages and associated conflicts between in-stream and off-stream uses, between agricultural and urban/industrial water uses, and between flood control and other water uses of reservoirs may be expected. Hydropower output also may decline due to lower riverflow.

Pacific Northwest

Climate change may alter the timing and volume of precipitation, increasing the risk of flooding, changing reservoir management practices, and affecting the output and reliability of hydroelectric power production and the availability of water for irrigation. Butcher et al. (1986) found that competition for water for irrigation, hydropower, and fisheries habitats is increasing in the Pacific Northwest.

California

The diversion of water from water-rich northern parts of the State and from the Colorado River to southern California via Federal and State systems of dams, aqueducts, and pumping stations has transformed California into the nation's leading agricultural state and has made possible the urbanization of southern California. Irrigation accounted for 83% of the total value of agricultural output in the State in 1982 (Bajwa et al., 1987). Because of this high economic dependence on water in an arid area, southern California is vulnerable to droughts and altered runoff patterns that may be caused by climate change.

Total annual runoff from the mountains surrounding the Central Valley is estimated to increase slightly under GCM scenarios, but runoff in the late spring and summer may be much less than today because of higher temperatures. The volume of water from the State Water Project may decrease by 7-16% (see Chapter IV). Current reservoirs do not have the capacity to increase storage of winter runoff and at the same time retain flood control capabilities. In addition, freshwater requirements to repel saline water near the major freshwater pumping facilities in the upper Sacramento-San Joaquin River Delta may have to be doubled under sea level rise scenarios, further reducing water available to southern California.

Decreases in water availability due to climate change may also cut hydroelectric power produced in California. In the 1976-77 drought, hydroelectric production in northern California dropped to less than 50% of normal, a deficiency relieved by importing surplus power from the Pacific Northwest and by burning additional fossil fuels at an approximate cost of a half billion dollars (Gleick, 1988).

Colorado, Rio Grande, and Great Basin River Basins

Total consumption is more than 40% of renewable supply in these river basins. The Colorado River Basin has huge reservoir storage, but demand exceeds supply in the lower half of the basin.

Ordinarily all of the Colorado River's water is consumed before it reaches the Gulf of California in Mexico. The Colorado River Compact of 1922, the 1963 Supreme Court decision in Arizona v. California, 1944 and 1973 treaties with Mexico, and other agreements define the allocation of Colorado River water to seven States and Mexico (Dracup, 1977). Some studies show that the Upper Colorado region will use all of its allocation by the year 2000, curtailing water available to lower Colorado and California (Kneese and Bonem, 1986). Climate change may further reduce the availability of water in these basins. Stockton and Boggess's (1979) model of a 2°C temperature increase and a 10% precipitation decrease cuts water supply in the upper Colorado by 40% and in the Rio Grande by 76%. Flaschka (1984) found that a decrease in precipitation of 25% combined with a temperature increase of 2°C could decrease annual runoff by 50% in the Great Basin.

Great Plains

The southern Great Plains States of Kansas, Nebraska, Oklahoma, and Texas produce almost 40% of the nation's wheat, 15% of its corn, and 50% of its fattened cattle (see Chapter 10). The region heavily depends on groundwater mining (when pumping exceeds aquifer recharge) for irrigation. The region was hard hit during the "dust bowl" years of the 1930s and suffered from severe drought in 1988 (see Chapter 10).

Because of greater stability in irrigated yields relative to dryland yields, a rise in commodity prices will make expansion of irrigated production economically feasible. Thus, while total agricultural acreage is expected to decrease, irrigated acreage and groundwater mining are expected to increase in the southern Great Plains (Chapter 10). Greater demand may be placed on the

Ogallala Aquifer, which underlies much of the region, causing further groundwater mining of the aquifer.

Water quality appears to be vulnerable to deterioration because of increased use of agricultural pesticides as a response to climate change in surface runoff and erosion and decreased water flows.

Great Lakes

According to the GCMs, higher temperatures may overwhelm any increase in precipitation and may evaporate lakes to below the lowest levels on record. The winter ice cover would be reduced but would still be present, especially in shallow areas and northern lakes. Navigation depths, recreation, erosion (e.g., shrinking Chicago shoreline), hydropower output, and water quality all would be affected (see Chapter 5).

Mississippi River

The Mississippi River has historically been affected by both spring floods and drought. In 1988, low flows due to drought received national attention. Low flows disrupt navigation, degrade the drinking water for southern Louisiana cities, and reduce the inflow of water to the vast Mississippi Delta.

Northeast

During periodic droughts in the Northeast, such as in 1962-65 and 1980-81, in-stream flow regulations ration water and threaten shutdowns of electrical powerplants (U.S. Army Corps of Engineers, 1977; Schwartz, 1977; Kaplan et al., 1981). Although the Northeast is humid, cities and powerplants demand large amounts of water at localized points in a watershed, necessitating storage and interbasin transfers. No new major storage has been built in the Northeast during the past 20 years, except the Bloomington Dam on the Potomac River. Water supply in lower New England, New York, and Pennsylvania, and power production in the Northeast remain vulnerable to the droughts that may occur, according to some climate change scenarios (Schwartz, 1977; Kaplan et al., 1981; and see Chapter 15).

Southeast

The drought of recent years is increasing demands for irrigation in the Southeast and is prompting farmers to consider shifting crops. However, studies for this report are unclear on whether the Southeast may become wetter or dryer.

Most reservoirs in the area have sufficient capacity to retain flood surges and maintain navigation, hydropower, water supply, and in-stream uses (e.g. dilution, wildlife) in both wet and dry climate change scenarios (see Chapter 6). However, dryer conditions would pose conflicts between recreational uses (which would be hurt by changes in reservoir levels) and all other in-stream and off-stream uses.

Finally, a possible decline in the flow of freshwater to the Gulf coast and estuaries would change the estuarine ecology.

NATIONAL IMPACTS OF CLIMATE CHANGE ON WATER RESOURCES

Climate change and regional models do not yet provide reliable data to indicate the net national or regional costs and benefits of climate change. Accurate quantitative predictions, such as dollar losses or numbers of people affected by climate change, cannot be made. With current information, however, we can indicate possible directions of impacts and sectors affected. For example, water shortages may stifle economic growth, impose hardships on existing users, and result in transfers of water to uses with the highest value.

A major national impact of climate change on water resources may be to increase conflicts between regional water uses. Decreased water availability and increased water demands may intensify conflicts in the following areas:

- between in-stream and off-stream uses;
- among off-stream uses, such as agriculture, domestic use, and thermoelectric power;
- between water supply and flood control in the West;
- between all uses and recreation in the Southeast; and
- between thermoelectric power production and in-stream uses, especially in the East.

Increased precipitation due to climate change may decrease conflicts in some areas, but only after management infrastructure is modified to account for large maximum floods.

The potential effects of climate change on supply and demand must be compared to existing water supply and demand and to projections for supply and demand without climate change. Figure

8-4 outlines the national impacts of climate change on water supply and demand. Table 8-3 highlights the vulnerability of major water uses to climate change impacts by region. The following sections outline the potential impacts of climate change on traditional categories of off-stream and in-stream water uses.

Off-Stream Uses of Water Resources

Water supply and use have changed significantly during the past decade. For the first time since 1950, when the United States Geological Survey began recording water withdrawals, there was a 10% national drop in total fresh and saline water withdrawals from 1980 to 1985 (443 billion gallons to 399 billion gallons) (Solley et al., 1988). Increased conservation and water recycling in agriculture, industry, and energy production, slower growth in energy demand, and decline in availability of new water supply reduced or tempered water use in all sectors (Solley et al., 1988). Withdrawals declined by 7% in irrigation, by 33% in industry, and by 13% in thermoelectric power during the period. Of the major users, only municipal/domestic water supply increased (by 7%).

Irrigation

Water-short western States are intensively exploring avenues for minimizing water requirements. Because of depleted groundwater supplies, the rising cost of obtaining groundwater, and high costs and limited sites for new surface water developments, irrigated agriculture is no longer expanding or is declining in some areas of the West (Solley et al., 1988). Groundwater pumping for irrigation has already started to decline in the southern Great Plains States and in Arizona, although the impacts on production have been mitigated by the adoption of more efficient irrigation systems and switching to crops offering higher returns to water (Frederick and Kneese, 1988).

In contrast, supplemental irrigation in the East is on the rise (Viessman and DeMoncada, 1980). But overall, in the absence of climate change, agricultural water use trends will be characterized by only moderate increases or, in some cases, decreases in the next 20 years (Viessman and DeMoncada, 1980). Most analysts predict greater agricultural output per unit of water input.

Climate change may severely affect agriculture. Summer drought in the West and Midwest and earlier runoff are likely to

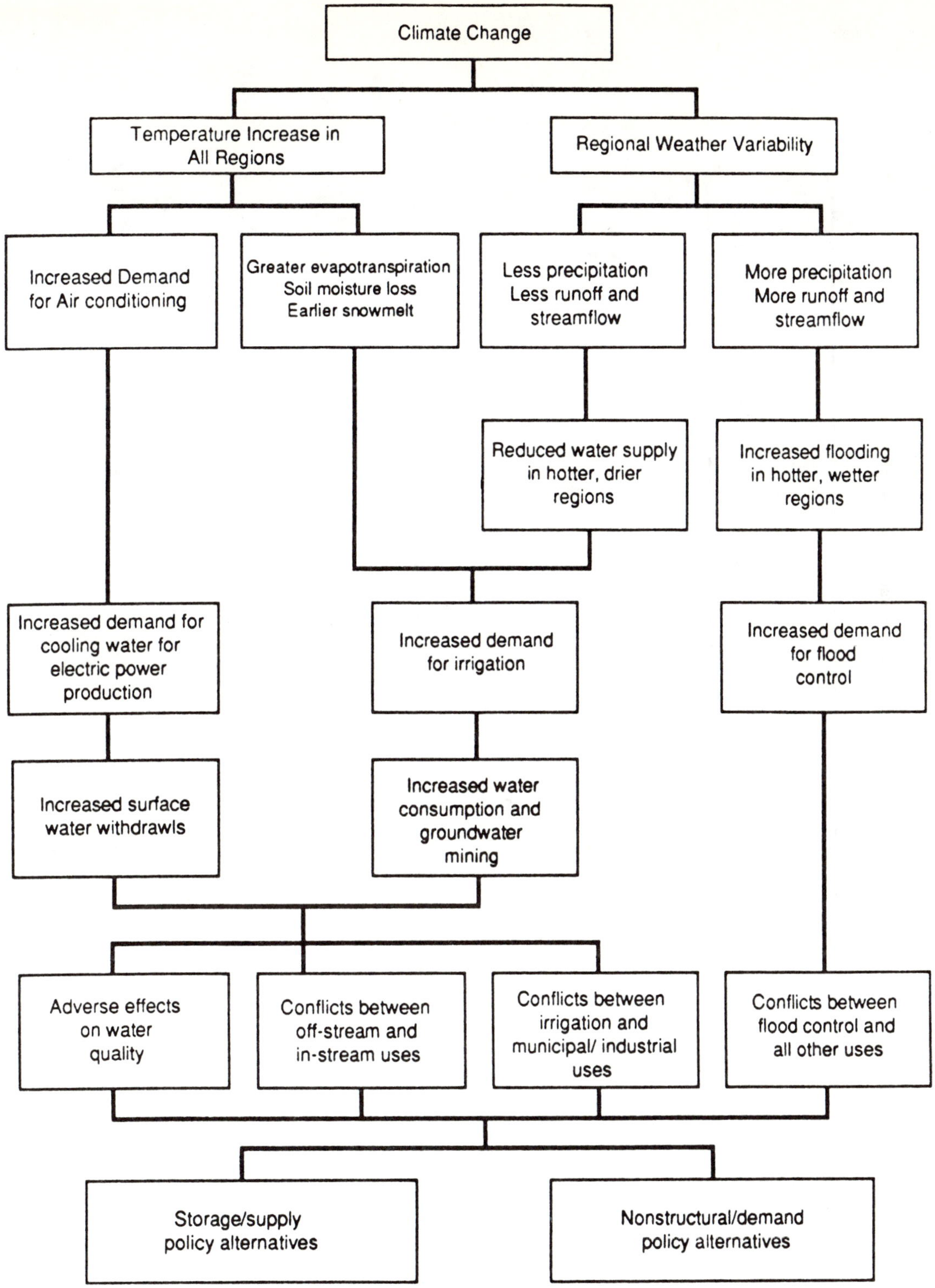

Figure 8-4. National impacts of climate change on water supply and demand.

Table 8-3. Potentail Regional Impacts of Climate Change on Water Uses: Areas of Vulnerability

	Pacific Northwest	California	Arid Western River Basins	Great Plains	Great Lakes	Mississippi	Southeast	Northeast
Irrigation	X	X	X	X			X	
Thermoelectric Power							X	X
Industrial		X	X					X
Municipal/ Domestic		X	X					X
Water Quality		X	X	X	X	X	X	X
Navigation					X	X	X	
Flood Control	X	X		X		X	X	
Hydropower	X	X			X		X	
Recreation					X		X	

change agricultural practices, cause dislocations in entire regions, and increase demands for irrigation. And even if the Southeast is wetter, higher temperatures may increase irrigation demand there (see Chapter 6).

Thermoelectric Power Generation

Future demand for water for power production will depend upon energy demand and Federal and State regulations governing in-stream water quality and thermal pollution. Due to minimum in-stream flow and maximum river temperature requirements of environmental regulations and the large amount of installed capacity along eastern rivers, freshwater withdrawals by steam electric powerplants have decreased, and siting of plants in coastal areas has increased. Thirty percent of installed capacity used saline surface water in coastal areas in 1985 (Solley et al., 1988). In addition, thermal regulations have effected a shift in new construction of cooling systems, from once-through to evaporative cooling (Miller, 1988). With once-through cooling, waste heat is discharged directly into the waste stream. Although evaporative cooling with towers and ponds alleviates thermal pollution, it increases water consumption.

During droughts, federal and state regulations protecting in-stream uses and limiting thermal discharges may constrain withdrawals for powerplant cooling (Kaplan et al., 1981; Miller, 1988). In addition, powerplant water needs on some eastern rivers are so large that during low-flow conditions, insufficient water is available to dissipate heat (Kaplan et al., 1981).

Demand for power may increase as warmer temperatures raise air-conditioning use (see Chapter 16). If streamflows are reduced as a result of climate change, powerplants using once-through cooling could be adversely affected (Miller, 1988). Increased demand for power may reinforce existing trends in powerplant design toward evaporative cooling and powerplant siting toward coastal sites. Purchases of water rights from other users by electric utilities in the West to meet cooling needs may increase. With less water available, low-flow conditions may interrupt power production and may increase power production costs and consumer electricity prices.

Industry

Since 1954, most major self-supplied industries steadily used less and less water per unit of production (Solley et al., 1988). This decline was partly due to efficiencies achieved by complying with federal and state water pollution legislation that restricts the discharge of untreated water. The trend toward more efficient industrial water uses is likely to continue. Industrial production across the country may be constrained by water quality if flow drops, and production in the West may be constrained by water supply curtailment.

Domestic Water Uses

Most municipal water supply systems are designed to provide reliable water at all times (safe yield). However, urban growth depends upon developed water supply which may be rapidly exhausted in some areas. In the Southeast and parts of the West, a large percentage of municipal water supply comes from groundwater (U.S. Water Resources Council, 1978; Solley et al., 1988). These regions have few groundwater recharge programs, so drought caused by climate change may decrease groundwater quality and quantity (see Chapters 6 and 7).

As relatively high-valued users of water, municipalities in the West are purchasing water rights to ensure adequate water supplies for urban growth. In the East, Midwest, and Southeast, municipalities may be able to increase safe yield by repairing and replacing existing leaking water delivery systems and by consolidating fragmented water supply districts.

In-stream Uses and Water Quality

The value of in-stream uses has risen relative to off-stream uses. Navigation and hydropower have retained their importance, as the values of wastewater dilution, ambient water quality, fish and wildlife habitats, and recreation have risen. Higher values placed on in-stream uses have made diversion of water for such uses as agriculture in the West and for powerplant cooling, more difficult. In general, climate change may put pressures on in-stream water uses and may degrade water quality.

<u>Water Quality</u>

The Federal Clean Water Act and its amendments ushered in a new era of water pollution control. Massive expenditures have decreased the amount of "conventional" water pollutants entering national water supplies. Total public and private, point and non-point, capital and operating water pollution abatement and control expenditures from 1972 to 1985 totaled $336 billion in 1982 dollars (Farber and Rutledge, 1987). Nevertheless, serious water quality problems remain.

One-third of municipal sewage treatment plants have yet to

comply with the provisions of the Clean Water Act (USEPA, 1987a). federal and state regulations governing previously unregulated toxic and hazardous water pollutants has just begun. And in the West, irrigation has caused the salinity levels in the water and soils in several river basins (the lower Colorado, the Rio Grande, and the San Joaquin) to rise so high that the viability of irrigation is being threatened (Frederick and Kneese, 1988).

With climate change, less freshwater may be available in some cases for diluting wastewater and recharging aquifers. Furthermore, increased thermal stratification accompanied by enhanced algal production may degrade the water quality of many lakes. Finally, the combination of declining freshwater availability and rising sea level would move salt wedges up estuaries, changing estuarine ecology and threatening municipal and industrial water supplies on the other hand, increased flows could allow for greater dilution of pollutants.

Navigation

Lower riverflow and lake levels would impede navigation. Systems that are particularly vulnerable are those with unregulated flow and high traffic, such as the lower Mississippi River and the Great Lakes.

Flood Control

The expected changes in the timing and magnitude of snowmelt and precipitation nationwide may restrict the flood control and multiple use capabilities of existing reservoirs and other structures. For example, drawing down the levels of reservoirs to contain floodwaters from anticipated increases in precipitation or earlier snowmelt may curtail water availability for other uses.

Because of the buffering and redundancy designed into large structures, major Federal flood control projects may be able to contain or mitigate the impacts of more frequent or severe floods. In addition, the majority of dams that provide substantial flood storage are in good condition. The National Dam Safety Inventory shows that the overall condition of the U.S. Army Corps of Engineers' more than 300 flood control reservoirs is sound (National Council on Public Works Improvement, 1988).

On the other hand, one-third of the non-Federal flood control dams inspected under the national non-Federal dam program were found to be unsafe, mostly due to inadequate spillways (National Council on Public Works Improvement, 1988). (Spillways are designed to prevent dam failure through overtopping.) These non-federal, smaller, mostly urban, flood control and stormwater structures may be more likely to have their capacity exceeded.

Some large dams may become susceptible to failure if storm intensity increases. The spillways of many large dams are designed to pass a "probable maximum flood" (an extreme flood event much greater than the 100-year flood). However, urbanization upstream from many dams is resulting in increased impervious surface (such as pavement) and increased peak runoff, and the probabilities of great floods, although remote, may increase due to climate change.

Hydropower

Because of the decline in water availability that could result from climate change, hydropower output and reliability could decline in the West and the Great Lakes. The Southeast could be facing the same problems unless it sacrificed or altered reservoir storage of streamflow increases in regions such as the Southeast hydropower production would increase also.

Recreation

Warmer and dryer conditions in the Southeast may increase the conflict between water uses, especially over reservoir releases and levels in the Tennessee Valley Authority region and the Lake Lanier, Georgia, systems. The conflicting uses may include releases for navigation, in-stream uses, urban water supply, groundwater recharge, hydropower and recreation (see Chapter 6).

NATIONAL POLICY IMPLICATIONS

Decreases in water availability, increased risk of flood damages, and the ensuing conflicts between water users competing for an increasingly scarce or difficult to manage resource are the major water resource issues raised by the possible impacts of a global warming trend. Some regions may restrict water use for lower value uses or may facilitate transfer of their water rights to higher value uses. How will we manage the transition to the new water resource regimes possible under climate change?

Current policy responses are complicated by the uncertainty attached to climate change forecasts. To be more helpful, the forecasts should be specific by area and should be accurate over the design-life of water resource structures. For example, they may depend in part on whether policymakers believe that the impacts will be felt as soon as 20 years from now or will occur gradually over 100 years. Forecasts of rapid climate changes may argue for measures to quickly prevent and avert climate change impacts; forecasts of gradual change may argue for policies that adapt and adjust to the change over time.

The implications of uncertainty include the following:

- Do not invest in irreversible, inflexible, large-scale, and high-cost measures.

- Design, modify, and rehabilitate structures and operating procedures that will provide robust and resilient water resource systems under different climate change scenarios.

- Implement a wide variety of measures for reducing demand, as long as they do not reduce the robustness and resilience of the systems.

Policy approaches to water resources may be grouped under supply (or structural) approaches and demand (or nonstructural) approaches. For example, water shortages may be addressed either by increasing developed supply (storage capacity) or decreasing water consumption and improving water quality.

Many or most of the policy approaches listed below have been recommended by water resource experts for 20 years. The potential impacts of climate change are an even greater impetus for implementing these policies.

Supply and Structural Policy Approaches

Developing Surface Water Structures

Surface water structures increase developed or available water supply or prevent flooding. These structures include dams, reservoirs, levees, and aqueducts.

Because of high costs of construction, adverse impacts on the environment and in-stream uses, limited sites available for new structures, and opposition by citizen groups, the trend during the past decade has been away from large excess-capacity, capital-intensive projects. Recent Federal financing for large structures has largely been limited to modifying, maintaining, and rehabilitating existing structures (Bureau of Reclamation, 1987). Only the Central Utah Project and the Central Arizona

Project have gone forward in recent years. Largely because of public opposition to new dams, only one major project in the Northeast has been completed in past 20 years -- the Bloomington Dam on the Potomac River. California citizens in 1982 voted down funds for the proposed Peripheral Canal that would increase diversion of water from north to south in the State. The trend toward local/state and away from federally financed projects has reduced funds available for large projects (National Council on Public Works Improvement, 1988).

These current trends in water resources management may be reevaluated in light of possible new demands for developed water supply under warm-dry climate change scenarios. Climate change may create pressure to build the Animas-La Plata and Narrows Projects proposed for Colorado, the Garrison Diversion in North Dakota, the Peripheral Canal in California, and structures to divert water from northern New England to southeastern Massachusetts.

Including Climate Change in Structural Design

Climate change should be a consideration in design and risk analysis for repair and replacement of existing projects and for constructing new projects (Hanchey et al., 1986). Current engineering and management approaches design for the mean, variability, and extremes of historical record. Large systems have substantial redundancy and robustness that enable them technologically and institutionally to adapt to large stresses (Matalas and Fiering, 1977). However, future climate may change hydrologic conditions from those on which the management of several major water systems is based.

In some cases, design includes both structural and nonstructural measures. Society may not be able to afford to build costly fail-safe water supply or flood control projects. For example, the design of Army Corps of Engineers' projects is increasingly becoming "safe-fail." This is a strategy of designing flood protection, for example, for less than the extreme event of record but for the 20- to 50-year flood, and then complementing the structural measure with emergency management plans that mitigate the damages (Hanchey et al., 1986).

Optimizing Water Resource Systems

Operational strategies for large-scale water systems and watersheds can have an enormous influence on the overall performance and resilience of the system and may provide additional water supplies to mitigate the impacts of climate change. Management practices may allow for substantial operating flexibility concerning releases of stored water. However, current design and management practices rely on historical data that may be superseded by new hydrologic averages and extremes.

Water yields may be substantially increased by operating existing facilities and surface and groundwater supplies as systems, rather than as independent projects (Sheer, 1985). For example, the Federal Bureau of Reclamation (1987) is adopting operational, management, or physical changes to gain more output from the same resources. Protection of groundwater recharge areas and methods for the conjunctive use of ground and surface waters are being implemented nationwide by water management agencies (Tripp and Jaffe, 1979; U.S. Water Resources Council, 1980).

Watershed management practices also affect water supply. For example, water yields can be significantly affected by timber harvest practices. Approximately 70-80% of water yield in the West is the result of snowmelt in high-elevation forests, much of which is under public jurisdiction (Troendle, 1983).

Demand Management and Nonstructural Policy Approaches

The limitations of supply and structural solutions to the nation's water problems may argue that greater attention should be given to demand management to resolve long-term water resource problems and to mitigate the potential adverse impacts of climate change. In the past, water was considered too essential a resource or too insensitive to price for its use to be allocated by market forces during shortages. Policy took the form of direct controls and appeals to conserve (Hrezo et al., 1986). In recent years, greater attention has been given to institutional, legal, and market policies to allocate and regulate limited water supplies among competing uses and to promote water conservation.

Reducing Irrigation Use

Relatively small reductions in the percentage of irrigation demand can make large amounts of water available for higher value urban and industrial uses. For instance, nearly 83% of withdrawals and 90% of the consumptive use of western water is for irrigation. Consequently, a 10% reduction in irrigation use may save 20 million acre-feet (maf) in water withdrawn and 10 maf in water consumed annually. This would save an amount approximately equal to the average annual flow of the Upper Colorado River Basin and would double the water available for municipal and industrial uses in the West (Frederick, 1986).

Estimates of water savings through improved efficiencies in irrigation management range between 20 and 50 % (Viessman and DeMoncada, 1980). Due in part to decreased availability, increased costs of pumping groundwater, and opportunities to market water rights to higher value users, irrigation declined 6 % from 1980 to 1985 (Solley et al., 1988).

Inexpensive water was a key factor in the settlement of the West and the expansion of agriculture (Frederick, 1986). The Bureau of Reclamation was established early in this century to promote the development of irrigation in the West. The Bureau provides irrigation for about 11 million acres, more than one-fifth of the total irrigated acreage. In 1975, it supplied 28.1 maf of water, 93% of which was used for irrigation (Bureau of Reclamation, 1976). Since the Bureau accounts for nearly one-third of all surface water deliveries and about one-fifth of total water deliveries in the 17 Western States, using this water more efficiently could be very important.

In the past, demand for Bureau water was not based on the real cost of the water, because more than 90% of the Bureau's irrigation projects have been subsidized, and payments on some projects no longer even pay for operation and maintenance (Frederick and Hansen, 1982). Irrigators fortunate enough to receive such inexpensive water may have little or no incentive to conserve. However, the Bureau's more recently stated objectives include revising their water marketing policy, promoting conservation, and pricing water to reflect its real cost (Bureau of Reclamation, 1987).

The 1985 Food Security Act and programs managed by the U.S. Department of Agriculture encourage water conservation and maintenance of water quality (Rosenberg, 1988). USDA programs focus on conserving soil and water, setting aside crops, and retiring marginal lands.

Municipal and Industrial Water Use

Municipalities throughout the country are finding it difficult and expensive to augment their supplies to meet the demands of population and economic growth. Traditional average-cost pricing designed to provide adequate service to customers and returns to water companies is being reevaluated (Congressional Budget Office, 1987). Average-cost pricing and declining-block rates tend to cause overinvestment in system capacity and overuse of existing capacity. Marginal-cost pricing or progressive-rate structures (charging for the actual amount of water used) can reduce domestic and industrial water consumption because water use is sensitive to price (Gibbons, 1986).

Water Markets and Transfers in the West

It is unlikely that the Federal Government will subsidize many more large irrigation projects. Instead, conservation incentives, prices, and water markets are becoming important management tools to balance water demand with supplies. Evidence shows that many water uses, including irrigation, are sensitive to price (Gibbons, 1986). However, while irrigators in may cases can adjust to higher prices and to water market opportunities, incentives for abandoning traditional water-intensive practices are still lacking for many areas in the West.

The "first in time, first in right" appropriation doctrine, which favors the longest standing water rights, governs much of the West's surface water and some groundwater. It encourages economic development by guaranteeing users a secure supply of water. This doctrine was potentially an important step for creating clear, transferable property rights to water -- a precondition for effective operation of water markets. The potential for water transfers to the highest value users has not yet been fully realized because the nature and transferability of the rights are obscured by legal and administrative factors (Trelease, 1977; Frederick, 1986; Saliba et al., 1987). Following are some examples.

- Rather than grant absolute ownership, States with prior appropriation rules grant rights to use water for beneficial purposes. Water rights not put to beneficial use may be forfeited. This encourages a use-it-or-lose-it attitude.

- Federal and Native American water rights remain unquantified in some areas, such as the Colorado River Basin.

- The U.S. Department of the Interior's Bureau of Reclamation supplies 30% of the irrigation water in the West. There are few incentives to conserve this water, because the Bureau's long-term, interest-free repayment contracts keep the price low and limit opportunities to transfer the water to users willing to pay more.

- Because of imperfect competition, third-party effects, uncertainty over administrative rules, and equity considerations, water market prices may not appropriately measure water values according to economic efficiency criteria (Gibbons, 1986; Saliba et al., 1987).

- It is possible to control groundwater withdrawals, but for a number of reasons it is difficult to establish market mechanisms for groundwater allocation. Because all groundwater users essentially draw from a shared pool, groundwater resources are treated as "common property." As a result, property rights are difficult to define, third- party impacts of transfers of groundwater rights are significant, and interstate agreements concerning allocation of interstate aquifer water are difficult to attain (Emel, 1987).

Despite the obstacles, transfer of water rights among users -- especially from irrigators to municipalities and power companies seeking water for urban expansion and electricity production -- is becoming common in many Western States (Wahl and Osterhoudt, 1985; Frederick, 1986). Methods include negotiated purchases, short-term exchanges during droughts, and water banks and markets (Wahl and Osterhoudt, 1985; Saliba et al., 1987; Wahl and Davis, 1986).

Legislation in many Western States has facilitated water transfers (Frederick, 1986; Frederick and Kneese, 1988). Arizona's new water law facilitates the purchase of land with water rights and the use of that water for urban development. Strict technical standards imposing conservation on municipal and industrial water uses, such as watering golf courses with wastewater, are also part of Arizona's laws (Saliba et al., 1987). California has applied the public trust doctrine to modify existing permits to require flow releases to protect the Sacramento-San Joaquin Delta from saltwater intrusion (Tarlock, 1987). Montana is increasingly basing water management plans on its in-stream flow requirements and is exploring ways to have these requirements for all future beneficial in-stream uses count as a bona fide use of the Missouri River in order to slow the rate of growth of diversion of water for off-stream uses (Tarlock, 1987). The Supreme Court decided in Colorado v. New Mexico (456 U.S. 176 (1982)) that equitable apportionment may be used to override prior appropriation priorities in cases of major flow reductions. The Supreme Court specifically mentioned climatic conditions in ruling that prior appropriation systems would otherwise protect arguably wasteful and inefficient uses of water at the expense of other uses.

Frederick and Kneese (1988) caution that water transfers occur gradually and are not likely to affect more than a small percentage of agricultural water rights for the foreseeable future. However, legal and institutional changes facilitating water markets and demand for water by high-value users may be accelerated under the stress of climate change (Trelease, 1977).

Interstate and Regional Compacts and Commissions

River basin and aquifer boundaries in many cases traverse or underlie portions of several States. Regional and interstate cooperation to manage water resources has a long tradition in some U.S. river basins. Implementing the agreements required for regional compacts and operating procedures and sharing of water supplies may require substantial and lengthy negotiations. However, the cost savings also can be substantial (Sheer, 1985).

Several interstate water authorities have significant allocation authority. For example, the Delaware River Basin Commission allocates water to users in the Delaware Basin and transfers it to New York City. It does so by the authority of a 1954 Supreme Court ruling (347 U.S. 995) and Federal legislation establishing the Commission in 1961 and granting it regulatory, licensing, and project construction powers. Similarly, water authorities in the Washington, D.C., metropolitan area operate Potomac River water supply projects as integrated systems under a 1982 agreement. Both the Delaware and Potomac regional compacts include provisions for drought allocations. (See Harkness et al., 1985, for management actions taken by the Delaware River Basin Commission during a 1984-85 drought.) Sheer (1985) estimated that coordinated water authority activities in the Potomac River Basin eliminated the need for new reservoirs, saving from $200 million to $1 billion. In the West, the most prominent interstate agreement is the Colorado River Compact, which divides river water allocations among seven Western States and Mexico (see Weatherford and Brown, 1986).

Numerous opportunities exist around the country for interstate, intrabasin, and interbasin coordination of water deliveries. In the East, consolidation of or coordination among fragmented urban water supply authorities can achieve economies of scale in water delivery, decrease the risk of shortage in any one subsystem within a region, and provide drought management procedures (Sheer, 1985).

Drought Management Policies

Integrating drought planning into water resource management may assume greater priority if climate change causes water shortages. The Model Water Use Act (Hrezo et al., 1986) advocates that states or water supply authorities integrate drought management and advance planning into their policies by designating a governmental authority for drought response and adopting mechanisms for automatically implementing and enforcing water use restrictions. In 1986, only seven States had comprehensive management plans for water shortages (Hrezo et al., 1986). Most States rely on general water rights, emergency responses, and litigation to allocate water during shortages.

Water Quality

Federal and State legislation and regulations for control of in-stream water quality have had a dramatic effect on reducing conventional water pollutants since the enactment of the 1972 Clean Water Act. However, some persistent water quality problems remain to be addressed (see USEPA, 1987b). In some cases, where river and lake levels are lower, more stringent controls on point and nonpoint sources may be required to meet water quality standards.

Many States have adopted measures to protect in-stream water uses. These include reserving flows or granting rights for particular in-stream uses and directing agencies to review impacts before granting new rights (U.S. Water Resources Council, 1980; Frederick and Kneese, 1988). Regulations limiting water use may have to be modified under some climate change scenarios that forecast increased incidence of low flows in some regions.

Policies for Floodplains

The National Flood Insurance Program was enacted in 1968, with major amendments in 1973. The program provides subsidized flood insurance for existing structures in flood-prone areas, provided that the community with jurisdiction regulates the location and construction of new buildings to minimize future flood losses. New structures that comply with the restrictions are eligible for insurance at full actuarial rates.

The program is second only to the Social Security Act in the magnitude of potential government liability. In 1979, the program took in $140 million in premiums and paid $480 million in payment claims. Recently, the program was authorized to apply claims payments to relocation expenses for structures exposed to repeated flood or erosion damage.

Where rainfall and flooding increase, the 100-year floodplain would expand, and rate maps would need revision. Premium payments and claims would rise.

RESEARCH NEEDS

1. Refine global climate change models so that water resource engineers and managers in regional planning can use their predictions more confidently.

2. Monitor atmospheric, oceanic, and hydrological conditions to detect evidence of water resource impacts of climate change.

3. Conduct research to identify implementable opportunities for adjusting and adapting to climate change. Types of research include the following:

   - Identify and quantify the relative vulnerabilities of water uses to climate change.

   - Focus research and monitoring on vulnerable water resource areas, such as the western edge of the Great Plains, where moisture is the limiting factor for agriculture, and river basins where demand is greater than developed safe yield supply under current conditions.

   - Quantify Federal and Native American water rights in the West.

   - Along with estimating how much water may be required for the various uses, develop information on the cost and worth or value of water in the various uses.

   - Examine how present institutions can better allocate water among users and provide incentives to conserve water and assess the extent to which laws and regulations exacerbate the effects of climate change (examples include thermal controls for rivers and Federal pricing and reallocation policies for irrigation water).

- Identify, project, and quantify economic, demographic, and institutional adjustments that may accompany climate induced impacts on water resources in the absence of concerted public action to provide data upon which to base policy decisions regarding action alternatives.

Chapter 8

BIBLIOGRAPHY

Beran, M. 1986. The water resource impact of future climate change and variability. In: Titus, J., ed. Effects of Changes in Stratospheric Ozone and Global Climate. Vol. 1: Overview. Washington, DC: U.S. Environmental Protection Agency, pp. 299-330.

Butcher, W.R., and N.K. Whittlesey. 1986. Competition between irrigation and hydropower in the Pacific Northwest. In: Frederick, K.D., ed. Scarce Water and Institutional Change. Washington, DC: Resources for the Future.

Cohen, S.J. 1986. Impacts of CO_2-induced climatic change on water Resources in the Great Lakes Basin. Climatic Change 8:135-53.

U.S. Congress, Congressional Budget Office. May 1987. Financing Municipal Water Supply Systems.

Dracup, J.A. 1977. Impact on the Colorado River Basin and Southwest water supply. In: The National Research Council, ed. Climate, Climatic Change, and Water Supply. Washington, DC: National Academy of Sciences, pp. 121-132.

Emel, J.L. 1987. Groundwater rights: definition and transfer. Natural Resources Journal 27:653-674.

Farber, K.D., and G.L. Rutledge. 1987. Pollution Abatement and Control Expenditure, 1982-85. Survey of Current Business. May 1987 pp. 21-26.

Frederick, K.D., and J.C. Hansen. 1982. Water for Western Agriculture. Washington, DC: Resources for the Future.

Frederick, K.D. 1986. Scarce Water and Institutional Change. Washington, DC: Resources for the Future.

Frederick, K.D., and A.V. Kneese. 1988. Reallocation by markets and prices. In: American Association for the Advancement of Science Report: Climatic Variability, Climate Change, and U.S. Water Resources. Washington, D.C.

Gibbons, D.C. 1986. The Economic Value of Water. Washington, DC: Resources for the Future.

Gleick, P.H. 1987. Regional Hydrologic Consequences of Increases in Atmospheric CO_2 And Other Trace Gases. Climatic Change 10: 137-161.

Gleick, P.H. 1988. The sensitivities and vulnerabilities of water supply systems to climatic changes. In: American Association for the Advancement of Science Report: Climatic Variability, Climate Change, and U.S. Water Resources. Washington, D.C.

Hanchey, J.R., K.E. Schilling, and E.Z. Stakhiv. 1987. Water resources planning under climate uncertainty. In: Preparing for Climate Change. Proceedings of the First North American Conference on Preparing for Climate Change. Government Institutes, Inc. pp. 394-405.

Hansen, J., A. Lacis, D. Rind, G. Russell, I. Fung, P. Ashcraft, S. Lebedeff, R. Ruedy, and P. Stone. 1986. The greenhouse effect: projections of global climate change. In: Titus, J., ed. Effects of Changes in Stratospheric Ozone and Global Climate. Vol. 1: Overview. Washington, DC: U.S. Environmental Protection Agency, pp. 199-218.

Harkness, W.E., H.F. Lins, and W.M. Alley. 1985. Drought in the Delaware River Basin. In: National Water Summary. 1985: Hydrologic Events and Surface-Water Resources. U.S. Geological Survey, pp. 29-40.

Hobbs, B.F., and P.M. Meier. 1979. An analysis of water resources constraints on power plant siting in the Mid-Atlantic States. Water Resources Bulletin 15(6):1666-1676.

Hrezo, M.S., P.G. Bridgeman, and W.R. Walker. 1986. Integrating drought planning into water resources management. Natural Resources Journal 26:141-167.

Kaplan, E., M. Rubino, and D. Neuhaus. 1981. Water Resource Development and Energy Growth in the Northeast. Brookhaven National Laboratory, Report BNL 51522; November.

Klemes, V., and J. Nemec. 1985. Assessing the impact of climate change on the development of surface water resources. In: Klemes V., ed. Sensitivity of Water Resource Systems to Climate Variations. Geneva: World Meteorological Office.

Kneese, A.V. and G. Bonem. 1986. Hypothetical shocks to water allocation institutions in the Colorado Basin. In: Weatherford, G., and F.L. Brown, eds. New Courses for the Colorado River: Major Issues for the Next Century. Albuquerque: University of New Mexico Press, pp. 87-108.

Linsley, R.K., and J.B. Franzini. 1979. Water Resources Engineering. New York: McGraw-Hill, Inc.

Manabe, S., and R.T. Wetherald. 1986. Reduction in summer soil wetness induced by an increase in atmospheric carbon dioxide. In: Titus, J., ed. Effects of Changes in Stratospheric Ozone and Global Climate. Vol. 1: Overview. Washington, DC: U.S. Environmental Protection Agency, pp. 249-256.

Matalas, N.C., and M.B. Fiering. 1977. Water-resource systems planning. In: The National Research Council, ed. Climate, Climatic Change, and Water Supply. Washington, DC: National Academy of Sciences, pp. 99-110.

National Council on Public Works Improvement. 1988. Fragile Foundations: A Report on America's Public Works. Final Report to the President and to the Congress. Washington, DC: U.S. Government Printing Office. February.

National Research Council. 1977. Climate, Climatic Change, and Water Supply. Washington, DC: National Academy Press.

National Research Council. 1988. Estimating Probabilities of Extreme Floods. National Research Council. Committee on Techniques for Estimating Probabilities of Extreme Floods, Water Science and Technology Board; and Commission on Physical Sciences, Mathematics, and Resources. Washington, DC: National Academy Press.

Nemec, J., and J.C. Schaake. 1982. Sensitivity of water resource systems to climate variation. Hydrological Sciences Journal 27:327-43.

Revelle, R.R., and P.E. Waggoner. 1983. Effects of the carbon dioxide induced climatic change on water supplies in the western United States. In: Changing Climate: Report of the Carbon Dioxide Assessment Committee. Washington, DC: National Academy Press.

Rind, D., and S. Lebedeff. 1984. Potential Climatic Impacts of Increasing Atmosphere CO_2 With Emphasis on Water Availability and Hydrology in the United States. U.S. Environmental Protection Agency Report. Washington, DC: U.S. Environmental Protection Agency.

Rosenberg, N.J. 1988. Global climate change holds problems and uncertainties for agriculture. In: Tutwiler, M.A., ed. U.S. Agriculture in a Global Setting. Washington, DC: Resources for the Future.

Saliba, B.C., D.B. Bush, W.E. Martinard, T.C. Brown. 1987. Do water market prices appropriately measure water values? Natural Resources Journal 27:617-652.

Schwarz, H.E. 1977. Climatic change and water supply: how sensitive is the Northeast? In: The National Research Council, ed. Climate, Climatic Change, and Water Supply. Washington, DC: National Academy of Sciences, pp. 111-120.

Sheer, D.P. 1985. Managing water supplies to increase water availability. In: National Water Summary. Hydrologic Events and Surface-Water Resources. U.S. Geological Survey, pp. 101-112.

Stockton, C.W., and W.R. Boggess. 1979. Geohydrological Implications of Climate Change on Water Resource Development. Fort Belvoir, VA: U.S. Army Coastal Engineering Research Center.

Strock, J.M. 1987. Adjusting water allocation law to meet water quality and availability concerns in a warming world. In: Preparing for Climate Change. Proceedings of the First North American Conference on Preparing for Climate Change. Rockville, MD: Government Institutes, Inc., pp.382-387.

Tarlock, A.D. 1987. Damning the dams and ditches: a review of D. Worster, rivers of empire: water aridity and the American West. Natural Resources Journal 27: 477-490.

Trelease, F.J. 1977. Climatic change and water law. In: The National Research Council, ed. Climate, Climatic Change, and Water Supply. Washington, DC: National Academy of Sciences, pp. 70-84.

Tripp, J.T.B. and A.B. Jaffe. 1979. Preventing ground water pollution: toward a coordinated strategy to protect critical recharge zones. Harvard Environmental Law Review 3.

Solley, W.B., C.F. Merck, and R.R. Pierce. Estimated use of water in the United States in 1985. U.S. Geological Survey. Circular 1004 Washington, D.C. U.S. Government Printing Office.

U.S. Army Corps of Engineers. 1977. Northeastern United States Water Supply Study, Summary Report. U.S. Army Corps of Engineers, North Atlantic Division; July.

U.S. Army Corps of Engineers. 1988. Lessons Learned from the 1986 Drought. U.S. Army Corps of Engineers, Water Resources Support Center, The Hydrologic Engineering Center; June.

U.S. Department of Agriculture. 1987. Economic Research Service. Agricultural Irrigation and Water Supply. Washington, DC: U.S. Government Printing Office. Agriculture Information Bulletin 532.

U.S. Department of the Interior, Bureau of Reclamation. 1987 Assessment '87 ...A New Direction for the Bureau of Reclamation. Washington, D.C.

U.S. Department of the Interior, Bureau of Reclamation. 1987 Implementation Plan ...A New Direction for the Bureau of Reclamation. Washington, D.C.

U.S. EPA. 1987a. U.S. Environmental Protection Agency. 1986 Needs Survey Report to Congress: Assessment of Needed Publically Owned Wastewater Treatment Facilities in the United States. EPA 430/9-87-001.

U.S. EPA. 1986b. U.S. Environmental Protection Agency. National Water Quality Inventory: 1986 Report to Congress. EPA-440/4-87-008.

U.S. Geological Survey. 1984. National Water Summary. Water-Supply Paper 2275. Washington, DC: U.S. Government Printing Office.

U.S. Geological Survey. 1985. National Water Summary. Water-Supply Paper 2300. Washington, DC: U.S. Government Printing Office.

U.S. Water Resources Council. 1978. The Nation's Water Resources 1975-2000: Second National Water Assessment. Vol. 1: Summary. Washington, D.C. U.S. Government Printing Office.

Viessman Jr., W., and C. DeMoncada. 1980. State and National Water Use Trends to the Year 2000. Congressional Research Service, The Library of Congress.

Wahl, R.W., and F.H. Osterhoudt. 1985. Voluntary transfers of water in the West. In: U.S. Geological Survey. National Water Summary. Hydrologic Events and Surface-Water Resources, pp. 113-124.

CHAPTER 9

SEA LEVEL RISE

FINDINGS

Global warming could cause a sea level rise of 0.5 to 2m by 2100. Such a rise would inundate wetlands and lowlands, erode beaches, exacerbate coastal flooding, and increase the salinity of estuaries and aquifers.

- A one-meter rise by the year 2100 could drown approximately 25 to 80 percent of the U.S. coastal wetlands. Their ability to survive would depend largely on whether they could migrate inland or whether levees and bulkheads block their path of migration.

- A one-meter rise could inundate 5,000-10,000 square miles of dryland if shores are not protected, and 4,000-9,000 square miles of dryland if only developed areas are protected.

- Most coastal barrier island communities would probably respond to sea level rise by raising land with sand pumped from offshore. Wide and heavily urbanized islands may use levees, while communities on lightly developed islands may adjust to a gradual landward migration of the islands.

- Protecting developed areas against such inundation and erosion by building of bulkheads and levees, pumping sand, and raising barrier islands would cost $73-111 billion (cumulative capital costs in 1985 dollars) for a one-meter rise by the year 2100.

- Developed barrier islands would likely be protected from sea level rise because of their high property values. However it would cost $50-75 billion (cumulative capital costs in 1985 dollars) to elevate beaches, houses, land, and roadways by the year 2100.

Policy Implications

- Many of the necessary responses to sea level rise need not be implemented until the rise is imminent, such as rebuilding ports, constructing levees, and pumping sand onto beaches. On the other hand, the cost of incorporating sea level rise into a wide variety of engineering and land use decisions would be negligible compared to the costs of not responding until sea level rises.

- Wetland ecosystems are likely to survive sea level rise only if appropriate measures are implemented soon. At the states and local levels, these measures include land use planning, regulation, and redefinitions of property rights. The State of Maine has already issued regulations to enable wetlands to migrate landward by requiring that structures be removed as sea level rises.

- The coastal wetlands protected under Section 404 of the Clean Water Act will gradually be inundated. The Act does not authorize measures to ensure survival of wetland ecosystems as sea level rises. Additional statutory authorities might be needed to protect migrating wetlands.

- The National Flood Insurance Program may wish to consider the implications of examining the sea level rise on potential liabilities. A recent HUD authorization act requires this program to purchase property threatened with erosion. The act may imply a commitment by the federal government to compensate property owners for losses due to sea level rise.

CAUSES, EFFECTS, AND RESPONSES

Introduction

The global warming from the greenhouse effect could raise sea level approximately one meter by expanding ocean water, melting mountain glaciers, and causing ice sheets in Greenland to melt or slide into the oceans. Such a rise would inundate coastal wetlands and lowlands, erode beaches, increase the risk of flooding, and increase the salinity of estuaries, aquifers, and wetlands.

In the last five years, many coastal communities throughout the world have started to prepare for the possibility of such a rise. In the United States, Maine has enacted a policy declaring that shorefront building will have to be moved to enable beaches and wetlands to migrate inland to higher ground. Maryland has shifted its shore-protection strategy from a technology that can not accommodate sea level rise to one that can. Seven coastal states have held large public meetings on how to prepare for a rising sea. Australia and the Netherlands are beginning to undergo a similar process.

Causes

Ocean levels have always fluctuated with changes in global temperatures. During the ice ages when the earth was 5°C (9°F) colder than today, much of the ocean's water was frozen in glaciers and sea level often was more than 100 meters (300 feet), below the present level (Donn et al., 1962; Kennett, 1982; Oldale, 1985). Conversely, during the last interglacial period

(100,000 years ago) when the average temperature was about 1°C (2°F) warmer than today, sea level was approximately 20 feet higher than current sea level (Mercer, 1968).*

When considering shorter periods of time, worldwide sea level rise must be distinguished from relative sea level rise. Although climate change alters worldwide sea level, the rate of sea level rise relative to a particular coast has greater practical importance and is all that monitoring stations can measure. Because most coasts are sinking (and a few are rising), the range of relative sea level rise varies from more than 3 feet per century in Louisiana and parts of California and Texas to 1 foot per century along most of the Atlantic and Gulf coasts, and to a slight drop in much of the Pacific Northwest (Figure 9-1). Areas such as Louisiana provide natural laboratories for assessing the possible effects of future sea level rise (Lyle et al., 1987).

Global sea level trends have generally been estimated by combining the trends at tidal stations around the world. Studies combining these measurements suggest that during the last century, worldwide sea level has risen 10 to 15 cm (4 to 6 inches) (Barnett, 1984; Fairbridge and Krebs, 1962). Much of this rise has been attributed to the global warming that has occurred during the last century (Meier, 1984; Gornitz et al., 1982). Hughes (1983) and Bentley (1983) estimated that a complete disintegration of West Antarctica in response to global warming would require a 200- to 500-year period, and that such a disintegration would raise sea level 20 feet. Most recent assessments, however, have focused on the likely rise by the year 2100. Figure 9-2 illustrates recent estimates of sea level rise, which generally fall into the range of 50 to 200 cm.

Although most studies have focused on the impact of global warming on global sea level, the greenhouse effect would not necessarily raise sea level by the same amount everywhere. Removal of water from the world's ice sheets would move the earth's center of gravity away from Greenland and Antarctica and would thus redistribute the oceans' water toward the new center of gravity. Along the U.S. coast, this effect would generally increase sea level rise by less than 10%. Sea level could actually drop, however, at Cape Horn and along the coast of Iceland. Climate change could also effect local influences on sea level by changing ocean currents, winds, and atmospheric pressure; no one has estimated these impacts.

---

*Other processes such as changes in ocean basins have also played a role, albeit over longer periods of time.

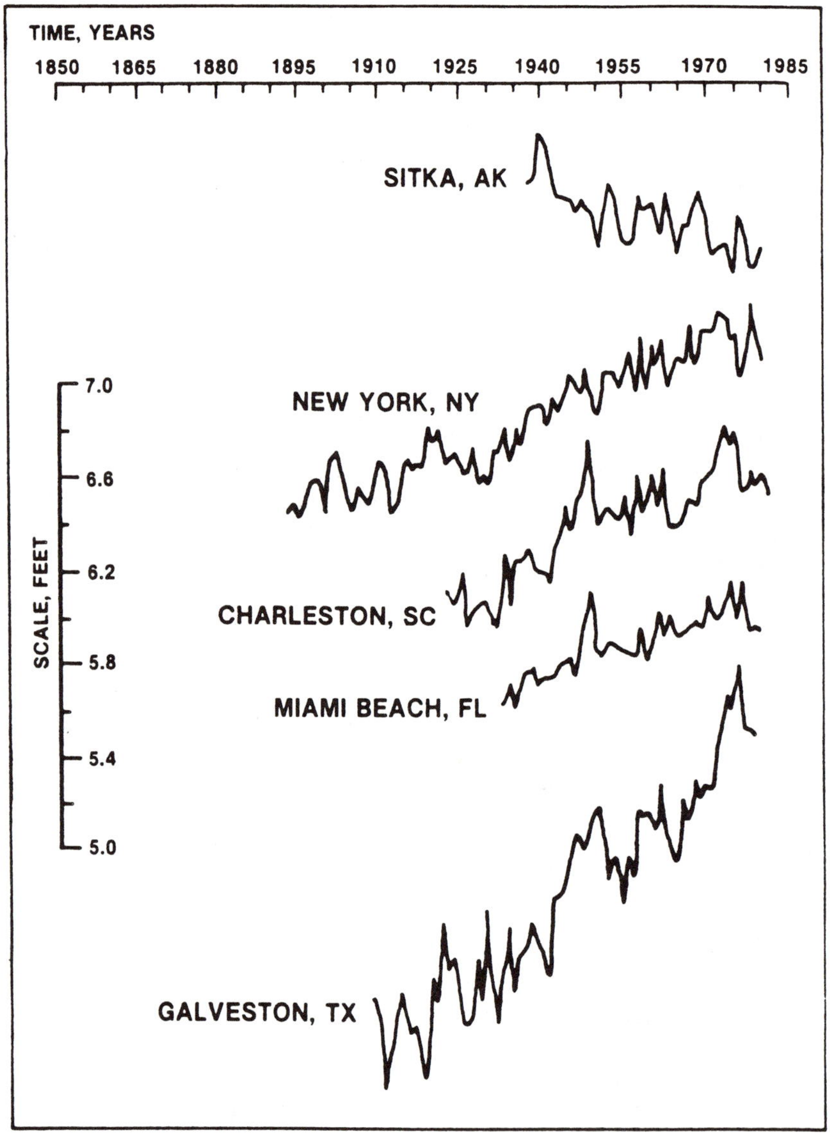

Figure 9-1. Time Series Graph of Sea Level Trends for New York, Charleston, Miami, Galveston, and Sitka

Source: Lyle et al, 1987.

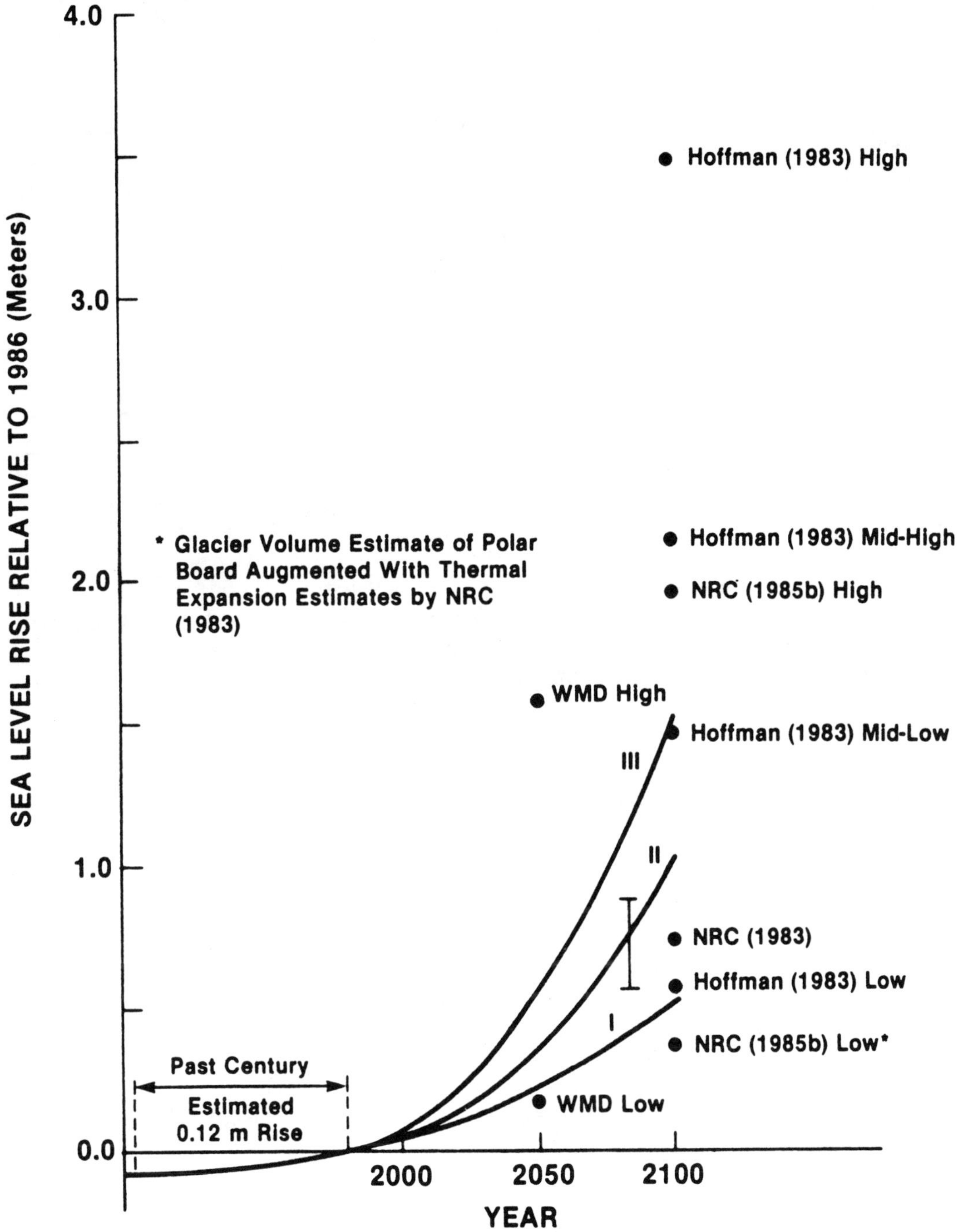

Figure 9-2. Estimates of Future Sea Level Rise

Source: Hoffman (1983, 1986); Meier (1985); Revelle (1983); Wigley (1987, Thomas

Effects

In this section and in the following sections, the effects of, and responses to sea level rise are presented separately. However, the distinction is largely academic and is solely for presentation purposes. In many cases, the responses to sea level rise are sufficiently well-established and the probability of no response is sufficiently low that it would be misleading to discuss the potential effects without also discussing responses. For example, much of Manhattan Island is less than 2 m above high tide; the effect of sea level rise would almost certainly be the increased use of coastal engineering structures and not the inundation of downtown New York.

A rise in sea level would inundate wetlands and lowlands, accelerate coastal erosion, exacerbate coastal flooding, threaten coastal structures, raise water tables, and increase the salinity of rivers, bays, and aquifers (Barth and Titus, 1984). Most of the wetlands and lowlands are found along the Gulf coast and along the Atlantic coast south of central New Jersey, although a large area also exists around San Francisco Bay. Similarly, the areas vulnerable to erosion and flooding are also predominately in the Southeast; potential salinity problems are spread more evenly along the U.S. Atlantic coast. We now discuss some of the impacts that would result if no responses were initiated to address sea level rise.

Destruction of Coastal Wetlands

Coastal wetlands are generally found between the highest tide of the year and mean sea level. Wetlands have kept pace with the past rate of sea level rise because they collect sediment and produce peat upon which they can build; meanwhile they, expanded inland as lowlands were undated (Fig 9-3). Wetlands accrete vertically and expand inland. Thus, as Figure 9-2 illustrates, the present area of wetlands is generally far greater than the area that would be available for new wetlands as sea level rises (Titus et al., 1984b; Titus, 1986). The potential loss would be the greatest in Louisiana (see Chapter 6, Southeast.)

In many areas, people have built bulkheads just above the marsh. If sea level rises, the wetlands would be squeezed between the sea and the bulkheads (see Figure 9-3). Previous studies have estimated that if the development in coastal areas is removed to allow new wetlands to form inland, a 1.5- to 2-m rise would destroy 30 to 70% of the U.S. coastal wetlands. If levees and bulkheads are erected to protect today's dry land, the loss could be 50 to 80% (Park, 1986; Titus, 1988; Armentano et al., 1988). Such a loss would reduce the available habitat for

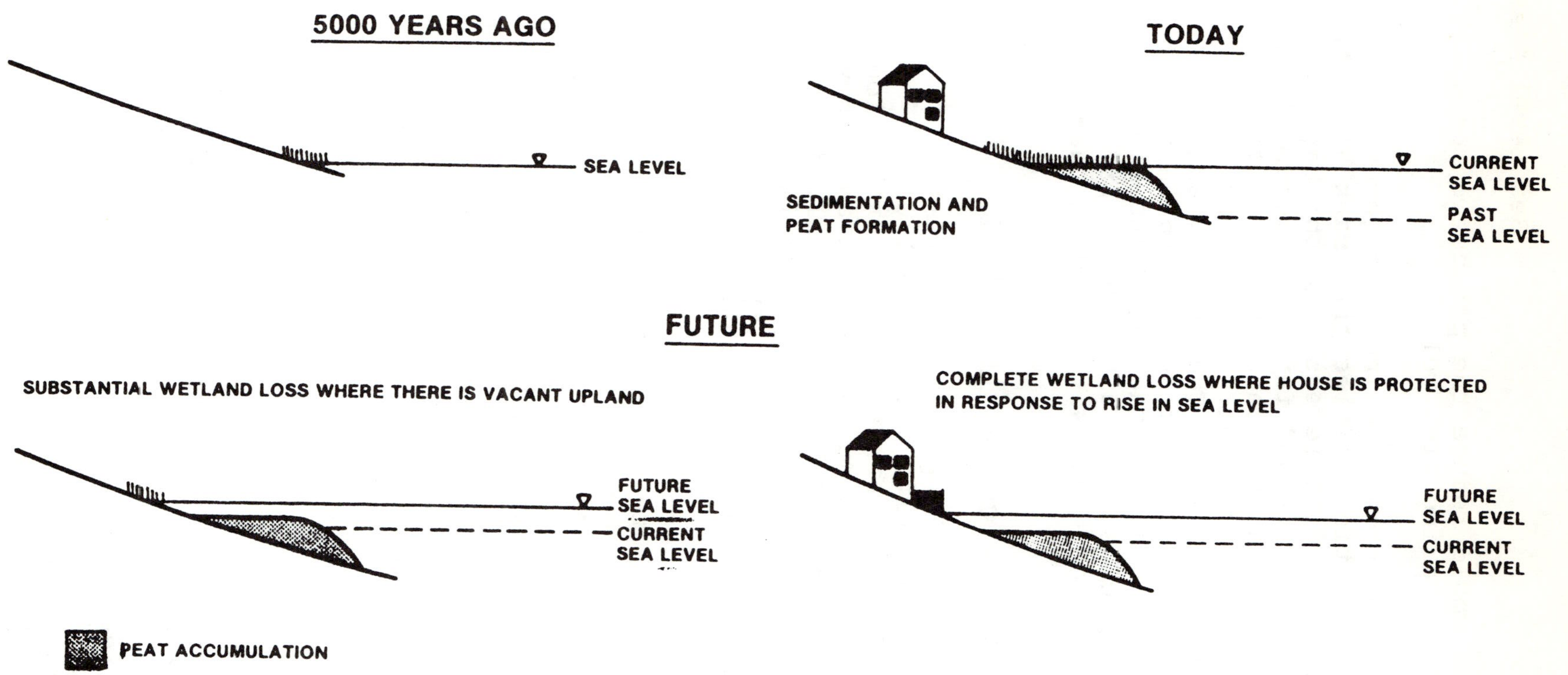

Figure 9-3. Evolution of marsh as sea rises. Coastal marshes have kept pace with the slow rate of sea level rise that has characterized the last several thousand years. Thus, the area of marsh has expanded over time as new lands were inundated. If in the future, sea level rises faster than the ability of the marsh to keep pace, the marsh area will contract. Construction of bulkheads to protect economic development may prevent new marsh from forming and result in a total loss of marsh in some areas.

Source: Titus, 1986.

birds and juvenile fish and would reduce the production of organic materials on which estuarine fish rely.

The dry land within 2 m of high tide includes forests, farms, low parts of some port cities, cities that sank after they were built and are now protected with levees, and the bay sides of barrier islands. The low forests and farms are generally in the mid-Atlantic and Southeast regions; these would provide potential areas for new wetland formation. Major port cities with low areas include Boston, New York, Charleston, and Miami. New Orleans is generally 8 feet below sea level, and parts of Galveston, Texas City, and areas around the San Francisco Bay are also well below sea level. Because they are already protected by levees, these cities are more concerned with flooding than with inundation.

Inundation and Erosion of Beaches and Barrier Islands

Some of the most important vulnerable areas are the recreational barrier islands and spits (peninsulas) of the Atlantic and Gulf coasts. Coastal barriers are generally long narrow islands and spits with the ocean on one side and a bay on the other. Typically, the oceanfront block of an island ranges from 5 to 10 feet above high tide, and the bay side is 2 to 3 feet above high water. Thus, even a 1-m sea level rise would threaten much of this valuable land with inundation.

Erosion threatens the high part of these islands and is generally viewed as a more immediate problem than the inundation of the bay sides. As Figure 9-4 shows, a rise in sea level can cause an ocean beach to retreat considerably more than it would from the effects of inundation alone. The visible part of the beach is much steeper than the underwater portion, which comprises most of the active "surf zone." While inundation alone is determined by the slope of the land just above the water, Bruun (1962) and others have shown that the total shoreline retreat from a sea level rise depends on the average slope of the entire beach profile.

Previous studies suggest that a 1-foot rise in sea level would generally cause beaches to erode 50-100 feet from the Northeast to Maryland (e.g., Kyper and Sorensen, 1985; Everts, 1985); 200 feet along the Carolinas (Kana et al., 1984); 100-1,000 feet along the Florida coast (Bruun, 1962); 200-400 feet along the California coast (Wilcoxen, 1986); and perhaps several miles in Louisiana. Because most U.S. recreational beaches are less than 100 feet wide at high tide, even a 1-foot rise in sea level would require a response. In many areas, undeveloped barrier islands could keep up with rising sea level by "overwashing" landward. In Louisiana, however, barrier

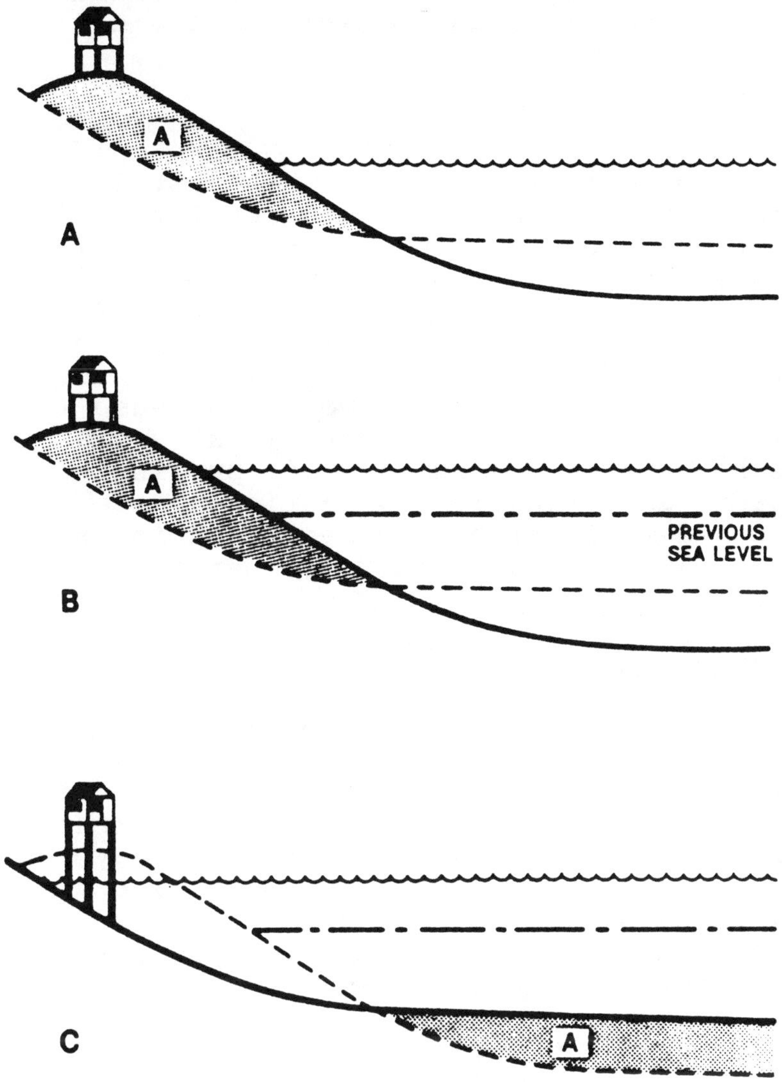

Figure 9-4. The Bruun Rule. A, initial condition; B, Immediate inundation when sea level rises; C, subsequent erosion due to sea level rise. A rise in sea level immediately results in shoreline retreat due to inundation, shown in the first two examples. However, a 1 m rise in sea level implies that the offshore bottom must also rise 1 m. The sand required to raise the bottom (A') can be supplied by beach nourishment. Otherwise, waves will erode the necessary sand (A) from upper part of the beach as shown in C.

Source: Titus, 1986.

islands are breaking up and exposing the wetlands behind them to gulf waves; consequently, the Louisiana barrier islands have rapidly eroded.

Flooding

Flooding would increase along the coast if sea level rises for three reasons: (1) A higher sea level provides a higher base for storm surges to build upon. A 1-m sea level rise would enable a 15-year storm to flood many areas that today are flooded only by a 100-year storm (e.g., Kana et al., 1984; Leatherman, 1984). (2) Beach erosion also would leave oceanfront properties more vulnerable to storm waves. (3) Higher water levels would reduce coastal drainage and thus would increase flooding attributable to rainstorms. In artificially drained areas such as New Orleans, the increased need for pumping could exceed current capacities. Finally, (4) a rise in sea level would raise water tables and would flood basements and in cases where the groundwater is just below the surface perhaps leach it above the surface.

Saltwater Intrusion

Finally, a rise in sea level would enable saltwater to penetrate farther inland and upstream into rivers, bays, wetlands, and aquifers. Salinity increases would be harmful to some aquatic plants and animals, and would threaten human uses of water. For example, increased salinity already has been cited as a factor contributing to reduced oyster harvests in the Delaware and Chesapeake Bays, and to conversion of cypress swamps in Louisiana to open lakes. Moreover, New York, Philadelphia, and much of California's Central Valley obtain their water from areas located just upstream from areas where the water is salty during droughts. Farmers in central New Jersey and the city of Camden rely on the Potomac-Raritan-Magothy aquifer, which could become salty if sea level rises (Hull and Titus, 1986). The South Florida Water Management District already spends millions of dollars every year to prevent Miami's Biscayne Aquifer from becoming contaminated with seawater.

Responses

The possible responses to inundation, erosion, and flooding fall broadly into three categories: erecting walls to hold back the sea, allowing the sea to advance and adapting to the advance, and raising the land. Both the slow rise in sea level over the last thousand years and the areas where land has been sinking more rapidly offer numerous historical examples of all three responses.

For over five centuries, the Dutch and others have used dikes and windmills to prevent inundation from the North Sea. By contrast, many cities have been rebuilt landward as structures have eroded; the town of Dunwich, England, has rebuilt its church seven times in the last seven centuries. More recently, rapidly subsiding communities (e.g., Galveston, Texas) have used fill to raise land elevations; the U.S. Army Corps of Engineers and coastal States regularly pump sand from offshore locations to counteract beach erosion. Venice, a hybrid of all three responses, has allowed the sea to advance into the canals, has raised some lowlands, and has erected storm protection barriers.

Most assessments in the United States have concluded that low-lying coastal cities would be protected with bulkheads, levees, and pumping systems, and that sparsely developed areas adapt to a naturally retreating shoreline (e.g., Dean etal., 1987; Gibbs, 1984; Schelling, 1983). This conclusion has generally been based on estimates that the cost of structural protection would be far less than the value of the urban areas being protected but would be greater than the value of undeveloped land.

Studies on the possible responses of barrier islands and moderately developed mainland communities show less agreement but generally suggest that environmental factors would be as important as economics. Some have suggested that barrier islands should use seawalls and other "hard" engineering approaches (e.g., Kyper and Sorensen, 1985; Sorensen, 1984). Others have pointed to the aesthetic problems associated with losing beaches and have advocated a gradual retreat from the shore (Howard et al., 1985). Noting that new houses on barrier islands are generally elevated on pilings, Titus (1986) suggested that communities could hold back the sea but keep a natural beach by extending the current practice of pumping sand onto beaches to raising entire islands in place.

Responses to erosion are more likely to have adverse environmental impacts along sheltered water than on the open coast (Titus, 1986). Because the beach generally is a barrier island's most important asset, economics would tend to encourage these communities to preserve their natural shorelines; actions that would prevent the island from breaking up also would protect the adjacent wetlands. However, along most mainland shorelines, economic self-interest would encourage property owners to erect bulkheads; these would prevent new wetland formation from offsetting the loss of wetlands that were inundated.

Most of the measures for counteracting saltwater intrusion attributable to sea level rise have also been employed to address current problems. For example, the Delaware River Basin

Commission protects Philadelphia's freshwater intake on the river and New Jersey aquifers recharged by the river by storing water in reservoirs during the wet season and releasing it during droughts, thereby forcing the saltwater back toward the sea. Other communities have protected coastal aquifers by erecting underground barriers and by maintaining freshwater pressure through the use of impoundments and injection wells.

HOLDING BACK THE SEA: A NATIONAL ASSESSMENT

The studies referenced in the previous section have illustrated a wide variety of possible effects from and responses to a rise in sea level from the greenhouse effect. Although they have identified the implications of the risk of sea level rise for specific locations and decisions, these studies have not estimated the nationwide magnitude of the impacts. This report seeks to fill that void.

It was not possible to estimate the nationwide value of every impact of sea level rise. The studies thus far conducted suggest that the majority of the environmental and economic costs would be associated with shoreline retreat and measures to hold back the sea, which can be more easily assessed on a nationwide basis, thus other impacts. Because the eventual impact will depend on what people actually do, a number of important questions can be addressed within this context:

- Would a gradual abandonment of moderately developed mainland areas significantly increase the amount of wetlands that survived a rise in sea level?
- Would the concave profiles of coastal areas ensure that more wetlands would be lost than gained, regardless of land-use decisions?
- Should barrier islands be raised in place by pumping sand and elevating structures and utilities?
- Would a landward migration of developed barrier islands or to encircling them with dikes and levees be feasible alternatives?
- How much property would be lost if barrier islands were abandoned?

Chapter 9

STRUCTURE OF STUDIES FOR THIS REPORT

Figure 9-5 illustrates the relationships between the various studies. As the top portion shows, investigators began with a case study of Long Beach Island, New Jersey. A case study was due to develop and evaluate methods for the national analysis. The Park and Leatherman studies performed the same calcualtion for the case study and the other sites included in the national assessment, with Park estimating the loss of wetlands and dryland, and Leatherman estimating the cost of dredging sufficient sand to raise recreational beaches and coastal barrier islands. However, Weggel and Yohe conducted detailed assessments for the case study that were not repeated for the other sites; the results were used in the Leatherman and Titus studies.

A central theme underlying these questions is that the implications of sea level rise for a community depend greatly on whether people adjust to the natural impact of shoreline retreat or undertake efforts to hold back the sea. Because no one knows the extent to which each of these approaches would be applied, this study was designed to estimate the impacts of sea level rise for (1) holding back the sea, and (2) natural shoreline retreat.

The tasks were split into five discreet projects:

1. Park et al. estimated the loss of coastal wetlands and dryland.
2. Leatherman estimated the cost of pumping sand onto open coastal beaches and barrier islands.
3. Weggel et al. estimated the cost of protecting sheltered shores with leaves and bulkheads.
4. Yohe began a national economic assessment by estimating the value of threatened property.
5. Titus and Greene synthesized the results of other studies to estimate ranges of the nationwide impacts.

Figure 9-5 illustrates the relationships between the various reports. As the top portion shows, the assessment began with a case study of Long Beach Island, New Jersey, which was necessary for evaluating methods and providing data for purposes of extrapolation. The Park and Leatherman studies performed the same calculations for the case study site that they would subsequently perform for the other sites in the nationwide analysis. However, Weggel and Yohe conducted more detailed assessments of the case study whose results were used in the Leatherman and Titus studies.

Because it would not be feasible for Leatherman to examine more than one option for the cost of protecting the open coast, Weggel estimated the cost of protecting Long Beach Island by

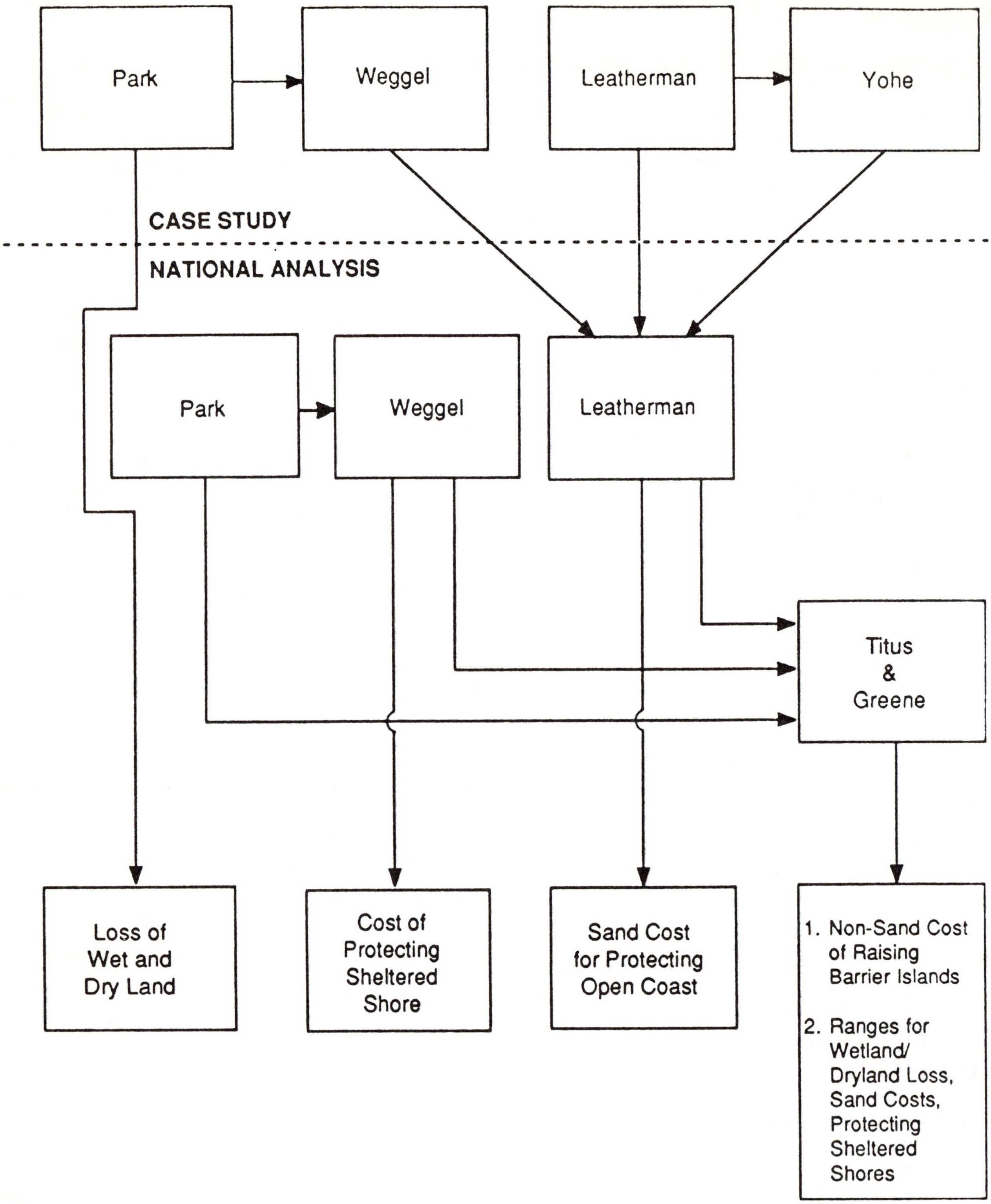

Figure 9-5. Overview of Sea Level Rise Studies-Authors

three approaches: (1) raising the island in place; (2) gradually rebuilding the island landward; and (3) encircling the island with dikes and levees; Yohe estimated the value of threatened structures. Titus analyzed Weggel's and Yohe's results and concluded that raising barrier islands would be the most reasonable option for the Leatherman study. They also noted that the cost of this option would be considerably less than the resources that would be lost if the islands are not protected as shown in Figure 9-6.

Once the case study was complete, Park, Leatherman, and Weggel proceeded independently with their studies (although Park provided Weggel with elevation data). When those studies were complete, Titus synthesized their results, developing a nationwide estimate of the cost of holding back the sea and interpolating Weggel's 200-cm results for the 50- and 100-cm scenarios.

In presenting results from the Park and Weggel studies, the sites were grouped into seven coastal regions, four of which are in the Southeast: New England, mid-Atlantic, south Atlantic, south Florida/gulf coast peninsula, Louisiana, other gulf (Texas, Mississippi, Alabama, Florida Panhandle), and the Pacific coast. Figure 9-7 illustrates these regions.

SCENARIOS OF SEA LEVEL RISE

Although the researchers considered a variety of scenarios of future sea level rise, this report focuses on the impacts of three scenarios: rises of 50, 100, and 200 cm by the year 2100. Following this convention of a recent National Research Council report (Dean et al., 1987), the rise was interpolated throughout the 21st century using a parabola, as shown in Figure 9-8. For each site, local subsidence was added to determine relative sea level rise. Along the coast of Florida (which is typical of most of the U.S. coast), sea level would rise 1 foot by 2025, 2040, and 2060 for the three scenarios and 2 feet by 2045, 2065, and 2100.

RESULTS OF SEA LEVEL STUDIES IN THIS REPORT

Loss of Coastal Wetlands and Dryland

Park sought to test a number of hypotheses presented in previous publications:

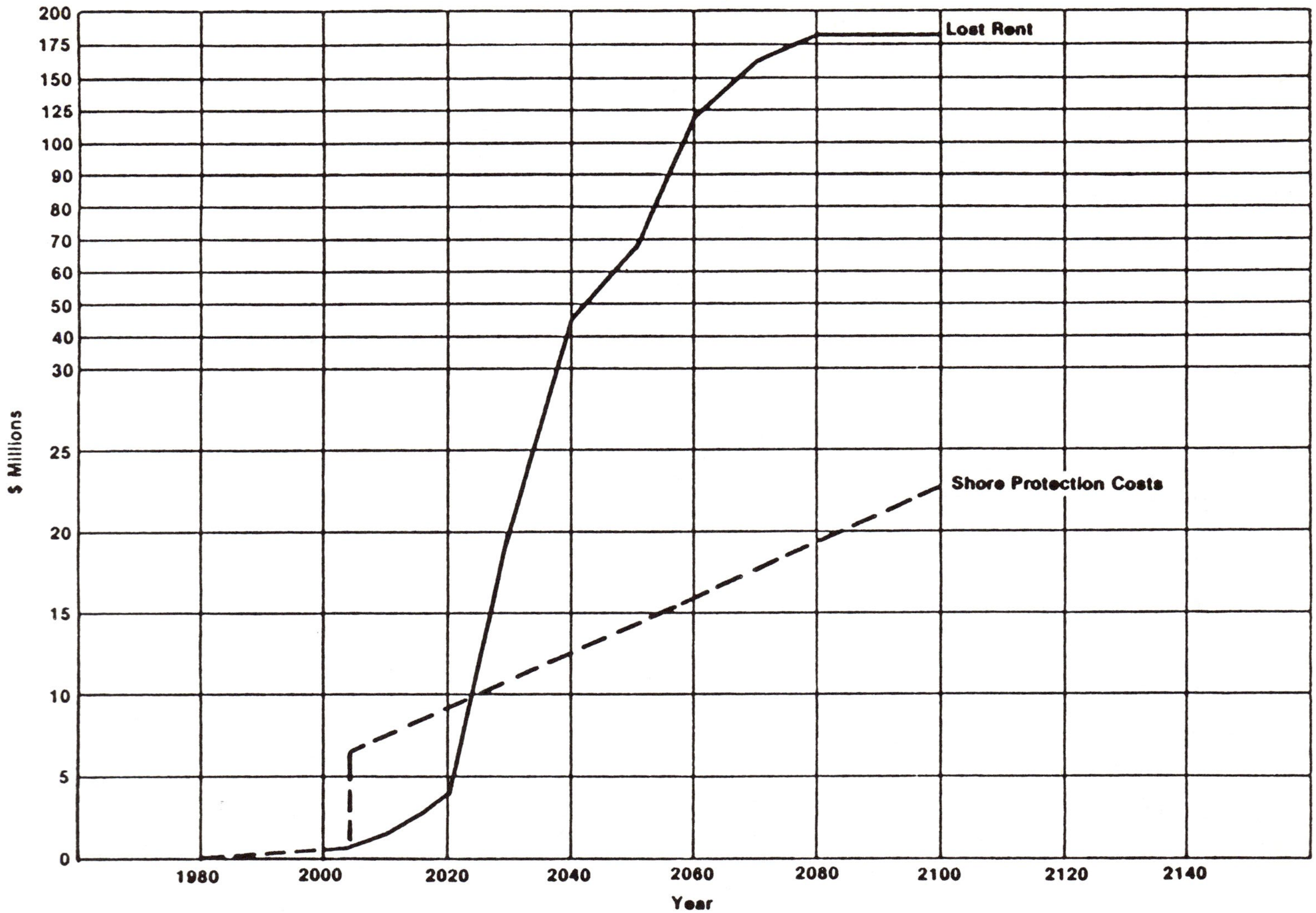

Figure 9-6. Annual Cost of Raising Island Versus Annual Costs (Lost Rent) From Not Protecting the Island

Source: Titus and Greene, Volume __.

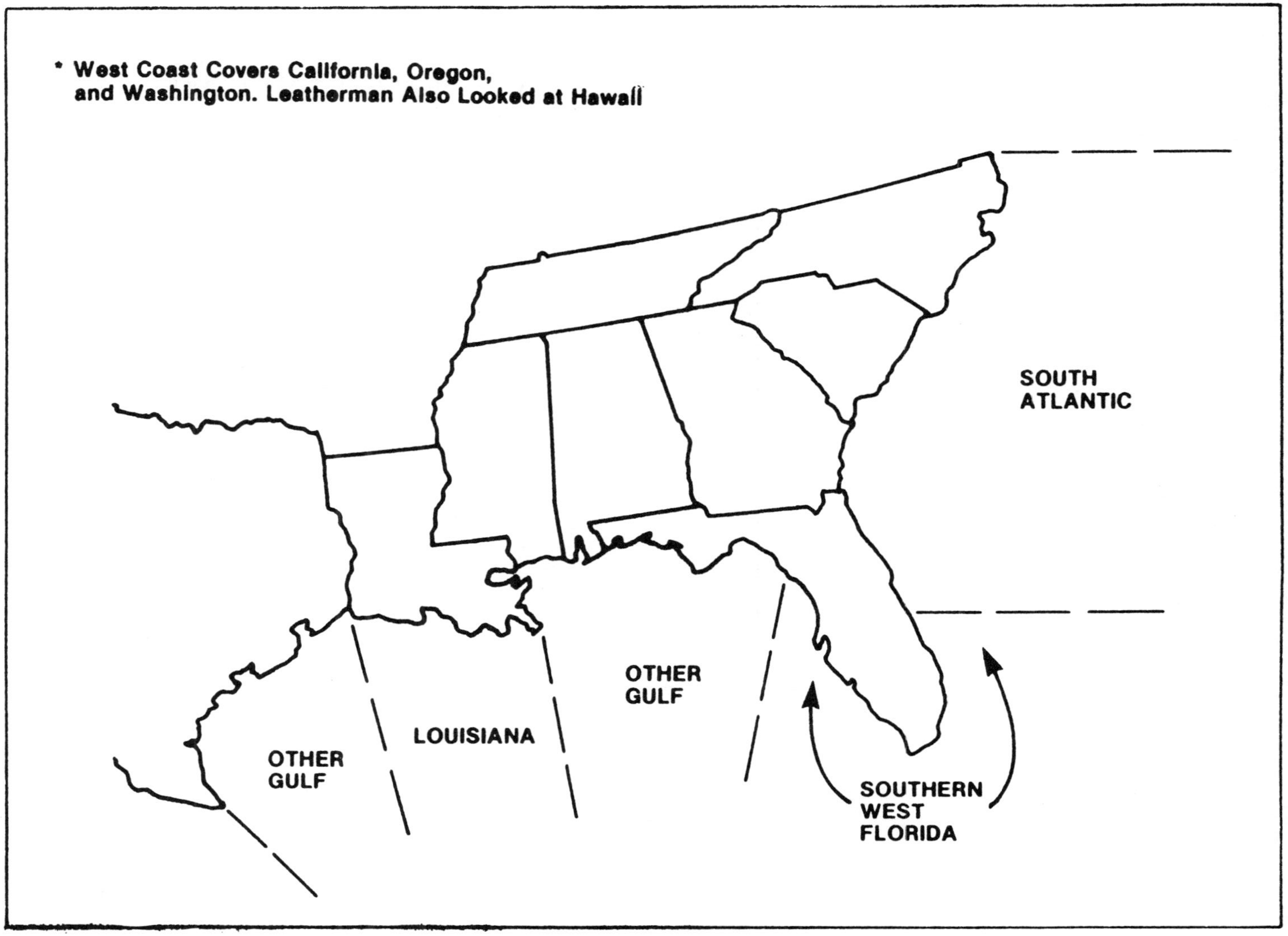

Figure 9-7. Coastal Regions Used In This Study

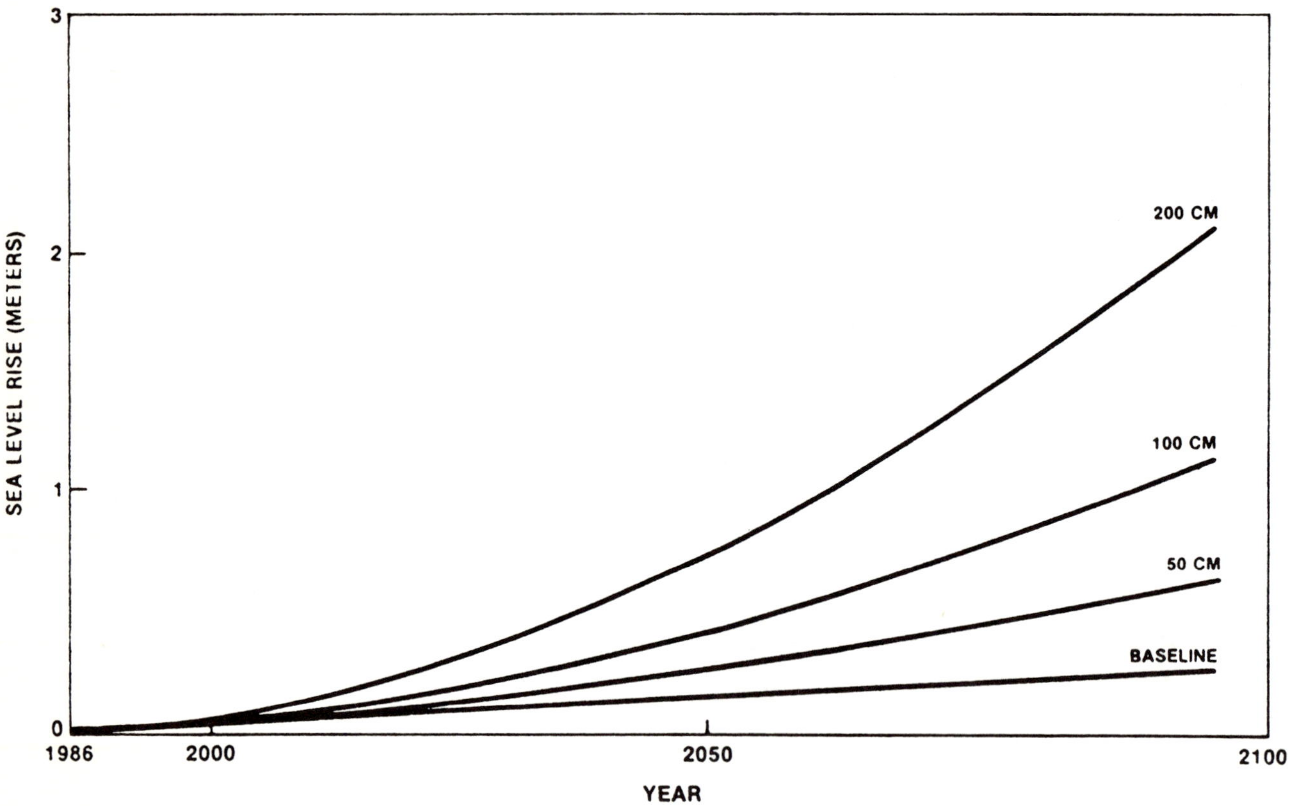

Figure 9-8. Sea Level Scenarios (Miami Beach)

- A rise in sea level greater than the rate of vertical wetland accretion would result in a net loss of coastal wetlands.

- The loss of wetlands would be greatest if all developed areas are protected, less if shorelines retreat naturally, and least if barrier islands are protected while mainland shores retreat naturally.

- The loss of coastal wetlands would be greatest in the Southeast, particularly Louisiana.

Study Design

Park's study was based on a sample of 46 coastal sites that were selected at regular intervals. This guaranteed that particular regions would be represented in proportion to their total area in the coastal zone. The sites chosen accounted for 10% of the U.S. coastal zone.

To estimate the potential loss of wet and dry land, Park first had to characterize their elevations. For wetlands, he used satellite imagery to determine plant species for 60- by 80-m parcels. Using estimates from the literature on the frequency of flooding that can be tolerated by various wetland plants, Park determined the percentage of time particular parcels currently under water. From this, Park inferred wet land elevation based on the known tidal range. For dry land, Park used spot elevation measurements to interpolate between contours on U.S. Geological Survey topographic maps.

Park estimated the loss of wetlands and dryland for no protection, protection of developed areas, and protection of all shores. For the no-protection scenario, estimating the loss of dryland is straightforward. However, for calculating wetland loss, Park used published vertical accretion rates (see Armentano et al., 1988). For accelerated sea level rise, he allowed for some acceleration of vertical accretion in areas with ample supplies of sediment, such as tidal deltas.

For sites in the Southeast, Park also allowed for the gradual replacement of salt marshes by mangrove swamps. The upper limit for mangroves is around Fort Lauderdale. Park used the GISS transient scenario to determine the year particular sites would be as warm as Fort Lauderdale is today and assumed that mangroves would begin to replace marsh after that year.

Limitations

The greatest uncertainty in Park's analysis is a poor understanding of the potential rates of vertical accretion. Although this could substantially affect the results for low sea level rise scenarios, the practical significance is small for a rise of 1 m because it is generally recognized that wetlands could not keep pace with the rise of 1 to 2 cm/year that such a scenario implies for the second half of the 21st century.

Errors can be made when determining vegetation type based on the use of infrared "signatures" that satellites receive. Park noted, for example, that in California the redwoods have a signature similar to marsh grass. For only a few sites Park was able to corroborate his estimates of vegetation type.

Park's study did not consider the potential implications of alternative methods of managing riverflow. This limitation is particularly serious regarding application to Louisiana, where widely varying measures have been proposed to increase the amount of water and sediment delivered to the wetlands. Finally, the study makes no attempt to predict which undeveloped areas might be developed in the next century.

At the coarse (500-m) scale Park used, the assumption of only protecting developed areas amounts to not protecting a number of mainland areas where the shoreline is developed but areas behind the shoreline are not. Therefore, Park's estimates for protecting developed areas should be interpreted as applying to the case where only densely developed areas are protected.

Results

Park's results supported the hypotheses suggested by previous studies. Figure 9-9 shows nationwide wetland loss for various (0- to 3-m) sea level rises for the three policy options investigated. For a 1-m rise, 66% of all coastal wetlands would be lost if all shorelines were protected, 49% would be lost if only developed areas were protected, and 46% would be lost if shorelines retreated naturally.

As expected, the greatest losses of wetlands would be in the Southeast, which currently contains 85% of U.S. coastal wetlands (Figure 9-9). For a 1-m sea level rise, 6,000-8,600 mi^2 (depending on which policy is implemented) of U.S. wetlands would be lost; 90-95% of these would be in the Southeast, and 40-50% would be in Louisiana alone. By contrast, neither the Northeast nor the West would lose more than 10% of its wetlands if only currently developed areas are protected.

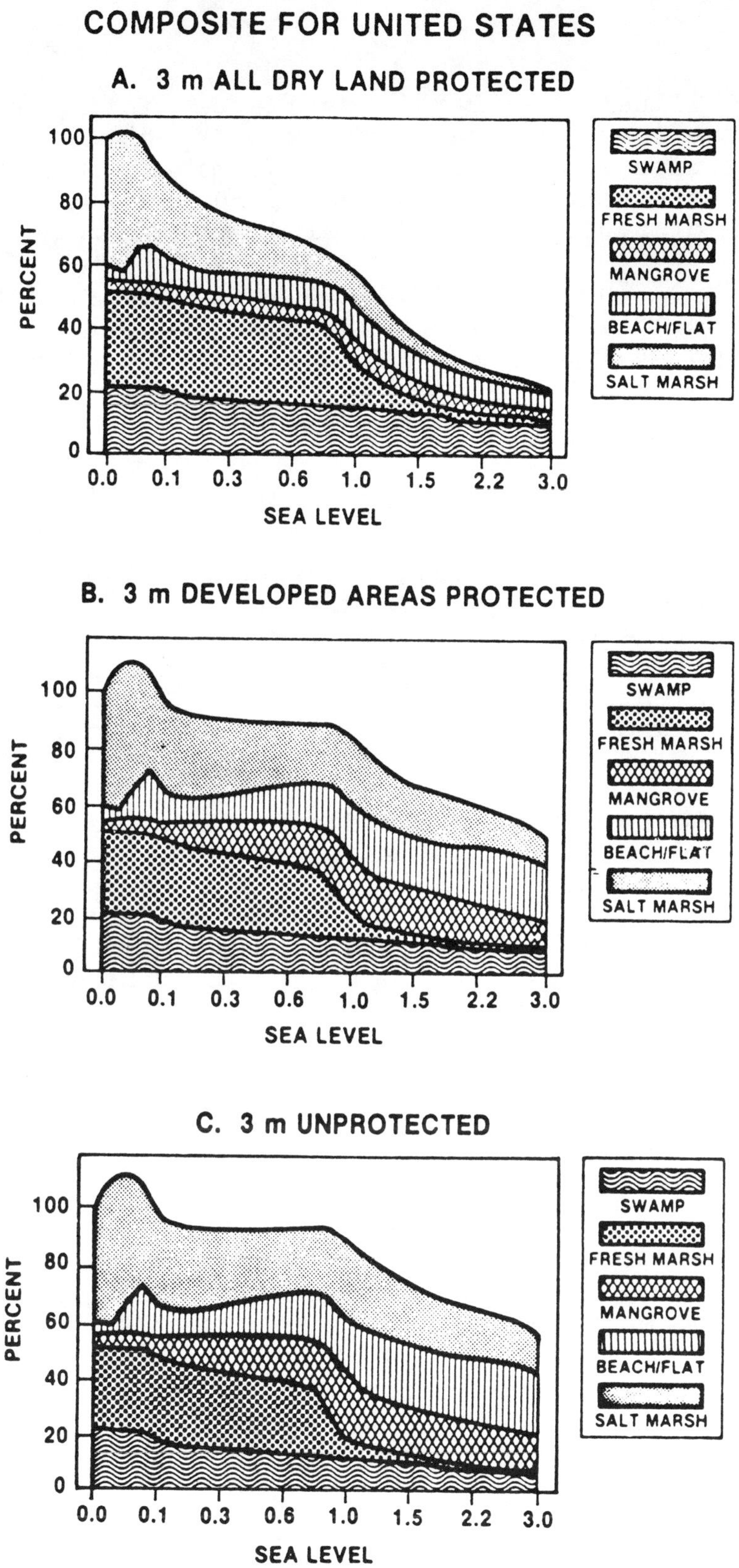

Figure 9-9. Nationwide Wetland Loss for Three Shoreline-Protection Options

Source: Park, Volume __.

Figure 9-10 illustrates Park's estimates of the inundation of dryland for the seven coastal regions. If shorelines retreat naturally, a 1-m rise would inundate 7,700 mi^2 of dryland, an area the size of Massachusetts. Rises of 50 and 200 cm would result in losses of 5,000 and 12,000 mi^2, respectively.

Approximately 70% of the dryland losses would occur in the Southeast, particularly Florida, Louisiana, and North Carolina. The eastern shores of the Chesapeake and Delaware Bays also would lose considerable acreage.

Costs of Defending Sheltered Shorelines

Study Design

This study began by examining Long Beach Island in depth. This site and five other sites were used to develop engineering rules of thumb for the cost of protecting coastal lowlands from inundation. Examining the costs of raising barrier islands required an assessment of two alternatives: (1) building a levee around the island; and (2) allowing the island to migrate landward.

After visiting Long Beach Island and the adjacent mainland, Weggel designed and estimated costs for an encirclement scheme consisting of a levee around the island and a drainage system that included pumping and underground retention of stormwater. For island migration, he used the Bruun Rule to estimate oceanside erosion and navigation charts to calculate the amount of sand necessary to fill the bay an equivalent distance landward. For island raising and island migration, Weggel used the literature to estimate the costs of elevating and moving houses and of rebuilding roads and utilities.

Weggel's approach for estimating the nationwide costs was to examine a number of index sites in depth and thereby develop generalized cost estimates for protecting different types of shorelines. He used the topographic information collected by Park for a sample of 95 sites to determine the area and shoreline length that had to be protected. He then applied the cost estimation factors to each site and extrapolated the sample to the entire coast.

After assessing Long Beach Island, Weggel conducted less detailed studies of the following areas: metropolitan New York; Dividing Creek, New Jersey; Miami and Miami Beach; the area around Corpus Christi, Texas; and parts of San Francisco Bay.

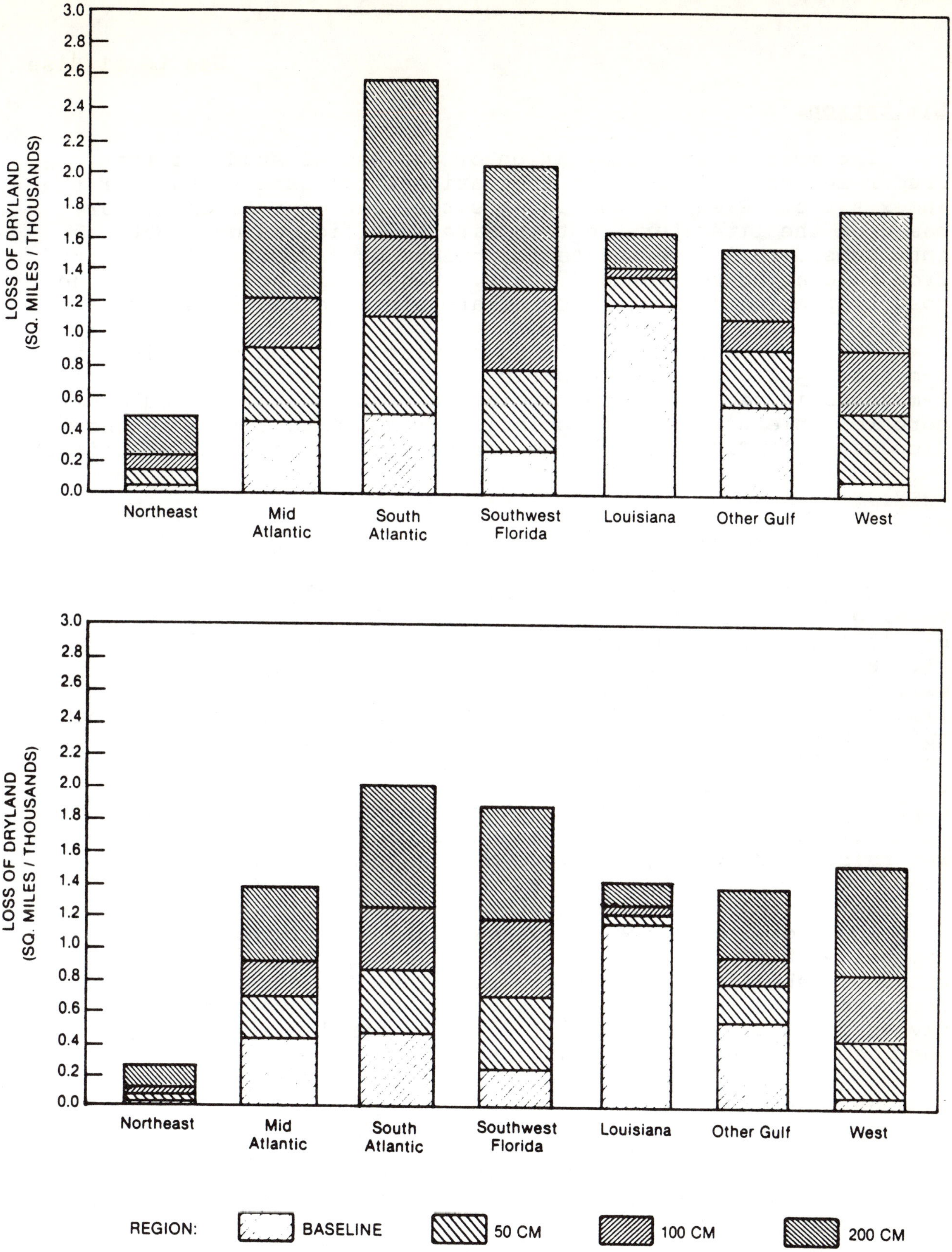

Figure 9-10. Loss of Dry Land by 2100 (a) without protection (b) with protection

Source: Derived from Park, Volume __; (see also Titus and Green, Volume __.)

Limitations

The most serious limitation of the Weggel study is that cruder methods are used for the national assessment than for the index sites. Even for the index sites, the cost estimates are based on the literature, not on site-specific designs that take into consideration wave data for bulkheads and potential savings from tolerating substandard roads. Weggel did not estimate the cost of pumping rainwater out of areas protected by levees.

Finally, Weggel was able to examine only one scenario: a 2-m rise by 2100. This scenario was chosen over the more likely 1-m scenario because an interpolation from 2 m to 1 m would be more reliable than an extrapolation from 1 m to 2 m. (See the discussion of Titus and Greene for results of the interpolation.)

Results

Case Study of Long Beach Island

Weggel's cumulative cost estimates clearly indicate that raising Long Beach Island would be much less expensive ($1.7 billion) than allowing it to migrate landward ($7.7 billion). Although the cost of building a levee around the island ($800 million) would be less, the "present value" would be greater (Table 9-1). Weggel concluded that the levee would have to be built in the 2020s, whereas the island could be raised gradually between 2020 and 2100. Thus, the (discounted) present value of the levee cost would be greater, and raising the necessary capital for a levee at any one time could be more difficult than gradually rebuilding the roads and elevating houses as the island was raised. Moreover, a levee would eliminate the waterfront view.

The Weggel analysis shows that landward migration is more expensive than island raising, primarily because of the increased costs of rebuilding infrastructure. Thus, migration might be less expensive in the case of a very lightly developed island. Levees might be more practical for wide barrier islands where most people do not have a waterfront view.

Nationwide Costs

Table 9-2 shows Weggel's estimates for the index sites and his nationwide estimate. The index sites represent two distinct patterns. Because urban areas such as New York and Miami would be entirely protected by levees, the cost of moving buildings and rebuilding roads and utilities would be relatively small. On the other hand, Weggel concluded that in more rural areas such as Dividing Creek, New Jersey, only the pockets of development would

Table 9-1. Total Cost of Protecting Long Beach Island From a 2-m Rise in Sea Level (millions of dollars)

Protective Measure	Encirclement	Island Raising	Island Migration
Sand Costs:			
Beach	290	290	0
Land creation/maintenance	NA	270	321
Moving/elevating houses	NA	74	37
Roads/utilities	0	1072	7352
Levee and drainage	542	0	0
Total	832	1706	7710

NA = Not applicable.

Source: Leatherman, Weggel.

Table 9-2. Cumulative Cost of Protecting Sheltered Waters (millions of 1986 dollars)

	New Bulkhead	Raise Old Bulkhead	Move Building	Roads/ Utils	Total
Index Sites					
New York	57	205	0.5	9.5	272.3
Long Beach	3	4	2.7	3.8	13.7
Dividing Creek	4	6	4.8	18.2	33.0
Miami area	11	111	0.3	8.3	130.7
Corpus Christi	11	29	2.8	40.9	83.4
San Francisco Bay	3	19	2.0	20.0	44.0

Nationwide Estimate

	Low	High
Northeast	6,932	23,607
Mid-Atlantic	4,354	14,603
Southeast	9,249	29,883
West	4,097	12,802
Nation	24,633	80,176

Source: Weggel et al.

be protected. The roads that connected them would have to be elevated or replaced with bridges, and the small number of isolated buildings would have to be moved.

Weggel estimates that the nationwide cost of protecting developed shorelines would be $25 billion, assuming bulkheads are built, and $80 billion assuming levees are built. Unlike wetland loss, the cost of protecting developed areas from the sea would be concentrated more in the Northeast than in the Southeast because a much greater portion of the southeastern coast is undeveloped.

Case Study of the Value of Threatened Coastal Property

Study Design

Yohe's objective was to estimate the loss of property that would result from not holding back the sea. Using estimates of erosion and inundation for Long Beach Island from Leatherman and Park et al., Yohe determined which land would be lost from sea level rise for a sample of strips spanning the island from the ocean to the bay. He then used the Ocean County, New Jersey, tax assessor's estimates of the value of the land and structures that would be lost, assuming that the premium associated with a view of the bay or ocean would be transferred to another property owner and not lost to the community. He estimated the annual stream of rents that would be lost by assuming that the required return on real estate is 10% after tax. Yohe assumed that a property on the bay side was "lost" whenever it was flooded at high tide, and that property on the ocean side was "lost" when the house was within 40 feet of the spring high tide mark. (See Titus and Greene, Volume x for discussion.)

Limitations

Yohe's results for a sea level rise of less than 18 inches are sensitive to the assumption regarding when a property would be lost. On the bay side, people might learn to tolerate tidal inundation. Unless a major storm occurred, people could probably occupy oceanfront houses until they were flooded at high tide. However, the resulting loss of recreational use of the beach probably would have a greater impact than abandoning the structure. Tax maps do not always provide up-to-date estimates of property values. However, the distinction between the tax assessor's most recent estimate of market value and the current market value is small compared with the possible changes in property values that will occur over the next century; hence, Titus and Greene used tax assessors estimates of market values.

Results

Yohe's results suggest that the cost of gradually raising Long Beach Island would be far less than the value of the resources that would be protected. Figure 9-6 compares Yohe's estimates of the annual loss in rents resulting from not holding back the sea with Weggel's estimates of the annual cost of raising the island for the 2-m scenario. With the exception of the 2020s, the annual loss in rents resulting from not holding back the sea would be far less than the annual costs of pumping sand and elevating structures.

Nationwide Cost of Pumping Sand Onto Recreational Beaches

Leatherman's goal was to estimate the cost of defending the U.S. ocean coast from a rise in sea level.

Study Design

Owing to time constraints, it was possible to consider only one technology. Based on the Long Beach Island results, Leatherman assumed that the cost of elevating recreational beaches and coastal barrier islands by pumping in offshore sand would provide a more representative cost estimate than assuming that barrier islands would be abandoned, migrate landward, or be encircled with dikes and levees.

The first step in Leatherman's analysis was to estimate the area of (1) the beach system, (2) the low bayside, and (3) the slightly elevated oceanside of the island. Given the areas, the volume of sand was estimated by assuming that the beach system would be raised by the amount of sea level rise. The bay and ocean sides of the island would not be raised until after a rise in sea level of 1 and 3 feet, respectively. Cost estimates for the sand were derived from inventories conducted by the U.S. Army Corps of Engineers.

Leatherman applied this method to all recreational beaches from Delaware Bay to the mouth of the Rio Grande, as well as California, which accounts for 80% of the nation's beaches. He also examined one representative site in each of the remaining States.

Limitations

Although the samples of sites in the Northeast and Northwest are representative, complete coverage would have been more accurate. Furthermore, Leatherman used conservative assumptions in estimating the unit costs of sand. Generally, a fraction of the sand placed on a beach washes away because the sand's grain is too small. Moreover, as dredges have to move further offshore to find sand, costs will increase.

For Florida, Leatherman used published estimates of the percentage of fine-grain sand and assumed that the dredging cost would rise $1 per cubic yard for every additional mile offshore the dredge had to move. For the other States, however, he assumed that the deposits mined would have no fine-grain sand and that dredging costs would not increase. (To test the sensitivity of this assumption, Titus and Greene developed an increasing-cost scenario.) Leatherman assumed no storm worse than the 1-year storm, which underestimates the sand volumes required.

A final limitation of the Leatherman study is that it represents the cost of applying a single technology throughout the ocean coasts of the United States. Undoubtedly, some communities (particularly Galveston and other wide barrier islands in Texas) would choose to erect levees and seawalls or to accept a natural shoreline retreat rather than to pump sand. This approach may overestimate the costs of pumping sand.

Results

Table 9-3 illustrates Leatherman's estimates. A total of 1,900 miles of shoreline would be nourished. Of 746 mi^2 of coastal barrier islands that would be raised for a 4-foot sea level rise, 208 mi^2 would be for a 2-foot rise. As the table shows, two-thirds of the nationwide costs would be borne by four southeastern states: Texas, Louisiana, Florida, and South Carolina.

Figure 9-11 illustrates the cumulative nationwide costs over time. For the 50- and 200-cm scenarios, the cumulative cost would be $2.3-4.4 billion through 2020, $11-20 billion through 2060, and $14-58 billion through 2100. By contrast, if current trends continue, the total cost of sea level rise for beach nourishment would be about $35 million per year.

Synthesis of the Three National Studies

Study Design

Although Weggel used Park's topographic data, the analysis in the three nationwide studies proceeded independently. Titus and Greene's primary objectives were to combine various results to estimate the nationwide cost of holding back the sea for

Table 9-3. Cost of Placing Sand on U.S. Recreational Beaches and Coastal Barrier Islands and Spits

State	Baseline	Sea Level Rise by 2100 50 cm	100 cm	200 cm
	($ millions)			
Maine[a]	22.8	119.4	216.8	412.2
New Hampshire[a]	8.1	38.9	73.4	142.0
Massachusetts[a]	168.4	489.5	841.6	1,545.8
Rhode Island[a]	16.3	92.0	160.6	298.2
Connecticut[a]	101.7	516.4	944.1	1,799.5
New York[a]	143.6	769.6	1,373.6	2,581.4
New Jersey[a]	157.6	902.1	1,733.3	3,492.5
Delaware	4.8	33.6	71.1	161.8
Maryland	5.7	34.5	83.3	212.8
Virginia	30.4	200.8	386.5	798.0
North Carolina	137.4	655.7	1,271.2	3,240.4
South Carolina	183.5	1,157.9	2,147.7	4,347.7
Georgia	25.9	153.6	262.6	640.3
Florida (Atlantic coast)	120.1	786.6	1,791.0[b]	7,745.5[b]
Florida (Gulf coast)	149.4	904.3	1,688.4[b]	4,091.6[b]
Alabama	11.0	59.0	105.3	259.6
Mississippi	13.4	71.9	128.3	369.5
Louisiana	1,955.8	2,623.1	3,492.7	5,231.7
Texas	349.6	4,188.3	8,489.7	17,608.3
California	35.7	174.1	324.3	625.7
Oregon[a]	21.9	60.5	152.5	336.3
Washington State[a]	51.6	143.0	360.1	794.4
Hawaii[a]	73.5	337.6	646.9	1,267.5
Nation	3,788.0	14,512.0	26,745.0	58,002.0

[a]Indicates States where estimate was based on extrapolating a representative site to the entire State. All other States have 100% coverage.

[b]Florida estimates account for the percentage of fine-grain sediment, which generally washes away, and for cost escalation as least expensive sand deposits are exhausted. All other estimates conservatively ignore this issue.

Source: Leatherman. (Baseline derived from Leatherman).

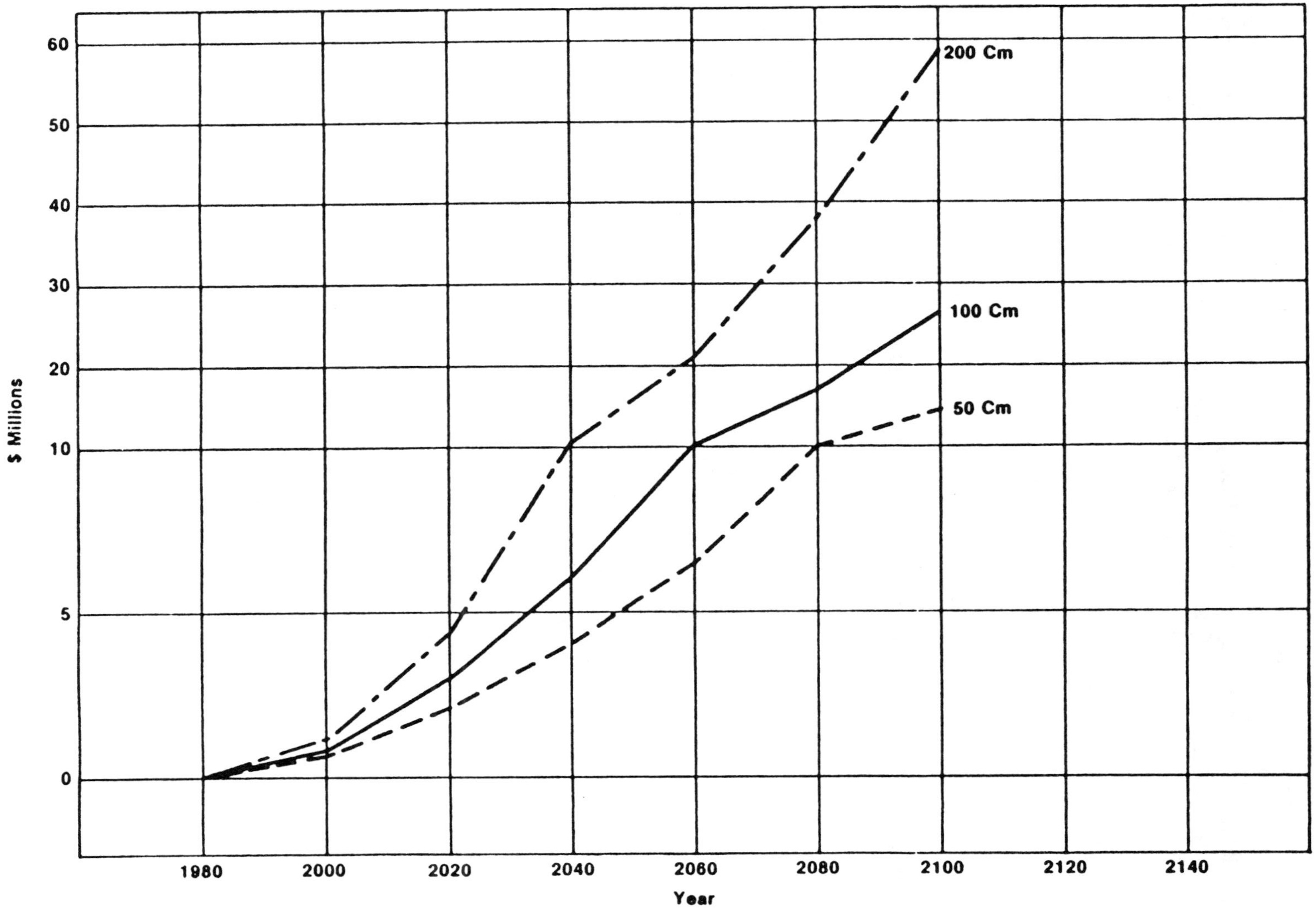

Figure 9-11. Nationwide Cost of Sand for Protecting Ocean Coast

Source: Leatherman, Volume -

various sea level rise scenarios and to derive ranges for the specific impacts.

Titus and Greene undertook the following primary tasks:

1. Use Park's results to weight Weggel's high and low scenarios according to whether levees or bulkheads would be necessary, and interpolate Weggel's cost estimate for the 2-m rise to rises of 50 and 100 cm.

2. Use results from Leatherman and Weggel, along with census data, to estimate the nationwide cost (other than pumping sand) of raising barrier islands.

3. Develop an increasing-cost scenario for the cost of protecting the open ocean coast.

4. Develop statistical confidence intervals for wetland loss, impacts of the various policy options, and costs of protecting developed shores.

<u>Cost of Protecting Sheltered Shores</u>

Titus and Greene developed a single estimate for protecting each site with bulkheads and levees by assuming that the portion of developed areas protected with levees would be equal to the portion of the lowlands that Park estimated would be inundated. They interpolated the resulting 2-m estimate to 50 and 100 cm estimates, based on Weggel's assumption that the cost of building bulkheads and levees rises as a function of the structure's height.

<u>Cost of Raising Barrier Islands Other Than Dredging</u>

Weggel's case study of Long Beach Island provided cost estimates for elevating structures and rebuilding roads, while Leatherman estimated the area that would have to be raised. Many barrier islands have development densities different from those of Long Beach Island because they have large tracts of undeveloped land or larger lot sizes. Therefore, Titus and Greene used census data to estimate a confidence interval for the average building density of barrier islands, and they applied Weggel's cost factors.

<u>Sensitivity of Sand Costs to Increasing Scarcity of Sand</u>

Titus and Greene used Leatherman's escalating cost assumptions for Florida to estimate sand pumping costs for the rest of the nation.

Confidence Intervals

The Park and Weggel studies involved sampling, but the researchers did not calculate statistical confidence intervals. Therefore, Titus and Greene developed 95% confidence intervals for the cost of protecting sheltered coasts, the area of wetland loss for various scenarios.

Limitations

Besides all of the limitations that apply to the Park, Leatherman, and Weggel studies, a number of others apply to Titus and Greene.

Cost of Protecting Sheltered Shores

Titus and Greene assumed that the portion of the coast requiring levees (instead of bulkheads) would be equal to the portion of lowlands that otherwise would be inundated. This assumption tends to understate the need for levees. For example, a community that is 75% high ground often would still have very low land along all of its shoreline and hence would require a levee along 100% of the shore. But Titus and Greene assume that only 25% would be protected by levees.

Cost of Raising Barrier Islands

The data provided by Weggel focused only on elevating roads, buildings, and bulkheads. Thus, Titus and Green do not consider the cost of replacing sewers, water mains, or buried cables. On the other hand, Weggel's cost factors assume that rebuilt roads would be up to engineering standards; it is possible that communities would tolerate substandard roads. In addition, the census data Titus and Greene used were only available for incorporated communities, many of which are part barrier island and part mainland; thus, the data provide only a rough measure of typical road density.

Sensitivity of Sand Costs to Increased Scarcity of Sand

Finally, Titus and Greene made no attempt to determine how realistic their assumption was that sand costs would increase by the same pattern nationwide as they would in Florida.

Results

Loss of Wetlands and Dryland

Table 9-4 illustrates 95% confidence intervals for the nationwide losses of wetlands and dryland. If all shorelines

Table 9-4. Nationwide Loss of Wetlands and Drylands

	Square Miles (%)			
	Baseline	50 cm	100 cm	200cm
Wetlands				
Total protection	N.C.	4944-8077 (38-61)	6503-10843 (50-82)	8653-11843 (66-90)
Standard protection	1168-3341 (9-25)	2591-5934 (20-45)	3813-9068 (29-69)	4350-10995 (33-80)
No protection	N.C.	2216-5592 (17-43)	3388-8703 (26-66)	3758-10025 (29-76)
Drylands				
Total protection	--	0	0	0
Standard protection	--	2180-6147	4136-9186	6438-13496
No protection	--	3315-7311	5123-10330	8791-15394

N.C. = Not Calculated

Source: Titus and Greene, Volume __.

were protected, a 1-m rise would result in a loss of 50-82% of U.S. coastal wetlands, and a 2-m rise would result in a loss of 66-90%. If only the densely developed areas are protected, the losses would be 29-69% and 61-80% for the 1- and 2-m scenarios, respectively. Except for the Northeast, no protection results in only slightly lower wetland loss than protecting only densely developed areas. Although the estimates for the Northeast, mid-Atlantic, the Gulf regions outside Louisiana, and the Florida peninsula are not statistically significant (at the 95% confidence levels), results suggest that wetland loss would be least in the Northeast and Northwest.

Costs of Holding Back the Sea

Table 9-5 illustrates the Titus and Greene estimates of the costs of holding back the sea. The low range for the sand costs is based on Leatherman's study, and the high range is based on the increasing cost scenario Titus and Greene developed. The uncertainty range for the costs of elevating structures reflects the uncertainty in census data regarding the current density of development. High and low estimates for the cost of protecting sheltered shorelines are based on the sampling errors of the estimates for the 46 sites that both Park et al. and Weggel et al. examined.

Titus and Greene estimated that the cumulative nationwide cost of protecting currently developed areas in the face of a 1-m rise would be from $73 to 111 billion, with costs for the 50- and 200-cm scenarios ranging from $32 to 309 billion.

POLICY IMPLICATIONS

Wetland Protection

The nationwide analysis showed that a 50- to 200-cm rise in sea level could reduce the coastal wetland acreage (outside Louisiana) by 17-76% if no mainland areas are protected, by 20-80% if only currently developed areas are protected, and by 38-90% if all mainland areas are protected. These estimates of the areal losses understate the differences in impacts for the various land-use options. Although a substantial loss would occur even with no protection, most of today's wetland shorelines still would have wetlands; the strip simply would be narrower. By contrast, protecting all mainland areas would generally replace natural shorelines with bulkheads and levees. This distinction is important because for many species of fish, the length of a wetland shoreline is more critical than the total area.

Table 9-5. Cumulative Nationwide Cost of Protecting Barrier Islands and Developed Mainland Through the Year 2100 (billions of dollars)

	Sea Level Scenario		
	50 cm	100 cm	200 cm
Open Coast			
Sand	15-20	27-41	58-100
Raise houses, roads, utils	9-13	21-57	75-115
Sheltered shores	5-13	11-33	30-101
Total*	32-43	73-111	119-309

Source: Titus and Greene (Volume ___).

*Note: Range for totals are based on the square root of the sum of squared ranges.

Chapter 9

Options for State and Local Governments

Titus (1986) examined three approaches for maintaining wetland shorelines in the face of a rising sea: (1) no further development in lowlands; (2) no action now but a gradual abandonment of lowlands as sea level rises; and (3) allowing future development only with a binding agreement to allow such development to revert to nature if it is threatened by inundation.

The first option would encounter legal or financial hurdles. The extent to which the "due-process" clause of the Constitution would allow governments to prevent development in anticipation of sea level rise has not been specifically addressed by the courts. Although purchases of land would be feasible for parks and refuges, the cost of buying the majority of lowlands would be prohibitive. Moreover, this approach requires preparation for a rise in sea level of a given magnitude; if and when the sea rises beyond that point, the wetlands would be lost. Finally, preventing future development would not solve the loss of wetlands resulting from areas that have already been developed.

Enacting no policy today and addressing the issue as sea level rises would avoid the costs of planning for the wrong amount of sea level rise but would probably result in less wetland protection. People are developing coastal property on the assumption that they can use the land indefinitely. It would be difficult for any level of government to tell property owners that they must abandon their land with only a few years' notice, and the cost of purchasing developed areas would be even greater than the cost of buying undeveloped areas.

Economic theory suggests that under the third alternative, people would develop a property only if the temporary use provided benefits greater than the costs of writing it off early. This approach would result in the greatest degree of flexibility, because it would allow real estate markets to incorporate sea level rise and to determine the most efficient use of land as long as it remains dry.

This approach could be implemented by regulations that prohibit construction of bulkheads as sea level rises or by the use of conditional long-term leases that expire when high tide falls above a property's elevation.

Maine (1987) has implemented this third approach through its coastal dune regulations, which state that people building houses along the shore should assume that they will have to move their houses if their presence prevents the natural migration of coastal wetlands, dunes, or other natural shorelines. A number of States also have regulations that discourage bulkheads, although they do not specifically address sea level rise. The option can be implemented through cooperative arrangements between developers, conservancy groups, and local governments. (See Titus and Greene for additional details.)

The Federal Role

Section 404 of the Clean Water Act discourages development of existing wetlands, but it does not address development of areas that might one day be necessary for wetland migration. This program will provide lasting benefits, even if most coastal wetlands are inundated. Although marshes and swamps would be inundated, the shallow waters that formed could provide habitat for fish and submerged aquatic vegetation. No one has assessed the need for a Federal program to protect wetlands in the face of rising sea level.

Coastal Protection

State and Local Efforts

State and local governments currently decide which areas would be protected and which would be allowed to erode. Currently, few localities contribute more than 10% of the cost of beach nourishment, with the States taking on an increasing share from the Federal Government. However, if global warming increases the costs of coastal protection over the next century by $50-300 billion, many coastal officials doubt that their States could raise the necessary funds. If State funds could not be found, the communities themselves would have to take on the necessary expenditures or adapt to erosion.

Long Beach Island, New Jersey, illustrates the potential difficulties. The annual cost of raising the island would average $200-1,000 per house over the next century (Titus and Greene, Volume __1). Although this amount is less than one week's rent during the summer, it would more than double property taxes, an action that is difficult for local governments to contemplate. Moreover, the island is divided among six jurisdictions, all of which would have to participate.

More lightly developed communities may decide that the benefits of holding back the sea are not worthwhile. Sand costs would be much less for an island that migrated. Although Weggel estimated that higher costs would be associated with allowing Long Beach Island to migrate landward than with raising the island in place, this conclusion resulted largely from the cost of rebuilding sewers and other utilities that would still be useful if the island were raised.

Regardless of how a barrier island community intends to respond to sea level rise, the eventual costs can be reduced by deciding on a response well in advance. The cost of raising an island can be reduced if roads and utilities are routinely elevated or if they have to be rebuilt for other reasons (e.g., Titus et al., 1987). The cost of a landward migration also can be reduced by discouraging reconstruction of oceanfront houses destroyed by storms (Titus et al., 1984a). The ability to fund the required measures also would be increased by fostering the necessary public debate well before the funds are needed.

Federal Efforts

While State governments generally are responsible for protecting recreational beaches, the U.S. Army Corps of Engineers is responsible for several major Federal projects to rebuild beaches and for efforts to curtail land loss in Louisiana. The long-term success of these efforts would be improved if the Corps were authorized to develop comprehensive long-term plans to address the impacts of sea level rise.

Beach Erosion

In its erosion-control efforts, the Corps has recently shifted its focus from hard structures (e.g., seawalls, bulkheads, and groins) to soft approaches, such as pumping sand onto beaches. This shift is consistent with the implications of sea level rise: groins and seawalls will not prevent loss of beaches due to sea level rise. Although more sand will have to be pumped than current analyses suggest, this approach clearly could ensure the survival of the nation's beaches.

Nevertheless, consideration of accelerated sea level rise would change the cost-benefit ratios of many Corps erosion control projects. As with the operations of reservoirs (discussed in Chapter 6: Southeast), the Corps is authorized to consider flood protection but not recreation. When they evaluated the benefits of erosion control at Ocean City, Maryland, the Corps concluded that less than 10% of the benefits would be for flood control (most were related to recreation). Had they considered accelerated sea level rise, however, the estimated flood protection benefits from having a protective beach would have constituted a considerably higher fraction of the total benefits (Titus, 1985).

Wetland Loss in Louisiana

By preventing freshwater and sediment from reaching the coastal wetlands, Federal management of the Mississippi River is increasing the vulnerability of coastal Louisiana to a sea level rise (e.g., Houck, 1983). For example, current navigation routes require the Corps of Engineers to limit the amount of water flowing through the Atchafalaya River and to close natural breaches in the main channel of the Mississippi; these actions limit the amount of freshwater and sediment reaching the wetlands. Alternative routes have been proposed that would enable water and sediment to reach the wetlands (Louisiana Wetland Protection Panel, 1987). These include dredging additional canals parallel to the existing Mississippi River gulf outlet or constructing a deepwater port east of the city.

Either of these options would cost a few billion dollars. By contrast, annual resources for correcting land loss in Louisiana have been in the tens of millions of dollars. As a result, mitigation activities have focused on freshwater diversion structures and on other strategies that can reduce current wetland loss attributable to high salinities but that would not substantially reduce wetland loss if sea level rises 50-200 cm (Louisiana Wetland Protection Panel, 1987).

The prospect of even a 50 cm rise in sea level suggests the solving the Louisiana wetland loss problem is much more urgent than is commonly assumed. Because federal activities are now a major cause of land loss, and would have to be modified to enable wetlands to survive a rising sea, the problem is unlikely to be solved without a Congressional mandate. A recent interagency report concluded that "no one has systematically determined what must be done to save 10, 25, or 50 percent of Louisiana's coastal ecosystem." (Louisiana Wetland Protection Panel, 1988). Until someone estimates the costs and likely results of strategies with a chance of protecting a significant fraction of the wetlands in face of rising sea level, it will be difficult for Congress to devise a long-term solution.

Flood Insurance

In 1968, Congress created the National Flood Insurance Program with the objective of reducing Federal disaster relief resulting from floods. The Federal Emergency Management Agency (FEMA), which already was responsible for administering disaster relief, was placed in charge of this program as well.

The National Flood Insurance Program sought to offer localities an incentive to prevent flood-prone construction. In return for requiring that any construction in a floodplain be designed to withstand a 100-year flood, the federal government would provide subsidized insurance to existing homes and a fair-market rate for any new construction (which was itself a benefit, since private insurers generally did not offer flood insurance). Moreover, as long as a community joined the program, it would continue to be eligible for federal disaster relief; if it did not join, it would no longer be eligible. As a result of this program, new coastal houses are generally elevated on pilings.

Although Congress intended to prevent coastal disasters, the National Flood Insurance Act does not require strategic assessments of long-term issues (see Riebsame, Volume _). Thus, FEMA has not conducted strategic assessments of how the program could be managed to minimize flood damage from shoreline retreat caused by both present and future rates of sea level rise.

Congress recently enacted the Upton-Jones Amendment (Public Housing Act of 1988), which commits the federal government to pay for rebuilding or relocating houses that are about to erode into the sea. Although the cost of this provision is modest today, a sea level rise could commit the federal Government to purchase the houses on all barrier islands that did not choose to hold back the sea. Furthermore, this commitment could increase the number of communities that decided not to hold back the sea.

The planned implementation of actuarially sound insurance rates would ensure that as sea level rise increased property risk, insurance rates would rise to reflect the risk. This would discourage construction of vulnerable houses, unless their value was great enough to outweigh the likely damages from floods. However, statutes limiting the rate at which flood insurance rates can increase could keep rates from rising as rapidly as the risk of flooding, thereby increasing the federal subsidy.

No assessment of the impacts of sea level rise on the federal flood insurance program has been undertaken.

Sewers and Drains

Sea level rise also would have important impacts on coastal sewage and drainage systems. Wilcoxen (1986) examined the implications of the failure to consider accelerated sea level rise in the design of San Francisco's West Side (sewerage) Transport, a large steel-reinforced concrete box buried under the city's ocean beach. He found that beach erosion will gradually expose the transport to the ocean. Although the structure probably could withstand wave attack, it would act as a seawall and eventually result in loss of the recreational beach or necessitate shore-protection projects that, over the course of a century, would cost more than the project itself. Wilcoxen concludes that had sea level rise been considered, the project probably would have been sited elsewhere.

The impacts of sea level rise on the construction grants program probably would be less in most other cases. As sea level rises, larger pumps would be necessary to transport effluents from settling ponds to the adjacent body of water. However, sea level rise would not necessarily require alternative siting. The projects serving barrier islands often are located on the mainland, and projects located on barrier islands are generally elevated well above flood levels. If barrier islands are raised in response to sea level as the nationwide analysis suggests, sewerage treatment plants will be a small part of the infrastructure that has to be modified.

Engineering assessments have concluded that it is already cost-beneficial to consider sea level rise in the construction of coastal drainage systems in urban areas. For example, the extra cost of installing the larger pipes necessary to accommodate a 1-foot rise in sea level would add less than 10% to the cost of rebuilding a drainage system in Charleston, South Carolina; however, failure to consider sea level rise would require premature rebuilding of the $4 million system (Titus et al., 1987).

RESEARCH NEEDS

A much better understanding of erosion processes is needed 1) to understand how much erosion will take place if no action is taken; and 2) to help identify the most cost-effective means for protecting sandy shores. An improved understanding of how wetland accretion responds to different temperatures, higher CO_2 concentrations, changing mineral content, and the drowning of adjacent wetlands is needed. This will refine our ability to project future wetland loss and, perhaps, devise measures for artificially enhancing their vertical growth.

This report did not examine the impacts of increased flooding because flood models have not been applied to the large numbers of coastal sites that would be necessary to conduct a nationwide assessment. Time dependant estuarine salinity models, such as that of the Delaware River Basin Commission should be applied to major estuaries for examining impacts on ecosystems and drinking water supplies.

Assessments of the impacts of global warming on coastal environments would be greatly improved by better estimates of future sea level rise. In addition to the improved ocean modelling that will be necessary for better projections of surface air temperatures (see Chapter 2: Global Climate Change), this will also require a substantial increase in the resources allocated for monitoring and modelling glacial processes. Finally, this report assumed that winds, waves, and storms remained constant; future studies will need estimates of the changes in these climatic variables.

REFERENCES

Armentano, T.V., R.A. Park, and C.L. Cloonan. 1988. Impacts on coastal wetlands throughout the United States. In: Titus, J.G. ed. Greenhouse Effect, Sea Level Rise, and Coastal Wetlands. Washington, DC: EPA.

Barnett, T.P. 1984. The estimation of "global" sea level change: a problem of uniqueness. Journal of Geophysical Research 89(C5):7980-7988.

Barth, M.C., and J.G. Titus, eds. 1984. Greenhouse Effect and Sea Level Rise: A challenge for this generation. New York: Van Nostrand Reinhold.

Bentley, C.R. 1983. West Antarctic ice sheets: diagnosis and prognosis. In: Proceedings of the Carbon Dioxide Research Conference: Carbon Dioxide, Science, and Consensus. Conference 820970. Washington, DC: Department of Energy.

Bruun, P. 1962. Sea level rise as a cause of shore erosion. Journal of Waterways and Harbors Division (ASCE) 1:116-130.

Dean, R.G. et al. 1987. Responding to Changes in Sea Level. Washington, DC: National Academy Press.

Dony, W.L., W.R. Farrand, and M. Ewing. 1962. Pleistocene ice volumes and sea level lowering. Journal of Geology 70:206-214.

Everts, C.H. 1985. Effect of sea level rise and net sand volume change on shoreline position at Ocean City, Maryland. In: Titus, J.G., ed. Potential Impact of Sea Level Rise on the Beach at Ocean City, Maryland. Washington, DC: U.S. Environmental Protection Agency.

Fairbridge, R.W., and W.S. Krebs Jr. 1962. Sea level and the southern oscillation. Geophysical Journal 6:532-545.

Gibbs, M. 1984. Economic analysis of sea level rise: methods and results. In: Barth, M.C., and J.G. Titus, eds. Greenhouse Effect and Sea Level Rise: A challenge for this generation. New York: Van Nostrand Reinhold.

Gornitz, V.S., S. Lebedeff, and J. Hansen. 1982. Global sea level trend in the past century. Science 215:1611-1614.

Hoffman, J.S., D. Keyes, and J.G. Titus. 1983. Projecting Future Sea Level Rise. Washington, DC: U.S. Environmental Protection Agency.

Hoffman, J.S., J. Wells, and J.G. Titus. 1986. Future global warming and sea level rise. In: Sigbjarnarson, G., ed. Iceland Coastal and River Symposium. Regkjavik/Iceland: National Energy Authority.

Houck, O.A. 1983. Land loss in coastal Louisiana: causes, consequences, and remedies. Tulane Law Review 58(1):3-168.

Howard, J.D., O.H. Pilkey, and A. Kaufman. 1985. Strategy for beach preservation proposed. Geotimes. 30(12):15-19.

Hughes, T. 1983. The stability of the West Antarctic Ice Sheet: What has happened and what will happen. In: Proceedings of the carbon Dioxide Research Conference: Carbon Dioxide, Science, and Consensus. Conference 820970. Washington, DC: Department of Energy.

Hull, C.H.J., and J.G. Titus. 1986. Responses to salinity increases. In: Hull, C.H.J., and J.G. Titus. 1986. Greenhouse Effect, Sea Level Rise, and Salinity in the Delaware Estuary. Washington, DC: U.S. Environmental Protection Agency and Delaware River Basin Commission.

Hull, C.H.J., and J.G. Titus. 1986. Greenhouse Effect, Sea Level Rise, and Salinity in the Delaware Estuary. Washington, DC: U.S. Environmental Protection Agency and Delaware River Basin Commission.

Kana, T.W., J. Michel, M.O. Hayes, and J.R. Jensen. 1984. The physical impact of sea level rise in the area of Charleston, South Carolina. In: Barth, M.C., and J.G. Titus, eds. Greenhouse Effect and Sea Level Rise: A challenge for this generation. New York: Van Nostrand Reinhold.

Kennett. 1982. Marine Geology. Englewood Cliffs, N.J: Prentice-Hall.

Kyper, T., and R. Sorenson. 1985. Potential impacts of sea level rise on the beach and coastal structures at Sea Bright, New Jersey. In: O.T. Magson, ed. Coastal Zone '85. New York: ASCE.

Leatherman, S.P. 1984. Coastal geomorphic responses to sea level rise: Galveston Bay, Texas. In: Barth, M.C., and J.G. Titus, eds. Greenhouse Effect and Sea Level Rise: A challenge for this generation. New York: Van Nostrand Reinhold.

Lyle et al. 1987. Sea Level Variations in the United States Rockville, MD: National Ocean Service.

Maine, State of. 1987. Dune Rule 355. August 9: Maine Department of Environmental Protection.

Meier, M.F. et al. 1985. Glaciers, Ice Sheets, and Sea Level. Washington, DC: National Academy Press.

Meier, M.F. 1984. Contribution of small glaciers to global sea level. Science 226(4681):1418-1421.

Mercer, J.H. 1968. Antarctic ice and Sangamon sea level. Geological Society of America Bulletin 79:471.

Oldale, R. 1985. Late quaternary sea level history of New England: a review of published sea level data. Northeastern Geology 7:192-200.

Revelle, R. 1983. Probable future changes in sea level resulting from increased atmospheric carbon dioxide. In Changing Climate. Washington, DC: National Academy Press.

Schelling, T. 1983. In: Changing Climate. Washington, DC: National Academy Press.

Sorensen, R.M., R.N. Weisman, and G.P. Lennon. 1984. Control of erosion, inundation, and salinity intrusion. In: Barth, M.C., and J.G. Titus, eds. Greenhouse Effect and Sea Level Rise: A challenge for this generation. New York: Van Nostrand Reinhold.

Thomas, R.H. 1986. In: Titus, J.G., ed. Effects of Changes in Stratosphere Ozone and Global Climate. Washington, DC: U.S.EPA and UNEP.

Thomas, R.H. 1985. Responses of the polar ice sheets to climatic warming. In: Meier, M.F. et al. Glaciers, Ice Sheets, and Sea Level. Washington, DC: National Academy Press.

Titus, J.G. 1988. Sea Level Rise and Wetlands Loss: An Overview. In: Titus, J.G. (ed). Greenhouse Effect, Sea Level Rise, and Coastal Wetlands. Washington, DC: U.S. Environmental Protection Agency.

Titus, J.G., ed. 1988. Greenhouse Effect, Sea Level Rise, and Coastal Wetlands. Washington, DC: U.S. Environmental Protection Agency.

Chapter 9

Titus, J.G. 1987. The Greenhouse Effects, Rising Sea Level, and Society's Response. In: Devoy, R.J.N. Sea Surface Studies. New York: Croom Helm.

Titus, J.G. 1986. Greenhouse effect, sea level rise, and coastal zone management. Coastal Zone Management Journal 14:3:147-171.

Titus, J.G. 1985. Sea level rise and the Maryland coast. In Potential impacts of sea level rise on the beach at Ocean City, Maryland. Washington, DC: U.S. Environmental Protection Agency.

Titus, J.G. 1984. Planning for sea level rise before and after a coastal disaster. In: Barth, M.C., and J.G. Titus, eds. Greenhouse Effect and Sea Level Rise: A challenge for this generation. New York: Van Nostrand Reinhold.

Titus, J.G., M.C. Barth, J.S. Hoffman, M. Gibbs, and M. Kenney. 1984. An overview of the causes and effects of sea level rise. In: Barth, M.C., and J.G. Titus, eds. Greenhouse Effect and Sea Level Rise: A challenge for this generation. New York: Van Nostrand Reinhold.

Titus, J.G., T. Henderson, and J.M. Teal. 1984. Sea level rise and wetlands loss in the United States. National Wetlands Newsletter Environmental Law Institute 6:4.

Titus, J.G., C.Y. Kuo, M.J. Gibbs, T.B. LaRoche, M.K. Webb, and J.O. Waddell. 1987. Greenhouse effect, sea level rise, and coastal drainage systems. Journal of Water Resources Planning and Management. ASCE 113(2):216-227.

Wilcoxen, P.J. 1986. Coastal erosion and sea level rise: implications for ocean beach and San Francisco's westside transport project. Coastal Zone Management Journal 14:3.

CHAPTER 10

AGRICULTURE

FINDINGS

Although climate change is not likely to threaten U.S. food supplies, it will affect crop yields and result in a northward shift in cultivated land, causing significant regional dislocations in agriculture with associated impacts on regional economies. It will expand crop irrigation requirements, stress livestock production, and increase infestations of agricultural pests and diseases.

Crop Yields

- The effects of climate change alone may reduce average yields of corn, soybeans, and wheat, both rainfed and irrigated, except in the northernmost latitudes where warmer conditions provide a longer frost-free growing season. Decreases in modeled yields result primarily from higher temperatures, which shorten a crop's life cycle.

- When the direct effects of CO_2 on crop photosynthesis and transpiration are approximated along with the effects of climate change, average rainfed and irrigated corn, soybeans, and wheat yields could overcome the negative effects of climate change in some locations. If climate changes are severe, yields could decline. Whether these modeled results will be reproduced under field conditions is uncertain.

- Even if the patterns of climate variability are unchanged, yield stability may decrease, particularly under rainfed conditions. This may occur because there would be more days above temperature thresholds for particular crops. The exact magnitude of change will be sensitive to changes in climatic variability.

Economic Impacts

- There may be a small to moderate aggregate reduction in the nation's agricultural output. Production appears to be adequate to meet domestic consumption needs.

- Assuming no change in export demand, reduced outputs would decrease exports, which could negatively affect global food supplies and the U.S. trade balance. This report did not analyze global changes in agriculture.

- Under the most severe climate change scenarios, continued technological improvements, similar to those in recent years, would have to be sustained to offset losses.

- The economic response of agriculture to shifts in regional productivity may be to shift crop production and associated infrastructure in a northward direction. It is not possible to determine if corn and soybean production could be sustained in northern areas, because soil conditions and other factors were not analyzed.

Irrigation Demand

- The demand for irrigated acreage is likely to increase in all regions. This is due to the reliability of irrigated yields relative to dryland yields and to higher commodity prices that make expansion of irrigated production more economically feasible. This result is sensitive to whether the cost of water to farmers increases.

- More irrigation may increase stress on and competition for regional water supplies. Increased irrigation also implies associated increases in surface and groundwater pollution and other forms of environmental degradation.

Agricultural Pests

- Climate warming could change the ranges and populations of agricultural pests. Temperature increases may enhance the survival of insect pests in the winter, extend their northward ranges, increase pest species with more than one generation per year, and allow pest establishment earlier in the growing season. These effects could result in a substantial rise in pesticide use, with accompanying environmental hazards.

Farm-Level Adjustments

- Farmers may adjust to climate change by using full-season and heat-resistant crop species or varieties, by altering planting dates, by planting two crops during one growing season, by increasing or altering their scheduling of irrigation, by using more pesticides, and by harvesting earlier. If climate change is not severe, these adjustments may mitigate losses in crop yields; more severe climate change is likely to make major adaptation necessary.

Livestock Effects

- Higher temperatures may increase disease and heat stress on livestock in some regions. Existing livestock diseases may shift north, while tropical diseases may extend their ranges into southern regions of the United States. Cold stress conditions may be reduced in the winter, but heat stress is likely to increase in the summer. Reproductive capabilities may also decrease.

Policy Implications

- Global climate change has important implications for all parts of the agriculture system. The agricultural research structure, which is dedicated to maintaining U.S. farm productivity, should enhance climate change research from the field level to the national policy level.

- Current U.S. Department of Agriculture (USDA) research on heat- and drought-tolerant crops and practices should be sustained and, possibly, enhanced to limit vulnerability to future climate change.

- The USDA should evaluate the implications of current legislation on the ability of agriculture to adapt to global warming by shifting the types of crops and location of farming. Such adaptation strategies should consider the impacts of soil erosion and water quality.

- The USDA, the Department of Commerce, the U.S. Trade Representative, and the State Department should consider the implications of potential long-term changes in the level of U.S. crop exports for the U.S. balance of trade and strategic interests.

- There is an urgent need for a national drought policy to coordinate Federal response to the possibility of increased frequency and duration of future droughts. Even without climate change, such a policy is needed not only for the agricultural sector, but also for other sectors.

SENSITIVITY OF AGRICULTURE TO CHANGES IN CLIMATE

Agriculture is a critical American industry, providing food for the nation's population and as much as $42.6 billion in exports for the nation's trade balance (Figure 10-1). The industry employs 21 million people -- more than other industry, when taking into account workers on farms and in the meat,

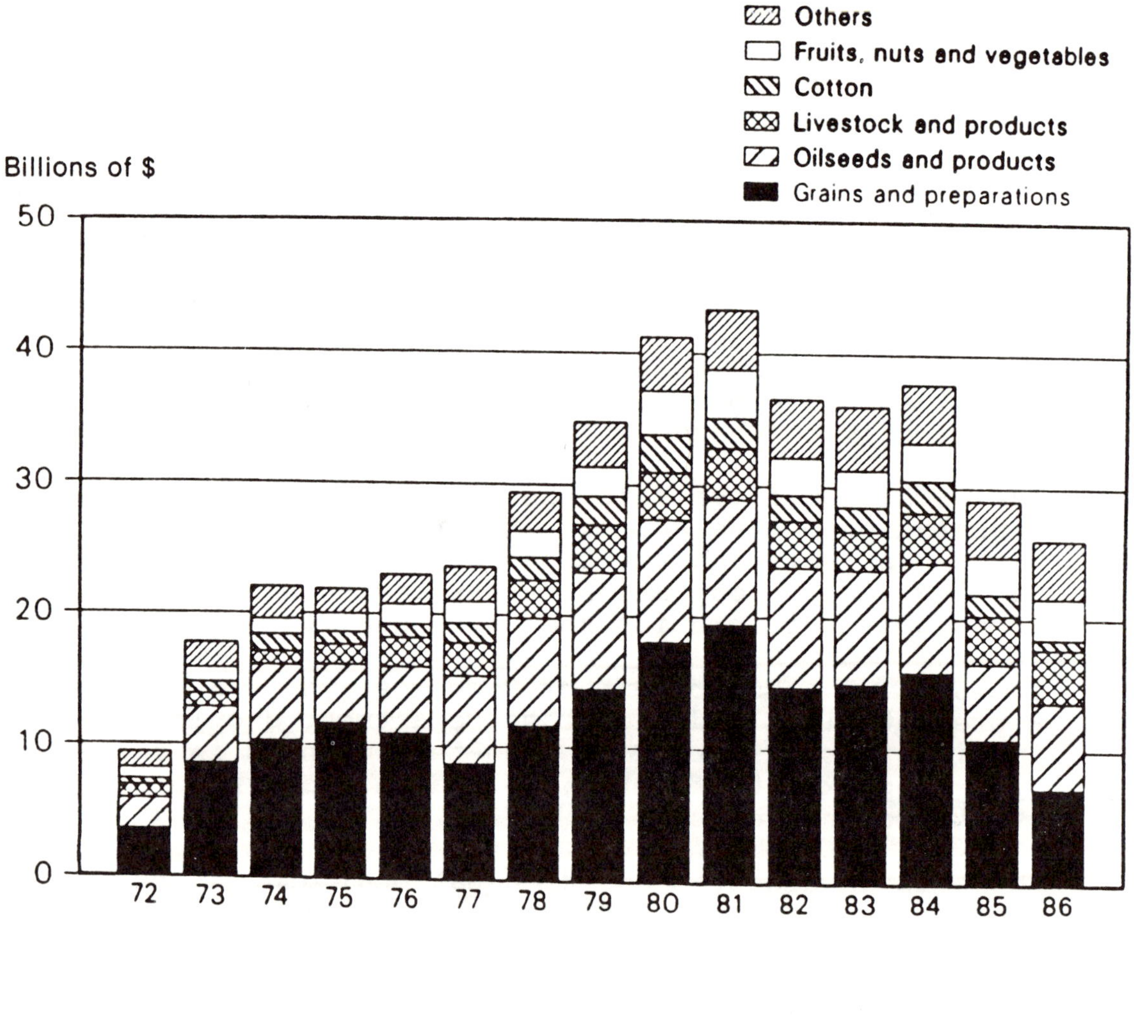

Figure 10-1. Value of U.S. agricultural exports by commodity.

Sources: The World Food Institute and U.S. Department of Agriculture, Economic Research Service, Foreign Agricultural Trade of the United States (Washington, DC), January-February 1987 and various other issues. Livestock excludes poultry and dairy products.

poultry, dairy, baking, and food processing industries (Council for Agricultural Science and Technology, 1988). The U.S. agricultural production system includes farm equipment manufacture, fertilizer and seed supplies, rural banking, and shipping. Total farm assets were $771 billion in 1985; food and fiber were 17.5% of the total gross national product in the same year. Wheat, corn, soybeans, cotton, fruits and vegetables, and livestock are among the most important U.S. agricultural commodities.

The study areas are major agricultural production regions (see Table 10-1). The Great Lakes and Southeastern States are major corn and soybean producers, while the Great Plains States grow mainly wheat and corn.

Worldwide, agricultural products must provide sustenance for the world's growing population, now estimated at about 5 billion and projected to rise to 8.2 billion by 2025 (World Bank, 1988). Global production and consumption of grain have grown steadily since 1960, although regional food shortages continue to occur due to climate variability and socioeconomic factors. Technological advances, such as improved hybrids and irrigation systems, have reduced the dependence of crop yields on the natural environment, but weather is still an important factor in agricultural productivity.

For example, during the Dust Bowl years of the 1930s, wheat and corn yields dropped by up to 50%. Midsummer 1983 saw an unpredicted drought in the U.S. Corn Belt and in the southeastern United States, causing the U.S. corn yields to fall by about a third, from over 7,000 kg/hectare (ha) to about 5,000 kg/ha. And failure of the monsoon season caused shortfalls in crop production in India, Bangladesh, and Pakistan in 1987. The 1980s have also seen the continued deterioration of food production in Africa, despite adequate world food supplies elsewhere, because of persistent drought, internal wars, poor distribution, weak infrastructure, and a deteriorating environment.

The 1988 drought demonstrated the impact that climate variability can have on agricultural productivity. This drought is expected to decrease corn yields by about 40%, and the cost of the 1988 Drought Relief Bill is estimated to be $3.9 billion (New York Times, 1988). The 1988 drought emphasizes anew the close link between agriculture and climate and the need to consider broadly the impacts of potential climate change on U.S. agriculture.

Future changes in climate due to increasing atmospheric trace gases, particularly CO_2, will affect agriculture because

Table 10-1. Crop Data by Region

	Corn	Wheat	Soybeans	Harvested Acres (thousands)
EPA Study Areas				
Southeast	311	272	306	29
Great Lakes	4,644	297	822	92
Great Plains	921	755	136	71
California	38	63	0.01	6
Total (48 States)	8,209	2,507	199	337

Units: Corn, wheat, and soybeans are in thousands of bushels; harvested acres are also in thousands.

Source: 1982 Census of Agriculture, Vol. 1, Geographical Area Series, Part 51, United States Summary and State Data. U.S. Department of Commerce, Bureau of the Census.

light from the sun, frost-free growing seasons, and the hydrologic cycle largely govern crop productivity. Livestock production will also respond to climate change through differing levels of heat and cold stress and altered ranges of disease-carrying vectors such as mosquitoes and ticks. Sea level rise, an associated impact of climate change, will threaten low-lying coastal agricultural regions with seasonal, and in some instances permanent, flooding, saltwater intrusion of freshwater aquifers and rivers, and salt contamination of soils. (See Water Resources chapter in this volume for linkages with agriculture.)

As predicted rises in temperature and other changes in climate variables such as precipitation occur, the higher levels of CO_2 in the air will also affect crops. Increasing CO_2 has boosted crop photosynthesis and has reduced the use of water by crops in experimental settings. Because experimental research has rarely dealt with both the climatic and the direct effects of CO_2 on plants simultaneously, it is difficult to assess the relative contributions of CO_2 and increased temperature to plant responses. This remains one of the most crucial questions in the analysis of impacts of climate change and increased CO_2 on agriculture.

Associated processes, such as changes in agricultural pests, may also play a role in responses of U.S. agriculture to climate change. The presence and abundance of pests affecting both crops and livestock are highly dependent on climate. The severity of the winter season, wind patterns, and moisture conditions determine in large part where pests will be prevalent. As these conditions change, the patterns, numbers, and types of insects also will change.

Much of U.S. agricultural production takes place under technologically advanced cropping systems that are primarily monocultural. Likewise, livestock production is highly specialized, both technically and geographically, and a high degree of integration exists between grain and livestock production. The economic robustness associated with general, multiple-enterprise farms has long since passed from the scene on any significant scale. Thus, the current vulnerability of our agricultural system to climate change may be greater in some ways than it was in the past.

Climate change and agriculture may affect the natural environment. As regional changes occur in crop and livestock production, these changes in agriculture may increase soil erosion, intensify the demand for water for irrigation, degrade water quality, reduce forested land, and destroy wildlife habitats. Other potential consequences are soil degradation, groundwater overdraft, loss of plant and aquatic communities, and reduced resilience in environmental resources in general. (See the Water Resources and Biological Diversity chapters of this report.)

Climate-driven changes in crop and livestock productivity cause changes in profitability and decisionmaking at the farm level, which in turn lead to shifts in farming systems at the regional level and in export-import supply and demand at the national level. Ultimately, impacts of climate change will reverberate throughout the global food economy, altering global food trade. Changes in U.S. agriculture will thus take place in a global context.

PREVIOUS STUDIES OF THE EFFECTS OF CLIMATE CHANGE AND AGRICULTURE

Relationships between climate and agriculture have been studied intensively for many years. However, relatively few studies have specifically addressed both the climatic and the direct effects that the growth in trace gases will have on agriculture. Even fewer studies have addressed these potential effects in an integrated approach that links both biophysical and economic spheres of analysis.

Most research attention in the United States has focused on the direct effects of CO_2 on crops, supported primarily by the U.S. Department of Energy. These studies are reviewed by Acock and Allen (1985) and Cure (1985) who found an average increase in yields of about 30% and increases in water-use efficiency for crops growing in favorable, current climate conditions between 330 and 660 ppm CO_2. Kimball (1985) and Decker et al. (1985) suggest that the potential effects of CO_2 and/or climate change on agricultural production systems may include shifts in production areas and changes in levels of livestock stresses, water availability, and pest control management.

Integrated approaches to the impacts of climate change on agriculture involving both biophysical and economic processes have been considered in studies by Callaway et al. (1982), the Carbon Dioxide Assessment Committee (1983), Warrick et al. (1986), and the Land Evaluation Group (1987). A benchmark international study on both the agronomic and economic effects of climate change on agriculture was conducted by the International Institute for Applied Systems Analysis (Parry et al., 1988). No study has as yet comprehensively examined the combined effects of climate change and the direct effects of CO_2 on U.S. agriculture.

Chapter 10

CLIMATE CHANGE STUDIES IN THIS REPORT

Structure of and Rationale for the Studies

The agricultural studies in this report involve the following research areas (see Table 10-2): (1) crop growth and yield, (2) regional and national agricultural economics, (3) demand for water for irrigation, (4) water quality, (5) pest-plant interactions, (6) direct effects of CO_2 on crop growth and yield, (7) impacts of extreme events, (8) potential farm-level adjustments, (9) livestock diseases, and (10) agricultural policy. The following studies were performed for this report:

Regional Studies

- Effects of Projected CO_2-Induced Climatic Changes on Irrigation Water Requirements in the Great Plains States -- Allen and Gichuki, Utah State University

- Climate Change Impacts upon Agriculture and Resources: A Case Study of California -- Dudek, Environmental Defense Fund

- Farm-Level Adjustments by Illinois Corn Producers to Climatic Change -- Easterling, Resources for the Future

- Impacts of Climate Change on the Fate of Agricultural Chemicals Across the USA Great Plains and Central Prairie -- Johnson, Cooter, and Sladewski, University of Oklahoma

- Impact of Climate Change on Crop Yield in the Southeastern U.S.A: A Simulation Study -- Peart, Jones, and Curry, University of Florida

- Effects of Global Climate Change on Agriculture: Great Lakes Region -- Ritchie, Baer, and Chou, Michigan State University

- Potential Effects of Climate Change on Agricultural Production in the Great Plains: A Simulation Study -- Rosenzweig, Goddard Space Flight Center/Columbia University

Table 10-2. Agriculture Projects for EPA Report to Congress on the Effects of Climate Change

Projects	Regional Studies				National Study
	California	Southeast	Great Lakes	Great Plains	
Crop Production	X	X	X	X	
Demand for Irrigation Water	X	X	X	X	
Water Quality			X	X	
Pest/Plant Interactions			X		
Farm-Level Management			X		
Regional/National Economics	X	X	X	X	X
Direct Effects of CO2	X	X	X	X	X
Livestock Pests & Diseases					X
Climatic Variability					X
Agricultural Policy					X

Chapter 10

National Studies

- The Economic Effects of Climate Change on U.S. Agriculture: A Preliminary Assessment -- Adams, Glyer, and McCarl, Oregon State University

- Analysis of Climate Variability in General Circulation Models -- Mearns, Schneider, Thompson, and McDaniel, National Center for Atmospheric Research

- Direct Effects of Increasing CO_2 on Plants and Their Interactions with Indirect (Climatic) Effects -- Rose

- Potential Effects of Climatic Change on Plant-Pest Interactions -- Stinner, Rodenhouse, Taylor, Hammond, Purrington, McCartney, and Barrett, Ohio Agricultural Research and Development Center

- Agricultural Policies for Climate Changes Induced by Greenhouse Gases -- Schuh, University of Minnesota

- Changing Animal Disease Patterns Induced by the Greenhouse Effect -- Stem, Mertz, Stryker, and Huppi, Tufts University

- Effect of Climatic Warming on Populations of the Horn Fly -- Schmidtmann and Miller, Agricultural Research Service

Production of corn, wheat, and soybeans is critical to the economic well-being of the nation's farmers and the national trade balance. These crops make up about two-thirds of the total U.S. agricultural acreage, and their economic value is equal to that of all other crops combined. These three crops were selected for the modeling studies on the effects of climate change on yields.

The results from the regional studies of crop production (not including California), hydrological predictions from the climate models, and an agricultural economics model were linked in an integrated approach, to enable translation of the estimated yield changes from the crop modeling studies and predicted changes in water availability into economic consequences (see Figure 10-2). Such a coordinated analytical framework is necessary to account for the effects of market forces on the total agricultural sector, including livestock, and

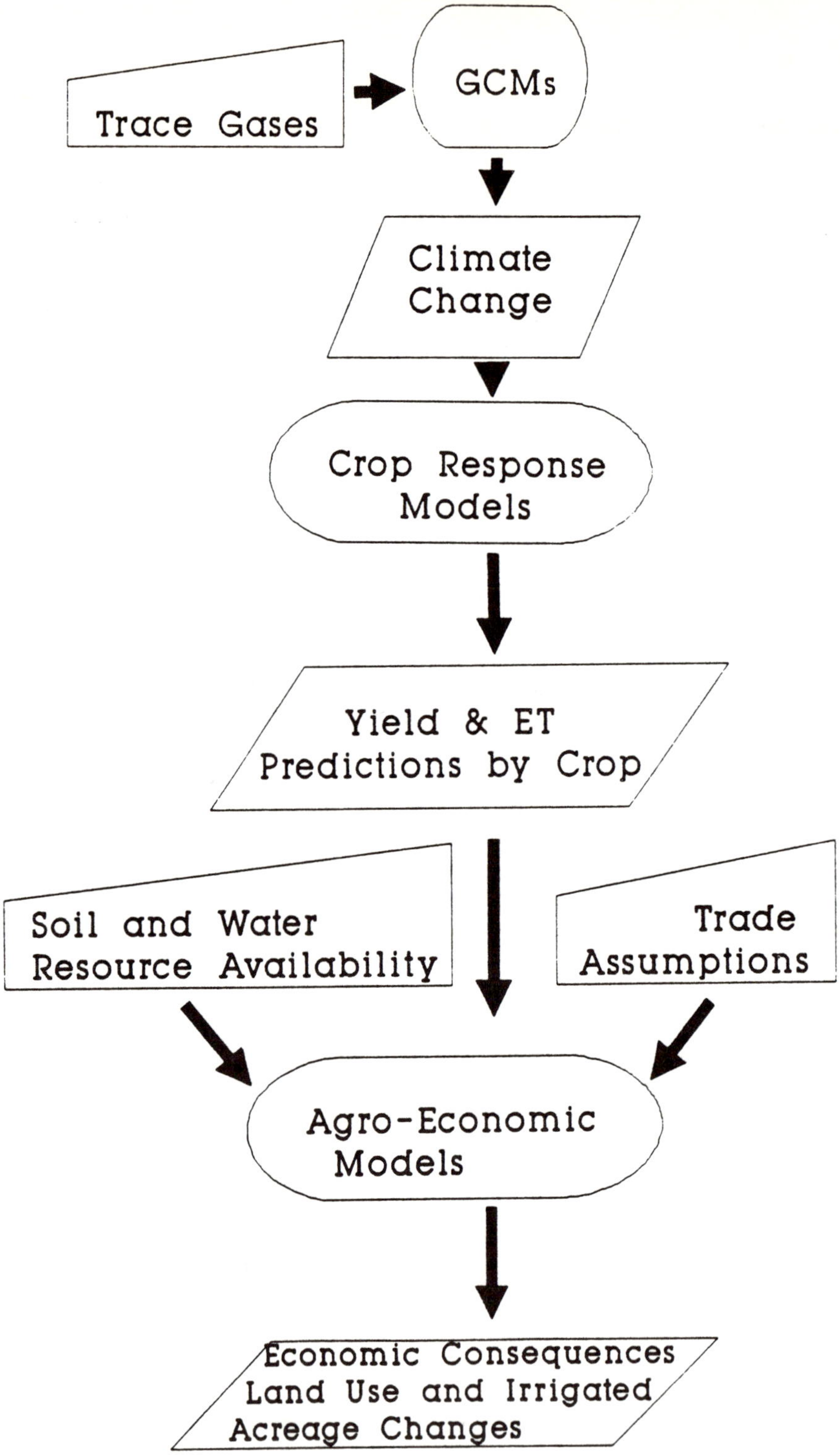

Figure 10-2. Flow chart of model interactions in EPA studies of the effects of global climate change on U.S. agriculture.

Source: Dudek (1987).

to evaluate the adequacy of the nation's resource base for agricultural production under climate change. Economic forces may lead farmers to grow more crops in areas with relatively high productivity and fewer crops in areas with relatively low productivity.

The studies of demand for irrigation water, water quality, and farm-level adjustment were also linked with the integrated modeling studies by common assumptions, sites, or outputs. Because California grows a large and diverse number of crop commodities, a simple approach was used to estimate crop yield changes for the California case study based on heat, sunlight, and photosynthetic response to CO_2. These yield changes were then used in a model of agricultural land and water use in California. Several adjustment experiments were included to test possible adaptation mechanisms, such as changes in planting dates and crop varieties.

The agricultural studies performed for this EPA report explore the sensitivities of the different parts of the agricultural system (shown in Table 10-1) to climate change scenarios. They are not meant to be predictions of what will happen; rather, they aim to define ranges and magnitudes of the potential responses as the systems are currently understood.

RESULTS OF AGRICULTURE STUDIES

Regional Crop Modeling Studies

Design of the Studies

Widely validated crop growth models -- CERES-Wheat and CERES-Maize (Ritchie and Otter, 1985) and SOYGRO (Jones et al., 1988) -- were used to simulate wheat, corn, and soybean yields at selected geographically distributed locations within the Great Lakes, the Southeast, and the Great Plains. California crop yield changes were predicted separately by using an agroclimatic index. (See the regional chapters of this report for specific results.) The crop models simulated both dryland and irrigated production systems. The crop modeling approach allowed for analysis of latitudinal gradients in changes in crop yields and provided compatible results for each climate change scenario to be used as input in the agricultural economics study. (See Ritchie et al., Peart et al., and Rosenzweig, Volume _.)

The direct effects of CO_2 -- i.e., increased photosynthesis and improved water-use efficiency -- were included with the climate change scenarios in some model runs to evaluate the combined effects. The direct effects were approximated by computing ratios of elevated CO_2 to ambient CO_2 values for daily photosynthesis and evapotranspiration rates (see Peart et al., Volume _).

Limitations

Uncertainties in the crop modeling studies reside in climate model predictions, locations of the climate stations (not always in production centers), crop growth models, and estimates of the direct effects of CO_2. In particular, the climate change scenarios did not include changes in climate variability, even though changes in the frequencies of extreme events may considerably affect crop yields. Technology and cultivars were assumed not to change from present conditions.

The direct effects of CO_2 in the crop modeling study results may be overestimated for two reasons. First, experimental results from controlled environments may show more positive effects of CO_2 than would actually occur in variable, windy, and pest-infested (weeds, insects, and diseases) field conditions. Second, because other radiatively active trace gases, such as methane (CH_4) also are increasing, the equivalent warming of a doubled CO_2 climate may occur somewhat before an actual doubling of atmospheric CO_2. A level of 660 ppm CO_2 was assumed for the crop modeling experiments, while the CO_2 concentration in 2060 is estimated to be 555 ppm (Hansen et al., 1988).

Results

Climate change scenarios alone, without the direct effects of CO_2, caused simulated yields of corn, soybeans, and wheat to decrease in the Great Lakes, Southeast, and Great Plains regions, except in the northernmost latitudes, where warmer conditions provide a longer frost-free growing season (Figure 10-3). The northern locations where yields increased included sites in Minnesota. Decreases in modeled yields resulted primarily from higher temperatures, which shorten the crop life cycle thus curtailing the production of usable biomass. Dryland modeled yields decreased more than irrigated yields.

When increased photosynthesis and decreased transpiration were included in the crop models along with the climate change scenarios, yields increased over the baseline in some locations but not in others (Figure 10-4). Particularly when combined with the hotter and drier GFDL climate change scenario, the direct effects of CO_2 did not fully compensate for changes in climate variables -- net yields decreased significantly from the base case.

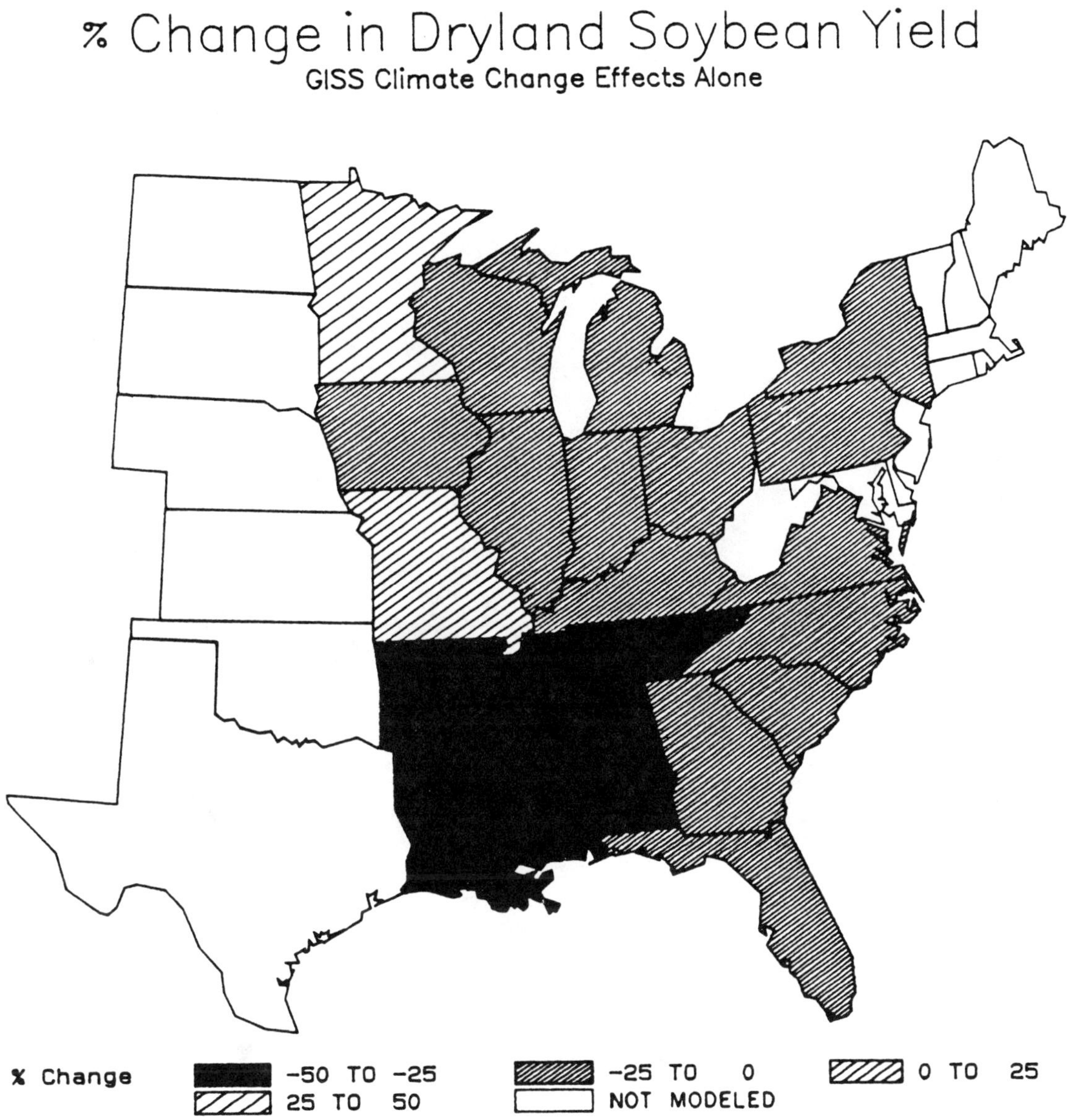

Figure 10-3. Percent change in rainfed soybean yields simulated by the SOYGRO model with the GISS climate change scenario.

Source: Peart et al. and Ritchie et al., Volume ___.

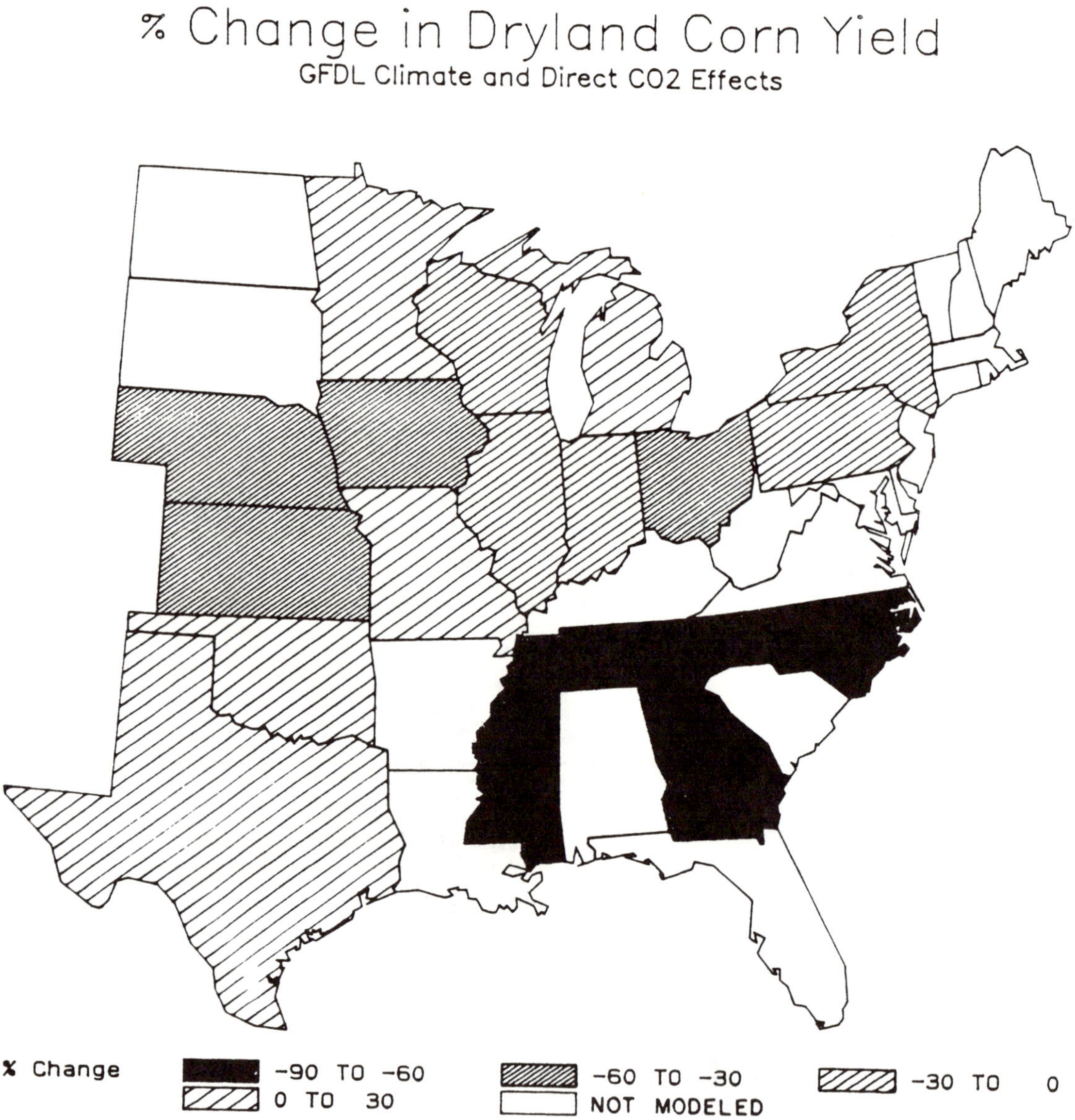

Figure 10-4. Percent change in rainfed corn yields simulated by the CERES-Maize model for the GISS climate change scenario with approximated direct effects of CO_2 on photosynthesis and water use.

Source: Peart et al. and Ritchie et al., Volume ___.

The crop models were also used to test several possible adaptations by farmers to the predicted climate changes. For example, a corn variety that is better adapted to longer growing seasons was tested in Indiana. This later maturing variety did not compensate entirely for the yield decreases caused by the warmer climate change scenarios.

Implications

An implication of the crop modeling studies is that pressure for increased irrigation may grow in many agricultural regions. If the major agricultural areas are to continue to provide a stable supply of food under the predicted changes in climate, supplemental irrigation may be required for many soils. This could further tighten water supply problems in some areas and increase pollution from nonpoint sources (i.e., pollution that is not traceable to any one distinct source, such as agricultural pollution).

The water supply for the irrigation needed to stabilize production would come from both surface and subsurface sources, depending on the region. However, considerable uncertainty exists regarding the future availability of surface water and groundwater supplies with climate change, and concerning the costs of using or extracting the water. (See the Water Resources chapter of this report.)

Regional and National Economics Study

The estimated yield changes from the crop modeling studies and projected changes in irrigation water demand and availability were introduced into an agricultural economic model to translate the physical effects of climate change into economic consequences. Adams et al. (see Volume _) estimated the regional and national economic implications of changes in yields of wheat, corn, soybeans, and other crops and in the demand for and availability of water associated with alternative global climate change scenarios. In addition, sensitivity analyses on critical assumptions were also performed -- particularly for technological change and yield benefits due to the direct effects of increased CO_2.

Study Design

A spatial equilibrium agricultural model developed by Adams et al. (1984) was used to represent production and consumption of numerous agricultural commodities for the U.S. farm production regions as designated by the USDA (see Figure 10-5). The model has been used to estimate agricultural losses due to increased

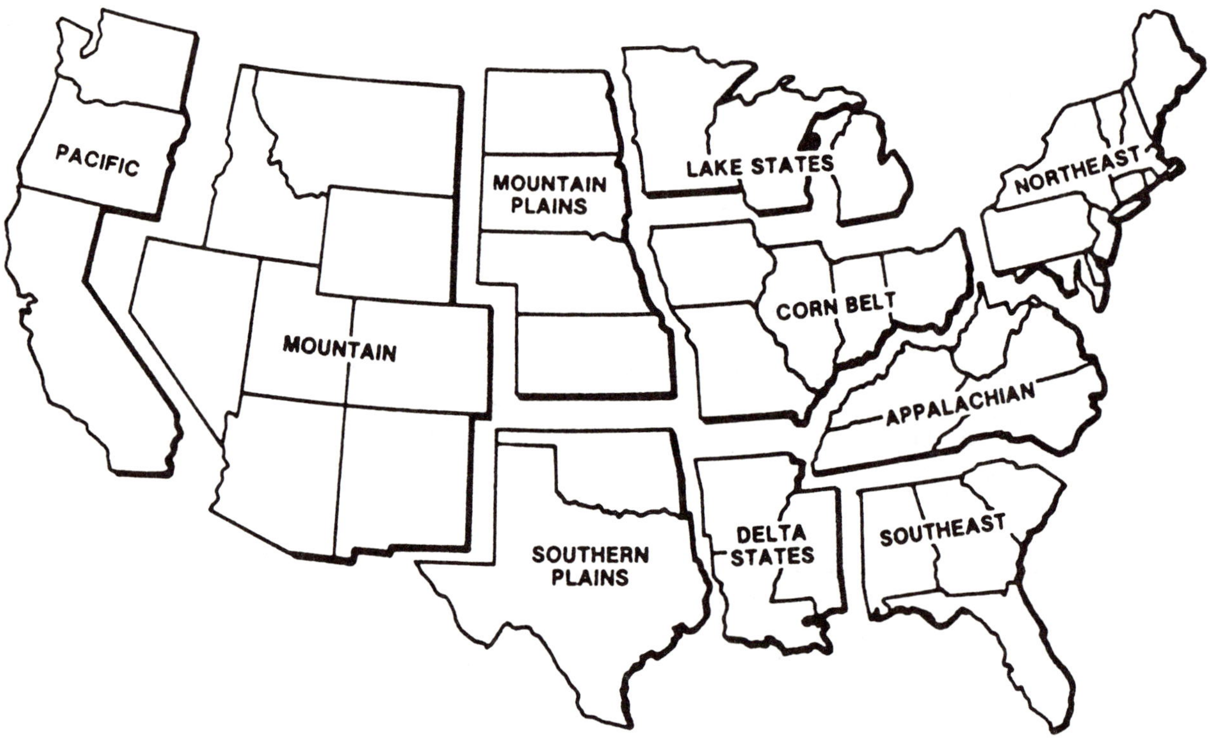

Figure 10-5. Farm production regions in the United States.

Source: USDA Economic Research Service (1987).

ultraviolet-B (UV-B) radiation caused by stratospheric ozone depletion (Adams et al., 1984). It consists of farm-level models for production regions, integrated with a national-level model of the agricultural sector. Both irrigated and nonirrigated crop production and water supply relationships are included for most regions. The model stimulates a long-run, perfectly competitive equilibrium, and was developed using 1980-83 economic and environmental parameters.

Limitations

The economic approach used in this study has several limitations. The economic model is static in the sense that it simulates an equilibrium response to climate change, rather than a path of future changes. Substitution of appropriate crop varieties and adjustments in farm management techniques were not included, thus possibly overestimating the negative effects of climate change. Estimates of the beneficial direct effects of CO_2 on crop yields may have biased results in the positive direction. Furthermore, changes in yields used as inputs to the economic model were only modeled for wheat, corn, and soybeans for a limited number of sites and regions, and national estimates were derived from these for all other crop commodities in the model. This introduces uncertainties into the results.

Potential changes in international agricultural supply, demand, and prices due to climate change are not explicitly included in the model. Such changes could have major impacts on U.S. agriculture.

Results

The economic model showed a small to moderate aggregate loss in economic welfare associated with the projected crop yield and hydrologic changes derived from the climate change scenarios (see Table 10-3). For the moderate climate change scenario, net losses were small; for the more extreme scenario, they were greater.

Production of most crops was reduced because of yield declines and limited availability of land and resources. The largest reductions were in sorghum (-20%), corn (-13%), and rice (-11%), because these estimated losses are equivalent to 5 and 28% of the 1982 value of crop and livestock commodities produced in the United States. The magnitudes of these changes, which are annual, may be compared to the estimated $6 billion cost of the Drought Relief Bill for 1988. In general, consumers lose and producers gain due to the increased prices of agricultural

Table 10-3. Aggregate Economic Effects of GISS and GFDL 2xCO_2 Global Climate Change on U.S. Agriculture (in billions of 1982 dollars)

Run	Economic Effects		
	Consumers	Producers	Total
GISS Analysis 4[a]	-7.3	+1.5	-5.0
GFDL Analysis 4	-37.5	+3.9	-33.6

[a]Analysis 4 includes the crop yield and irrigation water supply and demand consequences of climate change throughout the United States.

Source: Adams et al., Volume _.

commodities and inelastic demand (insensitivity to price changes) for agricultural crops.

Agricultural activity would shift northward with both GISS and GFDL climate change scenarios (see Figure 10-6). Land under production in Appalachia, the Southeast, the Mississippi Delta, and the Southern Plains, could decrease by 5 to 25%, while in the Lake States, the Northern Plains, and the Pacific it could increase by 5 to 17%. Irrigated acreage is estimated to increase in all areas, primarily because irrigation becomes economically feasible as agricultural prices rise (see Figure 10-7). Demand for additional water may be met in part by increased groundwater pumping, although this could be a matter of concern in areas where groundwater is already being overdrawn. Since changes in precipitation and recharge patterns are likely with climate change, the availability of groundwater and the costs to extract it might also be affected (see chapter on Water Resources).

The impact of higher CO_2 levels on crop productivity would probably reduce the negative economic impacts of climate change on the agricultural sector (Table 10-4). Under the less severe GISS climate scenario, the CO_2 direct effects would sufficiently counter the climatic effects in most regions, so that the net effect would be a gain in economic welfare. Consumers would be the primary beneficiaries, though (in aggregate) producers would also gain slightly. However, because the climatic effects on crops differ by region, the output of some regions would increase, while that of others would decrease. With the more severe GFDL climate change scenario combined with the direct effects of CO_2, economic welfare still would decrease, but not by as much as it did with the climate change scenarios alone. Here, consumers would be the losers.

Technological change, such as higher yielding crop varieties, chemicals, fertilizers, and mechanical power, has historically enabled agriculture to increase production with the same amount of or less land, labor, and other resources. When the effect of future technological change was modeled along with the less severe GISS climate change effects (without the direct effects of CO_2), most of the adverse climate effects were offset. Under the severe GFDL climate change scenario, continued and substantial improvements in yields would be required to overcome the climate change effects. Stated another way, the adverse effects of climate change would negate most of the higher output attributable to improved technology over the next 50 years.

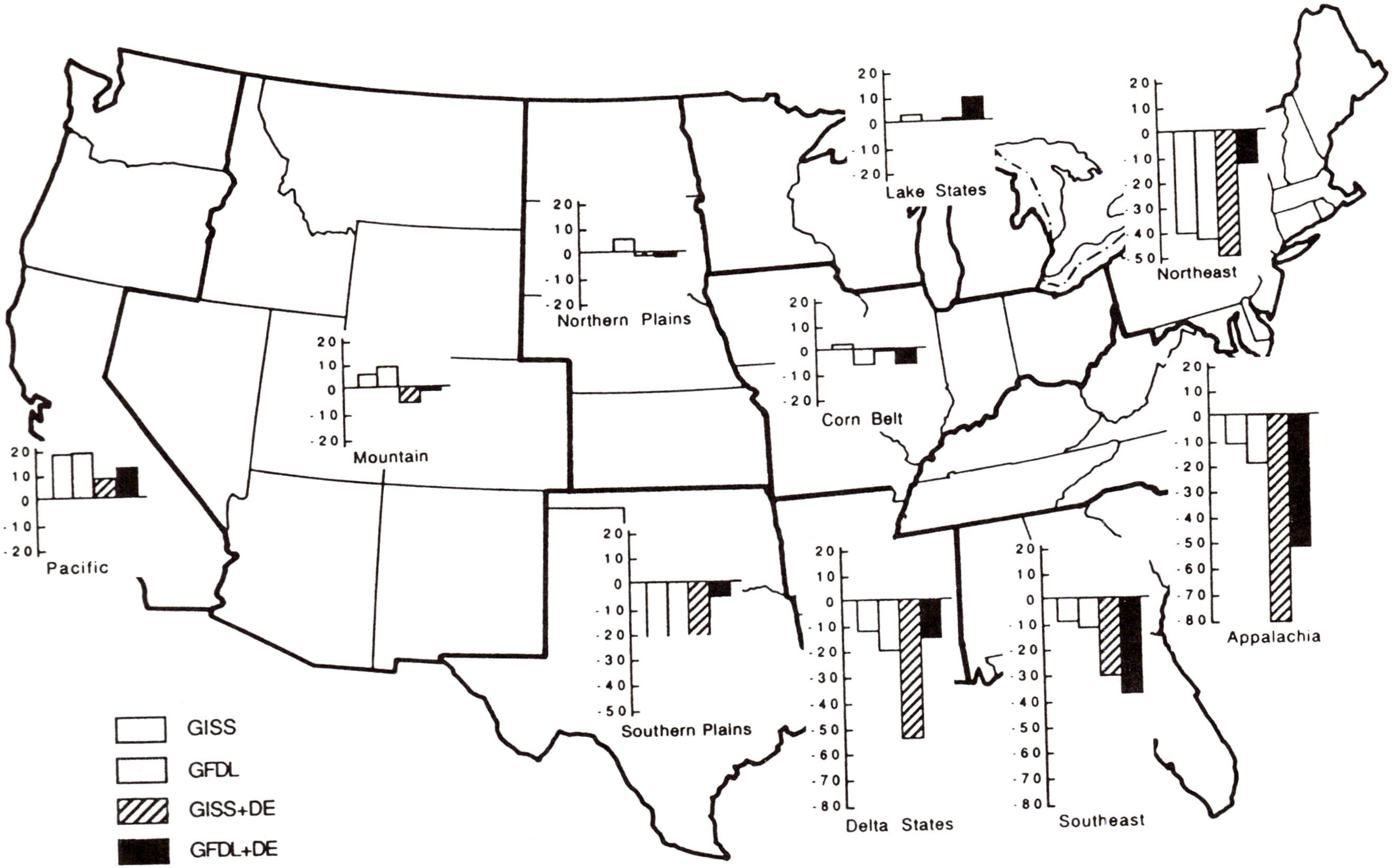

Figure 10-6. Percent change in regional agricultural acreage simulated by an economic model of the U.S. agricultural sector for the GISS and GFDL climate change scenarios with and without the direct effects of CO_2 on crop yields.

Source: Adams et al., Volume _.

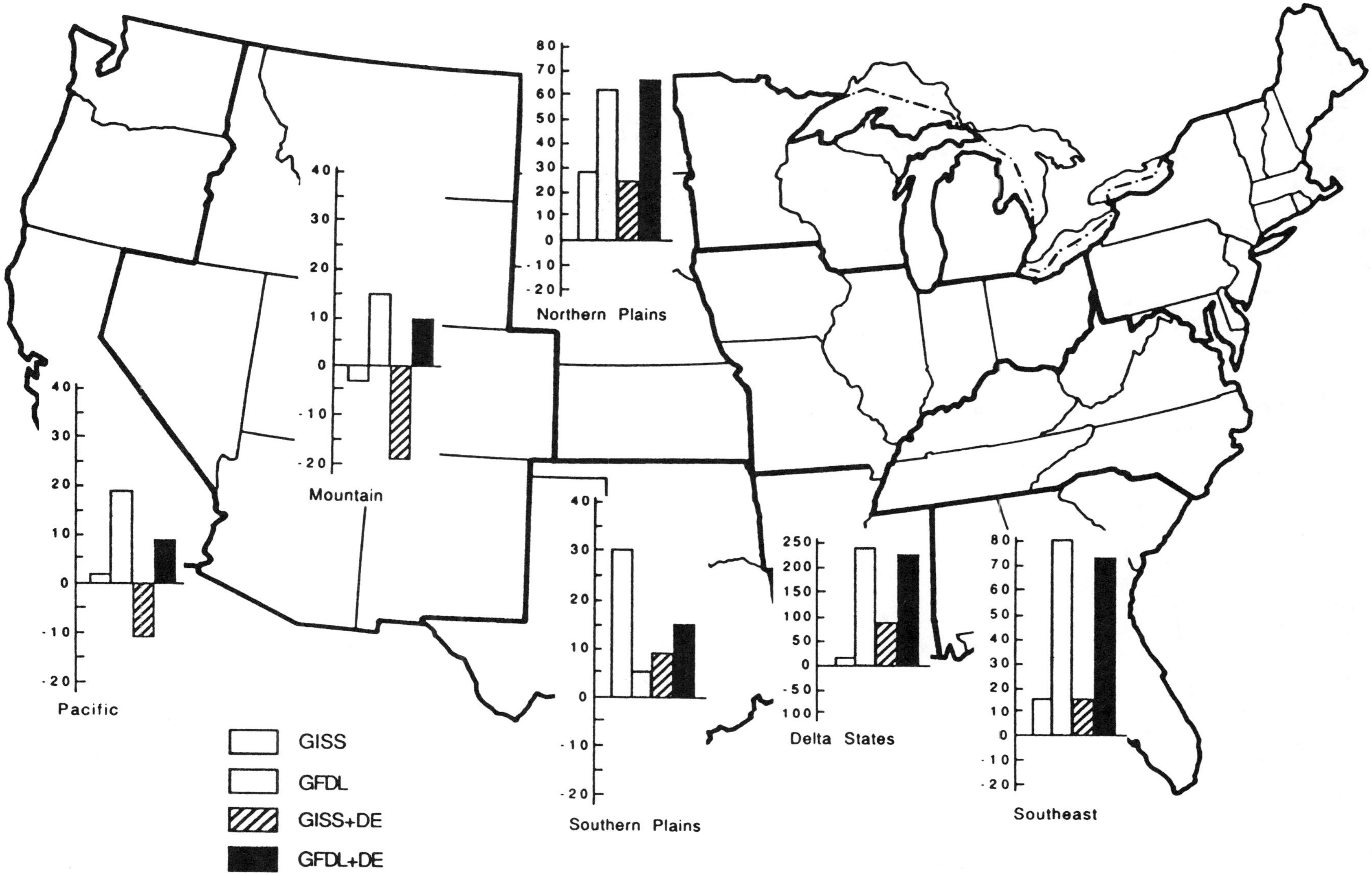

Figure 10-7. Change in regional irrigation acreage simulated by an economic model of the U.S. agricultural sector for the GISS and GFDL climate change scenarios with and without the direct effects of CO_2 on crop yields.

Source: Adams et al., Volume _.

Table 10-4. Combined Effects of Direct CO_2 and Climate Change on Agriculture: GISS and GFDL Analyses 4

Run	Economic Effects		
	Consumer	Producer ($ billion)	Total
GISS Analysis 4: without CO_2	-7.3	1.5	-5.9
GISS Analysis 4: with CO_2	9.4	1.3	10.6
GFDL Analysis 4[a]: without CO_2	-37.5	+3.9	-33.6
GFDL Analysis 4: with CO_2	-10.3	0.6	-9.7

[a]Analysis 4 includes the crop yield and irrigation water supply and demand consequences of climate change throughout the United States.

Source: Adams et al., Volume _.

Chapter 10

Implications

Food Supply

Although climate change could reduce the productive capacity of U.S. agriculture, the reduction does not imply major disruption in the supply of basic commodities for American consumers. Domestically, consumers would face slightly to moderately higher prices under some analyses, but supplies would be adequate to meet current and projected domestic demand. However, most exported commodities in some scenarios decline by up to 70%, assuming the demand for exports remains constant.

While climate change does not appear to be a food security issue within the United States, supply and demand for agricultural commodities could shift among international regions. Responses of U.S. agriculture would take place in this global context. These changes in global agriculture cannot be determined without analyzing the potential impacts of climate change on both major world agricultural production regions and potentially vulnerable food deficit regions.

Regional Economics and Land Use

Regional shifts in U.S. agricultural production patterns are highly likely, as all analyses showed a consistent northward shift in the production of the major commodities of wheat, corn, and soybeans (see Figure 10-6). Other crops, including specialty crops such as vegetables and fruits, could also shift. Where agricultural acreage declines, this might further weaken an economic base already under pressure from long-term structural changes in U.S. agriculture.

Environmental Concerns

Regional agricultural adjustments could place environmental resources at risk. Where agricultural acreage would increase, demands for natural resources, such as soil and water, might intensify current pressures on environmental elements, such as rivers, lakes, aquifers, wetlands, and wildlife habitats. Northern States, such as Minnesota and North Dakota, could become more productive for annual crops like corn and soybeans because of warmer temperatures and a longer frost-free growing season. Because of the presence of forests and wetlands in these regions, increased agricultural production in the area might threaten natural ecosystems, including wildlife habitats such as prairie potholes for ducks and flyways for bird migrations.

In addition, many of the glacial till soils in the northern latitudes are not as productive as Corn Belt soils. Thus, large increases in production of crops would most likely require increased applications of chemical fertilizers. The use of these fertilizers in humid regions on glacial till and sandy soils is now creating an environmental hazard to the underlying groundwater, receiving waters, and aquatic habitats in many areas. With climate change, water and fertilizer use would have to be carefully managed to minimize still more leaching of water-soluble nutrients such as nitrogen and potash, while continuing to provide enough nutrients for crop growth.

Demand for Water for Irrigation

Water is the single most critical factor in determining the development, survival, and productivity of crops. The amount of water that crops use and thus the demand for irritation water are governed largely by the evaporation process. The projected higher air temperatures due to increasing trace gases in the atmosphere could heighten evaporative demands. Farmers may irrigate more to satisfy these higher demands, which could accelerate depletion of groundwater and surface water resources. Also, the rate of evaporation might outstrip precipitation, thus decreasing crop yields.

Studies reported in the California and the Great Plains case studies (see Chapters 4 and 5) explicitly examined the potential changes in demand for water for irrigation. In these regions, water is a critical resource for agriculture; California and the parts of the Great Plains fed by the Ogallala Aquifer, in particular, depend very heavily on irrigation for crop production.

Irrigation Requirements in the Great Plains

Allen and Gichuki (see Volume _) computed irrigation water requirements for sites in the Great Plains for the baseline climate and the GISS and GFDL climate change scenarios. The direct effects of CO_2 and water use were also included. (See Chapter 7, Great Plains, for study design and limitations.) Major changes in irrigation water requirements were observed at all locations in the Great Plains, for all crops (see Figure 10-8). The most significant were the persistent increases in seasonal net irrigation water requirements for alfalfa, which were driven by the climate changes in temperature, wind, humidity, and solar radiation, and by the lengthening of the growing season. Decreases in irrigation requirements were observed for winter wheat and corn in most regions. These decreases were attributed to earlier planting dates and the shortening of the crop-growing

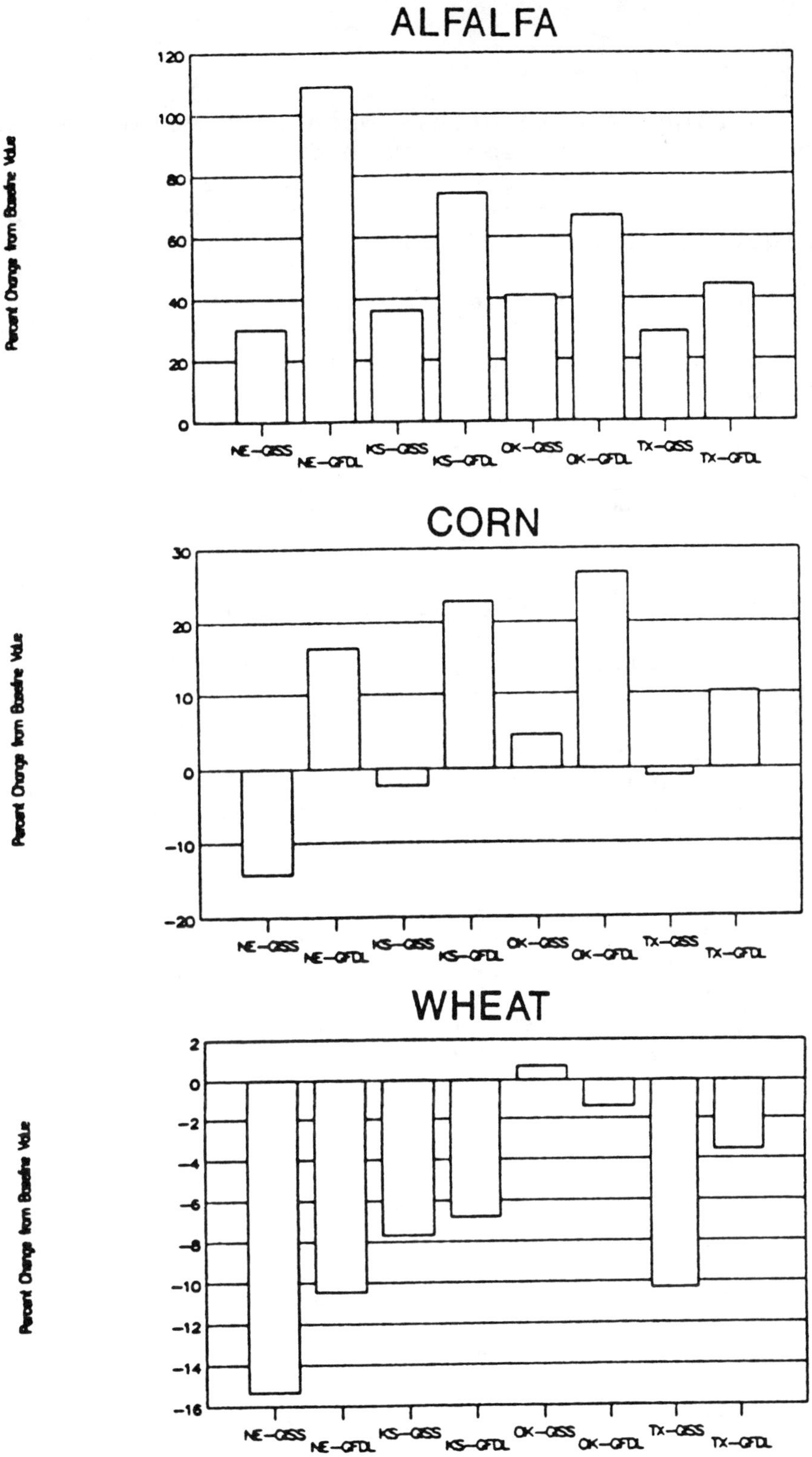

Figure 10-8. Percent change in net seasonal irrigation requirements with a 20% increase in bulk stomatal diffusion resistance.

Source: Allen and Gichuki, Volume __.

season by high temperatures. When crop varieties appropriate to the longer growing season were modeled, irrigation water requirements for winter wheat increased. Irrigation water requirements during peak periods increased in almost all areas (see Figure 10-9).

While farmers in the Great Plains would probably shift to longer season crops, the predicted climate conditions (warmer temperatures and predicted drying in some areas) during the later summer months could increase irrigation requirements and elevate leaf temperatures to a point that exceeds optimum temperatures required for high productivity. This might make it uneconomical to take full advantage of the predicted longer growing season, especially if the higher CO_2 levels increase photosynthesis and offset the effects of a shorter season to some degree.

Water Resources for Agriculture in California

In the California regional case study, Dudek (see Volume _) characterized the potential shifts in demand for water for agricultural production that would accompany shifts in cropping patterns driven by changing climate. (See California chapter for study design and limitations.) When climate change was considered alone, groundwater extraction and surface water use declined in California as a result of changes in both supply and demand. When the direct effects of CO_2 on crop yields were included, however, groundwater extraction increased due to improved yields of all crops except corn and to enhanced economic welfare. Institutional responses to changes in surface and groundwater use include water transfers, which improve irrigation efficiency. When water markets were included in the simulations, economic welfare was improved by 6-15% over the base, while crop acreage increased and groundwater extraction decreased.

Implications for Demand for Water for Irrigation

Expanded use of irrigation is implied from the regional crop modeling studies for the Great Lakes, the Southeast, and the Great Plains (see Chapters 5, 6, and 7). Increases in irrigated acreage are also predicted for most regions when the economics of crop production are factored in (see Adams et al., Volume _). When these results are considered along with the other Great Plains and California studies, it appears that climate change is likely to increase the demand for irrigation.

In the Great Plains, heightened evaporative demand and variability of rainfall may increase the need for irrigation in dryland farming regions. The projected changes in irrigation water requirements are varied, and specific crops and locations

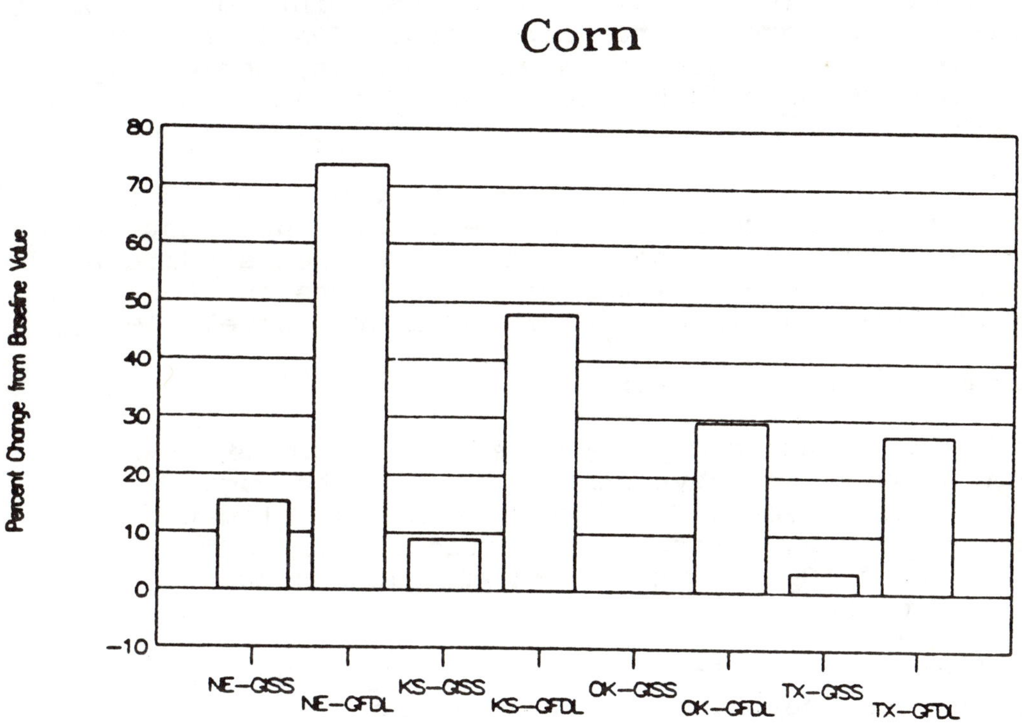

Figure 10-9. Percent change in peak irrigation requirements of corn with a 20% increase in bulk stomatal diffusion resistance.

Source: Allen and Gichuki, Volume _.

probably will be affected differently. Higher peak irrigation water requirements for some crops may require larger capacity irrigation systems and enlarge energy demands.

Intensified extraction of water poses serious environmental and economic problems, especially in areas where groundwater is being overdrawn. Streamflows also may slacken if more surface water is used for irrigation, thereby aggravating water quality problems. This in turn would harm fish, wildlife, and recreational activities.

Regional changes in cropping locations and patterns of water use also could exacerbate agricultural, nonpoint source pollution and further deplete groundwater resources. Institutional responses, such as markets for water transfers, could help improve irrigation water management and alleviate some of these negative effects.

The economic and social costs of shifting the location of irrigated agriculture could be considerable. The construction of irrigation systems consisting of reservoirs, wells, ditches, pipes, pumps, and sprinklers currently requires about $1,500-5,000 per hectare (ha) in capital investment (Postel, 1986).

Direct Effects of CO_2 on Crops

Global increases in CO_2 are expected to influence crop metabolism, growth, and development directly through physiological processes and indirectly through climate. Rose (see Volume _) reviewed recent experimental work performed on the direct effects of CO_2 on crops, with emphasis on wheat, corn, soybeans, and cotton.

Elevated concentrations of CO_2 directly affect plant processes such as photosynthesis and transpiration. CO_2 concentrations are also expected to influence these processes indirectly through predicted increases in temperature and other changes in climate variables such as precipitation. Because experimental research has rarely simultaneously studied both the direct and indirect effects of plant responses, it is difficult to assess the relative contributions of elevated CO_2 and climate changes to predictions of crop responses.

Research on the physiological effects has focused primarily on responses of rates of photosynthesis and transpiration to increasing concentrations of atmospheric CO_2. Photosynthesis rates have increased in these crops in relatively ideal experimental environments. At moderate temperatures, most crops will probably show increases in size and possibly yield as CO_2

concentrations rise. However, plants also have internal regulation mechanisms that may lessen these effects under field conditions.

Transpiration rates per unit leaf area decrease, while total transpiration from the entire plant sometimes increases because of greater leaf area. Drought-stressed plants exposed to high partial pressures of CO_2 should be better able to cope with water deficits. Leaf temperatures in all species are expected to rise even more than air temperatures, which may inhibit plant processes that are sensitive to high temperature.

Few studies have examined the interactive effects of CO_2, water, nutrients, light, temperature, pollutants, and sensitivity to daylength on photosynthesis and transpiration. Even fewer studies have examined the effects of these interactions on the growth and development of the whole plant. Therefore, considerable uncertainty exists concerning the extent to which the beneficial effects of increasing CO_2 will be seen in crops growing in the field under normal farming conditions with climate change.

Climate Impacts on Pest-Plant Interactions

Compared to the existing information on the potential effects of climate change on crop production, relatively little effort has been directed toward assessing the influence of climate change on plant-pest interactions. Atmospheric increases in temperature and CO_2, and changes in moisture regimes, all can directly or indirectly affect interactions between pests and crops.

Study Design and Results

Stinner et al. (see Volume _) conducted a literature survey and modeling experiments on the major mechanisms through which climate change may affect pest-plant interactions. This study emphasized the major insect pest and pathogen species of corn and soybeans. The survey indicated that temperature and precipitation patterns are the key variables that affect crop-pest interactions. The temperature increases associated with the climate change scenarios bring about the following trends: (1) increased survival for migratory and nonmigratory insect pest species in the winter, (2) northern range extensions of current pests in the higher latitudes and migration of southern species into the northern Grain Belt regions, (3) an increase in pest species with more than one generation per year in the northern

Grain Belt, (4) earlier establishment of pest populations in the growing season, and (5) increased abundance of pests during more susceptible growth stages of crops.

The potential changes in the overwintering ranges of four major pests were mapped for the GISS and GFDL climate change scenarios and were compared to present ranges. The overwintering capability of the four major pests may extend northward with both climate change scenarios. For example, the potato leafhopper, a serious pest on soybeans and other crops, at present overwinters only in a narrow band along the coast of the Gulf of Mexico (Figure 10-10). Warmer winter temperatures predicted by both the GFDL and GISS models suggest a doubling or tripling of the overwintering range in the United States, respectively. Increases in the winter population density of this size would increase the invasion populations in the northern States by similar factors. The invasions also would be earlier in the growing season, assuming planting dates do not change. Both features are likely to lead to greater insect density and damage. This pattern is repeated with the other three pests studied and indicates that these pests, and possibly others, may move northward and invade cropping systems earlier in the growing season under climate change conditions.

The Soybean Integrated Crop Management (SICM) model (Jones et al., 1986) was run with the GISS and GFDL climate change scenarios to estimate changes in damages caused by corn earworm. Modeling results show that earworm damage to soybeans will increase in severity in the Grain Belt under a warmer climate. Such damage could cause grain farmers in the Midwest to suffer significant economic losses. These results were particularly marked with the warmer and dryer GFDL scenario.

Study Limitations

Lack of knowledge about the physiological effects of CO_2 on crop plants and lack of experimental evidence on direct effects of CO_2 on insect-plant interactions make the study of pest-plant interactions particularly difficult. Only one cultivar is used in the modeling study under both the baseline and the climate change scenarios. In reality, farmers would probably switch to a more climatically adapted cultivar as climate changed.

Study Implications

Increased pest-related crop damage could intensify pesticide use. The economic and environmental ramifications of such an increase could be substantial, not only in current farming regions but also in new areas as agriculture shifts to the more

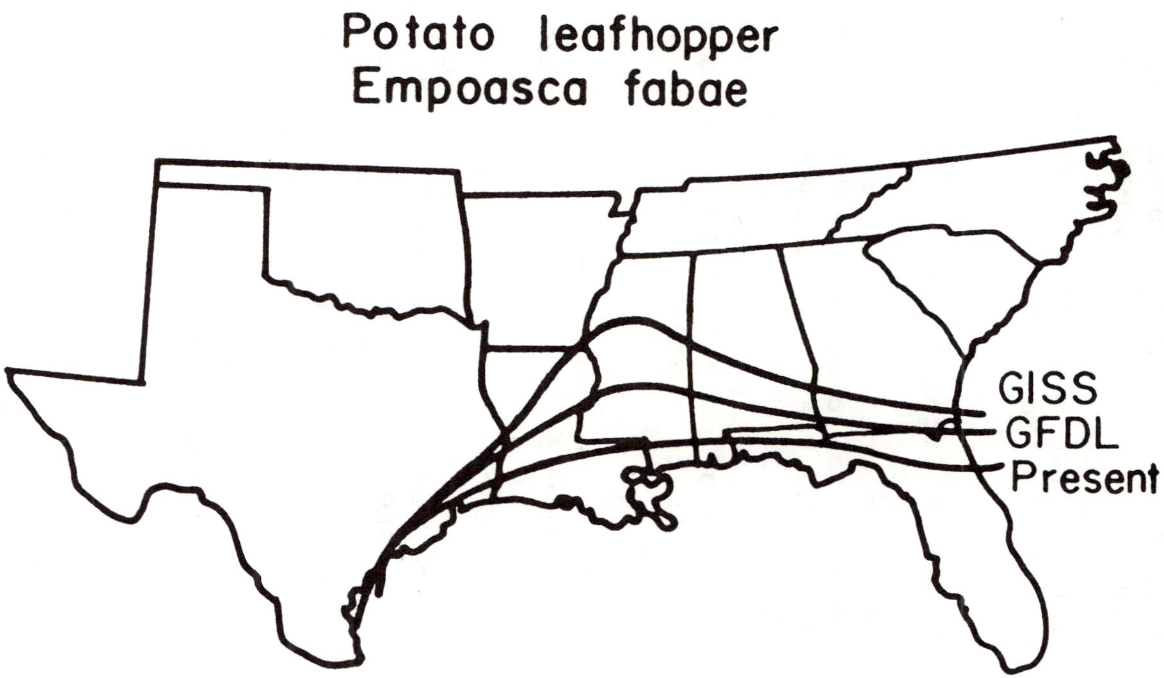

Figure 10-10. Present and potential (GISS and GFDL climate change scenarios) overwintering range of the potato leafhopper, Empoasca fabae, a major pest of soybeans.

Source: Stinner et al., Volume __.

northern regions such as the Northern Plains, the Great Lakes States, and the Pacific Northwest (see Figure 10-6).

Increased use of pesticides would create additional threats to the integrity of ecosystems through soil and water contamination and could increase risks to public health. If agricultural production is not to rely increasingly on chemicals that are potentially harmful to the environment, an increased need will exist for alternative pest management strategies such as biological control, genetic resistance, and innovative cropping systems.

Effects of Climate Change on Water Quality

Agricultural pesticides are ranked as a high priority pollution problem in many rural regions. Potentially toxic agricultural chemicals can be transported away from fields via runoff of surface soils and via downward leaching and percolation through the soil. An understanding of these processes is essential to the evaluation of potential threats to drinking water quality caused by climate change.

Study Design

Johnson et al. (see Volume _) modeled the partitioning of agricultural pesticides among uptake, degradation, surface runoff, and soil leaching for wheat, corn, and cotton production regions in the Great Plains and the Corn Belt. (See the Great Plains chapter of this report for details of the study.) They used the Pesticide Root Zone Model (PRZM) (Carsel et al., 1984), which simulates the vertical movement of pesticides in the soil. The model consists of hydrological and chemical transport components that simulate runoff, erosion, plant uptake, leaching, decay, foliar washoff, and volatilization of a pesticide. The interactions among soil, tillage, management systems, pesticide transport, and climate change were studied.

Limitations

The frequency and duration of precipitation remain the same in the climate change scenarios, even though these storm characteristics are critical factors in determining the transport of agricultural chemicals and may change. The scenarios assume that the number of days with rainfall does not change, but the intensity of rainfall increases or decreases. Runoff and leaching estimates would most likely be different if the number of days of rainfall changed and daily rainfall amounts were held constant.

The PRZM is a one-dimensional, point model that does not simulate the transport of water below the root zone. Thus, results on a regional basis must be extrapolated with care. The direct effects of CO_2 on crop growth, which may increase the size of the plants and the extent to which crops cover the soil, are not included.

Results

Regional changes in chemical loadings of water and sediment are likely but probably will not be uniform. There appears to be some consensus between the GCM scenarios concerning the estimated regional changes (Table 10-5). Modeled pesticides in runoff increase in the cotton production area, and pesticides carried by sediments decrease in the spring wheat and corn regions. Leaching of pesticides tends to be less everywhere owing to changes in seasonal precipitation and increased evaporation.

Implications

When the changes in water quality from the predicted climate change scenarios are considered in conjunction with the predicted increases in pests and implied higher applications of pesticides described in the study on pest-plant interactions, the potential for changes in the nation's water quality becomes apparent. Any deterioration in water quality could adversely affect public drinking water supplies and human health.

Climate Variability

The impacts of climate change result not only from a slow change in the mean of a climate variable but often from shifts in the frequency of extreme events. Droughts, freezes, and prolonged periods of hot weather have strong effects on agricultural production. While the agricultural modeling studies did not include the effects of potential changes in climate variability, Mearns and Schneider (see chapter on Climate Variability in this report) reviewed literature on agriculture and extreme events to determine the nature and magnitudes of significant impacts.

Corn, soybeans, wheat, and sorghum are sensitive to high maximum temperatures during blooming. Lower yields of corn, wheat, and soybeans have been found to be correlated with high temperatures. The damaging effect of runs of hot days on corn yields was particularly evident in 1983 in the U.S. Corn Belt.

Although low temperatures may be less of a problem with the predicted climate change, risks of frost damage to crops may

Table 10-5. Summary of GCM Model Consensus of PRZM Pesticide Transport by Cropping Region and Pesticide[a]

	Surface Pesticides Runoff Losses	Surface Pesticide Erosion Losses	Pesticide Leaching
SPRING WHEAT:			
Highly soluble/short-lived[b]	+	+	-
Highly soluble/long-lived		+	-
Slightly soluble/long-lived			-
WINTER WHEAT:			
Highly soluble/short-lived		+	+
Highly soluble/long-lived	+		-
Slightly soluble/long-lived			-
COTTON:			
Highly soluble/short-lived	+		+
Highly soluble/long-lived	+	+	-
Slightly soluble/long-lived	+	+	-
CORN:			
Highly soluble/short-lived	-	-	
Highly soluble/long-lived		-	-
Slightly soluble/long-lived	-	-	-

[a] + indicates that median values increase under climate change; - indicates that median values decrease under climate change; blank indicates no consensus among median values.

[b] ex. median value of all tillage, soil, weather site scenarios for highly soluble/short-lived pesticides in the spring wheat crop area.

Source: Johnson et al., Volume _.

change in the growing areas of certain crops. Citrus trees are very vulnerable to low minimum temperatures. Winter wheat is often damaged by low temperatures known as winter kill, especially in the absence of snow. (See Glantz, Volume _, for a discussion of how the Florida citrus industry adapted to increased frequency of freezes in the 1980s.) Even with warmer winters and fewer frosts, more damage can occur at less extreme temperatures. For example, the effect of freezing temperatures is exacerbated if the crops have not yet been hardened by the cold, or if they are no longer dormant and a cold snap occurs.

Drought is a major cause of yearly variability in crop production. In the Dust Bowl years of the 1930s, yields of wheat and corn in the Great Plains dropped to as much as 50% below normal. In 1988, agricultural disaster in areas of the northern Great Plains demonstrated a high vulnerability to drought; and nationwide a nearly 40% decrease in U.S. corn yields is predicted. Reduction in vegetative cover associated with drought also brings about severe wind erosion of soils, which will affect future crop productivity. Low yields of forage crops during droughts result in food shortages for livestock and premature selling of livestock.

Farm-Level Management and Adjustments to Climate Change

Adjustments to existing production practices are the first course of action in the face of climate change. The net effect of climate change with adjustment by farmers may be significantly different from the estimated effects of climate change alone.

Study Design

Several studies addressed possible adjustments that could modify the effects of climate change. These adjustments include changes in planting and harvesting dates, tillage practices, crop varieties, application of agricultural chemicals, irrigation technology, and institutional responses for management of water resources.

Results of Adjustment Studies

Ritchie et al. demonstrated that the yield reduction in corn in the Great Lakes can be partly overcome with selection of new cultivars that have a longer growing season (see Great Lakes chapter of this report). Rosenzweig (see Great Plains chapter of this report) showed that adjusting the planting date of winter wheat to later in the fall did not ameliorate the effects of climate change, but that changing to varieties more suited to the predicted climate overcame yield decreases at some locations.

Dudek's California study found that flexible institutional responses to climate change helped to compensate partly for negative climate change effects (see California chapter of this report). By allowing movement of water around the State by transferring water rights, California water resource managers can alleviate some groundwater extraction and compensate for surface water reductions.

Easterling (see Great Lakes chapter of this report) found that potential farmer adjustments to climate change include changes in tillage practices, increased fertilizer application, more full-season and heat-resistant varieties, changes in planting densities, higher pesticide use, earlier harvest, and less artificial drying. Different adjustments would occur at different times in the cropping season. With the hotter and drier GFDL scenario, farmers may be forced to adopt production practices that are different from those in use today. Climate changes that leave soils drier during summer than they are at present will most likely lead to increased use of irrigation in the Corn Belt. This increased use of irrigation is also supported by the projected price increases for all crops grown in Illinois.

Implications of Adjustment Studies

Although detrimental climate change effects on agriculture may be partly offset naturally by increased photosynthesis and water-use efficiency caused by higher levels of atmospheric carbon dioxide, farmers will no doubt use a variety of adjustments to adapt to climate change. Market forces also will aid adaptation to climate change because they help to allocate resources efficiently.

More severe climate changes are likely to necessitate major adaptations, such as farm abandonment in some regions. Prior investments may be lost as existing infrastructures become obsolete and new investments are needed for new locations. Some of these adjustments could be costly.

Improvements in agricultural technology also may be expected to ease adjustment by development of appropriate farming practices and crop variety and livestock species. Adjustment and adaptation to climate change should be included in agricultural research programs to enable this process to occur.

Livestock

Animal products are a critical source of protein, energy, vitamins, and minerals. U.S. livestock production, mainly from cattle, swine, sheep, and poultry, was estimated to be worth over $31 billion in 1986 (USDA, 1987).

Climate is known to significantly affect many aspects of animal health and production. The direct effects of climate change on animal health include differences in incidence of heat and cold stress, changes in weight gain, and decline in reproductive capabilities. Indirect effects may involve trends in agricultural commodities, such as the availability and prices of types of animal feeds and the expanded geographic distribution and activity of disease-carrying vectors.

Higher winter temperatures may lower the incidence of respiratory diseases in livestock (Webster, 1981). Conversely, warmer summers may necessitate more hours of indoor cooling during which pathogens are confined to housing structures. Climate warming may significantly increase the costs of air-conditioning in poultry housing. Changes in reproductive capabilities such as decreased ovulation rates, shortened intensity and duration of estrus, decreased fertility of males, and increased embryonic mortality also have been shown to occur with high temperatures (Ames, 1981).

Climate change may also affect the survivability, activity, and geographic distribution of vectors responsible for the transmission of infectious diseases in livestock. The activity and reproduction of disease-carrying vectors infecting livestock, humans, and crops are driven primarily by temperature, humidity, and precipitation. These impacts are likely to be similar to those on mortality and morbidity of disease in humans (see chapter on Human Health, this volume), and they also are similar to changes predicted for crop pests.

Design of Livestock Studies

Stem et al. (see Volume _) studied the available literature on four diseases to evaluate the range of potential changes in disease distribution and occurrence under climate change conditions. Schmidtmann and Miller (see Volume _) used a population dynamics simulation model to estimate the effects of the GFDL climate change scenario on the life cycles of the horn fly, a ubiquitous pest of pastured cattle throughout the United States.

Results of Livestock Studies

Stem et al. found that, under warmer conditions, livestock diseases currently causing serious economic losses in tropical countries could spread into the United States. Rift Valley fever is transmitted principally by mosquitoes, and danger of the disease may spread as rising winter temperatures become able to support an active increase in the mosquito population. African swine fever also may become more of a threat.

The ranges and activities of disease-carrying agents of blue tongue and anaplasmosis, diseases currently causing severe losses in cattle and sheep in the United States, may expand. If disease-carrying insects increase their winter survival and reproduce year-round in more States, the geographical distribution of blue tongue, which is caused by a virus, may expand northward and eastward. Anaplasmosis, a rickettsial infection of ruminants, is the second most important disease of cattle in the United States. The distribution of the habitat of the insect carrier of the disease could expand to northern States with climate change, and the day-to-day activity of the insects may increase, also causing an increase in disease transmission.

The horn fly causes annual losses of $730.3 million in the beef and dairy cattle industries (Drummond, 1987). Schmidtmann and Miller found that with the very warm GFDL climate change scenario, the horn fly season throughout most of the United States was extended by 8 to 10 weeks, horn fly populations increased (see Figure 10-11), and the average daily gain in beef cattle decreased. This horn fly increase will cause substantially greater reductions in the average daily gain of growing beef cattle. The GFDL simulation also increased pest activity in dairy cattle in the northern and northwestern States, a result that could significantly decrease milk production. Conversely, the summertime activity of the horn fly decreased in the southern States because the warmer climate exceeded the horn fly's tolerance to high temperatures.

<u>Limitations</u>

Horn fly data are based on counts taken at various times, under varying weather and management conditions. However, the prediction of current horn fly populations appears to be well correlated with observations. The model is not validated for the high temperatures predicted for the climate change. Schmidtmann and Miller used only the hottest scenario, GFDL; the other scenarios may have resulted in less of a geographic shift in the range of the horn fly.

It should be noted that this analysis is based on current livestock management breeds and distribution. Possible changes in these factors are beyond the scope of this study.

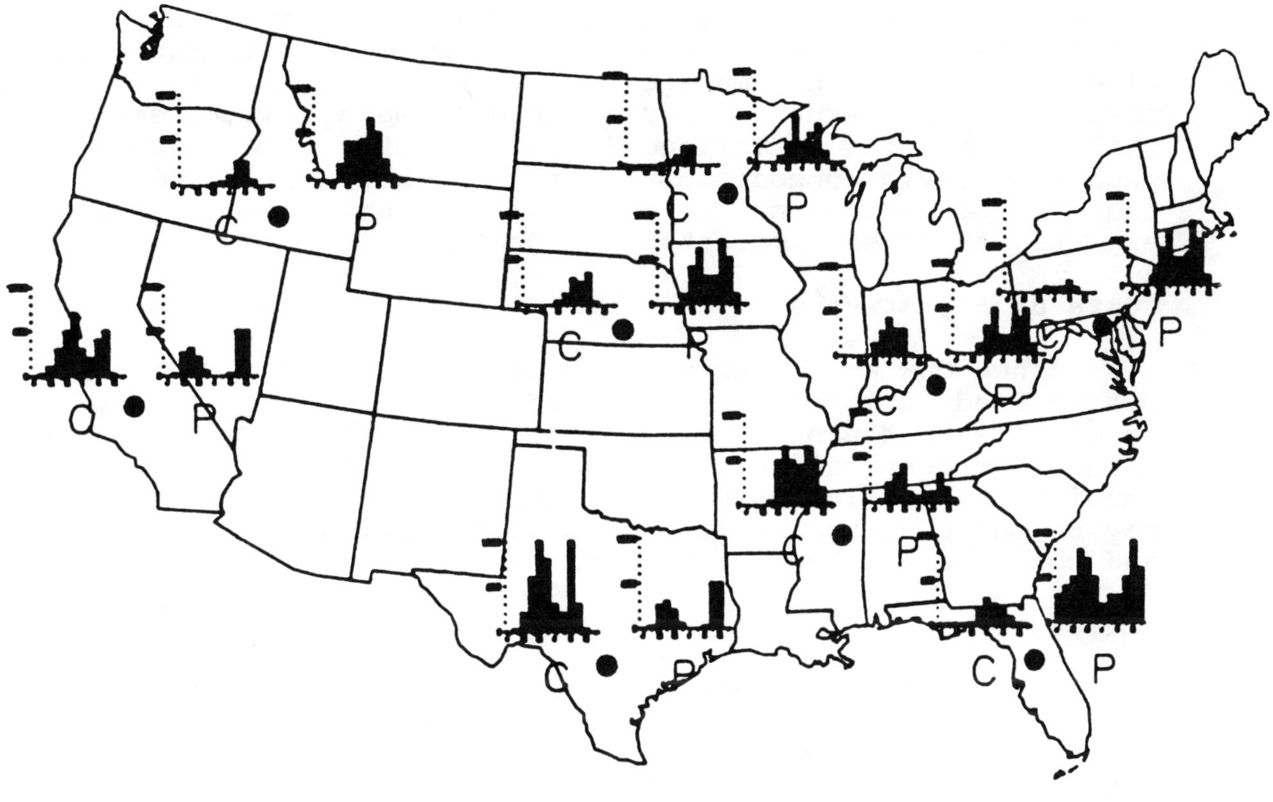

Figure 10-11. Current (C) and projected (P) populations of the horn fly, a major pest of beef and dairy cattle, using the GFDL climate change scenario.

Source: Schmidtmann and Miller, Volume _.

Implications of Livestock Studies

With climate change, patterns of livestock diseases and pests may change. Tropical livestock diseases may become more of a threat, because more geographical areas are potential ranges for the insect carriers of the diseases. Temperature conditions may improve in the winter but may be exacerbated in the summer. Reproductive capabilities may be lower.

ECONOMIC AND ECOLOGICAL IMPLICATIONS

Even though the U.S. agricultural system has historically shown considerable resiliency and may be less vulnerable to climate change than unmanaged ecosystems, many reasons to avoid complacency concerning the predicted climate change remain. While the studies summarized in this chapter show that the U.S. supply of basic commodities probably is not threatened by climate change, food prices may increase for consumers, and agricultural exports may be significantly reduced.

Costs of Adjustments

Since our agricultural production system consists mostly of specialized farms producing commodities in geographically specialized patterns of production, the costs of adjustments to changes in comparative advantage among agricultural regions with ensuing changes in resource use and agricultural infrastructure may be quite high. These shifts will also entail involvement of and costs to the Federal Government.

Effects of CO_2

It is also important to note that the crop modeling studies showed that the direct effects of CO_2 on crop photosynthesis and water-use efficiency ameliorate the negative effects of climate change in some locations but not uniformly and not everywhere. While much work remains to be done to improve both climate and crop models, reliance on overly optimistic assessments about the beneficial effects of CO_2 must be avoided when policy options are considered.

Water Quality

Changes in the agricultural production system are likely to have significant impacts on resource use and the environment. Many of the agricultural studies suggest that climate warming could result in changes in water quality from agricultural chemicals, increases in pesticide usage to control changes in pest vectors, and accelerated rates of demand for water for irrigation. Climate change could thus exacerbate many of the current trends in environmental pollution and resource use associated with agriculture as well as initiate new ones.

A northward migration of agriculture will increase the use of irrigation and fertilizers on sandy soils, thus endangering underlying groundwater quality. From South Dakota to southern Canada, critical prairie wetlands may be lost to drainage and conversion to cropland. Many of these areas are important wildlife habitats. Shifts in agricultural activities may increase the susceptibility of soils to wind and water erosion.

Global Agriculture

Finally, U.S. agriculture is an integral part of the global, international agricultural system. Consequently, the adjustment of U.S. agriculture to climate change cannot be considered in isolation from the rest of the world. The optimal configuration of U.S. adjustments will depend very much on how simultaneous changes in regional climate affect global agriculture and how other countries, in turn, respond to those changes.

NATIONAL POLICY IMPLICATIONS

Since climate change appears likely to reconfigure the agricultural activities and demographics of rural America, policies are needed to ease the inevitable adjustments and ensure the sustainability of our natural and human resources (see Schuh and Dudek, Volume _). Following are specific policy areas that policymakers could investigate to respond appropriately to the projected climate change.

Commodity Policies

Agricultural pricing and production policies should promote efficient adjustment to the changing conditions of global supply and demand induced by the greenhouse effect, which may include shifts in comparative advantage among regions and increased likelihood of droughts in some regions. Although these shifts may be slow, the cumulative effects are large and they deserve close monitoring. Market forces as well as government programs will play a crucial role in creating the flexibility to respond to climate changes by sending signals on the efficient use of resources, and in mitigating their ultimate impact as they have done in the past. Agricultural policies should be evaluated to ensure that they are appropriate to both current and future conditions.

Land-Use Programs

Federal legislation aimed at reducing the use of newly plowed grasslands, e.g., the "Sod-Buster Bill," and the related "Swamp-Buster Bill," which restricts agricultural encroachment into wetlands subject to flooding and water-logging, are examples of new policies meant to protect marginal lands. The basic goals of these new laws, which are part of the 1985 Farm Bill, are to protect the most erodible farmland by removing it from crop production and to use conservation as a tool for reducing overproduction.

Another program established in the 1985 Farm Bill that may help alleviate the negative effects of climate change is the Conservation Reserve Program. This program is aimed at removing from crop production the cropland classified as "highly erodible" by the Soil Conservation Service. The bill created a new form of long-term contract of up to 10 years and provides payments to farmers who apply conservation practices, such as maintaining a grass cover, on those acres. If successful, the Conservation Reserve Program may reduce the impact of climate fluctuations on total grain production by taking the most sensitive lands out of use.

The 1988 drought, however, demonstrates that the Conservation Reserve Program may be difficult to maintain in the face of climate stress. As the drought worsened during the summer, use of the set-aside lands was requested so that badly hit farmers could salvage some economic benefits from these acres. Such conflicts may be more common in the future, and land retirement strategies must be weighed against possible needed increases in production in the future.

Awareness of potential changes in agricultural land use due to regional climate change should be built into land-use planning programs, especially in regions where agricultural activities may expand into natural, unmanaged ecosystems. Large-scale drainage and water projects would need environmental impact studies to carefully assess this potential expansion of agricultural land (see Baldwin, Volume _).

Water-Resource Management Programs

Current water supply policies do not generally encourage optimum water-use efficiency. A greater degree of water efficiency should promote flexibility in light of the potential for increased irrigation demands with climate change. Policies such as water transfers and markets should be evaluated for irrigated areas.

Water Quality Policy

The increased use of agricultural chemicals, and changes in the hydrological cycle, threaten both soil and water supplies, and eventually, public health. These effects could be ameliorated by including potential climate change effects in water quality planning and by supporting alternative pest management strategies that use such techniques as biological control, genetic resistance, and innovative cropping systems.

Risk Management

Changes in the frequency, intensity, and location of extreme events are important for agriculture and the regional income that it produces. The adequacy of the private crop insurance and Federal disaster payment programs should be assessed in the face of climatic uncertainty. For example, only about 20-25% of potentially insurable acreage is currently covered by crop insurance. Farmers tend to rely on Federal disaster relief programs to bail them out of such disasters as droughts, floods, hail, and windstorms. Financial risk is also part of the credit structure that covers land, equipment, and production in modern farming.

The frequency and magnitude of climate extremes may be altered with climate change. Responding to the changes may be costly for the government if crops fail frequently. The Drought Relief Bill for the drought of 1988 is scheduled to cost $3.9 billion to cover just 1 year of a climatic extreme. On the other hand, some areas that currently suffer from climate extremes may benefit from climate change. Risk policy mechanisms for relief, recovery, and mitigation of climate change should be examined so that they will be ready to help farmers adjust.

International Trade Agreements

Policies designed to ease the adjustment to greenhouse effects must be global in scope because the effects, although varied, are global in nature. Comparative advantage will shift significantly both within the United States and in other countries. Population and economic activities also will change geographically with climate change, thus affecting the location of demand for agricultural products. It is already a goal of U.S. agricultural policy to incorporate global conditions of supply and demand into the agricultural sector. The potential seriousness of the impacts on the agricultural production system of the greenhouse effect may provide added incentive to establish such policies both nationally and globally. The vulnerability of current and potential food-deficit regions to climate change should also be considered.

Agricultural Contributions to the Greenhouse Effect

Agriculture itself is an active contributor to the greenhouse effect. Clearing of forested land for agriculture often involves burning of trees and shrubs that release CO_2. The biomass that is not burned tends to decay gradually, also emitting CO_2. Agricultural activities release other radiatively active trace gases. Flooded rice fields emit methane (CH_4) as a product of the anaerobic decomposition of organic matter. Ruminants also release methane as a consequence of their digestive processes. In addition, soils may volatilize some of the nitrogenous fertilizer applied to them in the form of nitrous oxide (N_2O). Finding effective ways to reduce these emissions presents a major challenge to the agricultural research community. In this regard, the Conservation Reserve Program and forestation efforts could provide a partial solution, since vegetation fixes CO_2 from the air.

Agricultural Research

The agricultural research community should enhance climate change research from the field level to the national policy level. It should continue to breed heat- and drought-resistant crop varieties and new crop species in preparation for global warming. Research in biotechnology may also be directed toward alleviating the negative effects of climate change. Improved water use and irrigation efficiency also take on renewed importance in the light of potential climate change.

Reduction of agricultural emissions of trace gases also deserves research attention. Development of nitrogen-fixing corn would help decrease fertilizer applications and thus agricultural emissions of N_2O. This would also mean a significant improvement in groundwater contamination.

RECOMMENDATIONS FOR FURTHER RESEARCH

1. Crop productivity -- Study the interactive effects of climate variability and change, CO_2, tropospheric ozone, UV-B from stratospheric ozone depletion, and other environmental and societal variables on agricultural productivity.

   Because of the significant changes in production indicated by these studies, the need for better simulation of the direct effects of CO_2 in the crop models, and the limited adjustment studies performed, further research should be conducted on a longer term basis. Necessary work includes validation of the crop models, resolving the differences in forecasts of the GCMs, and designing more appropriate scenarios including changes in climatic variability. Physiologically based submodels are needed for the effects of increased CO_2 on various crops. The effects on cotton also should be studied because cotton is a major crop, and because daylength may not allow it to shift northward. All the crop models should be improved in their simulation of the effects of increasing temperatures.

Research to this point has provided windows of knowledge concerning certain crops (especially soybeans) at specific stages of their life cycles. Both the direct and the climate change effects of high CO_2 are probably quite different at different stages of development. Research should be performed on the interactive effects of CO_2 and temperature over the whole life cycle of the plant, with realistic conditions of water and nutrition, rather than with "pampered" plants under optimal conditions. Then crop response to the combined climatic and physiological effects of CO_2 may be predicted more realistically.

2. <u>Adaptation strategies</u> -- The dynamic nature of climate change also needs to be studied: What is the rate of adaptation of regional agricultural systems compared to the rate of climate change? Evaluation of the thresholds of sensitivity of U.S. agriculture is needed. Studies should analyze the ability of various aspects of the agricultural production systems to adapt to various rates and degrees of climate change to determine these thresholds of sensitivity.

3. <u>Agricultural economics</u> -- Expand the national analysis to include crops and regions not now included (for example, cotton and grasslands, and pasture in the western regions of the United States). Conduct further analyses of regional implications of shifts in agriculture.

4. <u>International agriculture</u> -- Study the potential shifts in international comparative advantage and the vulnerability of food-deficit regions, and evaluate the implications of such shifts for the United States.

5. <u>Environmental impacts</u> -- Elucidate the impacts of climate change on water quantity, water quality, and other components of the environment caused by shifts in crop and livestock production and related industries.

REFERENCES

Acock, B., and L.H. Allen Jr. 1985. Crop respones to elevated carbon dioxide concentrations. In: Strain, B.R., and J.D. Cure, eds. Direct Effects of Increasing Carbon Dioxide on Vegetation. Washington, DC: U.S. Department of Energy. DOE/ER-0238. pp. 33-97.

Adams, R.M., S.A. Hamilton, and B.A. McCarl. 1984. The Economic Effects of Ozone on Agriculture. Corvallis, OR: U.S. Environmental Protection Agency. EPA-600/3-84-090.

Ames, David R. 1981. Effect of climate on livestock production data. In: Knapp, F.W., ed. Systems Approach to Animal Health and Production. Lexington, KY: University of Kentucky, pp. 148-148.

Callaway, J.M., F.J. Cronin, J.W. Currie, and J. Tawil. 1982. An analysis of methods and models for assessing the direct and indirect impacts of CO_2-induced environmental changes in the agricultural sector of the U.S. economy. Richland, WA: Pacific Northwest Laboratory, Battelle Memorial Institute. PNL-4384.

Carbon Dioxide Assessment Committee. 1983. Changing Climate. Washington, DC: National Academy of Sciences.

Carsel, R.F., C.N. Smith, L.A. Mulkey, J.D. Dean, and P. Jowise. 1984. Users Manual for the Pesticide Root Zone Model (PRZM). Athens, GA: U.S. Environmental Protection Agency. EPA-600/3-84-109.

Council for Agricultural Science and Technology. 1984. Long-Term Viability of U.S. Agriculture. Aces, Iowa. Report No. 114.

Cure, J.D. 1985. Carbon dioxide doubling responses: a crop survey. In: Strain, B.R., and J.D. Cure, eds. Direct Effects of Increasing Carbon Dioxide on Vegetation. Washington, DC: U.S. Department of Energy. DOE/ER-0238. pp. 99-116.

Decker, W.L., V. Jones, and R. Achutuni. 1985. The impact of CO_2-induced climate change on U.S. agriculture. In: White, M.R., ed. Characterization of Information Requirements for Studies of CO_2 Effects: Water Resources, Agriculture, Fisheries, Forests and Human Health. Washington, DC: U.S. Department of Energy. DOE/ER-0236. pp. 69-93.

Doorenbos, J., and A.H. Kassam. 1979. Yield Response to Water. Rome: Food and Agriculture Organization of the United States. FAO Irrigation and Drainage Paper 33.

Drummond, R.O. 1987. Economic aspects of ectoparasites of cattle in North America. In: Leaning, W.H.D., and J. Guerrero, eds. Proceedings of the MSD AGVET Symposium, The Economic Impact of Parasitism in Cattle. XXIII. World Veterinary Congress. Montreal, Quebec. pp. 9-24.

Hansen, J., I. Fung, A. Lacis, D. Rind, G. Russell, S. Lebedeff, R. Ruedy, and P. Stone. 1988. Global climate changes as forecast by the GISS 3-D model. Journal of Geophysical Research (In press.)

Horner, G., L. Putler, and S.E. Garifo. 1985. The Role of Irrigated Agriculture in a Changing Export Market. Washington, DC: U.S. Department of Agriculture, Economic Research Service. ERS Staff Report AGES850328.

Jones, J.W., K.J. Boote, S.S. Jagtap, G. Hoogenboom, and G.G. Wilkerson. 1988. SOYGRO V5.41: Soybean Crop Growth Simulation Model. User's Guide. Florida Agr. Exp. Sta. Journal No. 8304, IFAS. Gainesville: University of Florida.

Jones, J.W., J.W. Mishoe, G.Wilkerson, J.L. Stimac, and W.G. Boggess. 1986. Integration of soybean crop and pest models. In: Frisbie, P.E., and P. Adisson, eds. Integrated Pest Management on Major Agriculture Systems. Texas Agriculture Experiment Station. Publication No. MP-1616.

Kimball, B.A. 1985. Adaptation of vegetation and management practices to a higher carbon dioxide world. In: Strain, B.R., and J.D. Cure, eds. Direct Effects of Increasing Carbon Dioxide on Vegetation. Washington, DC: U.S. Department of Energy. DOE/ER-0238. pp. 185-204.

Land Evaluation Group. 1987. Implications of climatic warming for Canada's comparative position in agricultural production and trade. University School of Rural Planning and Development. University of Guelph. Guelph, Ontario.

Monteith, J.L. 1965. Radiation and crops. Experimental Agricultural Review 1(4):241-251.

New York Times. August 12, 1988. Drought cutting U.S. grain crop 31% this year, by K. Schneider. p. A1.

Parry, M.L., T.R. Carter, N.T. Konijn, eds. 1988. The Impact of Climatic Variations on Agriculture. Vol. 1. Assessments in Cool Temperate and Cold Regions. Kluwer, Dordrecht.

Postel, S. 1986. Altering the Earth's Chemistry: Assessing the Risks. Worldwatch Paper 71. Washington, DC: Worldwatch Institute.

Ritchie, J.T., and S. Otter. 1985. Description and performance of CERES-Wheat: A user-oriented wheat yield model. In: Willis, W.O., ed. ARS Wheat Yield Project. Washington, DC: U.S. Department of Agriculture, Agricultural Research Service. ARS-38. pp. 159-175.

USDA. 1987. U.S. Department of Agriculture. Agricultural Statistics. Washington, DC: U.S. Government Printing Office.

Warrick, R.A., R.M. Gifford, M.L. Parry. 1986. CO_2, climatic change and agriculture. Assessing the response of food crops to the direct effects of increased CO_2 and climatic change. In: Bolin, B., B.R. Doos, J. Jager, and R.A. Warrick, eds. The Greenhouse Effect, Climatic Change and Ecosystems. A Synthesis of the Present Knowledge, SCOPE 29. New York: John Wiley and Sons, pp. 393-473.

Webster, A.J.F. 1981. Weather and infectious disease in cattle. The Veterinary Record 108:183-187.

World Bank. 1988. World Population Projections. 1987-1988 edition. Short- and Long-term Estimates. Zachariah, K.C., and Vu, M.T. Baltimore, MD: Johns Hopkins University Press.

World Food Institute. 1987. World Food Trade and U.S. Agriculture, 1960-1986. Aces, IA: Iowa State University.

CHAPTER 11

THE POTENTIAL IMPACT OF RAPID CLIMATIC CHANGE ON FORESTS IN THE UNITED STATES

FINDINGS

Global warming could significantly affect the forests of the United States. Changes could be apparent in 30 to 80 years, depending upon the region, a site's quality, and the rate of climate change. There may be northward shifts in species ranges, dieback along the southern reaches of species ranges, and changes in forest productivity. The combined effects of climate change and other stresses may exacerbate these impacts. Different migration rates and climate sensitivities may result in changes in forest composition. Large reductions in the land area of healthy forests are probable during this century of adjustment to climatic changes.

Range Shifts

- The southern ranges of many forest species in the eastern United States could dieback as a result of higher temperatures and drier soils. The southern boundary could move several hundred to one thousand kilometers (up to six hundred miles) in a generally northward direction for the scenarios studied.

- The potential northern range of forests species in the eastern U.S. could shift northward as much as 600-700 km (370-430 mi) over the next century. Actual northward migration could be limited to as little as 100 km (60 mi), due to the slow rates of migration of forest species. It could take centuries for eastern forests to fully migrate to potential northern distributions. If climatic change is too fast, some tree species may not be able to form healthy seeds, thus halting migration.

- If elevated CO_2 concentrations substantially increase the water-use efficiency of tree species, the southern declines could be alleviated.

- If climate stabilizes, forests might eventually regain a generally healthy status (over a period of several centuries). In the meantime, declining forests could be subject to increased fires, pest attacks, and replacement with low-value trees, grasslands, and shrubs.

<u>Productivity Changes</u>

- Dieback along the southern limits of distribution of many species could result in productivity declines of 46 to 100

Productivity Changes

- o Dieback along the southern limits of distribution of many species could result in productivity declines of 46 to 100 percent, depending on how dry soils become.

- o Productivity could increase along the northern limits of some eastern tree species, particularly as slow-growing conifers are replaced by more rapidly growing hardwoods.

Combined Impacts with Other Stresses

- o Large regions of severely stressed forests, combined with possible increases in fires, pests, disease outbreaks, wind damage, and air pollution could produce major regional disturbances. These factors were not considered for this study.

- o Additional impacts of changes in forests include reductions in biotic diversity, increased soil runoff and soil erosion reduced aquifer recharge, changes in recreation, and changes in wildlife habitat.

Policy Implications

- o Institutions such as the U.S. Forest Service, state forest agencies, and private companies should begin to consider how to factor climate change in their long-term planning. Global climate change may need to be a factor in the Forest Service's 50-year plan.

- o Strategies that forest agencies may wish to consider include the feasibility of massive reforestation on a much larger scale than practiced now, and the possible introduction of subtropical species into the Southeast in order to maintain U.S. forest productivity. A coordinated public and private reforestation effort, together with development of new and adapted silvicultural practices, may be required. A large reforestation program would also reduce atmospheric CO_2 concentrations, slowing the rate of global warming.

EXTENT AND VALUE OF U.S. FORESTS

Forests occupy 33% of the U.S. land area and exist on some lands in all 50 States. In total, they occupy 299 million hectares (738 million acres) and are rich in such resources as water and wildlife.

Chapter 11

Many biotic and abiotic factors influence the condition of forests, but climate is the dominant factor (Spurr and Barnes, 1980). This chapter summarizes what is currently known and predicted regarding the effects of rapid climatic change on U.S. forests.

Distribution and Ownership

Eight major forest regions of the conterminous 48 States contain 84% of the forested ecosystems of the United States (Figure 11-1). The forested areas of Alaska and Hawaii represent the remaining 16% (Table 11-1). Each forest region includes one or more forest types distinguished by the major tree species present. As a general rule, some types in each region have predominantly coniferous tree species (i.e., evergreen, needle-leaved, and softwoods); other forest types are composed mostly of deciduous trees (i.e., tree species that are broad-leaved, have no winter foliage, and are hardwoods). Forest types with a mix of coniferous and deciduous trees, however, are not uncommon.

Superimposed over the natural distributions of trees, forests, and ecosystems in the United States is the human infrastructure. Ownerships include Federal, State, and private lands (Table 11-1). In broad terms, the United States contains public forests (28%), private industry lands (14%), and many small private forests or woodlands (58%) that mostly consist of parcels less than 200 hectares (ha) (about 500 acres) (Woodman, this report). The Federal Government ownership of 40 million ha (99 million acres) of commercial forest lands is primarily in the National Forest System managed by the U.S. Department of Agriculture's Forest Service (36 million ha or 91 million acres); most of the remainder is managed by the Department of Interior's Park Service, Fish and Wildlife Service, or Bureaus of Land Management and Indian Affairs. State ownerships total 9 million ha (23 million acres). Private lands are divided between those of industrial forest companies (27 million ha or 68 million acres) and small landowners, who collectively have 112 million ha (277 acres).

To the degree that all forest lands are owned by some individual or organization, all forest lands are under some form of management. A continuum of management policies exists, ranging from those lands intended to have minimal human intervention, except for protection from catastrophic wildfire (e.g., some parks and most wilderness areas), to lands where silvicultural practices are intensively applied (e.g., the most productive Federal, State, and industrial forest lands dedicated to growing tree crops, Table 11-2). The forests under government and industrial management constitute roughly one-fourth of the

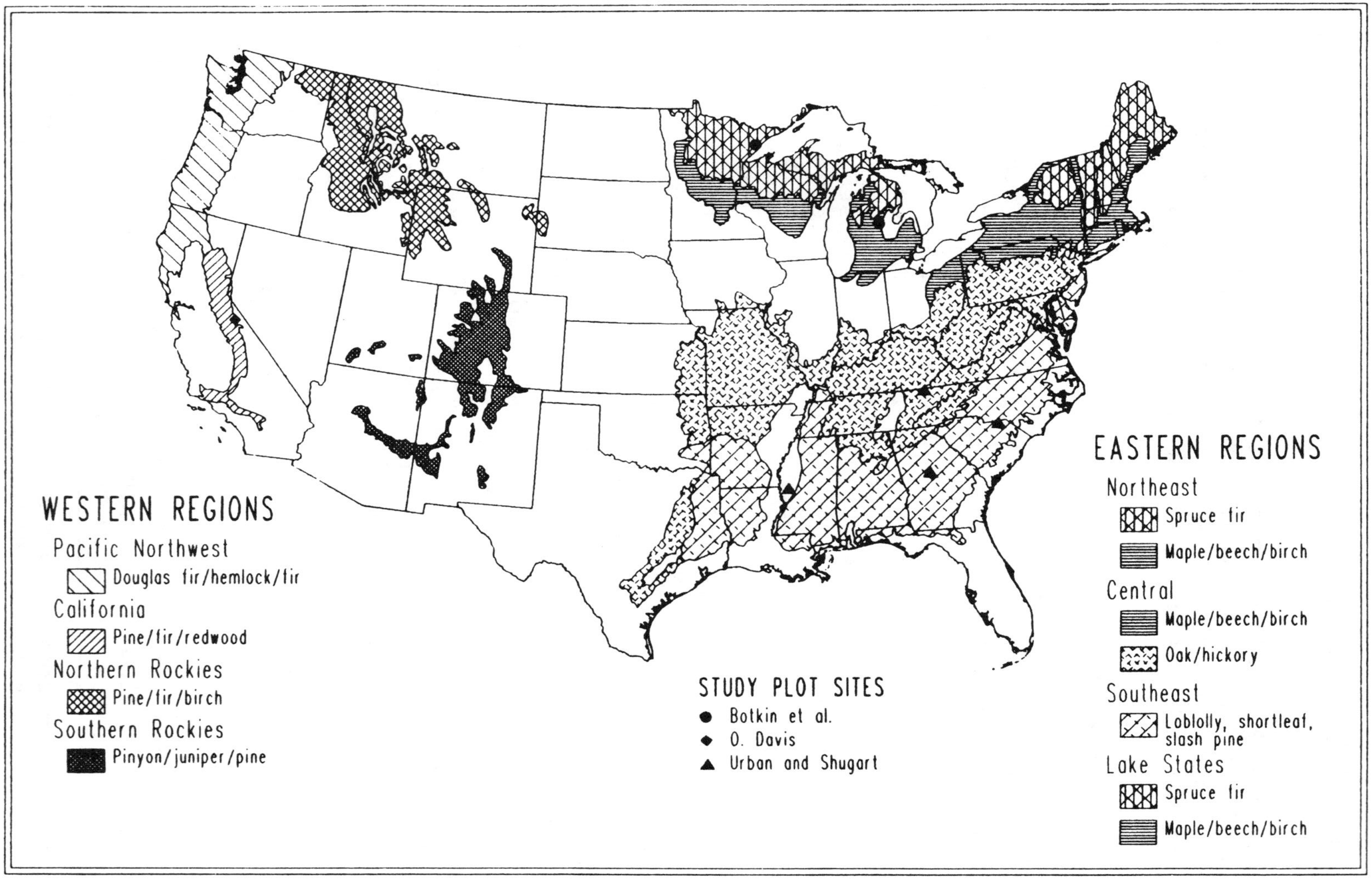

Figure 11-1. Major Forest Regions of the Conterminous United States and Their Primary Tree Groups

Table 11-1. Area of U.S. forest lands in 1977 by federal, state, private, and other ownerships (millions of hectares[1]; USDA, 1982).

	Region/States	Primary Tree Species	Commercial Forests[2] Federal	State	Private Industry	Private Non-Indus	Others[3]	Total	% Total
EAST	Northeast - CT, MA, ME, NH, RI, VT	spruce-fir maple-beech-birch	0.3	0.4	3.9	7.8	0.7	13.1	4.4
	Lake States - MI, MN, WI, ND, SD(E)	spruce-fir maple-beech-birch	2.3	2.8	1.7	9.9	4.2	20.9	7.0
	Central - DE, IA, IL, IN, KA, KY, MD, MO, NB, NJ, OH, PA, TN, WV	maple-beech-birch oak-hickory	1.8	2.0	8.6	22.9	2.6	37.9	12.7
	Southeast - AL, AR, FL, GA, LA, MS, NC, OK, SC, TN, TX, VA	loblolly, shortleaf slash pine	5.8	1.0	14.7	54.3	8.0	83.8	28.1
WEST	Northern Rockies - ID, MT, SD(W), WY	pine-fir-birch	9.1	0.6	0.8	2.7	9.3	22.5	7.6
	Southern Rockies - AZ, CO, NM, NV, UT	pinyon-juniper-pine	6.4	0.3	0.0	2.4	24.1	33.2	11.1
	Pacific Northwest - OR, WA	D. fir-hemlock-fir	7.8	1.2	4.0	3.2	5.3	21.5	7.2
	California - CA	pine-fir-redwood	3.4	.03	1.1	2.0	9.8	16.3	5.4
SEPARATE STATES	Alaska - AK	spruce-hemlock-hardwood	3.3	1.0	0.0	0.1	43.9	48.3	16.2
	Hawaii - HI	ohia	.01	0.2	0.0	0.2	0.4	0.8	0.3
		TOTAL	40.2	9.5	34.8	105.5	108.3	298.3	100
		% TOTAL	13.5	3.1	11.7	35.4	36.3	100	

[1] Hectare x 2.47 = acres

[2] Commercial forests are those capable of growing at least 1.4 m^3/ha/yr (20 ft^3/ac/yr) of industrial wood materials.

[3] Other forests include county and municiple forests and those federal lands withdrawn from industrial and wood production for use as parks, preserves, and wilderness.

Source: USDA (1982).

Table 11-2. Percentage of Forest Lands by Level of Management Within Four U.S. Regions. (Estimates for 1977

U.S. Region	% Forest Plantations[a]	% Other Commercial[b]	% Reserved/ Deferred[c]
East			
North	9	80	11
South	21	69	10
West			
Rocky Mountains	2	38	60
Pacific Coast	16	44	40

[a]Intensively managed plantations.
[b]Moderately managed forests.
[c]Other commercial.

Source: USDA (1982).

total and might be the easiest to manage under climatic impacts simply because they are larger blocks of lands already under strong management commitments.

Value of U.S. Forests

Most populated regions in the United States are located close to or within a forested region. For instance, the Boston-Washington corridor is within the eastern hardwoods. Atlanta and Southeast populations are interspersed among the southern pine forests. Chicago and nearby Great Lakes communities are surrounded by the mixed conifer-hardwood forests of that region, and the Los Angeles to San Francisco populations parallel the Sierra Nevadas to the east. Forests, therefore, are part of the environmental fabric and general habitability for the majority of U.S. citizens.

All forests shed water to some degree, and two-thirds of the water runoff in the contiguous 48 States comes from forested ecosystems. Precipitation passes through forested ecosystems as canopy throughfall and flows in or along tree stems, the ground surface, and the soil; eventually, some precipitation flows into streams. Water yields from U.S. forests provide about 750 billion liters (200 billion gallons) of water per day for major uses such as irrigation (47%), steam electric (26%), manufacturing (15%), and domestic use (6%). These levels of demand are projected to continue to the year 2030 (USDA, 1981).

A favorite use of forests is outdoor recreation. Activities include hiking, camping, hunting, sightseeing, boating, swimming, fishing, skiing, sledding, and snowmobiling. A 1977 survey of U.S. households indicated that a majority of people participated in outdoor recreation four or more times per year (USDA, 1981).

About 190 million ha (470 million acres), or 64% of the total U.S. forested ecosystems, are highly productive commercial forest lands. These lands represent about 10% of the world's forest area, but they supplied nearly a quarter of the world's industrial forest products in the late 1970s (USDA, 1982). In 1980, 1.7 million people were employed in timber-based occupations across the United States. Such employment is basic to the economic well-being of many small towns and communities (Schallau, 1988). The total value of timber products harvested in 1972 was about $6.4 billion, and the total value after such processes as manufacturing, marketing, transport, and construction amounted to $48 billion, or 4% of the nation's gross national product. In 1979, timber product exports and imports were valued at $7 and 9 billion, respectively. Looking ahead, the consumption of wood products in the United States is projected to increase between current levels and the year 2030 (USDA, 1982).

RELATIONSHIP BETWEEN FORESTS AND CLIMATE

Scientific understanding of forest ecosystems has greatly advanced with each decade of this century. Yet the literature contains little information concerning the direct or indirect effects of climate change on the complex of biological and physical processes in forest ecosystems. Some insights are gained from paleobotanical studies regarding past rates and magnitudes of change during glacial-interglacial cycles, as well as changes in forest composition. Similarly, observations of forest responses to unusual drought or other weather extremes provide some knowledge. Estimates of rate, magnitude, and quality of change have also been derived using computer models developed by plant ecologists or forest management scientists for other objectives. Their validation for understanding how a forest can adapt to climatic change is only in the initial stages.

Climate is a primary determinant of existing forests. The ranges of annual average temperature and rainfall variation determine global forest distributions relative to different biotic regions (Figure 11-2). Substantial increases in temperature or decreases in rainfall could, for example, produce a shift from a forest to a grassland type. Thus, accelerated climatic change resulting from human activities and related effects on U.S. forests is of high concern to citizens and policymakers alike.

Magnitude

As pointed out by Spurr and Barnes (1980), there is ". . . irrefutable evidence that vegetation (i.e., ecosystems) has been in an almost constant state of instability and adjustment due to an almost constantly changing climate over the past 10,000 years and even over the past several hundred years . . . ." Lines of evidence come from studies of fossils, tree rings, carbon-14 dating, plus peat and pollen analyses (Barbour et al., 1987).

Historical climatic changes appear to have been associated with such phenomena as fluctuations in solar radiation, earth orbit variations, and volcanic activity. Evidence of repeated continental glacial advances and contractions in the Northern

Hemisphere dramatically illustrates the large-scale effects of global climatic change.

In response to the glaciation, ecosystems shifted south. Evidence from fossil pollen, for example, indicates a southward shift of spruce into Georgia and east Texas during the last glacial advance and treeless tundra in the Great Lakes States (Spurr and Barnes, 1980). During the maximum interglacial warmth of 6,000 to Figure 11-2. Approximate Distributions of the Major World Bioass Based Upon Mean Annual Temperature and Precipitation (Hammond, 1972) 9,000 years ago, which was 1.5°C (2.7°F) warmer than the present temperature level, plant zones were one to several hundred km (60-250 miles) north of present distributions.

Rates

All forested ecosystems experience change on both spatial and temporal scales; each biological and physical forest component may respond to climatic variation on different spatial and temporal scales. For example, microorganisms, insects, and birds come and go with relatively short-term climatic variation; shrub species abundances vary within the timespan of decades; trees, once established, could persist for centuries. This understanding is important from the perspective of climatic change, since it implies that forested ecosystems do not respond as a unit, but in terms of parts. Different parts respond differently; consequently, future forested ecosystems under a rapidly changing climate could be quite different from those existing today.

At the expected rapid rate of climatic change, the potential rates of forest migration would become a major concern. Migration rates vary by species. Paleorecords of the Holocene (10,000 years ago to present) show that extension of ranges for tree species of eastern North American (in response to glacial retreat) varied from 10 to 20 km (6 to 12 miles) per century for chestnut, beech, maple, and balsam fir (Zabinski and Davis, this report). Other species within the oak and pine groups extended at faster rates, i.e., 30 to 40 km (19 to 25 miles) per century.

Mechanisms

Knowledge of causal links between weather patterns and forest response is fundamental to projecting growth and composition effects resulting from climate change. Another requirement is to understand the climatic influences on population processes of forest plants and animals. These include such phenomena as fires, windstorms, landslides, pest outbreaks,

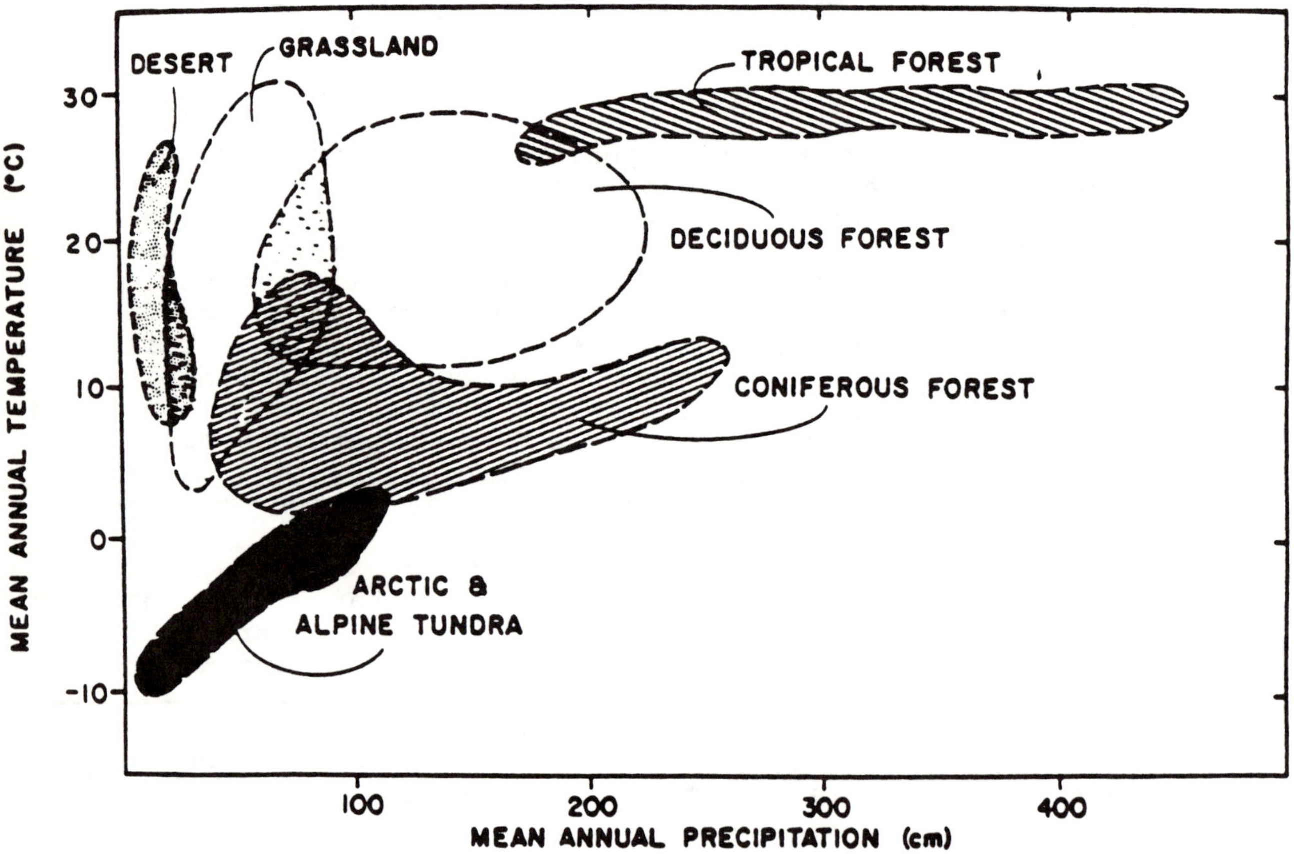

Figure 11-2. Approximate distributions of the major world bioass based upon mean annual temperatures and precipitation

Source: Hammond, 1972.

and other disturbances that affect survival and subsequent colonization by different species. Furthermore, the processes that control the dispersal of seeds through a mosaic of different ecosystem types (such as forest patches interspersed with agricultural lands, wetlands, grasslands, and other land-use groups) must be clearly defined.

Among the important factors now known to influence the growth and distribution of forests are the following.

Temperature

The optimum temperature for growth depends upon the tree species and other conditions. Warmer temperatures usually increase the growth of plants. However, very high temperatures decrease the growth of plants or kill those growing at optimum (or higher) temperatures. Cold temperatures can limit plant distributions by simply limiting growth at critical stages or by directly killing plants.

Precipitation

Too much or too little precipitation can limit forest production and survival. Too much rainfall in some areas can cause flooding or raise the water table, thus drowning roots that require oxygen or promoting fungal attack. Too little rainfall can reduce growth, cause susceptibility to fire or pestilence, and possibly kill plants. The seasonal timing of rainfall is more important than total annual rainfall, although forests also require some minimum total annual rainfall (see Figure 11-2).

CO_2 Concentration

High CO_2 concentrations could increase tree growth through increases in photosynthesis rates and water-use efficiency (primarily hardwood species) when water and other nutrients are not limited (Strain and Cure, 1985). Plant responses to CO_2 have been investigated largely in growth chambers and are difficult to extrapolate to the real world. Responses are varied and do indicate some measure of adaptive capability most likely imparted from ancestral exposure to much higher and lower levels in the geologic past. However, in natural situations, water nutrients or temperature usually are limiting factors in forest growth, thus making the impacts of CO_2 enrichment uncertain. If water use efficiency increases, then tolerance to drought might increase, ameliorating declines in southern parts of ranges. Unfortunately, the current state of knowledge does not allow generalizations on this subject.

Light

The amount of sunlight bathing an ecosystem sets the upper limit on net primary productivity. Thus, the tropics exhibit higher productivity than do the boreal regions. This potential productivity would, of course, be limited by other climatic effects such as drought, cold, heat, and natural disturbances, and the time required for forests to shift into new ranges. The length of day exerts considerable control on physiological processes such as release from and onset of dormancy. Significant northward shifts of forests would alter their day length regime, producing uncertain results.

Nutrient Status

In addition to climate, most forest growth is strongly influenced by availability of soil nutrients. Disturbances over vast regions, such as by drought followed by fire, can release large quantities of essential nutrients into the atmosphere and surface waters and can make soils deficient. Lengthy periods of soil development are usually required to replenish the soil nutrients before a large, mature forest stand can be supported. In turn, soils reflect properties of the forests that they support. This results from decades of nutrient uptake, litterfall, decomposition, and other processes.

Atmospheric Chemistry

Much of the nutrient budget of forests involves deposition of chemicals from the atmosphere as gases, aerosols, particles, and in solution. Although most of these act as nutrients, some produce acid deposition that can leach important soil nutrients (e.g., $SO_4^{=}$), produce a fertilizing effect (e.g., $NO_3^{-)}$, or damage leaf tissue (e.g., O_3). Climatic change will alter transport paths of air pollutants, and increased temperature could increase the rates at which gases convert to deleterious forms.

Disturbances

Almost continually, forests experience natural disturbances or stresses from biotic or abiotic agents alone or in combination. Examples are insects and disease outbreaks, plant competition, wildfire, drought, cold extremes, and windstorms.

These disturbances, which are among the primary factors controlling the successional processes in forests (Pickett and White, 1985), may range from an opening of small gaps in the canopy as the result of single tree death or of windthrow occurring when trees are blown down by heavy winds (predominant successional mechanisms in eastern hardwoods), to large clearings from fire, windthrow, or pestilence (predominant successional mechanisms in western forests).

Landscape Processes

The horizontal movement of material and disturbances through the landscape is critical to the processes controlling forest migration, species diversity in forests, and the spread of fire, windthrow, and pestilence effects. These processes are very poorly understood; quantification in the emerging field of landscape ecology is just beginning.

Multiple Stresses

In general, trees or forests stressed by one factor, e.g., accelerated climatic change, are more susceptible to natural stresses (secondary disturbances) such as insects, disease, or invading weed species. The concept of multiple stresses leading to forest declines is becoming more widely recognized (Manion, 1981). Regional climatic changes, even if temporary, frequently predispose forests to damage by other natural or anthropogenic stresses.

PREVIOUS STUDIES ON THE NATIONAL EFFECTS OF CLIMATIC CHANGE ON FORESTS

Concern regarding effects of climatic change on U.S. forests has prompted several excellent reviews. One of the most comprehensive (Shands and Hoffman, 1987) was the result of a conference sponsored by EPA, the National Forest Products Association, and the Society of American Foresters. While pointing out the high uncertainty associated with current predictions of climatic change, several authors suggested that if predictions are true, distributions of key forest species in the United States will change significantly.

Other recently produced compilations broadly consider forest effects, such as in agriculture, prairie land, and the Great Lakes (White, 1985; Titus, 1986; Meo, 1987; Tirpak, 1987). These reviews are largely pioneering efforts and some overlap occurs, but each presents some key points.

The methods used in the previous studies are quite similar to those used in this report. They include computer modeling of forest processes, literature surveys, studies of fossil evidence, and empirical relationships constructed by experts. The estimates of future change produced from these studies are generally based on the output of one or more of the General Circulation Models used for this report. Thus, the results of the previous studies are consistent with those reported here.

STUDIES IN THIS REPORT

Six studies on forest effects contributed to the regional case studies reported in this volume. The purpose was to use existing methods and data bases to protect U.S. forests from predicted climatic change. The selection of the six studies was based upon three criteria: use of available technology; testing hypotheses focusing on causal mechanisms; and a mix of studies that complemented each other, such that the strengths in one approach might overcome the weaknesses of another.

This report focuses primarily on forests within the lower 48 States. It is worth noting, however, that the largest magnitude of warming is expected in the northern latitudes encompassing the boreal forest and other forest types in Alaska and Canada. These large forests could be under significant risk from climatic warming.

Study Design

Characteristics of the six studies are briefly listed in Table 11-3. With the exceptions of the Overpeck and Bartlein study and the Woodman study, the methods are discussed in the regional case study chapters and will be mentioned only briefly here.

Two studies used correlations between tree distributions and climate (Overpeck and Bartlein; Zabinski and Davis). Overpeck and Bartlein's approach consists of correlating the modern pollen distributions of important tree species with January and July mean temperature and annual rainfall. The correlation is then tested by reconstructing past pollen distributions from General Circulation Model simulations of past climates (during the most recent glacial-interglacial cycle) for each species and comparing them to observed pollen distributions from those periods. Future pollen distributions are then constructed from the expected doubled CO_2 climate projected from the different model climate scenarios. The correlations are constructed on modern pollen distributions, rather than tree distributions, to allow the direct comparison to fossil pollen data. Modern pollen distributions are similar to, but not exactly the same as, modern

Table 11-4. Principal Investigators, Regional Focus and Method of Approach for the Regional Forested Ecosystem Studies

Principal Investigator	Region	Method
Overpeck and Bartlein	Eastern North America	Correlation and fossil studies
Urban and Shugart	Southeast Uplands	Forest dynamics model
Botkin et al.	Great Lakes	Forest dynamics model
Zabinski and M. Davis	Great Lakes	Correlation
Davis	California	Fossil studies
Woodman	Southeast, California, and National	Literature review

tree distributions. The verification studies indicate that the approach works reasonably well at a coarse spatial resolution. That is, northern trees are in the north and southern trees are in the south, with the regional patterns being reasonably well represented.

The approach of Zabinski and Davis is essentially the same as that of Overpeck and Bartlein, except that the correlations are constructed from the actual modern tree distributions rather than from the modern pollen distributions (see the Great Lakes chapter of this report).

Two of the studies used computer models of forest dynamics (Botkin et al.; Urban and Shugart). Both studies explored forest response on a range of soil types from well drained to poorly drained. Forest growth simulations from bare ground represent conditions after a fire, logging, or similar disturbance. Mature stand simulations are useful for investigating the potential response of present forests to gradual climatic change in the immediate future.

For the California case study, Davis reconstructed vegetation patterns in the Sierra Nevadas from fossil pollen studies for the interglacial warm periods between about 6,000 and 9,000 years ago. These reconstructions represent possible analogs of a future warm period at the lower magnitude of the predicted future warming.

A literature review was conducted by Woodman for the Southeast and California forested regions and peripherally for the nation as a whole. The purpose was to ascertain the attributes of the forest resource in terms of extent, ownership economic and recreational value, and policy considerations.

Study Results

The six studies conducted by EPA are consistent enough to indicate that climatic changes would significantly affect the natural forests of the United States. The distribution of healthy forests in the eastern United States appears to become greatly reduced from their areas during the next century (see Figure 11-3 and 11-4). This results from a very slow northward migration coupled with a fairly rapid decline in the southern and western parts of species ranges. Drier forest conditions in the United States, induced as much by increased temperature as by changes in rainfall, would mean less tree growth and therefore reduced forest productivity in general.

Figure 11-3. Present distribution of major forest genera and herbs vegetation of eastern N. America compared to potential future distributions after reaching euilibrium with the climate predicted for $2xCO_2$. The comparison is based upon simulated modern pollen data (a) and simulated future pollen abundances for each of the three $2xCO_2$ scenarios: (b) GISS; (c) GFDL; and (d) OSU (Overpeck and Bartlein, Volume _).

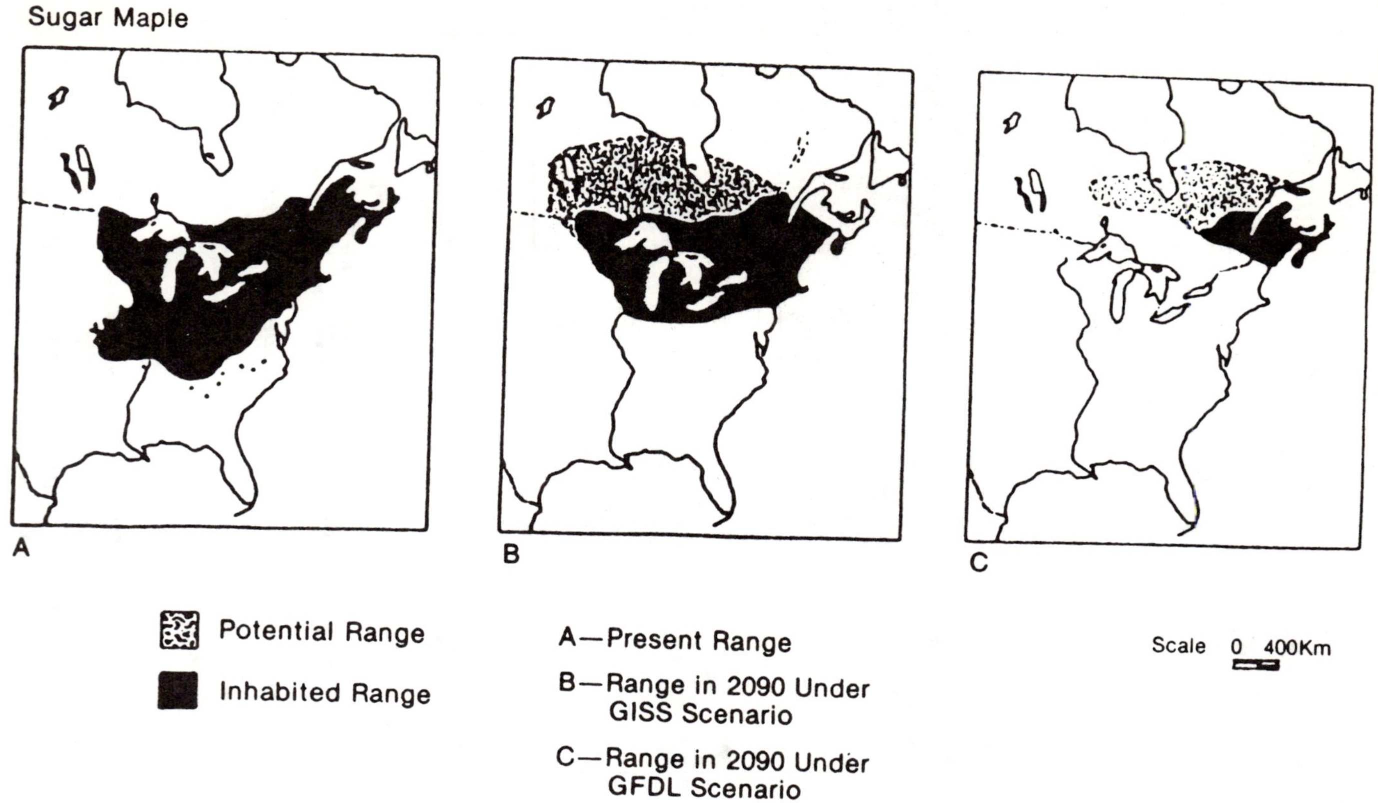

Figure 11-4. Present and future geographical range for sugar maple. a: Present range. b: $2XCO_2$ climate-space in 2090 A.D. under the GISS scenario. Black area is the predicted species range; stippled area is potential range. c: $2XCO_2$ climate-space in 2090 A.D. under the GFDL scenario. Black area is the predicted species range, stippled area is potential range (Zabinski and Davis, Volume _).

The forest simulation models provide an indication of the importance of uncertainties imparted by the climate scenarios. The climate scenarios differ primarily in their representation of regional rainfall patterns. The model results indicate that temperature has a large effect on forest health, either directly through cold and heat stress or indirectly by exaggerating drought effects. Thus, the overall characteristics of forest responses are remarkably similar among the three climate scenarios.

Magnitude

Eastern Forests - Northern Limits

All of the study results suggest a northward expansion of most eastern tree species (Figure __ displays results from Overpeck and Bartlein). That is, spruce, northern pine, and northern hardwood species would move north by about 600-700 km (375-440 miles) into the Hudson Bay region of the Canadian boreal forest (Overpeck and Bartlein; Zabinski and Davis). New England coniferous forests would be replaced by more hardwood forests and especially by the oak species from the eastern mid-United States (Botkin et al.; Overpeck and Bartlein; Zabinski and Davis). As the northern mixed forests shift from spruce-fir to sugar maple, some sites could actually triple their present productivity (Botkin et al.).

Additionally, southern pine species could shift about 500 km (310 miles) into the present hardwood forest lands of eastern Pennsylvania and New Jersey (Overpeck and Bartlein; Urban and Shugart; Solomon, 1986; Miller et al., 1987). Depending upon the severity of climate change, Urban and Shugart estimate that near the northern limits of slash pine in East Tennessee, above-ground woody biomass in 100 years could range from little change to extremely low biomass with no trees supported (i.e., a grassland, savanna, or scrub). However, even with little decrease in productivity, species shifts would alter the forest composition from shortleaf to loblolly pine, a more commercially valuable tree species.

Eastern Forests - Southern Limits

Ultimately, forest decline and mortality could truncate southern distributions of tree species by as much as 1,000 km (625 miles) in many northern hardwood species (Zabinski and Davis; Overpeck and Bartlein) or by as little as a few hundred km (60 miles) in southern pines and hardwoods (Urban and Shugart; Solomon and West, 1986). Under the driest scenario, Zabinski and

Davis predict local extinction in the Great Lakes region of many eastern tree species such as eastern hemlock and sugar maple (Figures 11-3 and 4). These estimates bear considerable uncertainty for all species.

These uncertainties are particularly true for the southern limits of southeastern species that border the continental margin. The actual southern climatic limitations of these species are not well known (Urban and Shugart). Nevertheless, under the most severe climate scenario in the Southeast with increased temperatures and decreased growing-season precipitation, Urban and Shugart's results suggest that the 18 tree species they considered would no longer grow in the southern half of the region. Present forest lands in the region would be replaced by scrub, savanna, or sparse forest conditions. Under the most mild scenario, much of the southern half of the Southeast would not support trees, and areas in South Carolina would be marginal, supporting about half the current biomass of forest.

Biomass accumulations in 100 years for mature natural forests in productive sites in the Lake States could be reduced to 23-54% of their present values (Botkin et al.; Solomon and West, 1986). On poor sites, forests could be converted to grassland or savanna with very low productivity, ranging from 0.4 to 28% of their present values.

Western Forests

Similar projections were made for six western coniferous species: ponderosa and lodgepole pine, Douglas-fir, western hemlock, western larch, and Englemann spruce (Leverenz and Lev, 1987). Predictions are mixed for the West. Because of the mountainous conditions in the West, upslope shifts are possible for Douglas-fir, ponderosa, and western hemlock in the northern Rocky Mountains. In the coastal mountains of California and Oregon, Douglas-fir should shrink in the lowlands and be replaced by western pine species (Davis, this volume; Leverenz and Lev, 1987). Overall, the western forest lands are predicted to favor more drought-tolerant tree species, such as the hard pine group, at the expense of fir, hemlock, larch, and spruce species.

If regional drought persisted, the frequency of fires could increase, significantly reducing total forested area. Also, with massive upslope movement, some species would be pushed off the tops of mountains into local extinction.

No quantitative estimates have been derived for productivity for the western forests under potential warming conditions. However, using the analog approach of Davis, under the most severe conditions projected for California, the species composition of the west-side Sierra Nevada forests would become more similar to that of the east-side forests. This would reduce the standing biomass to about 60% of current levels.

Rates of Decline and Migration

In the Great Lakes region, significant forest decline and forest compositional change could become evident within 30 to 60 years. In the Southeast region, forest declines could become evident in 60 to 70 years, with declines in the drier western portions occurring even earlier, perhaps in about 30 years (Figure 11-5) (Urban and Shugart).

These rapid declines, coupled with the expected magnitude of change, raise the question of how fast forests can migrate. Based upon fossil records, Zabinski and Davis have estimated that the maximum dispersal rate of several tree species in response to the last glacial retreat was roughly 50 km (30 miles) per century. Under the expected rapid warming, they estimated that a dispersal rate of about 100 km (600 miles) per century would be required to maintain species distributions near their current extent. Such migration routes are doubtful, suggesting greater reductions in species ranges under rapid climatic changes.
Mechanisms of Migration

Distribution changes suggested by these studies (i.e., migrations) must be considered carefully. Reproductive processes are essential for the migration of tree species across the landscape. For many tree species, climatic change could reduce natural regeneration in an existing location and introduce the species at different latitudes or altitudes. Reproductive processes in trees, such as flowering, pollination, seed set, seed germination, and seedling competitive success, are particularly sensitive to climate.

Specific regional climate scenarios vary as a function of the GCM. All scenarios predict increases in temperature; however, some predict increases in rainfall, and others predict decreases. The northward shifts of species appear to result from a release from cold temperature stress, which normally freezes flowers, seedlings, and even adult trees. However, the western and southern limits of eastern tree species appear to result from insufficient moisture stress and excessive heat stress, which

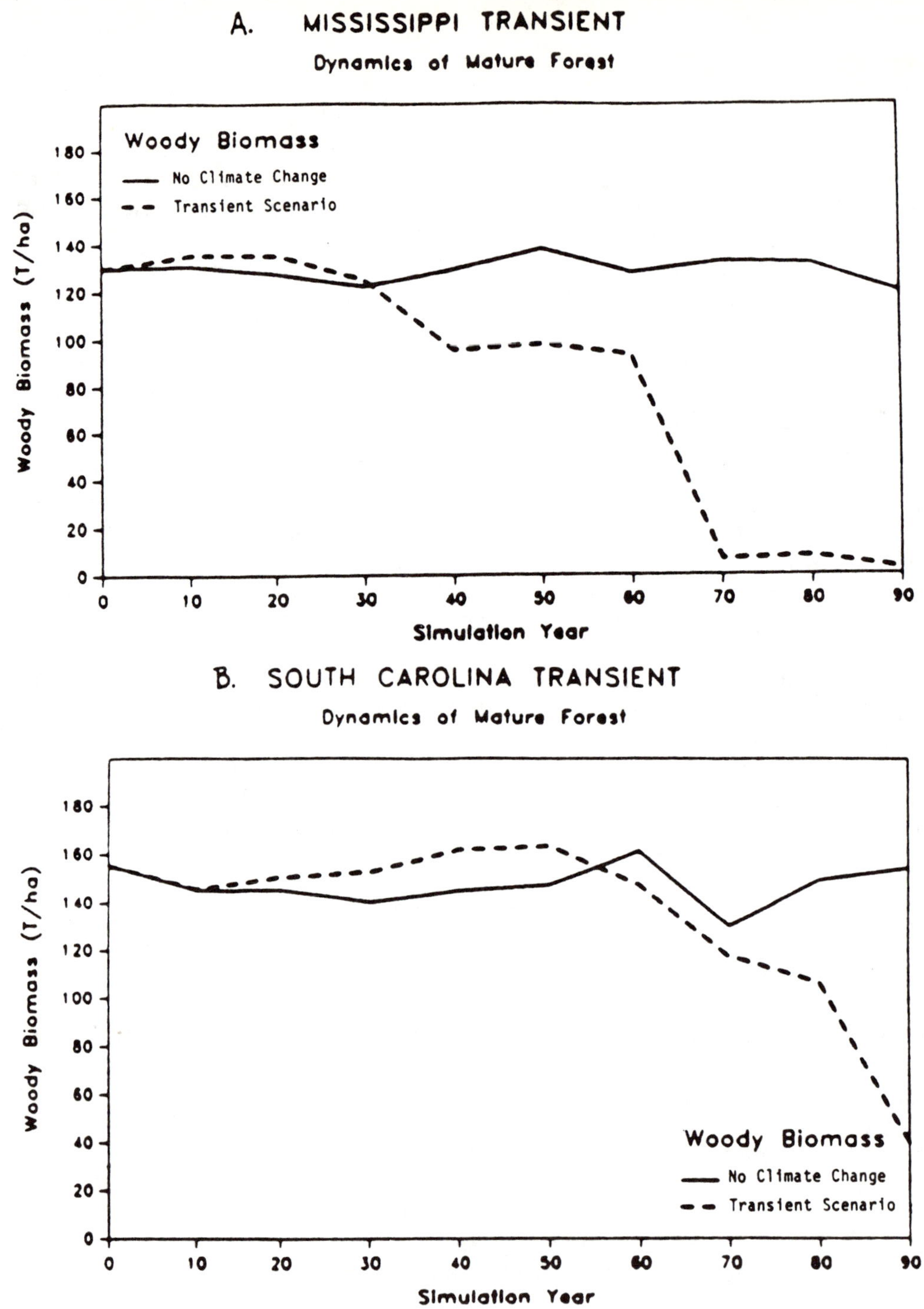

Figure 11-5. Projected changes in biomass (T/ha=2.24 t/acre) of mature forests in Mississippi (A) and South Carolina (B) under the GISS transient climatic change scenario (Chapter __; Urban and Shugart, Volume _), shifts toward drier conditions and the prairie advances to the East. Presumably changes in western forests would result from similar mechanisms. The drought-induced declines in forests might be partly offset if CO_2-induced increases in water-use efficiency were realized.

primarily affect sensitive life history stages but can also affect adult trees.

All of the study approaches used under all of the climatic scenarios predict major forest declines in the southern parts of species ranges and expansions to the north. These declines, resulting primarily from drought stress, would occur despite the differing rainfall predictions. Increases of temperature would severely reduce water intake, and therefore, would produce drought. Although precipitation could increase slightly in some U.S. locations, generally the increased evaporative demand caused by higher temperature would be a greater factor. Subsequent differences in rainfall between the various climate scenarios would modify the magnitude of regional drought patterns. The western limits of eastern forests could similarly retract as the climate. Existing forests probably would not shift intact, but would change in composition. Variations in migration rates and sensitivities to weather variables produce individual responses to climate change. These changes are consistent with the well-known dynamic nature of ecosystems and were projected for the forests of all regions. In the Great Lakes region, for example, beech would decrease in abundance, and birch and maple would increase (Zabinski and Davis; Botkin et al.). On some lands, forest productivity could remain about the same as today, but changes to less economically important species could be significant.

Not considered quantitatively in any of the studies are changes in forest disturbance regimes. These changes should not be considered lightly. Extreme and more frequent climatic variations (see Chapter 17) could cause much higher mortality in U.S. forests than the current experience. Although little is known as yet, some locations may experience an increase in the frequency of extreme weather events, for example, wind, ice, or snow storms, droughts, and flooding. Besides the direct damage these events can cause, they can cause drainage from insects, disease, and wildfires.

Study Limitations

Although predicted effects vary, these six analytical studies have results that are collectively consistent enough to advance our knowledge and justify concern regarding the future of U.S. forests under rapid climate change. The range of predicted effects is large; however, uncertainties exist regarding (1) the climate scenarios; (2) the kind and rates of responses of individual tree species; and (3) changes in forested ecosystems as a whole from environmental change. All of these factors significantly influence the precision and accuracy of the results.

A major uncertainty in the simulation model approach involves the rates of species dispersal into a region. The current generation of models has no dispersal mechanisms. A species is simply present or it is not present. For example, Botkin et al. excluded more southern tree species so that their dispersal was unrealistically nonexistent, and these southern species could never enter the region. But if they had been included in the simulations, these species would have entered the northern forests at the same rate as the climate change. This would have assumed dispersal rates far in excess of reality. This limitation can, in part, be overcome by studies, such as those of Zabinski and Davis, that provide some insight into actual dispersal rates and species migration.

The models also carry assumptions about the environmental controls of species limits. In most cases these assumptions are reasonable, given that the assumed control (for example, July temperature or annual rainfall) is usually related to the actual control, perhaps accumulated warmth or summer drought. However, large uncertainties exist in some instances. This is particularly true with regard to the climatic controls of the southern limits of southeastern forests, simply because of their association with the continental margin. Does the climate at that latitude represent the actual climatic limitation to the distributions, or are the species simply stopped by a geographic barrier? No one really knows. These uncertainties were partially addressed by Overpeck and Bartlein, who compared their fossil pollen approach to the modeling approach. The two approaches use similar relations to climate. Both approaches can be used to simulate forest distribution in the geologic past and both do reasonably well.

Several uncertainties with the pollen-climate correlation approach limit its precision and accuracy. First, many of the plant taxa used in the study are plant genera (e.g., pine, oak) rather than species, and thus the simulated results are not taxonomically precise. Second, the results are applicable only on a regional scale; local scale predictions are not made. Third, and very significant, the simulated results assume that all the plants are in equilibrium with the new climate. Rates of dispersal vary between species, and several hundred years may pass before plant communities are again in equilibrium with climate. How this lag would affect vegetation dynamics is not addressed in this study and is an important research question.

The paleoecological analysis of the past vegetation in the Sierra Nevadas (Davis) presents several uncertainties. First, differences with respect to weather variations (i.e., season to season and year to year) could produce strikingly different types of vegetation. Also, there is a large uncertainty about what the most appropriate analog period might be -- or if one even exists! Furthermore, the rate of climatic change in the future is predicted to be much faster than the rate of climatic change during the past 20,000 years. Lags in the response of species to the future climate could strongly affect the type of forest at any one location, whereas in the past, with a more slowly varying climate, lags in species response were not as important in determining forest composition. To the extent that climatic regimes predicted for the future are similar to the climates of certain periods in the past, a picture of future forests can be estimated from these past periods.

All of the studies are deficient in some very important processes controlling forest responses to climate, particularly disturbance regimes such as fires, windstorms, hurricanes, landslides, and pest outbreaks. In addition, the responses of mature trees to elevated CO_2 under conditions of moisture, temperature, or other nutrient limitations remain largely unexplored. Most research on elevated CO_2 on plants has been performed in controlled chambers using seedlings (Strain and Cure, 1985). However, the seedlings were not previously grown in or acclimatized to high CO_2 environments. Evidence has shown that plants grown under high CO_2 will respond much differently to changes in temperature, light, and moisture conditions (Strain and Cure, 1985). However, our ability to extrapolate CO_2 fertilization results from laboratory experiments to the natural world is limited, and our understanding of regional changes in water use efficiency is even less. Furthermore, complex interactions between fertilization effects and changes in water use efficiency can produce unexpected problems such as reduced heat tolerances due to effects on evaporation cooling. These interactions are least understood, but could produce major regional changes in forest responses. Therefore, it is not yet possible to quantitatively incorporate the direct effects of CO_2 on forests into studies such as these. Further, if water or nutrients are limiting to forest growth, then the effects of carbon fertilization may not be significant.

NATIONAL, ECOLOGICAL, AND SOCIOECONOMIC IMPLICATIONS

The effects of the predicted climate changes resulting from a doubling of CO_2 by 2030 may be considered from two perspectives: ecological and socioeconomic. Evidence for significant national implications is strong from both viewpoints.

Ecological Implications

Ecological implications for forests commonly start with tree response. But strong implications also exist in regard to other ecosystem components, e.g., animals, soils, water, secondary impacts, and as noted, the atmosphere through which climate change is mediated. Forest effects are described in terms of tree distribution changes and biomass production changes, but among the other major components, there is a host of interacting processes. Thus, significant changes in tree response would be accompanied by ecological reverberations throughout all the forested areas of these regions.

Tree distributions and biomass productivity

Forest migrations to the north in response to rapid warming in North America during the next century will be likely. However, significant lag is possible. Even under the maximum rates of species dispersal estimated by Zabinski and Davis, healthy forest areas may not redevelop for several centuries. Meanwhile, forest decline would not be under such constraints, so the extent of healthy forested regions in the United States probably would be greatly reduced. Though some locations may have increased productive potential from a biomass per hectare standpoint, the large reductions in areas with healthy forests would likely create a net reduction in forest productivity for the United States for several centuries.

Even if a massive reforestation effort were undertaken, the new forests resulting from species shifts may or may not be as productive as existing forests. More northern latitudes or higher elevations raise other considerations. Farther north, days are longer in the summer and shorter in the winter. At higher elevations, damaging ultraviolet light intensity is greater. All of these conditions could lower forest productivity below optimum and could hinder reproduction and survival.

Animals

A change in the size and relative homogeneity of forests could influence whether some animals can continue to live in habitats to which they have adapted.

Depending upon the species, animal migration can be hindered by boundaries between forests and other land types or facilitated as animals move along edges. In turn, some animals prefer young forests, and others prefer old forests. Some animals can exert a profound influence on forest structure and composition through selective browsing of seedlings, insect attack of different tree species, and other effects. All of these factors combine to influence the regional patterns of biotic diversity of both plants and animals (see Chapter___).

Soils

Soils under warmer climates also would change, though at a much slower rate than shifts in species distribution. Optimum soil conditions for high forest productivity supporting species at more northern latitudes or higher elevations might take several centuries to develop. Furthermore, it is not at all clear that some northern soils could even support more southerly species. The soils of the boreal forest are vastly different from those under the deciduous forests to the south.

Water

Where forests give way to drier conditions (e.g., in the Great Lakes and California), many lands now serving as watersheds might be used for a different purpose. Furthermore, regional-scale disturbances (such as fire) and applications of chemicals (such as fire retardants, fertilizers, and pesticides) could degrade regional water quality and increase airborne toxic chemicals (see chapter on Water Resources in this report).

Sea level rise is an additional water impact related to some forests. Projections for sea level changes are up to 1 meter (3 feet) in the next century (see Sea Level Rise chapter in this report). Many forest lands of high value for timber production (e.g., in the Southeast) or recreation (in the Northeast, Northwest, and California) are close to ocean coasts. Inundations, decreases in depth to the water tables, and saltwater intrusions could trigger rapid forest declines near these areas.

Secondary Impacts

As the southern bounds of forests tend to shift north, forest decline (sick and dying trees) could become extreme over vast areas. These areas would be highly susceptible to weed competition, pest outbreaks, or wildfire. As forests decline, species of lower economic value such as chokecherry, as well as weedy shrubs and herbs, could invade via wind dispersion. Under stressful environments, such species are severe competitors to most commercial tree species.

Trees experiencing less favorable growth conditions are more stressed and will be vulnerable to insect and disease attack. These secondary pest impacts could last "until the most vulnerable forest stands or tree species are eliminated" (Hedden, 1987). In addition, the incidence of catastrophic wildfires is predicted to increase in U.S. forests with higher temperatures. Simand and Main (1987) estimated an 8% rise in the occurrence of fires and a 20% increase in fire-suppression costs.

Socioeconomic Implications

The United States enjoys substantial economic and cultural benefits from its forests. Until recently, the nation's forest managers assumed that these benefits could be sustained by maintaining forests in a healthy condition (Fosberg, 1988). This was achieved, for example, by preventing fires or pest invasions, avoiding careless use, and enhancing productivity through good silviculture.

Beginning with the possibility of damage from regional air pollution damage to forests, suspected in the 1980s, alterations of the environment external to forests presented a new concern. Research and policy discussions to deal with this issue are ongoing.

If climate changes as rapidly as predicted, this additional external influence with its more global dimensions looms as a possible hazard to forests and their use. As can be imagined, a list of potential socioeconomic concerns would be large. To provide a brief perspective, three issues are considered.

<u>Quality of the Human Environment</u>

The forest amenities enjoyed by most U.S. citizens will be affected according to different forest responses. In the Boston-Washington corridor, a composition change from predominantly hardwood to more pine forests, though ecologically significant, would probably only superficially affect most people. In the Atlanta-Southeast region, the southern pine forests, while undergoing a gradual expansion of their northern boundaries, would have less vigor in the remaining stands. This could raise their vulnerability to damage by insects and disease, reducing esthetic values -- at least an intermediate impact for most of the local citizens. In contrast, drier climates in the Great Lakes and California regions may cause the loss of some forest lands to prairie or desert conditions, respectively -- a severe change for the people there, not only in their living environment but also in the whole spectrum of forest land use.

Recreation

Forests must be in a relatively healthy condition to support quality recreational use (Clawson, 1975). Forests undergoing gradual composition changes might remain healthy, but rapid changes would most likely cause stressed or declining forests. Such forest conditions would have less recreational appeal because of such factors as less pleasing appearance, greater threat of wildfire, and reduced hunting quality when game populations change or are diminished. Furthermore, drier conditions in U.S. forests would harm recreational opportunities that depend on abundant water or snow.

Wood Products

Altered U.S. forest productivity resulting from climatic change would have obvious major economic impacts. Significant yield reductions could lead to unemployment, community instability, industrial dislocation, and increased net imports of wood products.

Reforestation projects could make up for some losses in forest productivity and artificially advance migrations forced by climatic change. Reforestation technology has greatly improved in recent decades so that success rates also have increased greatly. Examples are high-vigor seedlings through improved nursery practices, genetic selection, and tissue culture. Improvements in the field include machine planting, fertilization, and weed control on selected sites. Results are evident from the large acreages of plantations established in the United States in recent decades, particularly with loblolly pine in the southeast and Douglas-fir in the Pacific Northwest (see Table 11-5).

Innovative manufacturing trends should prove to be timely during times of rapid forest change. High-strength and durable products from reconstituted wood (e.g., new particle board concepts, warp-proof hardwood lumber, paper products of fiber from multispecies) are now in use or well along in development. These new methods will lessen the present overdependency on a few commercial conifer species from stands above minimum size and quantity (Ince, 1987). The result will be an ability to use the timber resources of the future, however they change in composition.

Table 11-5. Levels of management within U.S. forests in 1977 as illustrated by areas in three broad uses within four major regions. Intensively managed are represented by plantations[2]; moderately managed are those forests naturally established but with commercial rates of growth[3]; and forests with protection as the primary level of management (i.e. deferred or reserved[4]; estimated from tables 3.1 and 9.2, USDA, 1982).

U.S. Regions	Management level (millions of hectares[1])			
	Intensive	Moderate	Protected	Total
East				
North	3	65	4	72
South	8	69	7	84
West				
Rocky Mountain	1	23	32	56
West Coast (WA, OR, CA)	3	22	13	38

1. Ha x 2.47 = acres

2. Plantations are forests established by seeding or planting, generally on the most productive lands, and followed by silvicultural tending such as thinning, fertilization and weed control until harvest at an economically optimum age.

3 and 4. Definitions in Table 11-1.

Chapter 11

NATIONAL FOREST POLICY AND CLIMATIC CHANGE

Historically, U.S. forest policies have undergone continued development to meet national change (Young, 1982). The earliest policies were adapted by the New England colonies in the 1600s to regulate overculturing near settlements. Wood was needed for fuel and buildings, but existing methods were not capable of long distance log transportation. Development of U.S. forest policies continued and has been particularly intense this century, as the National forests, National parks, and wilderness areas were established.

At present, forest managers are dealing with many additional policy issues. Five of these (Clawson, 1975) are important to climatic change/forest response:

- How much U.S. land should be devoted to forests?
- How much forest land should be withdrawn from timber production and harvest?
- How should the Federal forest lands be managed? (That is, the lands under the USDA Forest Service, USDI Park Service, Bureaus of Land Management and Indian Affairs, etc.)
- What constraints (e.g., mandatory forest practices) should be placed on forest managers to ensure national environmental goals?
- Who should pay the additional costs incurred in implementing new policies?

The large array of forest ownerships in the United States makes development and implementation of forest policy more complicated than in most countries, where nearly all forests belong to the government. The diversity of owners and managers results in widely divergent goals and objectives.

How Much Land Should Be Forested?

Changes in forest composition or regional boundaries induced by rapid climatic change would magnify the complexity of National forest policy even further. Lands in forests now would require review relative to such competing needs as agriculture and residential use (which would also be adjusting to climatic change).

How Much Should Be Withdrawn From Timber Production?

Where the productivity of wood is significantly reduced, increased, or shifted, a policy question that would surely arise concerns whether forest lands should be reallocated to maintain timber production. If so, how should competing forest uses, such as watersheds, parks, and wilderness, be treated? How much of each can the United States afford under changed climatic conditions? Should the Federal Government purchase more forest lands to support all public needs?

In the short term, forest managers could compensate for some loss of productivity by improved technology, though at increased costs. An example would be establishing more drought-tolerant plantations through genetic selections, improved nursery stock, and more intensive silvicultural practices (e.g., weed control and thinning). Introducing new species adapted to warmer climates might be possible in some locations, but this would call for developing new silvicultural regimes and use methods -- possible, but expensive in time and money. In the long term, if growing conditions become extremely difficult on some U.S. forest lands because of climatic changes, investment in tree crops at such locations could become unattractive altogether.

How Should We Manage Federal Forests?

The National forests under the USDA Forest Service are managed according to a series of complex legal directives and administrative procedures, beginning with the Organic Act of 1897 (Woodman). Ultimately, the objective became to manage the National forests for multiple uses, with timber on a sustained-yield basis and certain lands set aside as wilderness areas. The National Forest Management Act of the mid-1970s requires 15-year management plans for each National forest subject to public review.

Lands managed by the Department of Interior are under similar mandates. For example, a congressional act passed in 1976 charged the Bureau of Land Management to manage its 2.3 million ha of forest and range land according to multiple-use and sustained-yield principles. Similarly, the National Park Service is mandated to manage National parks, monuments, historic sites, and so forth, for the recreational enjoyment of people. Such activities as timber harvesting, hunting, mining, and grazing are not permitted. In addition to the Federal Government, most States, many counties, and some municipalities own forest lands.

The Forest and Rangeland Renewable Resources Planning Act of 1974 requires the Secretary of Agriculture to make periodic reviews of the nation's forest and rangeland resources. The act was amended in 1976 to also require management plans for the National forests. In the future, these assessments and planning efforts should include consideration of the possible effects of predicted climatic changes. One key issue is the level of priority given to maintaining forest health under changed climatic conditions. For instance, under more adverse environments, should National forests be left to decline as a natural process, thereby losing esthetic values in parks, water yields from watersheds, and highly productive timber crops? Or should silvicultural forest techniques, such as thinning, weed control, and fertilization, be employed to "save" them? This question and others will challenge the fundamental concepts of the benefits of multiple use and sustained yield of forests.

How Can We Ensure National Goals?

At the minimum, Federal agencies must plan and act in concert with the State forest organizations. In the first half of this century, the Federal Government attempted to regulate forest harvests on Federal and private lands. Development of this policy did not survive strong public concern and intense political debate against such policy (Worrell, 1970); the same sentiment would likely exist today and in the near term. However, under the influence of rapid climatic change, the nation may once again have to face the touchy issue of what restraints or forest practices must be regulated for all public and private lands.

Solomon and West (1985) point out that while climatic change might disrupt forest ecosystems in the future, it is uncertain whether forest managers could or would be able to apply silvicultural practices on a scale large enough to maintain the net productivity of commercial forest lands in the United States. Some States (e.g., Washington, Oregon, and California) have laws specifying fire protection requirements, control burn practices, and reforestation minimums following timber harvests. Zoning, permits, licenses, and various taxation measures also have been attempted with mixed results. It is much easier to prevent owners from destroying forests than to compel them to implement silvicultural practices.

To keep pace with the rapid climatic changes projected, the U.S. reforestation effort conceivably would need to be doubled or tripled in size. In recent years, about 800,000 ha (2 million acres)/year (approximately 700+ million seedlings) have been reforested in the United States (USDA, 1982). Costs range from \$200 to \$700 per ha (\$80 to \$280 per acre) depending upon species, site preparation, plantation density, and planting method. Using \$500 per ha (\$200 per acre) as a mode, the total annual expenditure is near \$400 million. About 0.4% of the commercial land base is reforested annually. At this rate, it would take 100 years to reforest 40% of the U.S. forest lands, assuming no repeat hectares to cover failures or harvests of the first plantations.

An expansion on the scale suggested above would require large investments in seed procurement, tissue culture capability, nursery capacity, and research to improve knowledge about the establishment and silviculture of drought-resistant plantations. Even if the dollar commitments were made, reforestation at this scale might only be possible if all forest lands were managed by one organization. The complex forest ownership pattern in the United States, therefore, would be an issue to overcome in a national reforestation program.

Who Should Pay?

Adjusting forest policies to address the issues arising from climatic change will most likely raise the costs of using the nation's forests -- whether for water, recreation, aesthetics, or timber. Additional research to answer many new questions will also require more funds. A major question will concern who should pay for these costs. Land owners? Forest users? Consumers? All taxpayers? The answers will come when better information is available on resulting forest effects, followed by public debate establishing new priorities for forest use in a changed climate.

RESEARCH NEEDS

The forest effects resulting from rapid climatic change are at present hypothetical. The change has not yet occurred, and many uncertainties are associated with the predictions. Effective policies to deal with new forest effects will require more information and fewer uncertainties that must come through forest ecosystem research. Four broad questions concerning U.S. forests frame the research needs for the 1990s: What will the effects be? How can they be measured reliably? How should they be managed? How can we ensure that research will be conducted in a timely fashion?

Effects

What will be the effect on the nation's forest ecosystems if climatic changes occur as predicted by the middle of the 21st century? While subsets of this question must include extent, magnitude, and risk considerations, additional knowledge is needed concerning the following factors:

1. Forest migration processes and rates, including the landscape processes that control the horizontal movements of forest, animals, and disturbances;

2. Interactions among the different landscape components and land-use practices that affect biodiversity, and water quantity and quality;

3. The impact of climate change alone and in combination with other natural or anthropogenic influences, such as insects, pathogens, CO_2 enhancement, air pollutants, UV-B radiation, and acid deposition on U.S. forests; and

4. The processes and mechanisms that play key roles in forest ecosystem effects -- both biologically as in photosynthesis and respiration, and physically as in flows of energy, carbon, water, and nutrients through ecosystems.

Methods

How can forest ecosystems be measured to reliably detect the effects of rapid climate change? Today, the response of ecosystems to environmental change is largely based upon extrapolating from field observations, from knowledge about seedlings or individual trees of a small number of commercially valuable species, and from computer models. The following must be accomplished:

1. A determination of the most useful integrating variables for forest ecosystems that indicate the effects of climatic change;

2. Effective sampling designs developed for experiments and long-term monitoring at the forest ecosystem scale; and

3. Improved models capable of projecting regional effects on forests across multiple spatial and temporal scales.

Management

What options are available to the public and private forest managers and owners in the United States to deal with the changes in the nation's forests that might occur in the next century? Research is needed to accomplish the following:

1. Understand the socioeconomic impacts of all forest ecosystem effects to clarify economic risks and alternatives; and

2. Develop technology to mitigate the adverse effects or to exploit the benefits of forest change, such as breeding, bioengineering, transplanting, fertilization, irrigation, and other management approaches.

Timing

The timing of the research is critical. The effects of climatic change may be some decades away, but this should not lessen the urgency to begin research toward better information and methods. The complexities of the science are very large. To build the required base of knowledge to deal with potential forest changes before they are upon the nation will require significant time and resources.

REFERENCES

Barbour, M.G., J.H. Burk, and W.D. Pitts. 1987. Terrestrial Plant Ecology, 2nd ed. Menlo Park, CA: Benjamin/Cummings Publishers.

Botkin, D.B. 1979. A grandfather clock down the staircase: stability and disturbance in natural ecosystems. In: Waring, R.H., ed. Proceedings of the 40th Annual Biological Colloquium. Forests: Fresh Perspectives From Ecosystem Analysis. Corvallis, OR: Oregon State University Press, pp. 1-10.

Clawson, M. 1975. Forests for Whom and for What? Resources for the Future. Baltimore, MD: Johns Hopkins University Press.

Cubbage, F.W., D.G. Hodges, and J.L. Regens. 1987. Economic implications of climate change impacts on forestry in the South. In: Meo, M., ed. Proceedings of the Symposium on Climate Change in the Southern U.S.: Future Impacts and Present Policy Issues. New Orleans, LA: U.S. Environmental Protection Agency.

Fosberg, M.A. 1988. Forest productivity and health in a changing atmospheric environment. In: Berger, A., ed. NATO Symposium on Climate and Geosciences. Dordrecht, The Netherlands: Reidel Publishing Company. In press.

Hammond, A.L. 1972. Ecosystem analysis: biome approach to environmental science. Science 175:46-48.

Hedden, R. 1987. Impact of climate change on forest insect pests in the southern U.S. In: Meo, M., ed. Proceedings of the Symposium on Climate Change in the Southern U.S.: Future Impacts and Present Policy Issues. New Orleans, LA: U.S. Environmental Protection Agency.

Ince, P.J. 1987. Technology, timber demand and timberland investment. In: A Clear Look at Timberland Investment, Milwaukee, WI, April 27-29, 1987. Conference proceedings. Forest Products Research Society.

Lavdas, L.G. 1987. The impact of climate change on forest productivity. In: Meo, M., ed. Proceedings of the Symposium on Climate Change in the Southern U.S.: Future Impacts and Present Policy Issues. New Orleans, LA: U.S. Environmental Protection Agency.

Leverenz, J.W., and D.J. Lev. 1987. Effects of carbon dioxide-induced climate changes on the natural ranges of six major commercial tree species in the western United States. In: Shands, W.E., and J.S. Hoffman, eds. The Greenhouse Effect, Climate Change, and U.S. Forests. Washington, DC: Conservation Foundation, pp. 123-156.

Manion, P.D. 1981. Tree Disease Concepts. New Jersey: Prentice Hall.

Meo, M., ed. 1987. Proceedings of the Symposium on Climate Change in the Southern United States: Future Impacts and Present Policy Issues. May 28-29, 1987. New Orleans, LA: University of Oklahoma and U.S. Environmental Protection Agency.

Miller, F.W., P.M. Dougherty, and G.L. Switzer. 1987. Effect of rising carbon dioxide and potential climate change on loblolly pine distribution, growth, survival and productivity. In: Shands, W.E., and J.S. Hoffman, eds. The Greenhouse Effect, Climate Change, and U.S. Forests. Washington, DC: Conservation Foundation, pp. 157-188.

Pickett, S.T.A., and P.S. White. 1985. The Ecology of Natural Disturbance and Patch Dynamics. Academic Press, Inc. Harcourt Brace Jovanovich.

Schallau, C.H. 1988. The forest products industry and community stability: the evolution of the issue. Montana Business Quarterly Summer: 1-8.

Shands, W.E., and J.S. Hoffman. 1987. The Greenhouse Effect, Climate Change, and U.S. Forests. Washington DC: Conservation Foundation.

Simand, A.J. and W.A. Main. 1987. Global climate change: the potential for changes in wildland fire activity in the Southeast. In: Meo, M., ed. Proceedings of the Symposium on Climate Change in the Southern U.S.: Future Impacts and Present Policy Issues. New Orleans, LA: U.S. Environmental Protection Agency.

Solomon, A.M., and D.C. West. 1985. Potential responses of forests to CO_2 induced climate change. In: White, M.R., ed. Characterization of Information Requirements for Studies of CO_2 Effects: Water Resources, Agriculture, Fisheries, Forests and Human Health. Washington, DC: U.S. Department of Energy. DOE/ER-0236. pp. 145-1709.

Solomon, A.M., and D.C. West. 1986. Atmospheric carbon dioxide change: agent of future forest growth or decline? In: Titus J.G., ed. Effects of Changes in Stratospheric Ozone and Global Climate, Vol. 3: Climate Change. Washington, DC: U.S. Environmental Protection Agency.

Spurr, S.H., and B.V. Barnes. 1980. Forest Ecology, 3rd Ed. New York: J. Wiley & Sons.

Strain, B.R., and J.D. Cure, eds. 1985. Direct Effect of Increasing Carbon Dioxide on Vegetation. Washington, DC: U.S. Department of Energy. DOE/ER-0238.

Tirpak, D.A., ed. 1987. Potential Effects of Future Climate Changes on Forest and Vegetation, Agriculture, Water Resources and Human Health, Vol. V. Assessing the Risks of Trace Gases That Can Modify the Stratosphere. Washington, DC: U.S. Environmental Protection Agency. EPA 400/1 - 87/001E.

Titus, J.G., ed. 1986. Climate Change, Vol. 3. Effects of Changes in Stratospheric Ozone and Global Climate. Washington, DC: U.N. Environmental Program and U.S. Environmental Protection Agency.

USDA. 1981. U.S. Department of Agriculture, Forest Service. An Assessment of the Forest and Range Land Situation in the U.S. Forest Resource Report No. 22.

USDA. 1982. U.S. Department of Agriculture, Forest Service. An Analysis of the Timber Situation in the U.S. 1952-2030. Forest Resource Report No. 23.

White, M.R., ed. 1985. Characterization of Information Requirements for Studies of CO_2 Effects: Water Resources, Agriculture, Fisheries, Forests and Human Health. Washington, DC: U.S. Department of Energy. DOE/ER-0236.

Worrell, A.C. 1970. Principles of Forest Policy. New York: McGraw-Hill.

Young, R.A., ed. 1982. Introduction to Forest Science. New York: J. Wiley & Sons.

CHAPTER 12

BIOLOGICAL DIVERSITY

FINDINGS

Unlike most other impacts, a loss of species and reduced biological diversity is irreversible. The ability of a natural community to adapt to changing climatic conditions will depend on the rate of climate change, the size of species ranges, the dispersal rates of the individual species, and whether or not barriers to species migration are present. Under rapid climate change, many species would be lost.

Species Diversity

- The effect of climate change on species and ecosystems will vary, with some species benefiting and others facing extinction. The uncertainties surrounding the rate of warming, individual species response, and interspecies dynamics, make impacts difficult to assess. However, climate change will alter competitive outcomes and destabilize natural ecosystems in unpredictable ways.

- In many cases, the indirect effects of climate change on a population, such as changes in habitat, in food availability, and in predator/prey relationships may have a greater impact than the direct physiological effects of climate change.

- Natural and manmade barriers to species migration, including roads, cities, mountains, bodies of water, agricultural land, unsuitable soil types, and habitat fragmentation, may exacerbate loss of species.

- The areas within the United States that appear to be most sensitive to changes in climate are those that have a number of threatened and endangered species, species sensitive to heat or drought stress, and species inhabiting coastal areas.

Marine

- The loss of coastal wetlands and coastal habitat resulting from sea-level rise and saltwater intrusion may profoundly affect the populations of all inhabitants of these ecosystems, including mollusks, shellfish, finfish, and waterfowl. However, there is no evidence to indicate these species would become extinct.

Freshwater

- Freshwater fish in large bodies of water, such as the Great Lakes, may increase in productivity, but some significant species could decline. Fish in smaller bodies of water may be more constrained in their ability to respond to climate change. They also may be adversely affected by reductions in water quality.

Migratory Birds

- Migratory birds are likely to experience mixed effects from climate change, with some arctic nesting herbivores benefiting and continental nesters and shorebirds suffering. The loss of wintering grounds due to sea level rise and changing climate could harm many species, as would the loss of inland prairie potholes due to potentially increased continental dryness.

Policy Implications

- Existing refuges, sited to protect a species or ecosystem under current climate, may not be properly located for this purpose under future conditions, or as species migrate.

- Wildlife agencies such as the Department of Interior, state government agencies and conservation organizations may wish to assess the feasibility of establishing migratory corridors to facilitate species migration.

- Areas that will become suitable future habitat for threatened and endangered species need to be identified and protected, such as lowland areas adjacent to current wetlands.

- The practice of restoration ecology will need to be broadened to restore ecosystems in new areas as climates shift.

- The increase in the number of species at risk as a result of climate change may require new strategies. Agencies such as the Fish and Wildlife Service's Endangered Species Program may wish to assess the relative risk of climate change and more current stresses on ecological systems.

- The museums and systematic collections of biological material in the United States may need to be mobilized to provide a comprehensive ability to describe the existing species of plants and animals and their ranges.

VALUE OF BIOLOGICAL DIVERSITY

Maintaining the biological diversity of our natural resources is an important goal for the nation. The preamble to the Endangered Species Act of 1973 formalizes this belief, stating that endangered and threatened species of fish, wildlife and plants "are of aesthetic, ecological, educational, historical, recreational and scientific value to the Nation and its people." We depend upon these biological resources for food, medicine, energy, shelter, and other important products. Reduced biological diversity could have serious implications for man, as untapped resources for research in agriculture, medicine, and industry are irretrievable lost.

The National Resource

Public and private lands in the United States provide sanctuary for an abundant diversity of plants and animals. About 650 species of birds reside in or pass through the United States annually. Over 400 species of mammals, 460 reptiles, 660 freshwater fishes, and tens of thousands of invertebrates can be found in this country, in addition to some 22,000 plants (U.S. Fish and Wildlife Service, 1981). These species compose a wide variety of ecosystem types within the United States, including coniferous and broad-leaf forest, grassland, desert, freshwater, marine, estuarine, inland wetland, and agricultural. Figure 12-1 shows the major ranges of natural vegetation in the United States.

The U.S. National parks, forests, wilderness areas, and fish and wildlife refuges are among the public lands that provide sanctuary for wildlife resources, including many endangered species. U.S. public lands, which encompass over 700 million acres (about 32% of the land area of the United States), support about 700 rare species and communities (Roush, 1986). Over 45% of the lands held by the Forest Service, Fish and Wildlife Service, NationalPark Service, and Bureau of Land Management are in Alaska, and over 48% are located in the 11 most western States (U.S. Department of the Interior, 1987). However, much of the nation's biological diversity lies outside these areas.

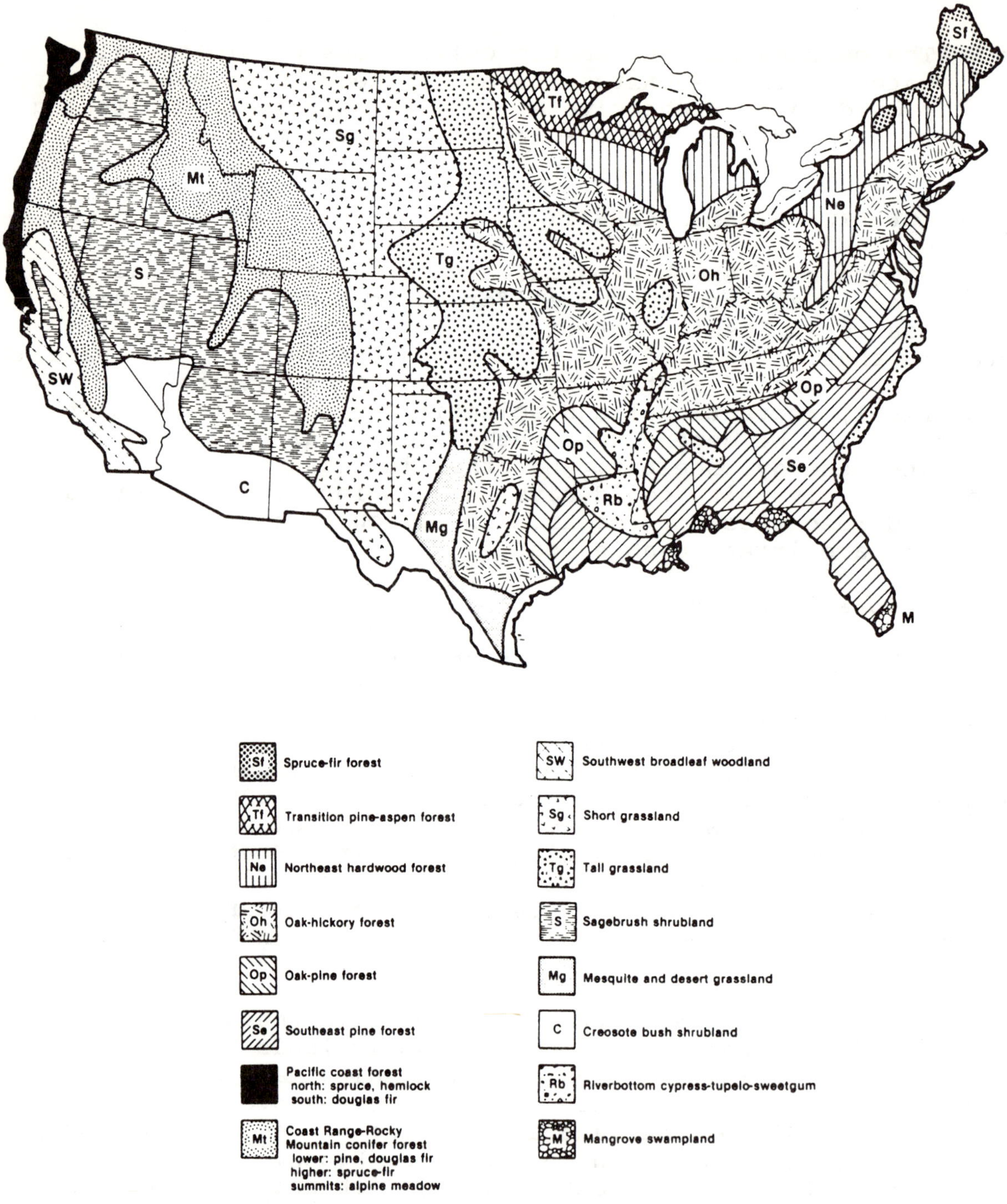

Figure 12-1. Natural vegetation in the United States.

Source: Hunt, 1972.

Private land holdings also account for a great deal of this nation's biological endowment. Private groups, such as the Nature Conservancy and the Audubon Society, manage 500,000 acres and 86,000 acres, respectively, for biological diversity.

GENERAL COMPONENTS OF BIOLOGICAL DIVERSITY

Biological diversity can be broadly defined as the full range of variety and variability within and among living organisms. It includes species diversity, genetic diversity, and ecosystem or community diversity.

Species Diversity

Each species occurs in a characteristic range or geographical area. The factors controlling species ranges are critical constraints on biological diversity. The presence of a species in an area suggests that the species must have successfully achieved the following: (1) dispersed into an area (no barriers to dispersal such as bodies of water or unsuitable soil types were present); (2) survived in that area (the physical characteristics of the area were suited to the animal's physiology, and food was available); and (3) became established in the area (the organism found an appropriate place in the food web in the absence of excessive competition and predation, and was able to reproduce).

The stresses brought about by development, overuse, and alteration of habitat have fragmented much of the world's natural habitat and have created many new barriers; consequently, for many species, dispersal has become much more difficult than it was in the past. For other species, man has inadvertently aided dispersal and has caused rapid spread in recent years. Such practices as clearcut logging prevent the dispersal of species adapted to dense forest conditions (e.g., flying squirrels) and promote the dispersal of species suited to open areas (e.g., deer).

Currently, 495 species are listed as endangered within the United States, and over 2,500 species await consideration for that status by the Fish and Wildlife Service. The list of endangered species is dominated by plants, birds, fishes, and mammals but also includes insects, amphibians, reptiles, mollusks, and crustaceans (U.S. Fish and Wildlife Service, 1988).

New species are created through the evolutionary process of speciation and are lost through extinction. Speciation generally requires a period of time considerably greater than centuries. However, extinction as a result of human activities is occurring extremely rapidly and at an increasing rate. Owing to its slowness, the process of speciation does little to offset species loss to extinction.

It is difficult to determine the exact rate of species extinction because the number of species on the Earth is known only to an order of magnitude. A recent estimate by Wilson (1988) places the total number of species between 5 and 30 million. Assuming 10 million species, Wilson made the rough calculation that one in every 1,000 species is lost each year. Wilson then compared this to estimates of extinction rates over geologic time, which ranged between 1 in every 1 million and 1 in every 10 million per year.

Species diversity is fundamental to an ecosystem's function and integrity. The significance of rare species should not be underestimated. A narrowly or sparsely distributed species may be a keystone in an ecosystem, controlling the structure and functioning of the community, or it may be a species of great and yet unknown value to humans.

Genetic Diversity

Each species that persists has a characteristic genetic diversity. That is, the individuals of that species have among them a collection of genes and gene combinations that may be more or less diverse.

The pool of genetic diversity within a species constitutes an adaptation to its present environment as well as a store of adaptive options for a new or changed environment. The loss of genetic diversity can contribute to the extinction of a species by reducing its ability to adapt to changing environmental conditions.

Generally, species with larger populations have greater genetic diversity. Species that are divided into partly connected subpopulations also tend to have greater genetic diversity. Species near extinction represented by few individuals in few populations have lower genetic diversity, a situation exacerbated by inbreeding.

Community and Ecosystem Diversity

Overall species diversity is also a function of the number of different types of areas in which communities can be built.* Ecosystem diversity refers to the distinctive assemblages of species and biotic processes that occur in different physical settings around the world.

FACTORS AFFECTING THE RESPONSE OF BIOLOGICAL DIVERSITY TO CLIMATE CHANGE

Natural communities are a significant indicator of climate change. Historically, biological communities have responded to changes in climate through large-scale geographic shifts, changes in species composition, physiological adaptation, and extinction (Peters and Darling, 1985). Local climate appears to be the primary factor defining an environmental setting and determining the species composition and spatial patterns of communities in terrestrial zones (Bolin, 1986). Temperature means, temperature extremes, and precipitation are the factors most often affecting the natural distributional limits of a species (Ford, 1982), while the actual distribution of a species is also affected by soil type, soil moisture, ecological dynamics, and regional isolation.

Rate of Climate Change

Predicting how a species or ecosystem might respond to a given environmental change is difficult. Adaptation to climate change will inextricably depend on the rate of climate change. For some species, migration rates may be inadequate to keep up.

The infinite number of combinations of dispersal range and reproductive age make the potential rate of migration different for every species. Paleorecords suggest migration rates between 10 and 20 km per century for chestnut, maple, and balsam fir, and between 30 and 40 km per century for some oak and pine species (see Forestry chapter in this report). On the other hand, cattle egrets have shown a much quicker migration rate by colonizing all of the North American tropics within approximately 40 years.

As species shift at different rates in response to climate change, communities may disassociate into new arrangements of

---

*The new discipline of landscape ecology attempts to relate topographic diversity to community and species diversity.

species. Local extinction can result either directly from physiological pressures or indirectly from changes in interspecies dynamics. Hence, the affect of climate change on an area will be to cause sorting and separation of species owing to the differential rate of migration and species retreat (Ford, 1982). Ecosystems, therefore, will not migrate as a unit.

The rate of climate change will be crucial to the survival of the species. A 3°C increase in temperature, for example, would effect a several hundred kilometer poleward shift in the temperate vegetation belts (Frye, 1983). If this change took place within a century, species would need to migrate several kilometers each year to adapt to this warming. Plants have a wide range of migration rates, and only some may be able to achieve this rate.

Many factors make evaluating the impact of climate change on ecosystems difficult. The great interdependencies among species in an ecosystem add considerable uncertainty to the effect that the various responses of individual species will have on the system. An impact upon a single species could significantly change the entire ecosystem. For example, elimination of a tree species in a region could have a significant effect on the whole forest ecosystem including birds, insects, and mammals.

Animal populations are generally much more mobile than plants. But animal distributions heavily depend on vegetation for food, protection, and nesting habitat. Species not directly dependent on vegetation ultimately depend on some other species that is. The ranges of the fig wasp and the fig depend entirely upon one another. In this case, the plant species depends on a single pollinator, and the insect species relies upon a single species of plant for food (Kiester et al., 1984).

Species do not immediately respond to changed and changing environmental conditions. A negative response, such as local extinction in an area, is usually quicker than the positive response of new species' colonization of a region (see Forestry chapter of this report). In the Arctic, the lag period between climate change and species response (colonization) may be several hundred years (Edlund, 1986). This lag period will leave areas open for weedy, opportunistic species that can quickly migrate and propagate in a region.

Barriers to Response

How quickly species migrate is affected by natural and manmade barriers and by competition. Peters and Darling (1985) examined potential species' responses to climate change, ecological interactions, and barriers to adaptation. Physical barriers include mountains, bodies of water, roads, cities, agricultural land, inappropriate soil type, and habitat heterogeneity (landscape patchiness).

Reserve and Island Species

Species living on reserves have additional constraints on their ability to respond to climate change that are similar to those of island species. This problem has become known as the island dilemma. Species on reserves are often remnants of larger populations and are more susceptible to environmental stress. Additionally, reserves that originally may have been well sited to protect a vulnerable population and its habitat may, after climate change and population response, exist outside the now suitable range. Figure 12-2 illustrates this problem.

Species on reserves are likely to be pressured from two directions as a result of climate change. A population isolated on a reserve surrounded by altered or unsuitable habitat receives little immigration from populations outside the reserve. Also, that population may not be able to colonize areas outside the reserve as these areas become suitable because of development or other alterations of habitat. Large reserves and buffer zones around reserves help to lessen this problem. Corridors between reserves lessen the problem of spatial isolation by allowing for some migration between reserves.

Habitat Area

Even without the added pressure of climate change, reserve populations are stressed because many reserves are not large enough to support a self-sustaining population (Lovejoy, 1979). The predictive theory of island biogeography (MacArthur and Wilson, 1967) showed that, other factors being equal, large islands accommodate larger numbers of land-dwelling species than do small islands. This held true for other ecologically isolated islands, such as mountaintops, woodlots, and lakes. Also, when large ecosystems become smaller through fragmentation, the number of species always declines. Figure 12-3 shows how mammalian extinctions have been inversely related to refuge area in North American parks.

Mountain Species

Just as species can migrate latitudinally, they can respond altitudinally to climate change by moving up a mountain slope.

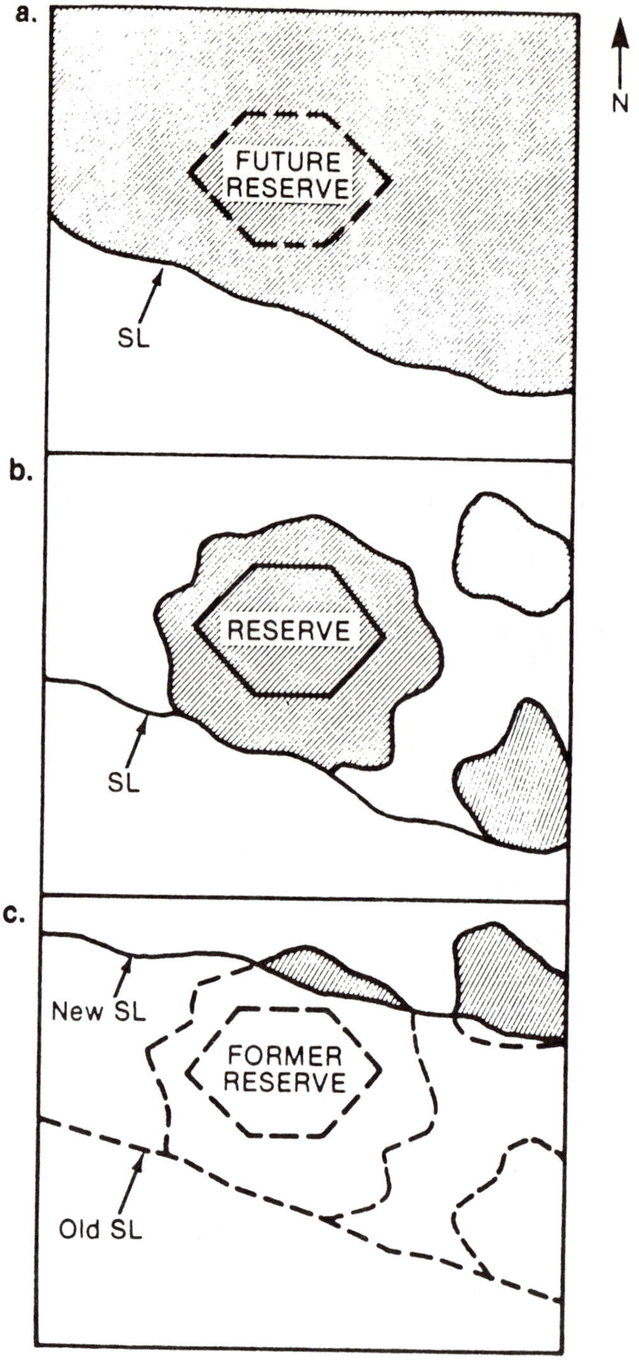

Figure 12-2. Effect of climate chante on biological reserves. Hatching indicates the following: (a) species distribution before human habitation (SL indicates southern limit of species range); (b) fragmented species distribution after human habitation; (c) species distribution after warming.

Source: Peters and Darling (1985).

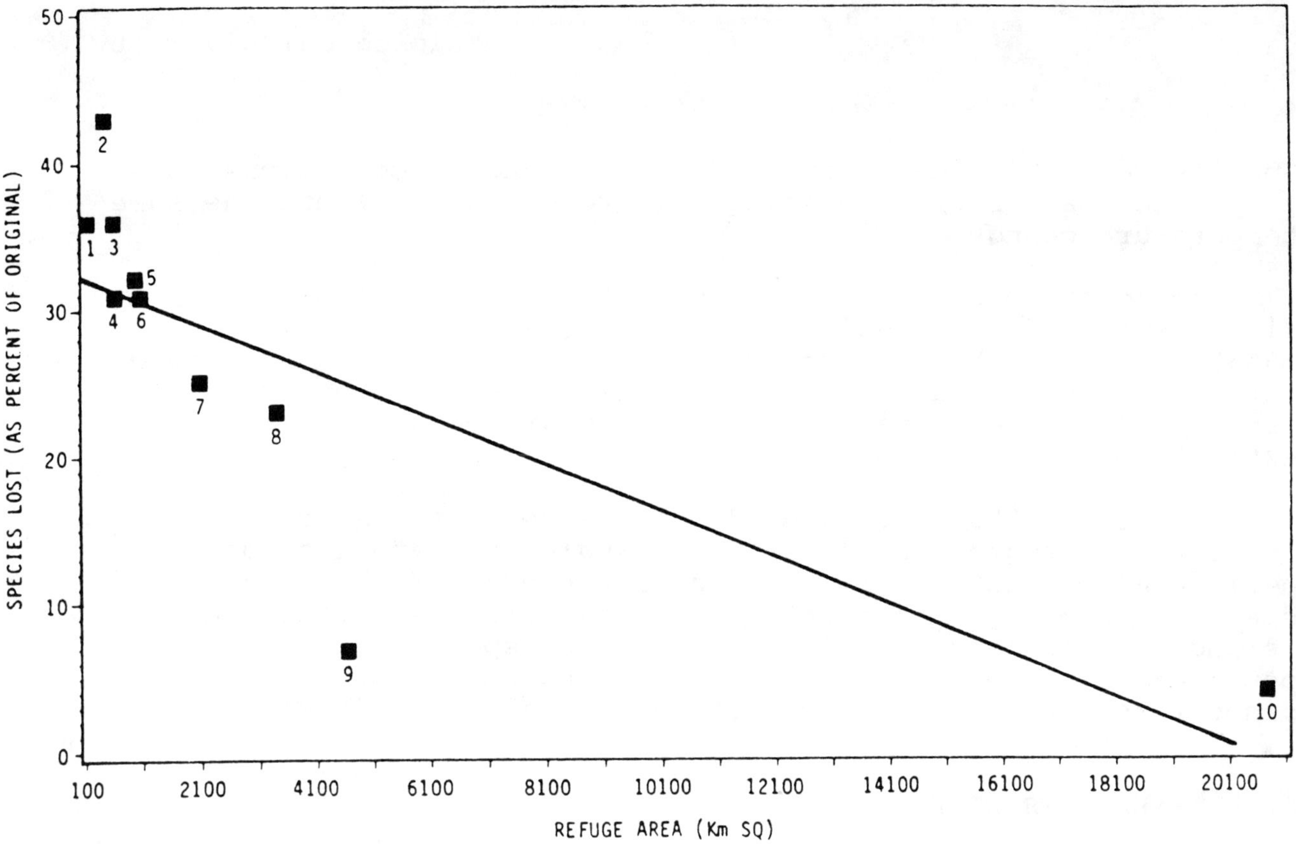

	PARK	AREA (km^2)	PERCENT OF ORIGINAL SPECIES LOST
1)	Bryce Canyon	144	36
2)	Lassen Volcano	426	43
3)	Zion	588	36
4)	Crater Lake	641	31
5)	Mount Rainier	976	32
6)	Rocky Mountain	1,049	31
7)	Yosemite	2,083	25
8)	Sequoia-Kings Canyon	3,389	23
9)	Glacier-Waterton	4,627	7
10)	Grand Teton-Yellowstone	20,736	4

Figure 12-3. Habitat area and loss of large animal species in North American parks (1986)

Source: Newmark (1987).

Species can often adapt more easily to changing conditions on a slope because a shorter migration is required to effect the same temperature change.

Among the problems associated with altitudinal migration are displacement of the species at the top (Peters and Darling, 1985). Also, with the increase in altitude, the area available for colonization usually becomes smaller, communities become very isolated, and these smaller populations are more prone to extinction.

Peters and Darling conclude that some of the species and communities at greatest risk from changing climate include peripheral populations (located near the edge of a contracting species range); geographically localized species (reserve and island species); genetically impoverished species (small populations); specialized species (living in narrow niches); poor dispersers; and alpine, arctic, and coastal communities.

CLIMATE EFFECTS RESEARCH

This section reviews some previous studies of ecological response to past changes in climate, recent studies of potential response to anticipated future climate change, and studies funded by EPA for this report, which use climate change scenarios from general circulation models for a doubled CO_2 environment (see Table 12-1).

In only the last few years, biological diversity has emerged as both a major scientific and policy issue. At present, much effort is being devoted to both the scientific and policy aspects of biological diversity. This effort should result in much new information is the relatively near future.

Forest Ecosystems

The tree species that make up any forest are major factors in determining the biological diversity found there. Treesprovide a multitude of habitats and are the basis of much of the food web in a forest.

The changes in forest composition resulting from climate change, (see Forests: Chapter 11), will have significant implications for biological diversity. Potential northerly range shifts of several hundred to a thousand kilometers may be limited by the tree species' ability to disperse. One possibility is that southern pine forests will move farther north into the

Chapter 12

Table 12-1. Studies Conducted for This Report. Cited in this Chapter

- Potential Responses of Great Lakes Fishes and Their Habitat to Global Climate Warming -- Magnuson, Regier, Shuter, Hill, Holmer and Meisner, University of Wisconsin

- The Effects of Global Climate Change on the Water Quality of Mountain Lakes and Streams -- Byron, Jassby and Goldman, University fo California at Davis

- The Effects of Climate Warming on Lake Erie Water Quality -- Blumberg and DiToro, HydroQual, Inc.

- Ecological Effect of Global Climate Change: Wetland Resources of San Francisco Bay -- Josselyn and Callaway, San Francisco State University

- Projected Changes in Estuarine Conditions Based on Models of Long-Term Atmospheric Alteration -- Livingston, Florida State University

regions currently occupied by mixed hardwood species. Some of these hardwood forests contain the highest tree species diversityfound anywhere in the United States (Braun, 1950). As they migrate north, species would inevitably be lost, and overall biological diversity would substantially decrease.

As forests are disrupted, the land will be invaded by weedy, opportunistic species, creating a system with very low diversity similar to that following logging. Ultimately these systems will not persist as succession takes place, but the pattern of succession following the removal of a forest by rapid climate change is unknown.

Tropical Forest Ecosystems

The greatest concentration of biological diversity in the world is in the rain forests of the Tropics (Wilson, 1988). Besides reducing diversity, deforestation contributes to disrupting the global carbon cycle by releasing CO_2 into the atmosphere and will directly affect the rate of climate change (Prance, 1986). Indeed, on a global scale the problems of tropical deforestation, rapid climate change through (among other factors) increased CO_2 production, and the loss of biological diversity can be seen as aspects of the same problem.

Tropical forests are also important as wintering grounds for migratory birds coming from the United States and as sources of new knowledge, because the patterns of interactions between species and climate are at their most sensitive and complex there. The Tropics may provide important leading indicators of climate change (Robinson, 1978; Janzen, 1986).

Freshwater Ecosystems

A study conducted by Magnuson et al. concludes that in most areas of the Great Lakes, climate warming would increase the amount of optimal thermal habitat for warm-, cool-, and cold-water fishes (see Chapter 5, Great Lakes); although overall productivity would increase, overall biological diversity could decrease through intensified species interactions.

A study by Byron et al. on mountain lakes suggests that climate change would cause a range of impacts, including higher productivity, changes in species composition, and decreased water quality resulting from an increase in algal growth (see Chapter 4, California). Blumberg (see Chapter 5, Great Lakes) found that thermal stratification in Lake Erie could decrease dissolved oxygen levels. This result, with possible loss of biological diversity, is likely to be repeated in many lakes.

The combined pressures of warmer waters, saltwater intrusion, and a rising sea level will significantly affect estuaries. The regional studies suggest that coastal estuaries are likely to see a growth in marine species and a loss of some estuarine species. A study by Josselyn on the San Francisco Bay estuary suggests a decline in species that use the delta for spawning (see Chapter 4, California). Livingston concluded that crabs, shrimp, oysters, and flounder in the Apalachicola estuary could not survive the warming projected by the GISS and GFDL scenarios (see Chapter 6, Southeast).

Saltwater Ecosystems

In general, a warmer global climate is expected to increase productivity in ocean fisheries, but the location and relative abundance of species are likely to change (Sibley and Strickland, 1985). Up to some threshold temperatures, such as 2°C, warmer ocean temperatures are likely to increase ocean productivity, but beyond that threshold, productivity could decline (Glantz, 1988). It is likely that as productivity decreases, biological diversity would decrease as well.

Coral Reef Ecosystems

Coral reefs provide the structural base for the very biologically diverse reef ecosystems. Coral reefs in the Caribbean and the Pacific may be severely stressed as a result of warmer water temperatures and the rising sea level associated with climate change. Extensive bleaching of coral (the expelling of symbiotic brown algae in response to environmental stress) occurred in the Pacific after the 1982-83 El Nino (Glynn, 1984) and in the Caribbean following a summer of elevated water temperatures in 1987 (Roberts, 1987). Loss of the brown algae, the primary food source of the coral, is thought to kill coral, making the reef ecosystem vulnerable to erosion and physical devastation.

Coral reefs also will be affected by sea level rise. Studies by Buddemeier and Smith (1987) and Cubit (1985) suggest that vertical accretion of reef flats eventually may be unable to keep up with an accelerating rise in sea level. Reef flats also may be subject to the stress of increasingly large waves, erosion, and sedimentation, which can inhibit coral growth (Buddemeier, 1988).

Arctic Ecosystems

Within the North American arctic, plant size, vigor, and reproduction could be expected to increase with higher temperatures. Some low-lying plants would most likely become upright, and there would be a northerly movement of the tree line and all vegetative zones (Edlund, 1986).

However, in the Arctic, rising temperatures may be a mixed blessing. Overall biological productivity is likely to increase, and some species may be able to increase their range. But many species specially adapted to the Arctic will face the same type of problem that mountaintop species face: they will have nowhere to go once they reach the Arctic Ocean. Thus, native Arctic species may be specially at risk. Other Arctic species may face their own problems. For example, caribou will be severely harmed if rivers do not freeze for long enough periods to allow for migration.

Migratory Birds

Migratory waterfowl are likely to experience very mixed effects as a result of warmer temperatures (Boyd, 1988). Herbivorous, Arctic nesting species, such as geese, are likely to benefit from the shortened winter season and from the increases in vegetation, in nesting habitat, and in ecosystem productivity (Harington, 1987). Smaller Arctic nesting shorebirds, on the other hand, are likely to be harmed by the encroachment of taller vegetation, potentially eliminating the preferred low-lying tundra breeding ground. Other effects on shorebirds could result from changes in ecosystem predator/competitor relationships and changes in the seasonal timing of such events as larval blooms, upon which these birds depend for nourishment while in a flightless stage and during migration (Myers, 1988).

Waterfowl that breed in the continental interior may suffer more than arctic nesters. Over half of all waterfowl in North America originate in the prairie pothole region, a large agricultural area riddled with highly ecologically productive permanent and semipermanent wetlands. Increased temperature and changes in seasonal precipitation could reduce the highly variable number of potholes (wetlands) in the area and could significantly impair the productivity of breeding ducks (Burke, 1988). Over 35% of the seasonal wetlands within the prairie pothole region were dry during the breeding season owing to the drought of 1988 (U.S. Fish and Wildlife Service, 1988). The Fish and Wildlife Service is forecasting that only 66 million ducks will migrate during the fall of 1988, a total of 8 million fewer than in 1987 and the second-lowest on record (Irion, 1988). The productivity index for mallards (number of young per adult) is 0.8, which is down by over 20% from the historical average (U.S. Fish and Wildlife Service, 1988).

Waterfowl and other migratory birds are likely to be affected on both ends of their migratory journey and at staging areas along the way. The loss of coastal wetlands, already an area of great concern in the United States, reduces the amount of habitat available to waterfowl, creating population pressures on a limited resource. Of the 215 million acres of wetlands in the coterminous United States at the time of settlement, fewer than 99 million acres (46%) remain (U.S. Fish and Wildlife Service, 1987). Loss of an additional 26-82% of existing coastal wetlands is expected over the next century as a result of a 1-meter rise in sea level, saltwater intrusion, and human development (see Sea Level Rise: Chapter 9). Loss of wintering habitat along the Gulf of Mexico would affect many waterfowl, including mallards, pintails, and snow geese.

The Tropics, which may be significantly altered by rapid climate change, is the winter home for many species of migratory birds. The need to protect a species in all parts of its range underscores the truly global nature of the effects of rapid climate change on biological diversity (Terborgh, 1974).

Endangered Species

Hundreds of species are currently listed as endangered in the United States, and several thousand await consideration for that status. These species are likely to be stressed further as a result of climate change.

Threatened and endangered species of the Southeast are very susceptible to the impacts of sea level rise. Some species potentially at risk in that region include the Key deer, manatee, Florida panther, and Everglades kite (Breckenridge, 1988). Climate change could also greatly increase the number of rare, threatened, and endangered species in the United States.

Other Direct and Indirect Stresses

As plant and animal species experience increasing pressures from changes in temperature, precipitation, and soil moisture, so too will agriculture and urban water supplies. The changes that result from the human response to climate change may have the greatest impact on biological diversity. If the continental interior of North America dries, for example, habitats may be cultivated, and our current uses of water resources may change. These secondary effects may also prove detrimental to biological diversity.

NATIONAL POLICY IMPLICATIONS

Climate change presents new challenges for policymakers, regulators, and resource managers. Planning for climate change may help to minimize the disruption to natural systems and facilitate adaptation under changing conditions. Decisions will need to be made in an environment of increased pressure on many other resources.

Policies regarding rare and endangered species are likely to change as the number of species at risk greatly increases. As more species become stressed and potentially threatened by climate change, reevaluation of protection policies may be required. The tradeoffs between protection of individual species and species' habitat and the broader protection of biodiversity will have to be reexamined. As a part of this question, decisions concerning whether to protect existing communities or foster the establishment of new ones will need to be made.

Management Options to Maintain Biological Diversity

Only a limited number of techniques are available for maintaining biological diversity. However, these techniques can be adapted and intensified to meet the potentially great impacts of rapid climate change.

Maintenance of Native Habitats

The most direct way to maintain biological diversity is to manage land to retain ecosystems, communities, and habitats. This has been successfully undertaken on a broad scale by Federal and State governments and by private organizations. Ecosystem conservation, especially as represented by the National Parks and other large reserves, maintains much of our national biological diversity. This will continue to be a useful first step for maintaining biological diversity in the face of climate change.

Land acquisition and management policies will need to take climate change into account. Climate change and the future requirements of whole ecosystems must be considered in siting and managing reserves. To preserve functioning ecosystems, large areas of land will be required. Preserves need to be at least large enough to support self-sustaining populations. Lands that will be more important as future plant and animal habitat need to be identified, evaluated, and set aside. Although identification of appropriate future habitat is difficult, and highly dependent on the future rate and extent of climate change, some areas, such as lowland areas adjacent to current wetlands, hold good potential for habitat protection.

To protect a species, alternative sites should be considered with regard to the ecological needs of target species under changing conditions. Siting preserves in mountainous areas is beneficial because it allows for the shorter-distance altitudinal shifts of adjustment to changing climate. Stream corridors, which can be effective avenues of dispersal for terrestrial as well as aquatic organisms, should be protected wherever possible. Providing corridors for migration between reserves also will enhance the ability of wildlife to adapt to climate change. Ideally, these corridors should be wide enough to maintain the ecosystem characteristics of the reserve in their center. Some species do not find the habitat conditions of narrow corridors suitable for migration.

Maintenance of Species in Artificial Conditions

When individual species are threatened with extinction, a possible option is to ensure that the species is propagated in captivity. Indeed, some rare species, such as the Pere David deer and the California condor, now exist only in captivity. This technique can be made to work for a variety of species, depending on their biology and the degree to which they successfully adapt to captive conditions. As more species become threatened with extinction due to climate change, the effort applied in this area may have to increase dramatically. However, only a tiny fraction of the nation's species can be maintained in this way.

Restoration of Habitat

Restoration ecology is a new discipline whose goal is to develop methods to restore damaged ecological communities to their prior unaltered state. Except in forestry, where reforestation has a longer tradition, restoration ecology has been in existence for only a few years. Nonetheless, it offers some real promise for ameliorating the effects of rapid climate change.

Normally, restoration is done at the site where the community previously existed and was altered or damaged. Historical and baseline information is used to manage the species in such a way as to eliminate unwanted new species and to encourage and possibly reintroduce native species.

The next step in restoration ecology is to restore natural communities on sites where they have not previously existed. This activity has not yet been attempted but may be necessary to save communities displaced by climate change. If the climate changes so that many of the key species of a community can no longer survive in their original range, and if the species are incapable of dispersing and establishing themselves elsewhere, then the artificial transplantation of entire communities may become necessary. This transplantation of communities would be a monumental task and could help to save much biological diversity, but it cannot possibly be undertaken on the scale necessary to preserve all species threatened by climate change. Restoration ecology can be useful for extending reserve boundaries and providing migratory corridors.

Planning Options

While there are only a few management techniques to maintain biological diversity, many different groups in our society can implement them. These groups can be divided into the private and public sectors.

Many different groups in the private sector, ranging from private individuals to large conservation organizations, will have an interest in maintaining biological diversity. However, all will need information about the current and probable future state of biological diversity. The Federal Government may be able to play a role here by providing information on the study of biological diversity.

Resource managers and policymakers should incorporate climate change into their long-term planning. Periodic reviews of the management plans for public lands should include consideration of the possible effect of climate change.

The four major Federal land management agencies develop plans intended to lay out a comprehensive framework and direction for managing Federal land. Land Resource Management Plans, required for each National forest, define the direction of management in the forest for the next 10 to 15 years. In addition, the Forest Service prepares 50-year plans, as required by the Resource Conservation Act. The National Park Service prepares a General Management Plan for each unit in the system that defines a strategy for achieving management objectives within a 10-year time frame. A Statement for Management is also prepared for each National park and is evaluated every 2 years; this includes a determination of information needs. The Bureau of Land Management's (BLM's) Resource Management Plans and the Fish and Wildlife Service's Refuge Master Plans are prepared and revised as needed for BLM Resource Areas and Wildlife Refuges (U.S. Department of the Interior, 1987).

The Federal Government manages an enormous amount of land and should consider management options to preserve biological diversity on much of that land. The major management techniques of habitat maintenance and restoration ecology could be applied by the agencies actively responsible for managing the nation's public lands.

RESEARCH AND DEVELOPMENT NEEDS

The ability to protect biological diversity is severely restricted by a lack of knowledge regarding the rate of climate change, the precise nature of the change, how individual species will respond, and how ecological balances will shift. Increased research effort is needed in three areas: identification of biological diversity, species interactions and biological diversity, and management options for biological diversity.

Identification of Biological Diversity

The first and most pressing need is for more and better coordinated research to simply identify the biologically diverse resources of our country. This involves both systematics (organism classification) and ecology. The nation needs a more coordinated ability to identify its plants and animals (see H.R. 4335), and it needs range maps and habitat requirement information for those species.

The apparently simple task of identifying the species of plants and animals that exist in a given area is actually a major barrier to further understanding. Common species are usually relatively easy to identify, but even they present serious problems when attempts are made to decide whether a widespread group is, for example, one or two species. The nation's museums and systematic collections needs to be mobilized to solve this problem. For example, there is currently no federally sponsored Flora (listing of all known plants) of the United States. In this regard, the United States lags behind many Third World countries, such as Panama and Peru.

Species Interactions and Biological Diversity

The second area needing research is the direct effects of climate on species and the indirect interactions of species with other species dependent on climate. Comprehensive mapping of species ranges along temperature and moisture gradients would provide valuable information. The direct effect of climate change on vegetation needs to be better assessed, and more estimates of species dispersal rates would significantly improve our ability to identify species at greatest risk.

A variety of ecosystems within a diversity of climatic regions and terrains need to be intensively studied using analogue climate regions under changed climate conditions. Although an ecosystem's response under changing climate conditions will not be wholly predictable, modeling individual ecosystem responses would enhance knowledge of the likely effects. Further research on how species interact and how trophic structures might change with climate would help predictive capabilities.

It will be impossible to study in detail even a fraction of the nation's species. Those groups chosen for study either must be representative of many species or must possess some special properties (such as extreme sensitivity to climate change). The method of deciding which group to study is itself a major outstanding research question.

Management Options for Biological Diversity

Management options are limited in both number and scope. Research is needed to increase the scope of the current management options and to discover new options.

National parks, forests, and wildlife refuges are suitable areas for study. In-depth evaluation of 5 to 10 of these areas could provide valuable information regarding the range of impacts and management options.

Chapter 12

REFERENCES

Bolin, B., B. Doos, J. Jager, and R. Warrick, eds. 1986. The Greenhouse Effect, Climatic Change, and Ecosystems (SCOPE 29). Chichester: John Wiley & Sons.

Boyd, H. 1988. Impact of Climate Change on Arctic-Nesting Geese. Paper presented at Climate Institute Wildlife Symposium, Washington, DC; January 21.

Braun, E.L. 1950. Deciduous forests of eastern North America. New York: Hafner Publishing Company.

Breckenridge, R. 1988. Personal Communication. (Phone conversation).

Brown, L., W. Chandler, A. During, C. Flavin, L. Heise, J. Jacobson, S. Postel, C. Shea, L. Starke, and E. Wolf. 1988. State of the World 1988. Worldwatch Institute. New York: W.W. Norton and Co., Inc.

Buddemeier, R. 1988. Impacts of Climate Change on Coral Reefs, Islands and Tropical Coasts. Paper presented at the Climate Institute Workshop on the Impact of Climate Change on the Third World, Washington, DC; March 24-25.

Buddemeier, R., and D. Hopley. 1988. Turn-Ons and Turn-Offs: Causes and Mechanisms of the Initiation and Termination of Coral Reef Growth. Paper presented for submittal to the Proceedings of the Sixth International Coral Reef Symposium, James Cook University, Townsville, Australia; August 8-12.

Buddemeier, R., and S. Smith. 1987. Coral Reef Growth in an Era of Rapidly Rising Sea Level: Predictions and Suggestions for Long-Term Research. Coral Reefs, Printer's Proof.

Burke, L. 1988. The Impact of Climate Change on Mallard Productivity in the Prairie Pothole Region. Work in progress.

CEQ. 1980. Council on Environmental Quality. Environmental Quality: Eleventh Annual Report. Washington, DC: U.S. Government Printing Office.

Cubit, J. 1985. Possible Effects of Recent Changes in Sea Level on the Biota of a Caribbean Reef Flat and Predicted Effects of Rising Sea Levels. Preprint from the Proceedings of the Fifth International Coral Reef Congress, Tahiti.

Edlund, S. 1986. Modern Arctic Vegetation Distribution and Its Congruence With Summer Climate Patterns. Environment Canada, Proceedings: Impact of Climatic Change on the Canadian Arctic.

Ford, Michael. 1982. The Changing Climate. London: George Allen and Unwin.

Frye, R. 1983. Climatic change and fisheries management. Natural Resources Journal 23:77-96.

Glantz, M. 1988. Personal Communication. (Phone conversation).

Glynn, Peter. 1984. Widespread Coral Mortality and the 1982-83 El Nino Warming Event. Environmental Conservation 11(2):133-140.

Harington, C. 1987. The Impact of Changing Climate on Some Vertebrates in the Canadian Arctic. Climate Institute, Proceedings of the First North American Conference on Preparing for Climate Change.

Irion, R. 1988. Washington Post, Monday, August 1, 1988. Drought helps 2 Endangered Species Rebound, but It's a Dismal Year for Ducks.

Janzen, D. 1986. The future of tropical ecology. Annual Review of Ecology and Systematics 17:305-324.

Jordan, W., R. Peters, and E. Allen. 1988. Ecological restoration as a strategy for conserving biological diversity. Environmental Management 12(1):55-72.

Kiester, A.R., R. Lande, D.W. Schemske. 1984. Models of coevolution and speciation in plants and their pollinators. American Naturalist 124(2):220-243.

Leatherman, S. 1987. Effects of Sea Level Rise on Beaches and Coastal Wetlands. Climate Institute, Proceedings of the First North American Conference on Climate Change, Washington, DC, pp. 140-146.

Lovejoy, T. 1979. Refugia, refuges and minimum critical size: problems in the conservation of the neotropical herpetofauna. In: Duellman, W., ed. The South American Herpetofauna: Its Origin, Evolution and Dispersal. University of Kansas Museum Natural Hist. Monograph 7:1-485.

MacArthur, R., and E. Wilson. 1967. The Theory of Island Biogeography. Princeton, NJ: Princeton University Press.

Myers, J.P. 1988. Remarks to Climate Institute Wildlife Symposium, Washington, DC; January 21.

Myers, N. 1979. The Sinking Ark. New York: Pergamon Press.

Naveh, Z., and A. Lieberman. 1984. Landscape Ecology. New York: Springer-Verlag.

Newmark, W.D. 1987. A land-bridge island perspective on mammalian extinctions in western North American parks. Nature 325:430-432.

Peters, R., and J. Darling. 1985. The greenhouse effect and nature reserves. Bioscience 35(11):707-717.

Prance, G.T. 1986. Tropical Rain Forests and the World Atmosphere. AAAS Selected Symposium No. 101. Boulder, CO: Westview Press, pp. xxi-105.

Regier, H., J. Holmes, J. Meisner. 1987. Likely Effects of Climate Change on Fisheries and Wetlands, With Emphasis on the Great Lakes. Climate Institute, Report for the First North American Conference on Preparing for Climate Change.

Roberts, L. 1987. Coral bleaching threatens atlantic reefs. Science 238:1228-9.

Robinson, M.H. 1978. Is tropical biology real? Tropical Ecology 19(1): ____.

Roush, J. 1986. Private action for public lands. The Nature Conservancy News 36(4):4-7.

Sibley, T., and R. Strickland. 1985. Fisheries: some relationships to climate change and marine environmental factors. In: DOE. Characterization of Information Requirements for Studies of CO_2 Effects, Ch. 5.

Strain, B. 1986. The biosphere and links to climate. In: Rosenzweig, C., and R. Dickinson, eds. Climate-Vegetation Interactions. Proceedings of a NASA/Goddard Space Flight Center Workshop, January 27-29, 1986. Published by the Office for Interdisciplinary Earth Studies, University Corporation for Atmospheric Research, Boulder CO.

Strain, B. 1987. Direct effects of increasing atmospheric CO_2 on plants and ecosystems. Tree 2(1):18-21.

Terborgh, J. 1974. Preservation of natural diversity: the problem of extinction prone species. BioScience 24(12):715-722.

Tiner, R. Jr. 1984. Wetlands of the United States: Current Status and Recent Trends. U.S. Fish and Wildlife Service. Newton Corner, Massachusetts: Habitat Resources.

Toppins, J.C., and J. Bond. 1988. The Potential Impact of Climate Change on Fisheries and Wildlife in North America. Report of the Climate Institute to the U.S. EPA.

U.S. Department of the Interior. 1987. National Park Service and National Recreation and Park Association. Trends 24(2):2-43.

U.S. Fish and Wildlife Service. 1981. Endangered Means There's Still Time. Department of the Interior, U.S. Fish and Wildlife Service, Washington, DC.

U.S. Fish and Wildlife Service. 1988. Endangered Species Technical Bulletin; May.

Wilcox, B. 1982. Biosphere Reserves and the Preservation of Biological Diversity. Towards the Biosphere Reserve: Exploring Relationships Between Parks and Adjacent Lands. Parks Canada and U.S. National Park Service, Kalispell, Montana; June.

Wilson, E.O. 1988. Biodiversity. Washington, DC: National Academy Press.

CHAPTER 13

IMPACT OF CLIMATE CHANGE ON AIR QUALITY

FINDINGS

- Changes in regional temperatures, precipitation patterns, clouds, windspeed and direction, and atmospheric water vapor that will accompany global climate change will affect future air pollution levels and episodes in the United States.

- While uncertainties remain, it is likely that an increase in global temperatures will have the following effects on air quality, if other variables remain constant.

  - Ozone levels in many urban areas would increase because higher global temperatures would speed the reaction rates among chemical species in the atmosphere.

  - Manmade and natural emissions of hydrocarbons and manmade emissions of nitrogen and sulfur oxides would increase with a temperature rise. Natural emissions of sulfur would also change, but the direction is uncertain. The hydrocarbons and nitrogen oxides participate in reactions that produce ozone.

  - The formation of acidic materials (such as sulfates) would increase with warmer temperatures because sulfur and nitrogen oxides would oxidize more rapidly. The ultimate effect on acid deposition is difficult to assess because of changes in clouds, winds, and precipitation.

  - Visibility may decrease because of the increase in hydrocarbon emissions and in the rate at which sulfur dioxide is oxidized to sulfate.

- Preliminary analyses of 4°C temperature increase in the San Francisco Bay area suggest that maximum ozone concentrations would increase by approximately 20%, that the area in which the National Ambient Air Quality Standard (NAAQS) would be exceeded would almost double, and that the number of people-hours of exposure would triple. The Midwest and Southeast also would incur high concentrations and an increase in the area of high ozone by a factor of three.

- The additional costs for air pollution controls that would be caused by global warming cannot be estimated at this time. Increases in ambient ozone levels resulting from climate change could increase the number of nonattainment areas and make attainment more expensive in many regions. Preliminary estimates suggest that an expenditure of several million dollars per year may be necessary for additional volatile organic compound (VOC) controls.

- Because of the close relationship between air pollution policies and global climate change, it is appropriate for EPA to review the impact of global climate change on air policies and the impact of air pollution regulations on global climate change.

RELATIONSHIP BETWEEN CLIMATE AND AIR QUALITY

The summer of 1988 provided direct evidence regarding the importance of weather to pollution episodes in the United States. Despite significant progress in reducing emissions of many pollutants over the last decade, the extended stagnation periods and high temperatures caused ozone levels in 76 cities across the country to exceed the national standard by at least 25%. Whether this recent summer is an appropriate analogue for the future cannot be determined with certainty, but scientists have recognized for some time that air pollution does vary with seasons and is directly affected by ventilation, circulation, and precipitation, all of which could be affected by future global climate changes.

Ventilation

Two major factors, referred to as "ventilation" when considered together, control the dilution of pollutants by the atmosphere: windspeed and the depth of the atmospheric mixing layer (frequently called the mixing depth). If windspeed is high, more air is available to dilute pollutants, thus lowering pollutant concentrations. The mixing layer (the distance between the ground and the first upper-layer inversion) tends to trap pollutants because the inversion on top of it acts as a barrier to vertical pollutant movement. Thus, pollutant concentrations decrease as mixing depth increases, providing greater dilution.

The ventilation characteristics of an area change, depend on whether a high- or low-pressure system is present. Low-pressure systems usually produce good ventilation because they normally have greater mixing depths and windspeeds, and precipitation is often associated with them. High-pressure systems, on the other hand, generally produce poor ventilation conditions because they frequently have smaller mixing depths on their western sides and lower windspeeds. They also tend to move more slowly than lows, so more emissions can enter their circulation patterns. In addition, they are frequently free of clouds, resulting in maximum sunlight and therefore more photochemical ozone production during the day. Also, during the evenings, the clear skies allow surface-based (see below) inversion layers to form, concentrating pollutants in a small volume of air and often creating very high air pollution levels.

Climatologically, certain places in the country are frequently windy (Figure 13-1), and others frequently have large mixing depths (Figure 13-2). These areas will have cleaner-than-average air if they do not contain not too many pollutant sources. Areas frequently affected by high-pressure systems -- causing lower windspeeds and smaller mixing depths -- will have more major air pollution episodes.

Circulation

Two semipermanent high-pressure systems are important to the global circulation pattern and greatly influence U.S. air pollution climatology: the large Pacific high, which is often situated between the Hawaiian Islands and the west coast of North America, and the Bermuda high, located over the western Atlantic Ocean.

The Pacific high often results in extended periods of air stagnation over the western United States from Oregon and California to over the Rockies, and is responsible for many severe ozone episodes in southern California. Air stagnation associated with the westward extension of the Bermuda high occurs most often during the summer months and affects the eastern United States from southern Appalachia northward to New England; it is an important cause of the air stagnation that occurs in the heavily populated eastern part of the country. Within the Bermuda high, pollutants are slowly transported from the industrial areas of the Ohio River Valley into the populated areas of the Northeast. It is also responsible for the general southwest-to-northeast airflow in the summer, carrying pollutants along the metropolitan corridor from Richmond to Boston and exacerbating the ozone problem in the Northeast.

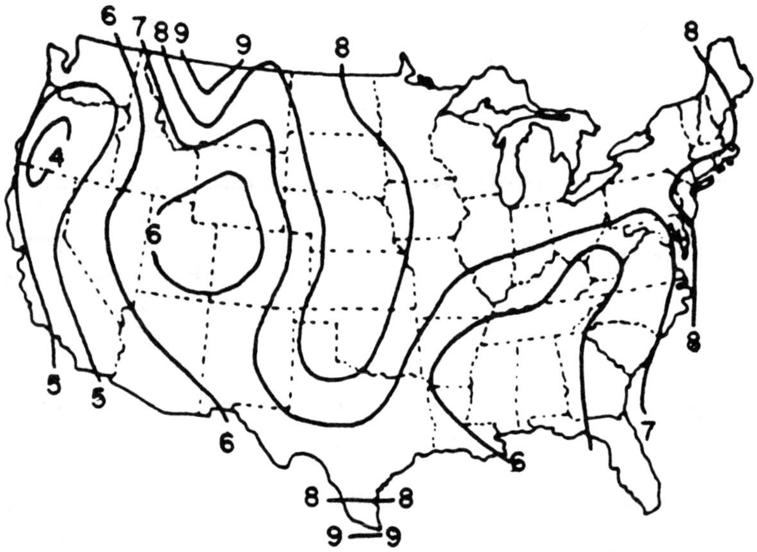

Figure 13-1. Mean Annual Windspeed Averaged Through the Afternoon Mixing Layer (speeds are in meters per second)

Source: Adapted from Holzworth (1972).

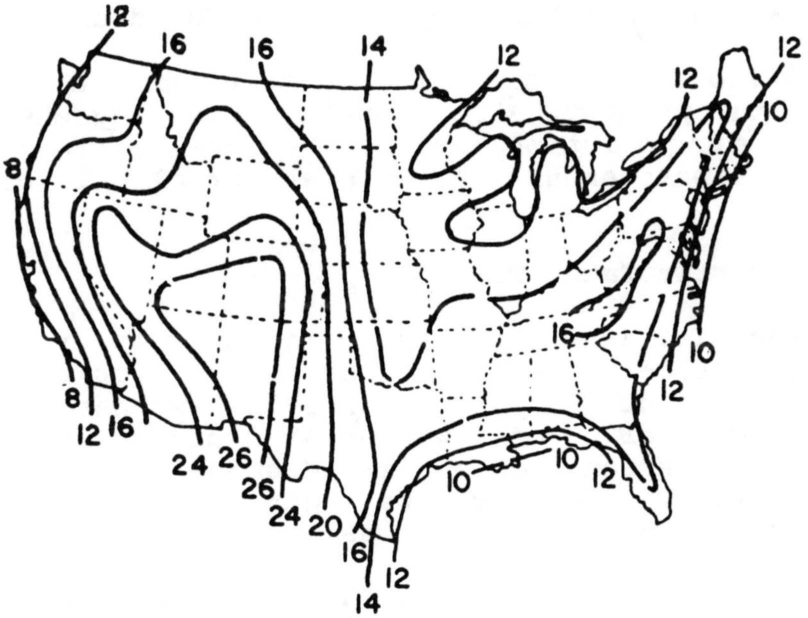

Figure 13-2. Mean Annual Afternoon Mixing Height (in hundreds of meters)

Source: Adapted from Holzworth (1972).

Precipitation

Atmospheric pollutants in both particulate and gaseous forms are incorporated into clouds and precipitation. These pollutants can then be transported to the ground through rainfall (wet deposition). Cloud-formation processes and the consequent type of precipitation, together with the intensity and duration of precipitation, are important in determining wet deposition of pollutants.

PATTERNS AND TRENDS IN AIR QUALITY

To protect the public health and welfare, the U.S. EPA has promulgated National Ambient Air Quality Standards (NAAQS). In 1986, 75 million people lived in counties with measured air quality levels that violated the primary NAAQS for ozone (O_3) (Figure 13-3) as compared with 41.7 million people living in countries exceeding the primary NAAQS for total suspended particulates (TSP), 41.4 million people for carbon monoxide (CO), 7.5 million people for nitrogen dioxide (NO_2), 4.5 million people for lead (Pb), and 0.9 million people for sulfur dioxide (SO_2).

Although millions of people continue to breathe air that is in violation of the primary NAAQS, considerable progress is being made in reducing air pollution levels. Nationally, long-term 10-year (1977-86) improvements can be seen for all six of the above pollutants.

Total Suspended Particulate (TSP)

Annual average TSP levels decreased by 23% between 1977 and 1986, and particulate emissions decreased by 25% for the same period. The more recent TSP data (1982-86) show that concentrations are leveling off, with a 3% decrease in ambient TSP levels and a 4% decrease in estimated emissions during that time. Note that TSP air quality levels generally do not improve in direct proportion to estimated emission reductions, because air quality levels are influenced by factors not included in the emissions estimates, such as natural dust, re-entrained street dust, and construction activity. In Figure 13-4, 1986 annual geometric mean TSP concentrations are plotted for the 89 largest metropolitan statistical areas (MSAs). The highest concentrations generally were found in the industrial Midwest and arid areas of the West.

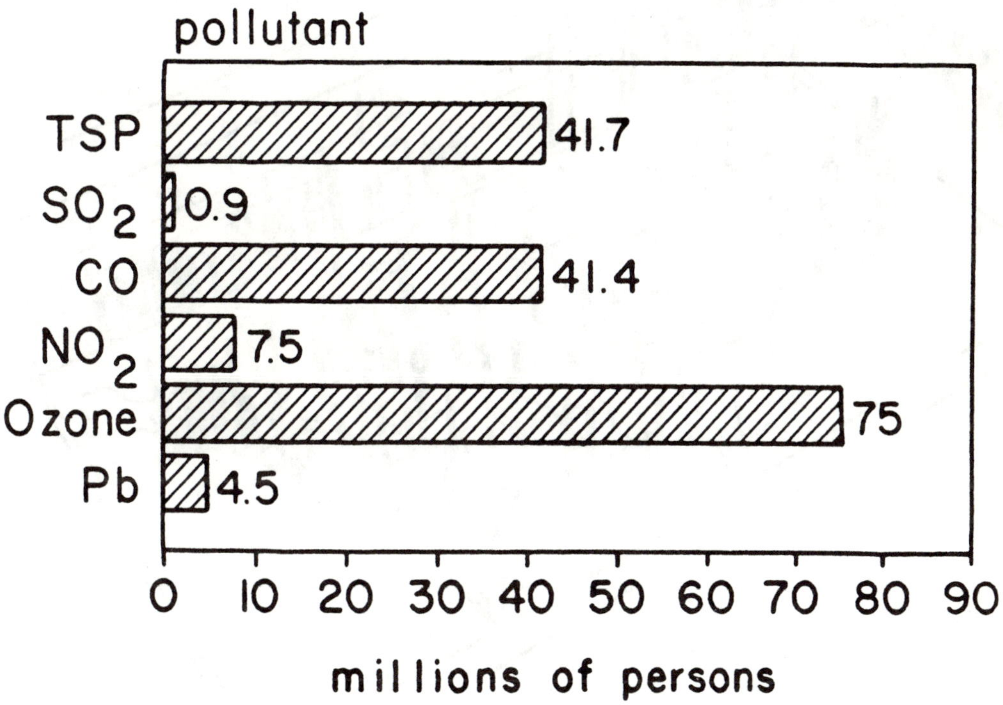

Figure 13-3. Number of Persons Living in Counties With Air Quality Levels Above the Primary National Ambient Air Quality Standards in 1986 (based on 1980 population data)

Source: U.S. EPA (1988).

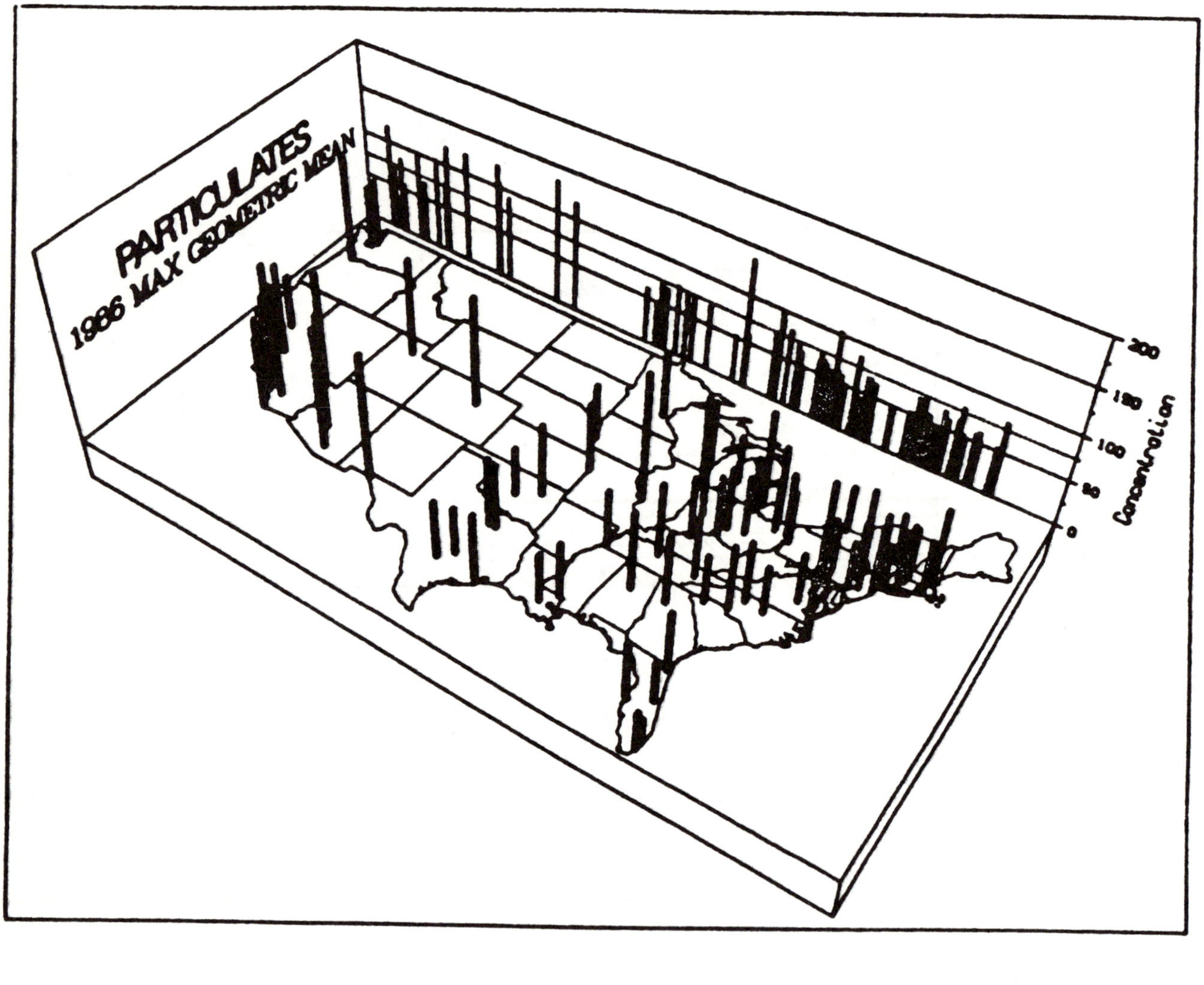

Figure 13-4. United States Map of the Highest Annual Geometric Mean TSP Concentration by Metropolitan Statistical Areas for 1986

Source: U.S. EPA (1988).

In the future, air quality may decrease as the benefits of current pollution control measures are affected by increases in population and economic growth.

Sulfur Dioxide (SO_2)

Annual average SO_2 levels decreased 37% from 1977 to 1986, improving at a rate of approximately 4% per year. An even greater improvement was observed in the estimated number of exceedances of the 24-hour standard for SO_2 concentration, which decreased by 98%. These decreases correspond to a 21% drop in sulfur dioxide emissions during this 10-year period. However, most of the exceedances and the improvements occurred at source-oriented sites, particularly a few smelter sites.

In Figure 13-5, the 1986 annual arithmetic mean SO_2 concentration is plotted for the 89 largest MSAs. The higher concentrations were found in the heavily populated Midwest and Northeast. All urban areas had ambient air quality concentrations lower than the current annual primary standard of 0.03 ppm.

Ozone (O_3)

Nationally, between 1977 and 1986, the annual average of the daily maximum ozone values decreased by 20%; between 1979 and 1986, O_3 levels decreased by 13%. Emissions of volatile organic compounds (VOCs), ozone precursors, decreased by 19% from 1977 to 1986, and by 20% from 1979 to 1986. The estimated number of exceedances of the ozone standard decreased by 38% between 1979 and 1986. In Figure 13-6, the 1986 annual average of the daily maximum O_3 concentrations is plotted for the 89 largest MSAs. Many of these areas did not meet the 0.12 ppm primary NAAQS. The highest concentrations were in southern California, but high levels also persisted in the Texas gulf coast, the northeast corridor, and other heavily populated regions.

<u>Acid Deposition</u>

Widespread concern exists concerning the effects of acid deposition on the environment. With the present monitoring network density in eastern North America, it is now possible to quantify regional patterns of concentration and deposition of sulfate, nitrate, and hydrogen ions, primary constituents of acid deposition. In Figures 13-7 through 13-9, isopleth maps show the

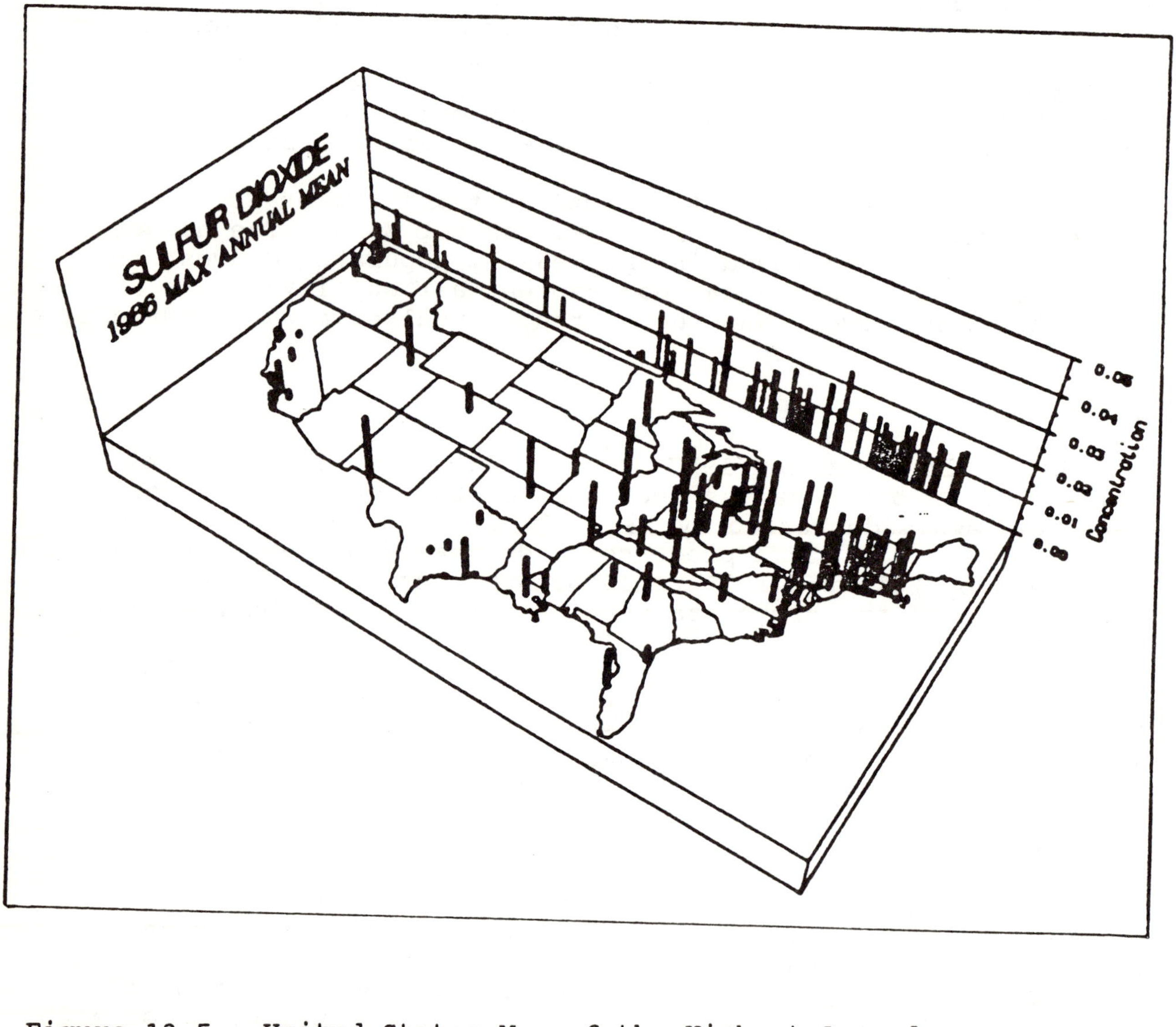

Figure 13-5. United States Map of the Highest Annual Arithmetic Mean SO_2 Concentration by Metropolitan Statistical Areas for 1986

Source: U.S. EPA (1988).

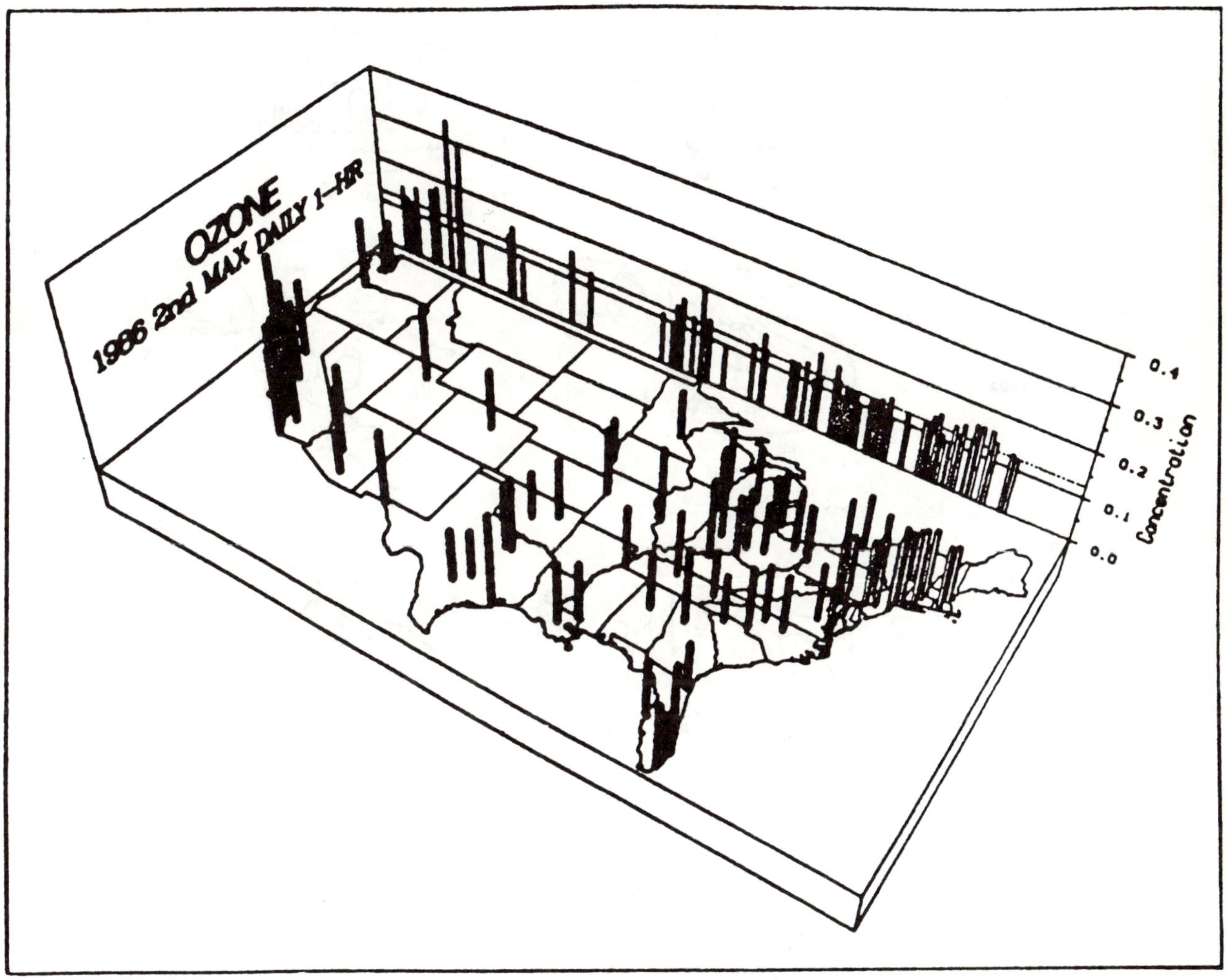

Figure 13-6. United States Map of the Highest Second Daily Maximum 1-hour Average O_3 Concentration by Metropolitan Statistical Areas for 1986

Source: U.S. EPA (1988).

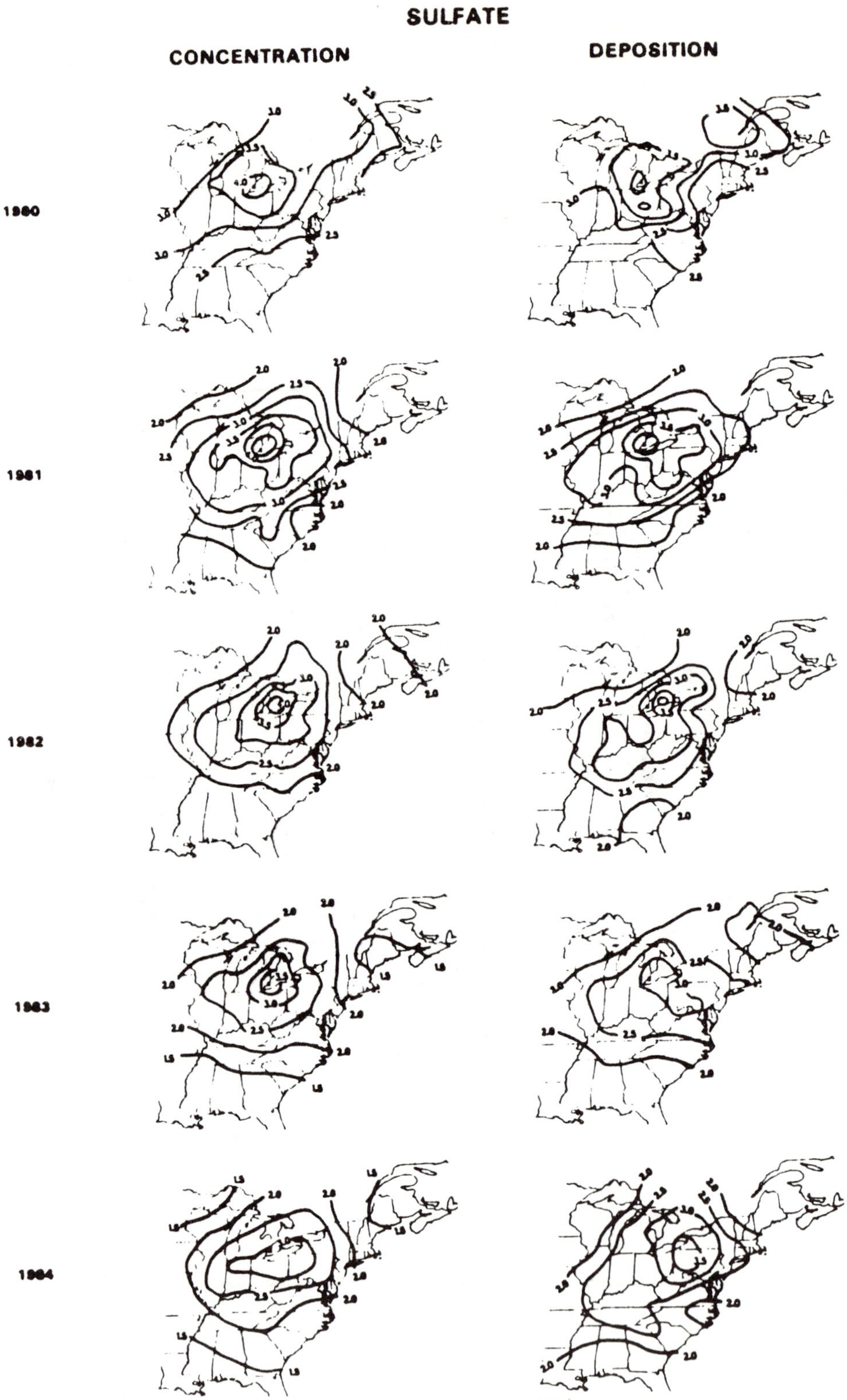

Figure 13-7. Isopleth Maps of Average Annual Concentrations (mg l^{-1}) and Total Annual Deposition (g m^{-2}) of Sulfate in 1980-84

Source: Seilkop and Finkelstein (1987).

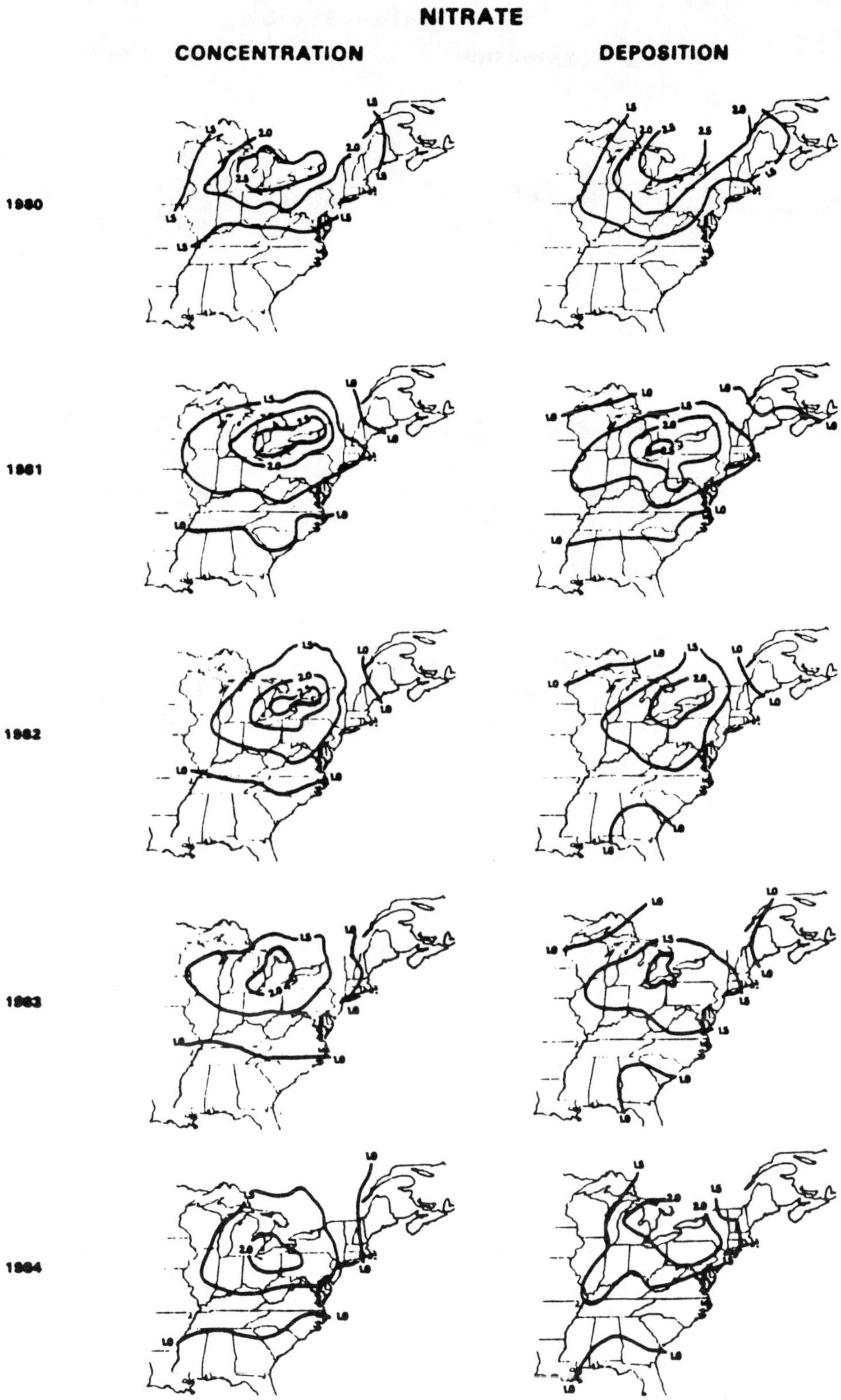

Figure 13-8. Isopleth Maps of Average Annual Concentration (mg l^{-1}) and Total Annual Deposition (g m^{-2}) of Nitrate in 1980-84

Source: Seilkop and Finkelstein (1987).

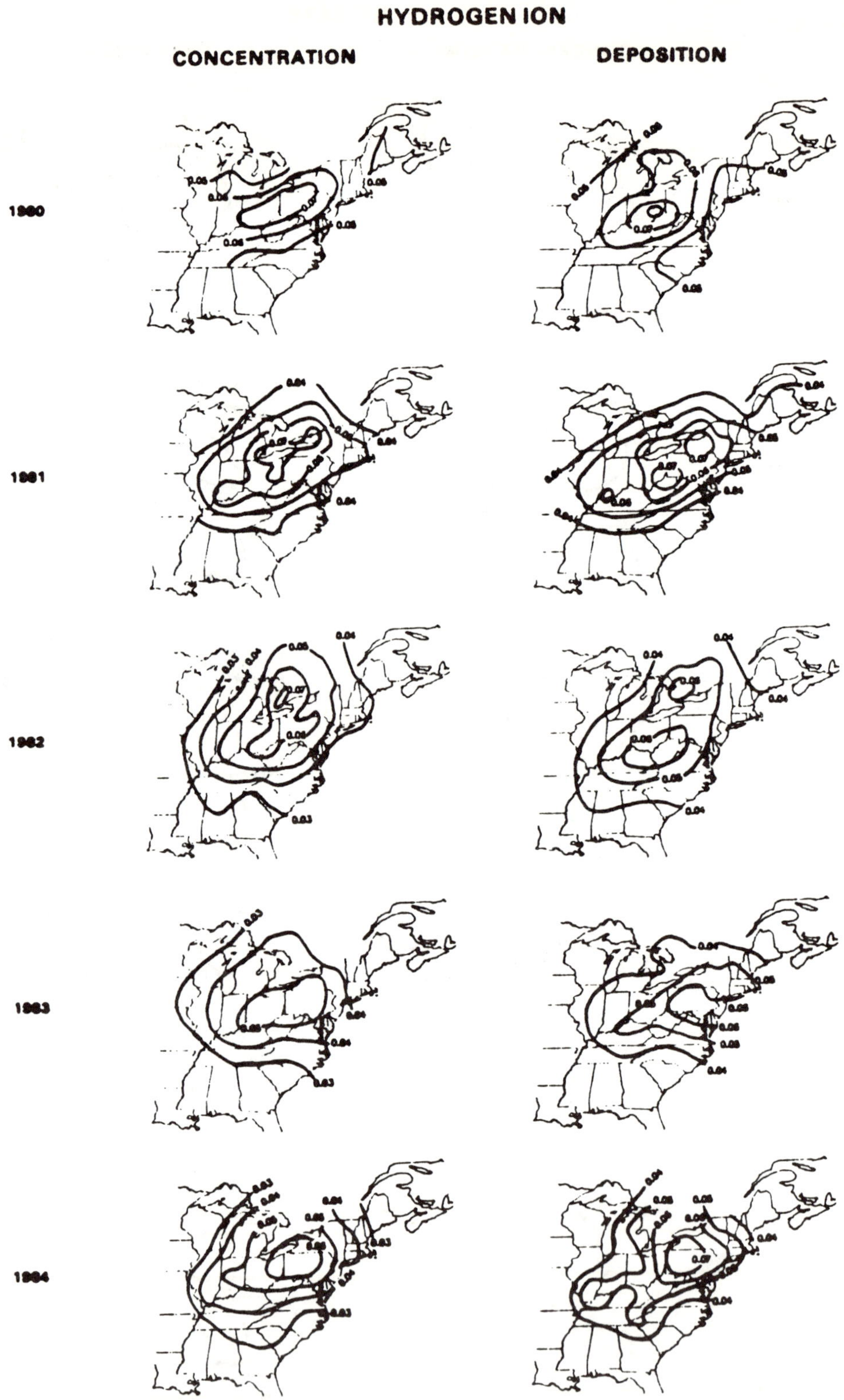

Figure 13-9. Isopleth Maps of Average Annual Concentration (mg l^{-1}) and Total Annual Deposition (g m^{-2}) of Hydrogen Ion in 1980-84

Source: Seilkop and Finkelstein (1987).

geographic pattern of acid deposition, as reflected by the concentration and deposition of these three species (Seilkop and Finkelstein, 1987).

For the relatively short period of 1980 through 1984, evidence indicates a 15 to 20% decrease in the total deposition and average concentration of sulfate, nitrate, and hydrogen ions in precipitation falling over eastern North America. These decreases occurred in the regions of maximum deposition and were most dramatic between 1981 and 1983. Evidence indicates increased levels of these species in 1984, although these concentrations were still less than those observed in 1980 and 1981. The observed decreases correspond with reported reductions in the U.S. emissions of sulfur oxides (SOx) and nitrogen oxides (NOx), and sulfate and nitrate precursors. However, the emission figures are subject to estimation error and should be used cautiously (Seilkop and Finkelstein, 1987).

STUDIES OF CLIMATE CHANGE AND AIR QUALITY

Some of the climate factors that could affect air quality are listed in Table 13-1. To explain these relationships for this report, EPA sponsored two projects to identify the potential impacts of climate change on air quality:

1. Climate Change and Its Interactions with Air Chemistry: Perspectives and Research Needs - Penner, Connell, Wuebbles, and Covey, Lawrence Livermore National Laboratory.

2. Examination of the Sensitivity of a Regional Oxidant Model to Climate Variations - Morris, Gery, Liu, Moore, Daly, and Greenfield - Systems Applications, Inc.

Climate Change and Its Interactions With Air Chemistry

Penner et al. conducted a literature review of studies on the relationship of climate and air quality. They also organized a workshop on the issue.

Effect of Climate Change on Ozone Formation

Changes in ventilation, circulation, precipitation, and other aspects of climate affect the concentrations of the ozone precursors VOCs and NOx. Climate changes can also increase or decrease the rates at which these precursors react to form ozone. The effects of change in global temperature and in stratospheric ozone concentration on tropospheric ozone precursor

Table 13-1. Climate Change Factors Important for Regional Air Quality

A CHANGE IN...

1. the average maximum or minimum temperature and/or changes in their spatial distribution leading to a change in reaction rates and the solubility of gases in cloud water.

2. stratospheric O_3 leading to a change in reaction rates.

3. the frequency and pattern of cloud cover leading to a change in reaction rates and rates of conversion of SO_2 to acid deposition.

4. the frequency and intensity of stagnation episodes or a change in the mixing layer leading to more or less mixing of polluted air with background air.

5. in background boundary layer concentrations of water vapor, hydrocarbons, NOx, and O_3 leading to more or less dilution of polluted air in the boundary layer and altering the chemical transformation rates.

6. the vegetative and soil emissions of hydrocarbons and NOx that are sensitive to temperature and light levels, leading to changes in their concentrations.

7. deposition rates to vegetative surfaces whose absorption of pollutants is a function of temperature, light intensity, and other factors, leading to changes in concentrations.

8. energy usage, leading to a change in energy-related emissions.

9. aerosol formation, leading to changes in reaction rates and the planetary albedo (reflectivity).

10. circulation and precipitation patterns leading to a change in the abundance of pollutants deposited locally versus exported off the continent.

[a]Adapted from Penner et al. (1988).

concentrations, reaction rates, and tropospheric ozone concentrations are discussed below.

Temperature Change

Studies of the Effects of Temperature on Ozone. Smog chamber and modeling studies have shown that ozone levels increase as temperature increases. Kamens et al. (1982) have shown in an outdoor smog chamber study that the maximum ozone concentration increases as the daily maximum temperature increases (holding light intensity constant). Their data show that there is no critical "cut-off" temperature that eliminates photochemical ozone production. Instead, a general gradient is observed as a function of temperature.

Temperature-dependent modeling studies were conducted by Gery et al. (1987). Results for New York in June 1980 are shown in Table 13-2. In general, ozone concentration increased with increasing temperature. The concentration of hydrogen peroxide (H_2O_2), a strong oxidant that converts SO_2 to sulfuric acid, was also observed to increase with higher temperatures. This is compatible with the increase in ozone because the entire photochemical reaction process is accelerated when temperature rises. As a result, cities currently violating the ozone NAAQS will be in violation to a greater degree in the future, and cities that are complying with the NAAQS now could be forced out of compliance just by a temperature increase. Figure 13-10 shows the predicted increase in ozone for two temperatures increases in Los Angeles, New York, Philadelphia, and Washington.

Modeling studies by Penner et al. have shown that the effect temperature has on ozone formation also depends on the ratio of volatile organic compounds to nitrogen oxides, both of which are ozone precursors. Figure 13-11 shows that ozone levels will generally go up, except in areas where the ratio of VOCs to NOx is low.

Temperature change has a direct effect on ozone concentrations because it increases the rates of ozone-forming reactions. However, temperature rise can also affect ozone formation by altering four other aspects of climate or the atmosphere: cloud cover, frequency and intensity of stagnation periods, mixing layer thickness, and reactant concentrations.

Effect of Changes in Cloud Cover. The reduction in light intensity caused by increased cloud cover can reduce ozone production. Penner et al. (1988) calculate that a reduction in

Table 13-2. Maximum Hourly Concentrations and Percentage Changes for Ozone, H_2O_2, and PAN for the Future Sensitivity Tests Using an EKMA Model for the Simulation of June 24, 1980, New York

	Concentration			Percent change (from base)		
Ozone						
Change in Temp (C)	Concentration (ppm) 0	+2	+5	0	+2	+5
Stratospheric Ozone[a]						
Base	0.125	0.130	0.138	--	4.0	10.4
-16.6%	0.150	0.157	0.167	20.0	25.6	33.6
-33.3%	0.165	0.170	0.178	32.0	36.0	42.4
Hydrogen Peroxide (H_2O_2)						
Change in Temp (C)	Concentration (ppb)					
Stratospheric Ozone[a]						
Base	0.05	0.06	0.08	--	20.0	60.0
-16.6%	0.43	0.58	0.84	760.0	1060.0	1580.0
-33.3%	3.08	3.31	3.60	6060.0	6520.0	7100.0
Peroxyacetyl Nitrate (PAN)						
Change in Temp (C)	Concentration (ppb)					
Stratospheric Ozone[a]						
Base	3.98	3.50	2.79	---	-12.1	-29.9
16.6%	5.85	5.26	4.34	47.0	32.2	9.1
-33.3%	7.59	6.73	5.49	90.7	69.1	37.9

[a]Base refers to the present stratospheric ozone column. The -16.6 and -33.3% refer to a depletion of the base value. Ultraviolet light will increase with the depletion.

Source: Gery et al. (1987).

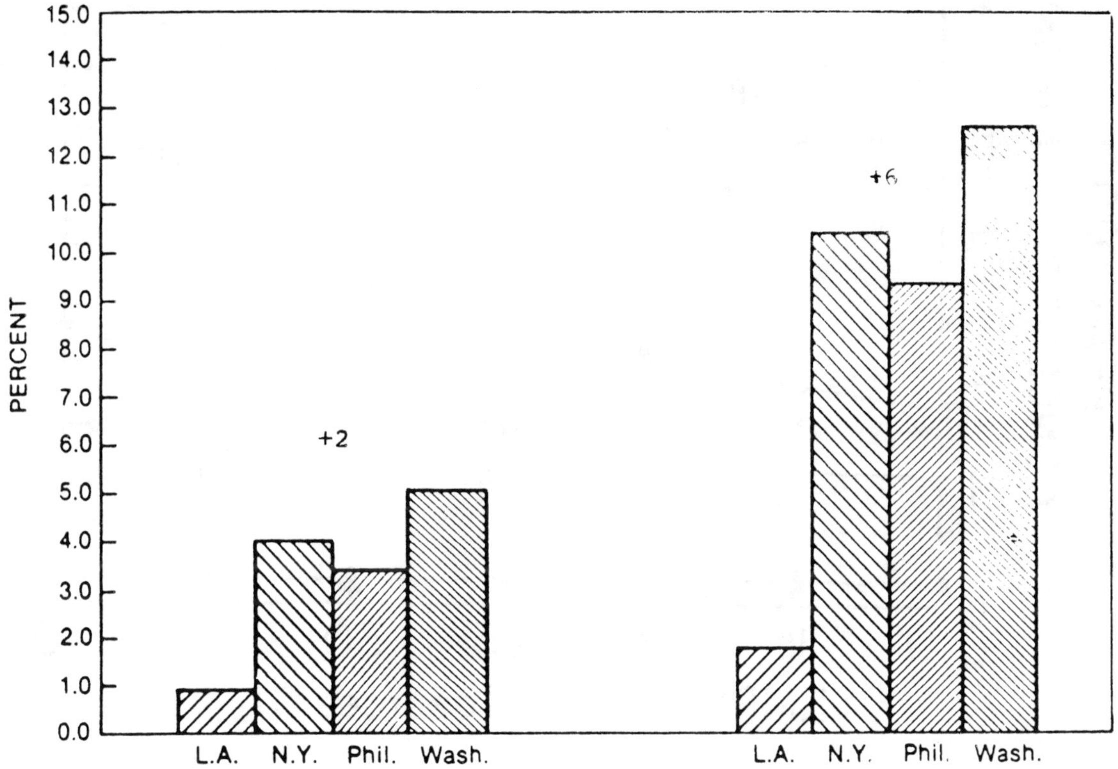

Figure 13-10. Percent increase in predicted ozone over future base case (0.12 ppm) for tow temperature increases in four cities.

Source: Gery (1987).

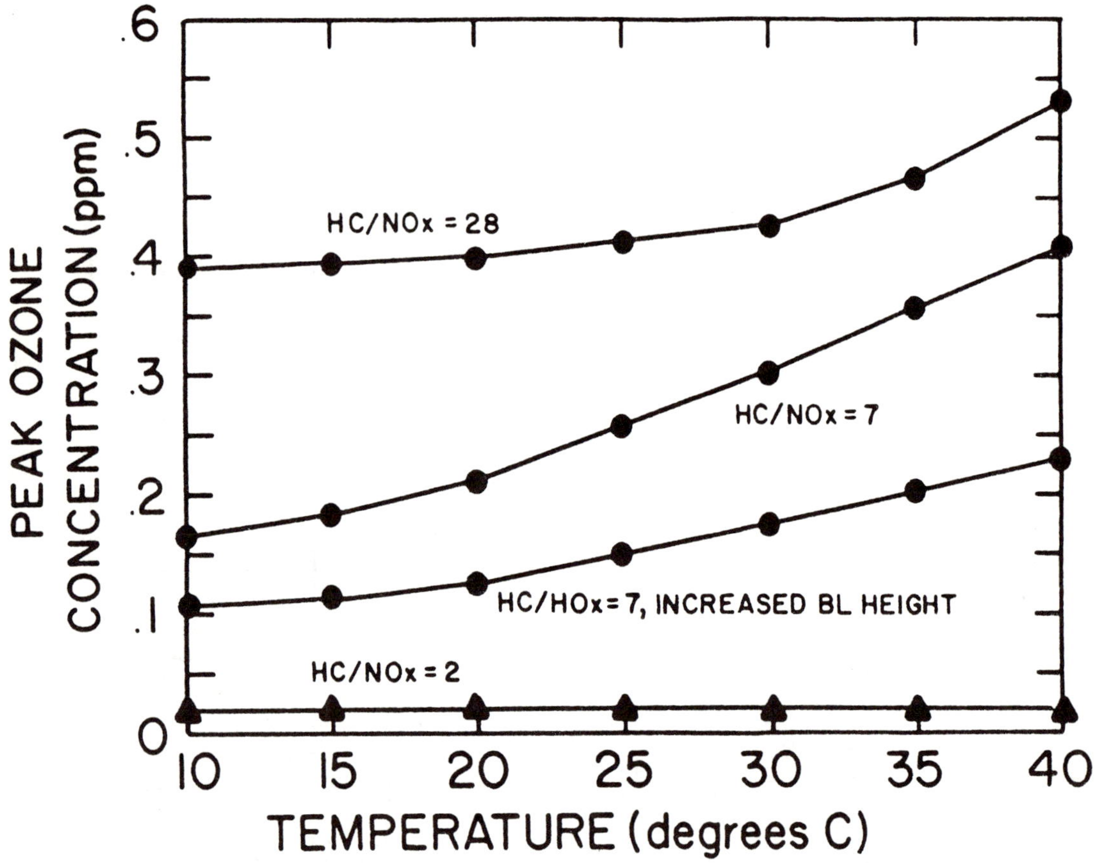

Figure 13-11. The effect of temperature on the peak O_3 concentrations predicted in a box model calculation of urban O_3 formation. Calculations are shown for three hydrocarbon to NOx ratios. The effect of increasing the boundary layer depth for the case with a hydrocarbon to NOx ratio of 7 is also shown

Source: Penner et al. (1988).

light intensity of 50% throughout the day will reduce the ozone formation. However, the magnitude of ozone reduction depends on the time of day when the cloud cover occurs. If clouds occur in the afternoon or evening, little effect is observed in the ozone production, but if clouds occur during the morning hours, photochemical reactions are slowed, and less ozone is produced. Although a global temperature change would affect cloud cover, the type and direction of the change are unknown.

The Penner et al. study assumes that cloud cover causes an equal decrease in all wavelengths of solar radiation. However, clouds are not expected to cause an equal decrease at all wavelengths. Solar radiation is needed to form ozone. Since Penner et al. may have underestimated the intensity of some wavelengths of light, they may have overestimated the decrease in ozone production.

Effect of Water Vapor. Water vapor is involved in the formation of free radicals (reactive compounds) and hydrogen peroxide, which are necessary for the formation of ozone. Global increases in temperature are expected to raise tropospheric water vapor levels.

If sources of water vapor are not perturbed by vegetative changes, and if global circulation patterns do not significantly affect precipation events (an unlikely assumption), then global water vapor levels are expected to increase with increasing temperature. A temperature increase of 2° could raise the water vapor concentration by 10-30% (Penner et al., 1988). This change should affect both oxidant formation and sulfur dioxide oxidation (acid deposition).

Smog chamber studies have shown that at high pollutant levels, increases in water vapor can significantly accelerate both the reaction rates of VOCs and the rate of oxidant formation (Altshuller and Bufalini, 1971). In the boundary layer of the troposphere, Walcek (1985) has shown with the use of a regional acid deposition model (RADM) that the ozone, hydrogen peroxide, and sulfate production rates all increase with increasing water vapor.

Effect of Changes in Frequency and Intensity of Stagnation Periods. As noted previously, high-pressure systems significantly enhance ozone formation potential. During a high-pressure episode, pollutants are exposed to high temperatures and prolonged irradiation (Research Triangle Institute, 1975), resulting in high levels of ozone. If the intensity and frequency of high-pressure episodes increase with global warming, then ozone levels can be expected to be even higher.

Effect of Changes in Mixing Layer Thickness. As shown in Figure 13-10, increases the mixing layer height decreases ozone formation, presumably because there are less ozone precursors per volume of atmosphere. An increase of global temperature would probably lead to a increase in average mixing depths due to greater convection, which raises the mixing depth and increases mixing.

Effect of Changes in Reactant Concentrations. The concentrations of ozone precursor pollutants (VOCs, NOx) play a large part in determining the amount of ozone produced. With increasing temperature, both natural and anthropogenic emissions of these precursors are expected to increase, so ozone production should increase.

Lamb et al. (1983) have shown that natural hydrocarbon (VOC) emissions from deciduous forests would increase by about a factor of three. Emissions of NOx from powerplants would grow due to a greater demand for electricity during the summer months. Evaporative emissions of VOCs from vehicles and refueling would also be expected to rise with warmer temperatures. However, exact predictions of the effects of all these factors on ozone formation are difficult to make because the relationship between precursor emissions and ozone is extremely complex and not fully understood and because increases in emissions are difficult to quantify.

An example of this complex relationship between ozone and its precursors is shown in Figure 13-12. At high VOC levels and low NOx, adding or reducing VOCs has very little effect on ozone formation. Likewise, when NOx concentrations are high and VOC concentrations are low, increasing NOx reduces ozone formation while lowering NOx increases ozone formation. Thus, VOCs and NOx must be examined together when considering any ozone reduction strategy based on controlling ozone-forming precursors.

Stratospheric Ozone Change

Changes in stratospheric ozone concentration can also affect tropospheric ozone formation because stratospheric ozone regulates the amount of ultraviolet (UV) radiation available for producing ozone in the troposphere. Stratospheric ozone absorbs UV light from the sun and decreases the UV energy striking the Earth's surface. When stratospheric ozone is depleted by the chlorofluorocarbons (CFCs) generated by man, more UV reaches the

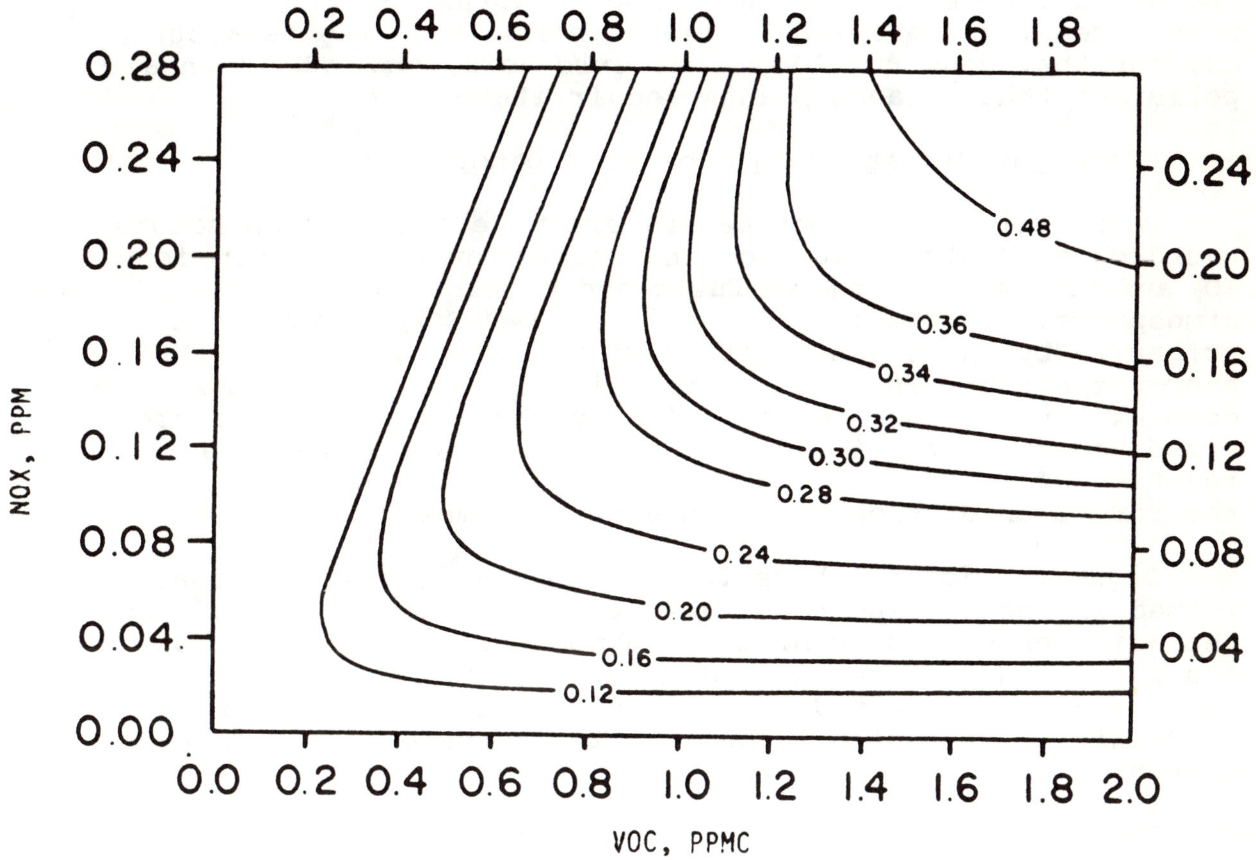

Figure 13-12. Ozone Isopleths as a Function of NOx and Volatile Organic Compounds (VOCs)

Source: Dodge (1977).

Earth's surface, which increases the photolysis rates of compounds that absorb solar radiation (nitrogen dioxide, formaldehyde, acetaldehyde, ozone, and hydrogen peroxide).* Faster photolysis produces more free radicals (high-energy species) that increase the amount of smog. Thus, less stratospheric ozone will lead to enhanced ozone formation in the troposphere.

Modeling results for New York from Gery et al. (1987) show that tropospheric ozone increased when stratospheric ozone decreased (Table 13-2). They also show that H_2O_2 and peroxyacetyl nitrate (PAN) yields increased. H_2O_2 is a strong oxidant that converts SO_2 to sulfuric acid, and PAN is an air pollutant that damages plants and irritates eyes.

<u>Effect of Climate Change on Acid Deposition</u>

Rainwater and surface waters are more acidic than natural background levels because of industrial and mobile emissions of SO_2 and NOx, which form sulfuric and nitric acids in the atmosphere. In the air, sulfuric acid (H_2SO_4) is produced primarily by the reaction of SO_2 with hydroxyl radicals (high-energy species); in clouds, the oxidation of SO_2 to H_2SO_4 is more complex, involving reactions with hydrogen peroxide and other dissolved oxidants. Nitric acid (HNO_3) is produced in air by the reaction of hydroxyl radicals with NOx. The exact pathways for the formation of these acids are quite complex.

Organic acids, such as formic and acetic acids, are also formed in the atmosphere. However, their relative importance to the acid deposition problem is unknown at present. Because they are weak acids (compared to H_2SO_4 and HNO_3), their contribution to the problem is expected to be much less than that of the inorganic acids (Galloway et al., 1982; Keene et al., 1983, 1984; Norton, 1985).

The acids produced in the atmosphere can be "dry deposited" to the earth's surface as gases or aerosols, or they can be "wet deposited" as acid rain. Changes in total acid levels depend on changes both in atmospheric chemistry and changes in precipitation. Wet deposition is affected most by the amount, duration, and location of precipitation. Since the direction of regional precipitation changes is not known, it is not known whether acid rain will increase or decrease in the future.

---

*Photolysis is the breakdown of chemicals as a result of the absorption of solar radiation.

However, many of the same factors that affect ozone formation will also affect the total deposition of acids.

Temperature Change

Higher temperatures accelerate the oxidation rates of SO_2 and NOx to sulfuric and nitric acids. Gery et al. (1987) have shown that a temperature rise would also speed the formation of H_2O_2, further increasing the conversion of SO_2 to sulfuric acid (Table 13-2). Considering only the chemistry occurring with a 10°C temperature rise, SO_2 oxidation increased 2.5 times. It is likely that oxidation would also increase with a smaller temperature. The limiting factor in the oxidation of SO_2 appears to be the availability of H_2O_2. The model also suggested that a temperature increase would cause more sulfuric acid to form near the sources where SO_2 is emitted.

Effect of Global Circulation Pattern Changes. Changes in global circulation patterns will greatly affect local acid deposition, because they will alter ventilation and precipitation patterns. Galloway et al. (1984) have calculated that over 30% of the sulfur emissions from the eastern United States are transported to the north and farther east. Changes in circulation patterns would affect this transport, although the direction or magnitude of the effect is unknown.

Effects of Changes in Emissions. If electricity demand rises with rising temperatures (see Chapter 16, Electricity Demand) and if more fossil fuels are burned, SO_2 and NOx emissions would increase. An approximately 10% growth in use of electricity in the summer could increase SO_2 emissions during the summer by approximately 30%. This, in turn, would increase acid deposition.

Effects of Reduced Stratospheric Ozone. A decrease in stratospheric ozone due to CFCs will increase acid deposition because more UV radiation will be available to drive the chemical reactions. As discussed above, a modeling study by Gery et al. (1987) showed an increase in the yield of H_2O_2 when stratospheric ozone was reduced by 33%. Because H_2O_2 is a strong oxidant, SO_2 will probably also be oxidized more quickly into sulfate aerosols and acid rain, but this depends the availability of liquid water (e.g., clouds, rain).

Reduced Visibility. The growth in natural organic emissions sulfates resulting from warmer temperatures should reduce visibility. This assumes that the frequency of rain events, wind velocity, and dry deposition rates remain the same. If rain events increase, washout/rainout should increase and visibility would be better than predicted.

MODELING STUDY OF CLIMATE AND AIR QUALITY

Study Design

Morris et al. (Volume X) applied a regional transport model RTM-III to an area covering central California and a region covering the midwestern and the southeastern United States. The model was run for the present-day conditions and for a future climate with higher temperatures. For California, Morris et al. used input data from August 5-10, 1981; for the Midwest and the Southeast, they used input data from July 14-21, 1981.

The RTM-III is a three-dimensional model that represents point sources embedded in a grid framework. The model has three prognostic vertical layers and a diagnostic surface layer. This means that the surface layer is represented by actual observations. The other three layers are predicted by using the surface layer data. The photochemical reactions are based on the latest parameterized chemical mechanism.

Study Limitations

The estimates for ozone are only coarse approximations. Morris et al. used the National Acid Precipitation Assessment Program (NAPAP) emissions data of 1980, but appear to underestimate actual ratios of VOCs to NOx as measured in urban areas. Ching et al. (1986) state that for most cities, the NAPAP data underestimate VOC emission values by a factor of three or more.

The model simplified some reactions of the hydrocarbons (VOCs) because the chemistry is not well known. The model ignored future increases in emissions that would occur with increased temperatures.

This study did not estimate climate induced alterations in most meteorological variables, except temperature and water vapor, which is an oversimplification. For example, this study assumed that the mixing heights remain unchanged for the temperature increase scenario; in reality, mixing heights could increase with rising temperature. Holding the mixing heights constant probably overemphasized the importance of temperature in oxidant production, because an increased mixing layer depth might have had a dilution effect.

Study Results

Central California Study

Table 13-3 summarizes the results from the base case scenario and a climate sensitivity scenario that used a 4°C temperature increase and an attendant increase in water vapor concentration. All of the days studied show a larger area exposed to high levels of ozone. An increase in temperature may lengthen the duration of high ozone levels, although the maximum levels may be the same. Figure 13-13 illustrates the August 6 base case and climate sensitivity case. The temperature change increased the August 6 maximum ozone concentration from 15 parts per hundred million (pphm) to 18 pphm, a 20% increase in ozone. The area in which the NAAQS was exceeded almost doubled from 3,700 to 6,600 km^2.

The temperature increases in the two main cities in the San Joaquin Valley (Fresno and Bakersfield) resulted in an approximately 0.5 pphm increase (approximately 8%) in maximum daily ozone concentration. In regions farther away from the emissions, such as the Sierra Nevada Mountains, little change in ozone levels was observed with the increased temperature.

Midwest and Southeast Study

The results from applying RTM-III to the midwestern and southeastern areas are shown in Table 13-4. On one particular day (July 16) raising the temperature caused maximum ozone to increase from 12.5 pphm to 13.0 pphm (Figure 13-14). Although this is only a slight increase (0.5 pphm), the predicted area of exceedance of the ozone NAAQS increased by almost a factor of three, from 9,800 to 27,000 km^2. The differences occurred mainly in the upper Midwest.

Population Exposure

As discussed above, both study domains show a significant area that is exposed to higher levels of ozone when the temperature is increased. Data taken from the 1980 census from central California and the midwestern and southeastern areas were used to determine the number of people exposed to ozone for the base case and temperature rise scenarios. Table 13-5 presents the number of people-hours of exposure to ozone concentrations exceeding 8, 12, and 16 pphm. These estimates of human exposure were generated by multiplying the number of people in the grid cells by the total number of hours that the predicted hourly

Table 13-3. Maximum Daily Ozone Concentrations Predicted by the RTM-III for Each Day of the Central California Modeling Episodes for the Base Case and the Case of Climate Sensitivity to Increased Temperature

	Maximum Daily Ozone Concentration (pphm)		
Day of Episode	Base Case	Increased Temperature	Percent Increase
August 5, 1981	11.8	12.1	3
August 6, 1981	15.0	18.0	20
August 7, 1981	11.7	13.1	12
August 8, 1981	13.5	13.7	2
August 9, 1981	10.5	11.2	7
August 10, 1981	9.1	9.18	8

Source: Morris et al. (1988).

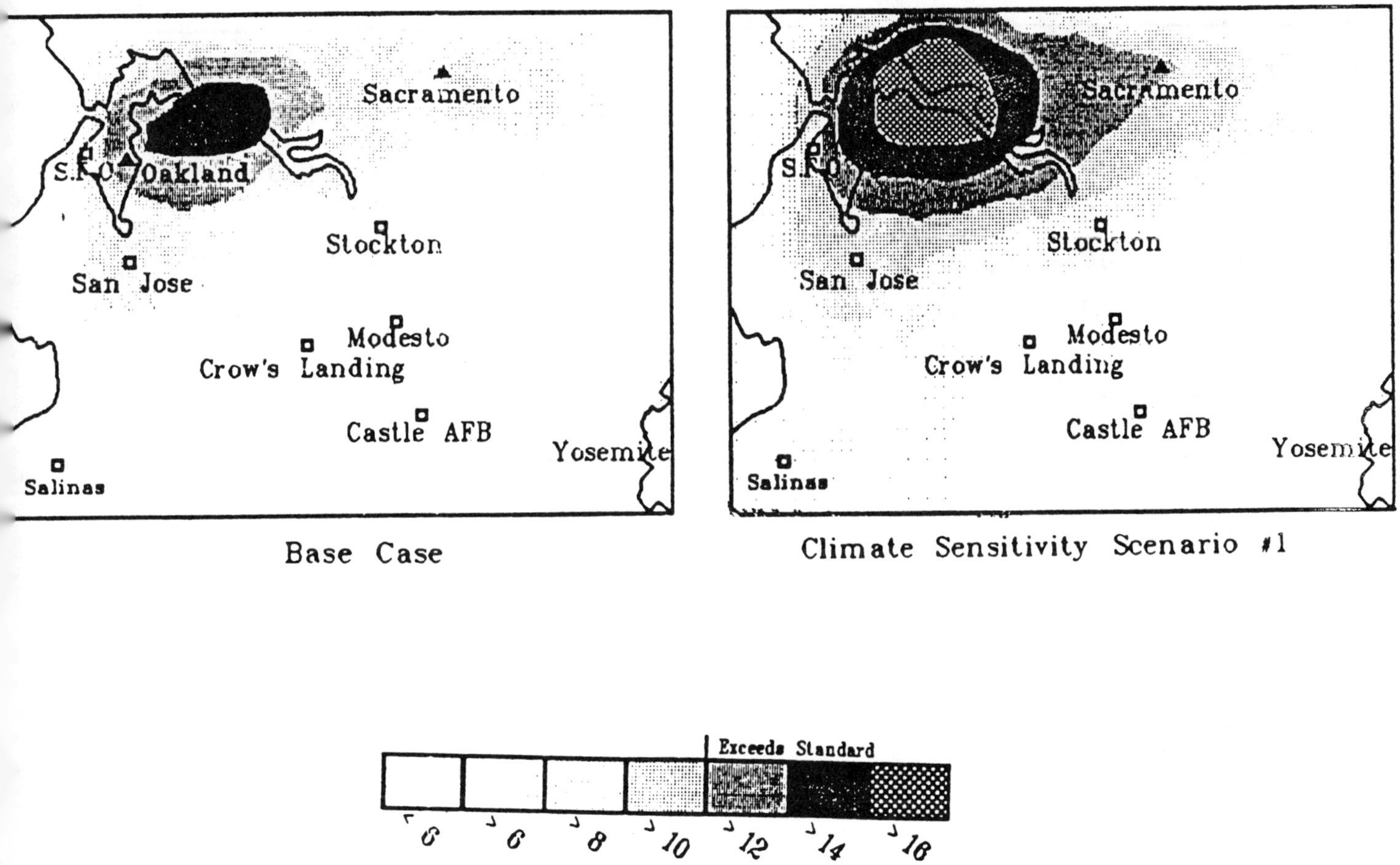

Figure 13-13. Comparison of Predicted Maximum Daily Ozone Concentrations (pphm) for the Base Case and Climate Sensitivity Scenario No. 1 (temperature and water increase) for August 6, 1981

Source: Morris et al. (1988).

Table 13-4. Maximum Daily Ozone Concentrations Predicted by the RTM-III for Each Day of the Midwestern/Southeastern Episode for the Base Case and the Case of Increased Temperature (Morris et al. 1988).

Date	Maximum Daily Ozone Concentrations (pphm)		
	Base Case	Increased Temperature	Percent Increase
14 July 1980	11.3	11.3	0.0
15 July 1980	11.5	11.9	3.5
16 July 1980	12.5	13.0	4.0
17 July 1980	11.7	12.0	2.6
18 July 1980	11.2	12.1	8.0
19 July 1980	13.8	14.8	7.2
20 July 1980	11.1	11.2	0.9
21 July 1980	12.6	12.3	-2.4

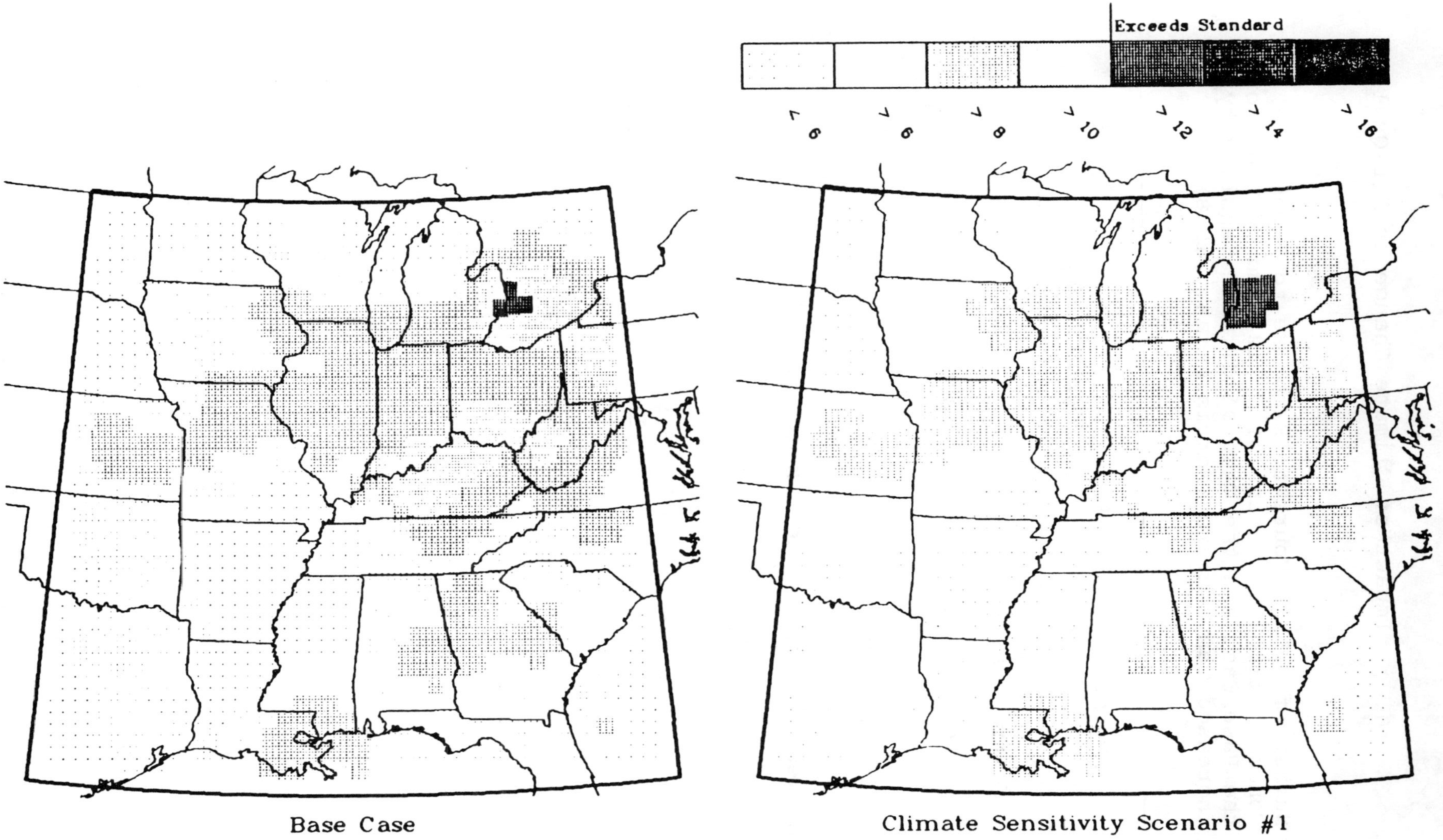

Figure 13-13. Comparison of Predicted Maximum Daily Ozone Concentrations (pphm) for the Base Case and Climate Sensitivity Scenario No. 1 (temperature and water increase) for July 16, 1980

Source: Morris et al. (1988).

Table 13-5. Number of People-Hours of Exposure to Ozone Concentrations in excess of 8, 12, and 16 pphm for the Base Case and the Case of Climate Sensitivity to Increased Temperature (Morris et al., 1988).

Scenario	Exposure to $O_3 \geq 8$ pphm	Exposure to $O_3 \geq 12$ pphm	Exposure to $O_3 \geq 16$ pphm
	Central California Modeling Episode		
Base Case	70,509,216	660,876	0
Increased Temperature	102,012,064	2,052,143	92,220
	Midwestern/Southeastern Modeling Episode		
Base Case	1,722,590,208	29,805,348	0
Increased Temperature	1,956,205,568	47,528,944	0

ozone concentration in those grid cells exceeded the 8, 12, or 16 pphm levels. Actual exposure levels may be less because indoor levels are generally lower than ambient air levels.

ECONOMIC, ENVIRONMENTAL, AND ECOLOGICAL IMPLICATIONS

Ozone

A national standard for ambient levels of ozone was established with the original Clean Air Act in 1972 along with standards for five other pollutants. While headway has been made in meeting all these national air quality standards, progress in meeting the ozone standard has been particularly slow and frustrating for concerned lawmakers and environmental officials at all levels of government. At the end of 1987, the date anticipated in the Act for final attainment of the ozone standard more than 60 areas had not met the standard. In recent years, the number of nonattainment areas has fluctuated with meteorology often overwhelming the progress being made through reduced emissions. Thus "bad" weather (summertime conditions favorable to ozone formation) in 1983 led to an increase in the number of nonattainment areas while "good" conditions in 1986 led to a decline in the number of areas.

An increase in ozone levels due to climate change is important for several reasons:

- Ozone levels in many areas are just below the current standard. Any increase in ozone formation is likely to push levels above the standard.
- Many inexpensive controls for ozone are already in place in nonattainment areas. Increases in ozone levels will require relatively expensive measures to sufficiently reduce ozone precursors to attain the standard.
- The standard itself is defined in terms of the highest levels of ozone experienced in an area, not average levels. (As a yearly average, no area of the country would exceed the standard of 0.12 ppm.) Thus, a factor such as temperature that may have a modest effect on average levels of ozone formation may have a much more significant effect on peak levels.

A rough estimate of each of these factors can illustrate the potential policy problems created by a rising temperature scenario. The data in Figure 13-11 suggest that 5°C degree rise in temperature may lead to an increase in peak ozone concentrations of around 10%. We can then ask what impact a 10% increase in peak ozone levels would have on the number of potential ozone violations. In the 1983-1985 period for example, 68 areas showed measured exceedances of the ozone air quality standards (for technical and legal reasons not all these areas were officially designated nonattainment areas.) A 10% increase in ozone levels in that period doubled the number of nonattainment areas to 136. This would include 41 new MSAs added to the list and 27 nonMSAs. These new nonattainment areas would add most midsize and some small cities in the in the midwest, south and east to the list of nonattainment areas.

The policy implications of this should not be overstated. The temperature is predicted to rise over a number of years, and the full effect will not be felt until well into the next century. Over the next several decades, various national measures to reduce ozone precursors, such as a reduction in the volatility of gasoline, may go into effect. These would provide a cushion to marginal areas and could offset a temperature effect. However, other factors suggest that the temperature problem will persist.

Ozone levels and ozone precursors are closely related to economic expansion and population growth. Consumer solvents (e.g., paints, sprays, and even deodorants) are a major source of ozone precursors. These are very difficult to control and will undoubtedly increase in the future in areas currently attaining the standards. Growth in other sources of ozone precursors would bring many areas relatively close to the limits of the ozone standard. Gradual increases in temperature would make remaining in compliance with the standard more difficult. Although any sudden change in the number of nonattainment areas as a result of a secular trend toward increased temperature is unlikely, a number of small to midsize cities eventually may be forced to develop new control programs.

The implications of warmer temperature rise for existing nonattainment areas can also be estimated. In these areas existing and planned control measures may not be adequate to reach the standard, if additional ozone forms. In the past, EPA has attempted to project the emission reductions and costs associated with the attempts of existing nonattainment areas to reach the ozone standard. Using the same modeling approach, the effects of a temperature increase were analyzed to estimate the additional tons and costs associated with a projected temperature rise. Extrapolations of existing inventories to the year 2000 suggest that higher temperatures could require an additional reduction of 700,000 tons of VOC from an inventory of about 6 million tons. Given that most current nonattainment areas will have already implemented the most inexpensive measures, these additional reductions may cost as much as $5,000 per ton year. Their aggregate cost could be as much as $3.5 billion each year.

These conclusions should be viewed as very preliminary. Nonetheless, they demonstrate that the potential economic consequences for an already expensive program to combat ozone could be significant.

Acid Rain

The global climate change and acid rain issues are likely to collide in the near future for several reasons:

First, emissions from fossil fuel powerplants both influence acid rain and contribute to global warming. In the future, global warming may increase energy demand and associated emissions. Because the growth of northern electricity demand may be less than anticipated, regional shifts in the growth emissions may occur in the future.

Second, global climate change will influence atmospheric reaction rates and the deposition and form of acidic material. It is conceivable that regions of high deposition may shift or that more acid rain may be transported off the North American continent. Strategies that seek to control powerplants in regions near sensitive areas may or may not be as effective, as global climate change occurs.

Third, global climate change will alter the impacts of acid rain on ecological and other systems in as yet unpredictable ways. For example:

- o Changes in the amount of rainfall may dilute the effect of acid rain on many sensitive lakes.
- o Changes in clouds will alter the fertilization of high elevation forests.
- o Changes in humidity and frequency of rain may alter degradation rates for materials.
- o Increased midcontinental dryness would alter the amount of calcium and magnesium in dust, neutralizing impacts on soils.

- o Increased numbers of days without frost would decrease forest damage associated with frost and overfertilization by atmospheric nitrogen.

- o Changes in snowpack and the seasonality of rainfall will change acid levels in streams and alter the timing and magnitude of spring shocks an aquatic species.

Finally, solutions to both problems are inextricably linked because the implementation of acid rain control measures and ? life extension programs of many utilities may not occur until after the year 2010, that is when global warming may be fully recognized. Some solutions, e.g., SO_2 scrubbers and clean coal technologies, may abate acid rain levels, but may do little to improve or may increase global warming. Other solutions, including increased energy efficiency and fuel switching to natural gas or renewable energy sources, may provide positive solutions to both problems.

In summary, an examination of the time horizons of importance to both acid rain and global climate change problems suggests that these two issues can no longer be viewed in isolation. Emissions, atmospheric reaction rates, pollutant transport, and environmental impacts will be altered by climate change. This suggests that a more holistic approach must be taken to air pollution problems and that proposed solutions should be evaluated on the basis of their contributions to solving both problems.

Carbon Monoxide

While global warming as a general rule will exacerbate major pollution problems, it may actually have a modest positive effect on the problem of carbon monoxide (CO) nonattainment. This is because most CO violations occur during the winter months, and that automobile CO emissions (the primary source of CO) are greatest under cold start conditions. With slightly warmer conditions overall, there should be some lessening of the CO problem.

POLICY IMPLICATIONS

The Environmental Protection Agency issues air pollution regulations to improve air quality and to protect public health and welfare. In general, current regulations to reduce oxidant levels will also provide positive benefits toward a goal of limiting the rate of growth in global warming. Other programs aimed at reducing carbon monoxide levels, particularly from mobile sources, or CFCs to protect the stratospheric ozone layer, also positively affect greenhouse gases and the rate of global warming. However, the regulatory activities of the Agency have not been retrospectively reviewed to determine their impacts on global warming. In some cases there may be important benefits; for example, current emission standards for automobiles do not encourage more efficient use of gasoline. A different form of standard, while potentially disruptive to air pollution efforts, might produce positive greenhouse gas benefits via reduced energy consumption. These types of analyses will have to be performed in the future.

Because of the climate change issue, the following are some of the more important policy issues:

- Air pollution control agencies such as EPA should undertake a broad review to determine the impact of global climate change on air pollution policies. In particular, the cost of added controls resulting from climate change should be determined, perhaps as each significant regulation is proposed or reevaluated.

- The impact of EPA regulations, particularly the impact on energy use and greenhouse gases, should be a more important weight in future regulatory decisions. Since EPA regulations often serve as models for other countries, the energy penalty, while sometimes small in the United States, may be important on a global basis.

- Future reports to Congress and major assessments of ecological effects, e.g., the 1990 Acid Deposition Assessment document, should include sensitivity analyses of alternative climates. Risk management decisions of the Agency could then be made with improved knowledge of climate impacts.

Recommendations for Further Research

Some of the key questions which need to be resolved regarding climate change and air quality include the following: How important will climate change be relative to other factors such as population growth to future air pollution problems? Is the impact of climate change likely to be significant enough so as to require totally different air pollution strategies? What mix of control strategies could be most cost effective in reducing acid rain, global warming, tropospheric ozone and other pollution problems? The research elements needed to address these issues include basic research, sensitivity analyses, full scale atmospheric modeling, and cost effectiveness studies. Examples are presented below:

Basic Research: There is an important need to understand how manmade and natural emissions of hydrocarbons and other pollutants might change in the future when temperatures, CO_2, and UV-B radiation increase and other climate parameters vary.

Sensitivity Analyses: Analyses of ozone concentrations are dependent on boundary layer height, clouds, water vapor, wind speed, UV-B radiation and other parameters. Sensitivity tests using single models could improve our understanding of the relative importance of these variables and could provide important information for general circulation modelers.

Full Scale Modeling - Complete understanding of the interactions of climate change and air quality will ultimately require that general circulation models and mesoscale chemistry models be linked in some direct or indirect manner. This will require the development of innovative approaches between the general circulation and air pollution modeling communities.

Cost Effectiveness Studies - There are currently a number of Congressional proposals to improve the Clean Air Act and to reduce global climate change. In order to assume that both air quality and global climate change goals are achieved, analyses of the cost effectiveness of alternating strategies will be necessary.

REFERENCES

Altshuller, A.P., and J.J. Bufalini. 1971. Photochemical aspects of air pollution. Environmental Science and Technology 5:39-64.

Ching, J.K.S., J.H. Novak, K.L. Schere, and F.A. Schiermeier. 1986. Reconciliation of Urban Emissions and Corresponding Ambient Air Concentrations Using Mass Flow Rate Technique. Draft report prepared for EPA/ORD; April.

Dodge, M.C. 1977. Effect of Selected Parameters on Predictions of a Photo-chemical Model. EPA-600/3-77/048; June.

Galloway, J.N., G.E. Likens, W.C. Keene, and M.M. Miller. 1982. The composition of precipitation in remote areas of the world. Journal of Geophysical Research 87:8771-8788.

Galloway, J.N., D.M. Whelpdale, and G.T. Wolff. 1984. The flux of sulfur and nitrogen eastward from North America. Atmospheric Environment 18:2595-2607.

Gery, M.W., R.D. Edmond, and G.Z. Whitten. 1987. Tropospheric Ultraviolet Radiation: Assessment of Existing Data and Effect on Ozone Formation. EPA Report 600/3-87/047; October.

Hales J. 1988. In: Sensitivity of Urban/Regional Chemistry to Climate Change: Report of the Workshop. Chapel Hill, NC; Feb. 17-18.

Holzworth, G.C. 1972. Mixing Heights, Windspeeds and Potential for Urban Air Pollution Throughout the Contiguous United States. EPA, Research Triangle Park, NC. Office of Air Programs Publication No. AP-101.

Kamens, R.M., H.E. Jeffries, K.G. Sexton, and A.A. Gerhardt. 1982. Smog Chamber Experiments to Test Oxidant-Related Control Strategy Issues. U.S. EPA Report 600/3-82-014; August.

Keene, W.C., J.N. Galloway, and J.D. Holden Jr. Measurement of weak organic acidity in precipitation from remote areas of the world. Journal of Geophysical Research 88:5122-5130.

Keene, W.C., and J.N. Galloway. 1984. Organic Acidity in precipitation of North America. Atmospheric Environment 18: 2491-2497.

Lamb B.K., H.H. Westberg, T. Quarles, and D.L. Flyckt. 1985. Natural Hydrocarbon Emission Rate Measurements From Selected Forest Sites. EPA-600/3-84-001; October.

Norton, R.B. 1985. Measurements of Formate and Acetate in Precipitation at Niwot Ridge and Boulder, Colorado. Geophysical Research Letters 12:769-772.

Ramanathan, V., R.J. Cicerone, H.B. Sing, and J.T. Kiehl. 1985. Trace gas trends and their potential role in climate change. Journal of Geophysical Research 90:5547-5566.

Research Triangle Institute. 1975. Investigation of Rural Oxidant Levels as Related to Urban Hydrocarbon Control Strategies. EPA-450/3-75-036; March.

Schertz, T.L., and R.M. Hirsch. 1985. Trend analysis of weekly acid rain data--1978-1983. U.S. Geological Survey, Water Resources Investigations Report 85-4211.

Seilkop, S.K., and P.L. Finkelstein. 1987. Acid precipitation patterns and trends in eastern North America, 1980-84. Journal of Climate and Applied Meterology 26(8):980-994.

U.S. EPA. 1988. U.S. Environmental Protection Agency, Office of Air Quality Planning and Standards. National Air Quality and Emissions Trends Report, 1986. Research Triangle Park, NC. EPA report No. 45014-88-001.

Walcek C. 1985. In: Sensitivity of Urban/Regional Chemistry to Climate Change: Report of Workshop. Chapel Hill, NC; Feb. 17-18.

CHAPTER 14

HUMAN HEALTH

FINDINGS

Global warming may lead to increases in morbidity and mortality during the summers, particularly among the elderly. These effects may be more pronounced in some regions than in others, with northern regions more vulnerable to the effects of higher temperature episodes than southern regions. Milder winters may offset increases in morbidity and mortality, although net mortality may increase. If people become acclimatized to the warmer temperatures by using air conditioning, changing work place habits, and altering the construction of homes and cities, the impact on summer mortality rates may be substantially reduced. The full scope of the impacts of climate change on human health remains uncertain and the subject for future research.

- o Although there may be increases in weather-related summer deaths due to respiratory, cardio-, acular, and cerebrovascular disease, there may be decreases in weather-related winter deaths for the same diseases. In the United States, however, on average, weather-related deaths are greater in summer than in winter. Thus, global warming could produce a net increase in deaths.

- o Sudden changes in temperature are correlated with increases in deaths. So if climatic variability increases, morbidity and mortality are also likely to increase. A decrease in the frequency or intensity of climate extremes is likely to be associated with a decrease in mortality and morbidity.

- o Seasonal variation of fetal and infant mortality (higher in the summers, lower in the winters) is thought to be due to summertime increases in infections. The longer and hotter summers that may accompany climate change may exacerbate this effect at least in some regions.

- o Vector-borne diseases, such as those carried by ticks, fleas and mosquitos, could increase in certain regions. In addition, climate change may alter habitats. For example, some forests may become grasslands, thereby modifying the incidence of vector-borne diseases.

- o While uncertainties remain about the magnitude of other effects, it's climate change could have the following impacts:

-- Increases in summer rainfall would multiply the amount of ragweed growing on cultivated land.

-- Changes in humidity may heighten the incidence and severity of skin infections and infestations such as ringworm, candidiasis, and scabies.

-- Increases in the persistence and level of air pollution episodes associated with climate change will have adverse health effects.

CLIMATE-SENSITIVE ASPECTS OF HUMAN HEALTH

Human illness and mortality are linked in many ways to the environment (Figure 14-1). Mortality rates, particularly for the aged and very ill, are influenced by the frequency and severity of extreme temperatures. The life cycles of disease-carrying insects, such as mosquitoes and ticks, are affected by changes in temperature and rainfall, as well as by modifications in habitat that result from climate change. Air pollution, frequently associated with climate change, is known to increase the incidence or severity of respiratory diseases such as emphysema and asthma. Climate variables have been implicated in affecting such chronic diseases as stroke and heart disease, such contagious diseases as influenza and pneumonia, such allergic diseases as asthma and hay fever, and such vector-borne diseases as encephalitis and Lyme disease.* In addition, adverse effects on reproduction, such as increased incidence of premature births, are associated with certain seasons. Table 14-1 lists the number of deaths and the number of physician visits (used to estimate the amount of illness associated with a given effect) associated with major causes of mortality and illness in the United States.

---

*Vector-borne diseases are those that are spread to humans or animals by an insect, such as a mosquito or tick. The agent is the disease-causing organism, such as a virus, that is carried and transmitted by the vector. Some vectors, such as ticks, live on other animals, such as deer and birds, which are called hosts. For example, Lyme disease is caused by a bacteria (the agent), which is carried by a certain specie of tick (the vector), which lives on deer and mice (the hosts).

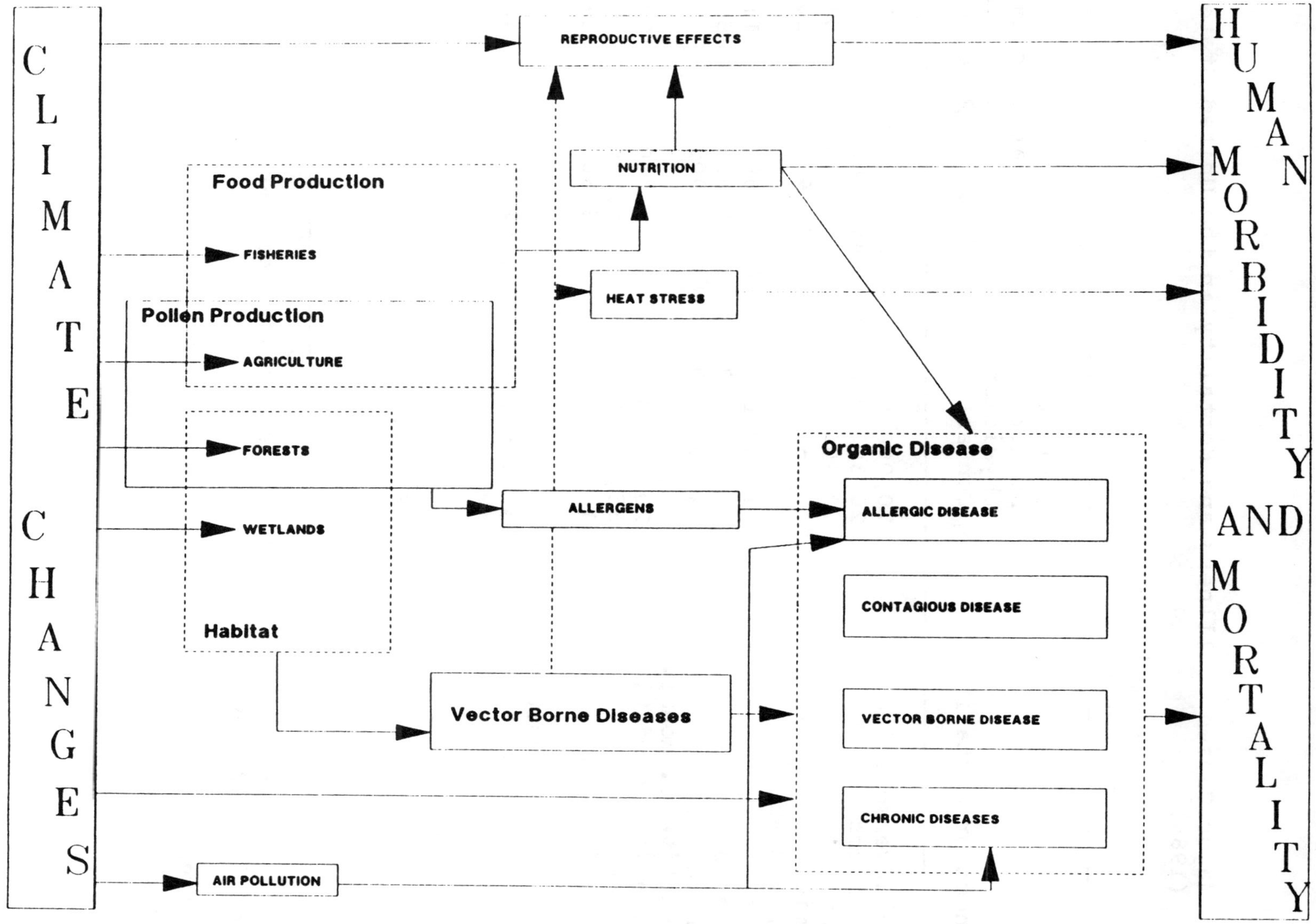

Figure 14-1. Schematic of How Climatic Changes Can Impact on Human Health

Table 14-1. Major Causes of Illness and Mortality in the United States (1984)[a]

Cause of Illness or Mortality	Estimated Number of Physician Contacts	Estimated Mortality	
		Number	Rate/100,000
Accidents and adverse effects	70,000,000	93,520	39.6
Malignant neoplasms	20,300,000	453,660	192.1
Heart diseases	72,400,000	763,260	323.2
Suicides, homicides	-	47,470	20.1
Chronic liver disease and cirrhosis	1,400,000	26,690	11.3
Cerebrovascular diseases	9,100,000	154,680	65.5
Congenital anomalies	4,300,000	12,900	5.5
Chronic obstructive pulmonary diseases and allied conditions	20,500,000	70,140	29.7
Diabetes mellitus	35,600,000	35,900	5.2
Pneumonia and influenza	14,500,000	58,800	24.9
Total for **weather-sensitive diseases**	152,100,000	1,082,780	448.5
Total all causes	248,100,000	1,717,020	717.1

[a]Conditions that can be influenced by changes in weather and climate are indicated in **bold type**.

Source: CDC (1986).

General Mortality and Illness

The relationship between mortality and weather has been studied for over a century with the (Kutschenreuter, 1959; Kalkstein, Vol. _) relationship between mortality and temperature receiving the most attention. Kutschenreuter (1959) observed "mortality is higher during cold winters and hot summers and lower during warm winters and cool summers." As indicated in Figure 14-2, the people most sensitive to temperature extremes are the elderly. Picciotto, 1985). One possible explanation is that for individuals already stressed by the circulatory problems associated with vascular and heart disease, heat waves "overload" the thermoregulatory system, which is struggling to maintain the appropriate body temperature. This results in heat stress, heatstroke, and often mortality as well (White and Hertz-Picciotto, 1985).

In addition to the elderly, people working in hot environments, such as steel mills and construction sites, are at special risk from heat waves (Dukes-Dobos, 1981). These workers face even greater risk if they have underlying medical problems such as the following:

- impaired circulation;
- lack of acclimatization/physical conditioning;
- higher than normal body temperature due to disease or exercise;
- chronic diseases such as alcoholism, diabetes, and obesity;
- reduced ability to sweat due to medications; and/or
- previous heat-related illness.

Cardiovascular, Cerebrovascular, and Respiratory Diseases

Although more information exists on the relationship of climate to total mortality from all causes, there is substantial information indicating that changes in climate variables are associated with impacts on the cardiovascular, cerebrovascular, and respiratory systems. As previously shown in Table 14-1, diseases or dysfunction of these three systems cause the majority of deaths observed on a yearly basis in the United States, as well as significant illness. Incidences of these diseases rise as climate extremes increase.

Changes in temperature may affect a number of heart problems including heart attack, coronary heart disease, and possibly coronary atherosclerosis and rheumatic heart disease (Vuori, 1987). Various stroke-related diseases are also thought to be

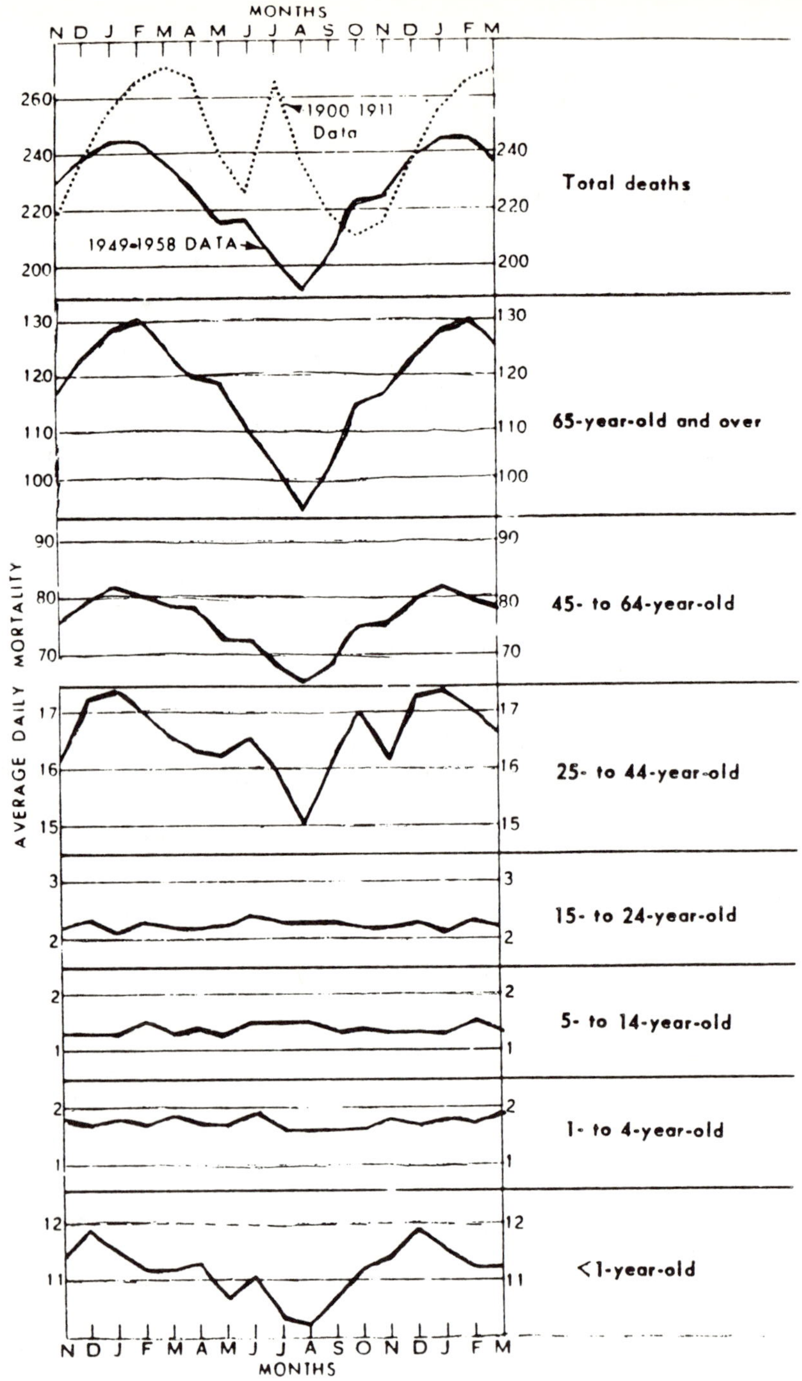

Figure 14-2. Annual Mortality for All Age Groups for 1949-58 a,b

[a]Temperature curves are based on 1921-50 monthly normal temperatures.

[b]Mortality curves are based on daily average mortality for each month computed from 1949 to 1958 monthly totals of mortality reported for all age groups.

Source: Kutschenreuter (1959).

affected by weather (Gill et al., 1987). Respiratory diseases sensitive to weather include pneumonia, bronchitis, influenza and other respiratory virus infections, asthma, and hay fever. A number of respiratory ailments have been linked to air pollution, which varies seasonally because it is affected by changes in weather. Increases in the incidence of lung cancer, emphysema, chronic bronchitis, and asthma have been attributed to the presence of industrially and photochemically produced pollutants present in urban smog (Lopez and Salvaggio, 1983).

The exact relationship between temperature changes and illness from diseases such as heart attack and stroke is not as clear cut as the relationship reported for mortality from these diseases and temperature because mortality has national reporting procedures whereas illness has to be estimated from such data as hospital admission figures. In the few studies that have evaluated illness from cardiovascular or cerebrovascular disease, admissions have shown a relationship to weather changes similar to that observed for mortality (Sotaniemi et al., 1970; Gill et al., 1988).

Illness from respiratory diseases is somewhat easier to estimate, principally because two such diseases, asthma and hay fever, affect as much as 3 and 6% of the U.S. population, respectively, causing significant losses in work time. The most common seasonal pattern for the allergic type of asthma and for hay fever is an increased occurrence during the spring in response to grass pollens. A nonseasonal form of allergic asthma may also occur in response to allergens such as molds, which are affected by changes in precipitation and temperature.

Vector-Borne Diseases

Two tick-borne diseases currently posing a public health problem in the United States, Rocky Mountain spotted fever and Lyme disease, induce similar initial symptoms: high fever, chills, headache, backache, and profound fatigue. Rocky Mountain spotted fever can eventually result in hemorrhagic areas that ulcerate, and Lyme disease may cause permanent neurologic, cardiac, and rheumatologic abnormalities (APHA, 1985). The ticks that spread these diseases, and therefore the geographic distribution of the diseases themselves, are affected both directly and indirectly by climate variables. Such environmental factors as temperature, humidity, and vegetation directly affect tick populations and the hosts of the tick populations, e.g., deer, mice, and birds.

Mosquito-borne diseases, such as malaria and certain types of encephalitis (inflammation of the brain), are not a major health problem in the United States today because occurrences are relatively rare. However, mosquitoes are also weather-sensitive insects favoring a warm, humid climate. The spread of mosquito populations and the diseases they carry depends in part upon such climate factors as temperature, humidity, and upon vegetation, which is also influenced by the climate.

Human Reproduction

Preterm delivery (i.e., premature birth) and perinatal mortality (i.e., death just before, during, or just after birth) are two adverse reproductive outcomes that are associated with particular seasons and, thus, might be affected by climate change. Statistically significant increases in preterm births and in perinatal mortality in the summer months have been documented (Keller and Nugent, 1983; Copperstock and Wolfe, 1986) (see Figure 14-3). The data on total perinatal deaths correspond closely with those on perinatal deaths associated with infection in the mother or infant, suggesting that the seasonality in perinatal death observed is linked to a seasonality of reproductive infections (Keller and Nugent, 1983).

POTENTIAL HUMAN HEALTH EFFECTS OF CLIMATE CHANGE

To assess the effects of climate change on human health, EPA sponsored three studies for this report (Table 14-2). Wiseman and Longstreth (1988) reviewed the literature on the role of climate, season, and weather variables in the incidence of, and mortality due to, vector-borne diseases, and they evaluated the potential impact of climate change on these diseases in the United States. They also conducted a workshop of eminent scientists to evaluate the potential impacts of global climate change on vector-borne infectious diseases in the United States. Following the workshop, Haile (1988) conducted modeling studies of the potential impact of climate change on 1) the distribution of the American dog tick, vectors of Rocky Mountain spotted fever, and 2) the potential for malaria transmission in the United States. The third study, by Kalkstein (1988), was an extension of an earlier modeling study that assessed the potential effects of global climate change on the elderly and on total mortality in New York (Kalkstein et al., 1986). In Kalkstein (1988), the New York analysis was expanded to include 14 other cities. A detailed review of these three studies, supplemented with other information from the literature, is presented in this section.

[a]

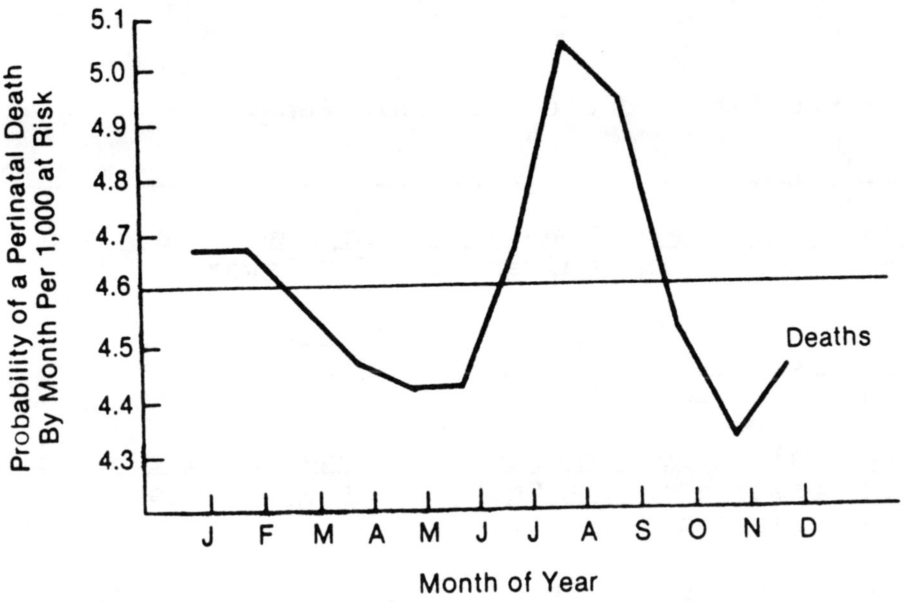

[b]

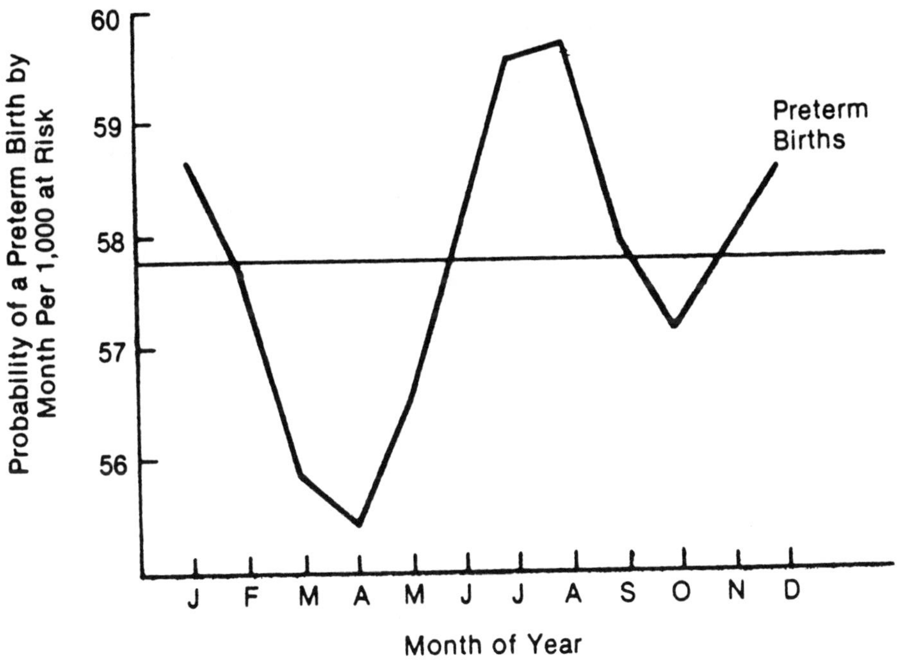

Figure 14-3. Probabilities of Perinatal Death [a] or Preterm Delivery [b]

Source: Keller and Nugent (1983).

Table 14-2. Studies Conducted for This Report.

- "The Impact of CO_2 and Trace Gas - Induced Upon Human Mortality" Kalkstein, University of Delaware.
- "Computer Simulation of the Effects of Changes in Weather Pattern on Veetor-Borne Disease Transmission" Haile, U.S. Department of Agriculture.
- "The Potential Impact of Climate Change on Patterns of Infectious Disease in the United States" Longstreth, ICF/Clement Associates, Inc.
- "Effect of Climatic Warming on Populations of the Horn Fly" Schmidtmann and Miller, Agricultural Research Service
- "Changing Animal Disease Patterns Induced by the Greenhouse Effect" Stenn, Mertz, Stryker and Huppi, Tufts University.

General Mortality

Preliminary analyses suggest that unless the U.S. population becomes fully acclimatized* to higher temperatures, climate change will be associated with a rising number of summer deaths. In addition, if climate change causes more intense heat waves, mortality is likely to increase. In the absence of complete acclimatization to warmer winters, the number of winter deaths is also likely to decline. With acclimatization, mortality may not decrease. It is not clear what the net effect of these two offsetting trends may be.

There have been very few studies of the effects of global climate change on human mortality. All of the identified effects have come from a single group of investigators: Kalkstein and his colleagues from the Center for Climate Research at the University of Delaware. This group (Kalkstein et al., 1986) developed a regression equation involving nine weather elements, such as temperature, windspeed, and humidity, to give the best algorithm for describing the current impact of weather on mortality. The algorithm used mortality data from New York City for 1964-66, 1972-78, and 1980.

The analysis revealed the existence of a summertime "threshold temperature" -- the maximum temperature above which mortality increases -- of 92°F for total deaths. This information was then used to assess the potential impact of climate change due to the emission of greenhouse gases, under the assumption that the population would not acclimatize, as well as under the assumption that it would acclimatize. Unacclimatized impacts were estimated by combining the climate scenarios and the historical weather algorithm described above, and acclimatized impacts were estimated by developing analogue cities that have weather variables today that look like those New York is estimated to have under climate change.

Under full acclimatization and a scenario predicting that New York will be 5 to 7°F warmer than it is today, no additional deaths were predicted. However, assuming no acclimatization, the

*Estimations of the impact of warming on future mortality must address the question of whether humans will acclimatize (physiologically adapt) to changing weather. How quickly humans may become acclimatized is a topic of considerable controversy, so it is difficult to predict whether the climate changes being predicted will occur sufficiently slowly to allow acclimatization.

number of summertime deaths attributable to temperatures above the threshold increased seven to ten-fold. Changes in winter weather were not estimated to affect mortality.

For this report, Kalkstein extended the New York analysis to cover 14 additional metropolitan areas and to evaluate the impact of two climate scenarios: the GISS doubled CO_2 scenario, and the GISS transient A scenario, evaluated at 30 years and at 60 years past the base period. Threshold temperatures were calculated for each city for summer and winter. Historical relationships between mortality and weather were derived independently for each of these 15 cities for both summer and winter. Table 14-3 summarizes the results for total mortality, by city and by season (summer or winter), for the doubled CO_2 scenario with and without acclimatization.

The cities with the highest estimated number of weather-associated summer deaths historically were New York City, Chicago, and Philadelphia: each averaged over 100 weather-induced deaths per summer. All of the cities with the highest average number of summer deaths are in the Midwest or Northeast, and those with the lowest number of summer deaths are in the South.

In the absence of any acclimatization, total summertime mortality in the United Stated under conditions of doubled CO_2 is estimated to rise from an estimated current total of 1,156 deaths to 7,402 deaths, with deaths in the elderly (aged 65 or over) subset contributing about 70% of each figure (727 and 4,605, respectively). Currently, the percentage of elderly in the U.S. population is increasing. Thus, the number of mortalities estimated to result from climate change may be larger than that found by Kalkstein because his analysis is predicated on today's age distribution. Even with full acclimatization, the number of weather-associated summer deaths approximately doubles to 2,198, possibly because hot weather increases physiological stress. Kalkstein's analysis also estimates a drop in the number of wintertime deaths. Historically, however, the number of weather-related deaths during the winter in the United States is much smaller (243) than that observed for the summer, and winter deaths were estimated to fall to 52 without acclimatization and to 159 with acclimatization. The net result for the United States is an increase in yearly mortality associated with doubled CO_2.

able 14-3. Estimated Future Mortality Under Doubled CO_2 Climate Conditions ithout and With Acclimatization

	Number of Deaths per Season					
	Summer			Winter		
ity	Current	Without	With	Current	Without	With
tlanta	18	159	0	2	2	0
hicago	173	412	835	46	2	96
incinnati	42	226	116	14	6	0
allas	19	309	179	16	1	0
etroit	118	592	0	16	2	37
ansas City	31	60	138	21	5	0
os Angeles	84	1,654	0	0	0	0
emphis	20	177	0	0	0	0
inneapolis	46	142	235	5	1	0
ew Orleans	0	0	0	0	0	0
ew York	320	1,743	23	56	18	25
klahoma City	0	0	47	0	0	0
hiladelphia	145	938	466	10	1	1
aint Louis	113	744	0	47	7	0
an Francisco	27	246	159	10	7	0
Total	1,156	7,402	2,198	243	52	159

ource: Kalkstein (1988).

Cardiovascular, Cerebrovascular, and Respiratory Diseases

Overall global warming and climate change may exacerbate the effects of cardiovascular, cerebrovascular, and respiratory diseases. Data show an inverse relationship between mortality and temperature (i.e., deaths go down as temperature goes up) for the range between -5°C and about +25°C, with sharp increases at temperatures above and below this range, particularly for the elderly and for hot weather (White and Hertz-Picciotto, 1985).* Data illustrating this relationship for coronary heart disease and stroke are shown in Figures 14-4 and 14-5, respectively (Rogot and Padgett, 1976). This complex relationship precludes simple prediction of the net effect of climate change. For example, while it seems likely that hot weather-associated mortality from these diseases will increase in many localities, this trend is likely to be offset, at least in part, by a decrease in cold weather-associated mortality.

Just as higher summer temperatures are associated with increases in mortality from cardiovascular, cerebrovascular, and respiratory diseases, they are also likely to be associated with increases in the occurrence of these diseases. Particular stress may be put on the respiratory system because climate change can increase pollen, urban smog (discussed below), and heat stress.

For example, if, as has been suggested in the chapter on forests, climate change encourages a transition from forest to grassland in some areas, grass pollens could increase. This, in turn, may increase cases of pollen-induced hay fever and allergic asthma. (However, the switch from forest to grassland would reduce the amount of tree pollens that also cause allergic responses in some individuals.) Rises in humidity also may affect the incidence of mold-induced asthma and hay fever.

As indicated in the chapter on air pollution, global warming may modify global and regional air pollution because it is likely to be associated with changes in the concentrations of ozone and may also have impacts on acid deposition and general oxidant formation. Increasing occurrences of numerous respiratory diseases, such as lung cancer, emphysema, bronchitis, and asthma, have been attributed to the pollutants in urban smog (Lopez and Salvaggio, 1983). Many of the trace gases implicated in global warming contribute to these problems; other chemicals are created

---

*The exact range appears to depend on the city.

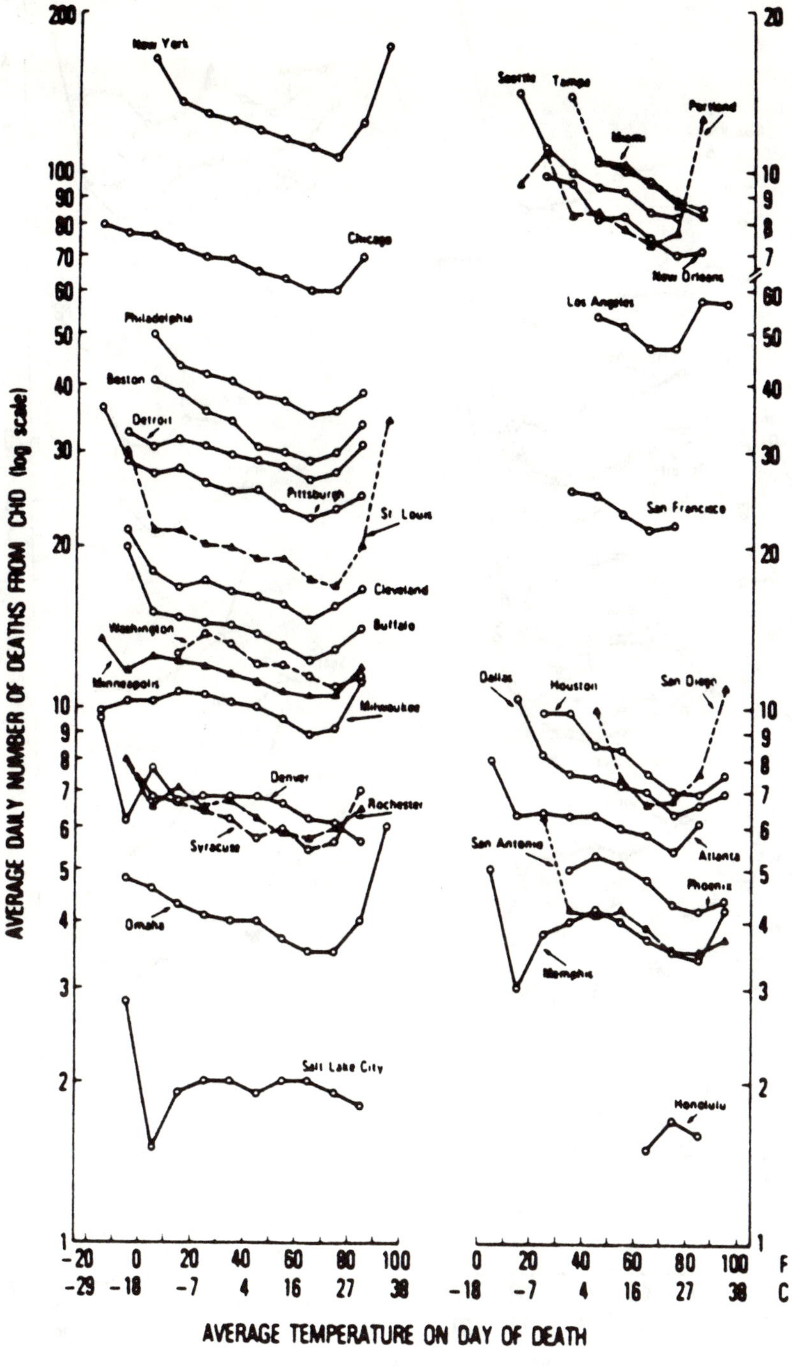

Figure 14-4. Relationship of Temperature to Heart Disease Mortality

Source: Rogot and Padgett (1976).

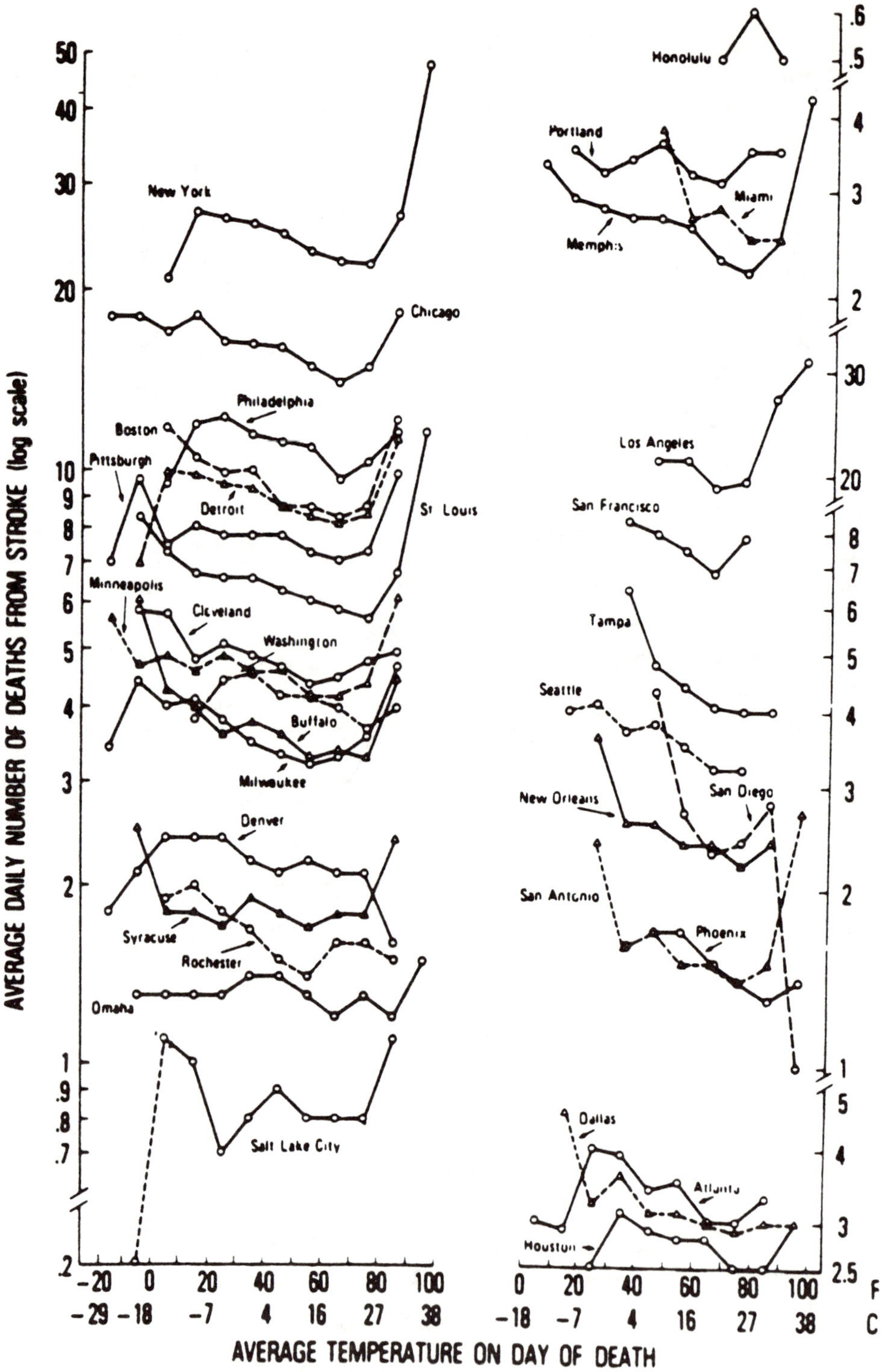

Figure 14-5. Relationship of Temperature to Mortality From Stroke

Source: Rogot and Padgett (1976).

from the interaction of ultraviolet light with these and other chemicals present in the atmosphere.

The component that causes the greates concern is urban smog (Grant, 1988). If global warming causes an increase in tropospheric ozone, serious consequences could result for adult asthmatics and people who suffer from acute or chronic bronchitis.

Vector-Borne Diseases

The changes in humidity and temperature caused by the increase in greenhouse gases are likely to alter the geographic ranges and life cycles of plants, animals, insects, bacteria, and viruses. (For further discussion of forestry and agriculture, see Chapter 11, Forestry, and Chapter 10, Agriculture.) For example, the range of many plant pests may move northward by several hundred miles. Such changes are also likely to occur for insects that spread diseases to both humans and animals. Vector-borne diseases are relatively rare in the United States, although most of those found in the United States are increasing in incidence. Some, such as Lyme disease, are increasing dramatically (CDC, 1986).

<u>Tick-Borne Diseases</u>

Both Rocky Mountain spotted fever and Lyme disease are considered to be public health problems in the United States. Although these two diseases are spread by different species of ticks, some overlap exists in their geographic distribution (Figure 14-6). Because tick populations appear to be limited by the size of their host populations, the spread of tick-borne diseases may be particularly sensitive to any change that may affect the geographic range of these hosts and, consequently, the range of the vector, or carrier.

In addition to the presence of the host, tick populations also depend upon the seasonality of environmental factors such as temperature, humidity, and vegetation. Optimally, climate must be warm enough to promote progression through the life cycles, humid enough to prevent the drying out of eggs, and cold enough in winter to initiate the resting stage.

As for many tick-borne diseases, the opportunity for a tick to acquire the infective agent from an infected animal is limited to the short period when the level of the agent in the blood of the host is high enough for the tick to receive an infective dose. Higher temperatures may increase the amount of the agent

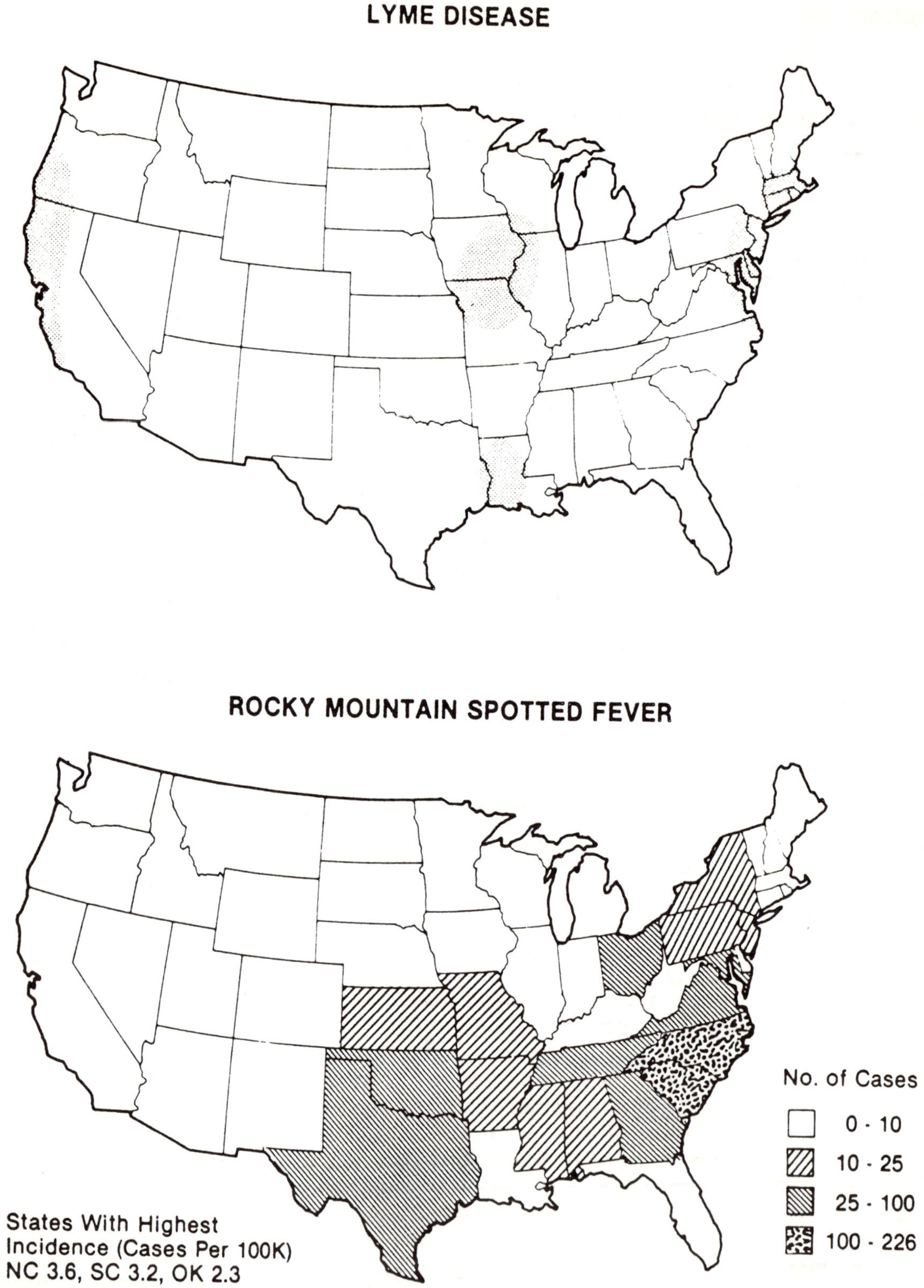

Figure 14-6. Geographic Distribution of Lyme Disease and Rocky Mountain Spotted Fever

Source: Wiseman and Longstreth (1988).

(the organism that is transmitted by the carrier, such as a virus) and the time it remains lodged on the host animal. Both of these mechanisms will increase the rate of infection of the carrier. However, although higher temperatures may favor the presence of the agent, there is some indication that they could disrupt the life cycle of some tick species. In these cases, warmer temperatures would reduce both tick survival and the spread of diseases they carry.

Tick populations also vary with the natural vegetation of an area. The incidence of Rocky Mountain spotted fever, in particular, has been linked to natural vegetation and changes in climate.

In examining the potential impact of climate change in the United States on Rocky Mountain spotted fever, Haile used a weather-based model, ATSIM, to evaluate the impact of the scenario climate changes on the distribution of the American dog tick, the primary carrier of this disease (Haile, 1988; Mount and Haile, 1988). The model uses data inputs from the three GCMs to estimate population dynamics, growth rate, and generation time under the various scenarios. Haile assumed that habitats and host density did not change in response to global warming. Sample results for six cities representing the most southern, the most northern, and the two middle latitudes are presented in Figure 14-7. The results indicate that under all scenarios, tick populations would shift from south to north and would be virtually eliminated from the most southern locations (Jacksonville and San Antonio). However, in the middle latitude cities, the results are mixed and depend on the scenario evaluated.

In this analysis, the only model inputs that were changed to simulate climate change were the weather inputs. Other important parameters in the model are the distribution of habitat between forests and meadows and the presence of suitable hosts. Both parameters are likely to be changed relative to current conditions under climate change. As indicated in the chapter on forests, a change from forests to meadow may occur in certain areas of the country; this would depress the tick population. However, the distribution of small mammals also may change. If small mammal populations increased, tick populations would grow.

In a sensitivity analysis of their model, Mount and Haile (1988) found that the model predictions could vary sixteenfold, depending on the inputs used for host density, whereas the variability conferred by changes in the weather inputs is about fourfold. Based on the sensitivity analysis, host densities are

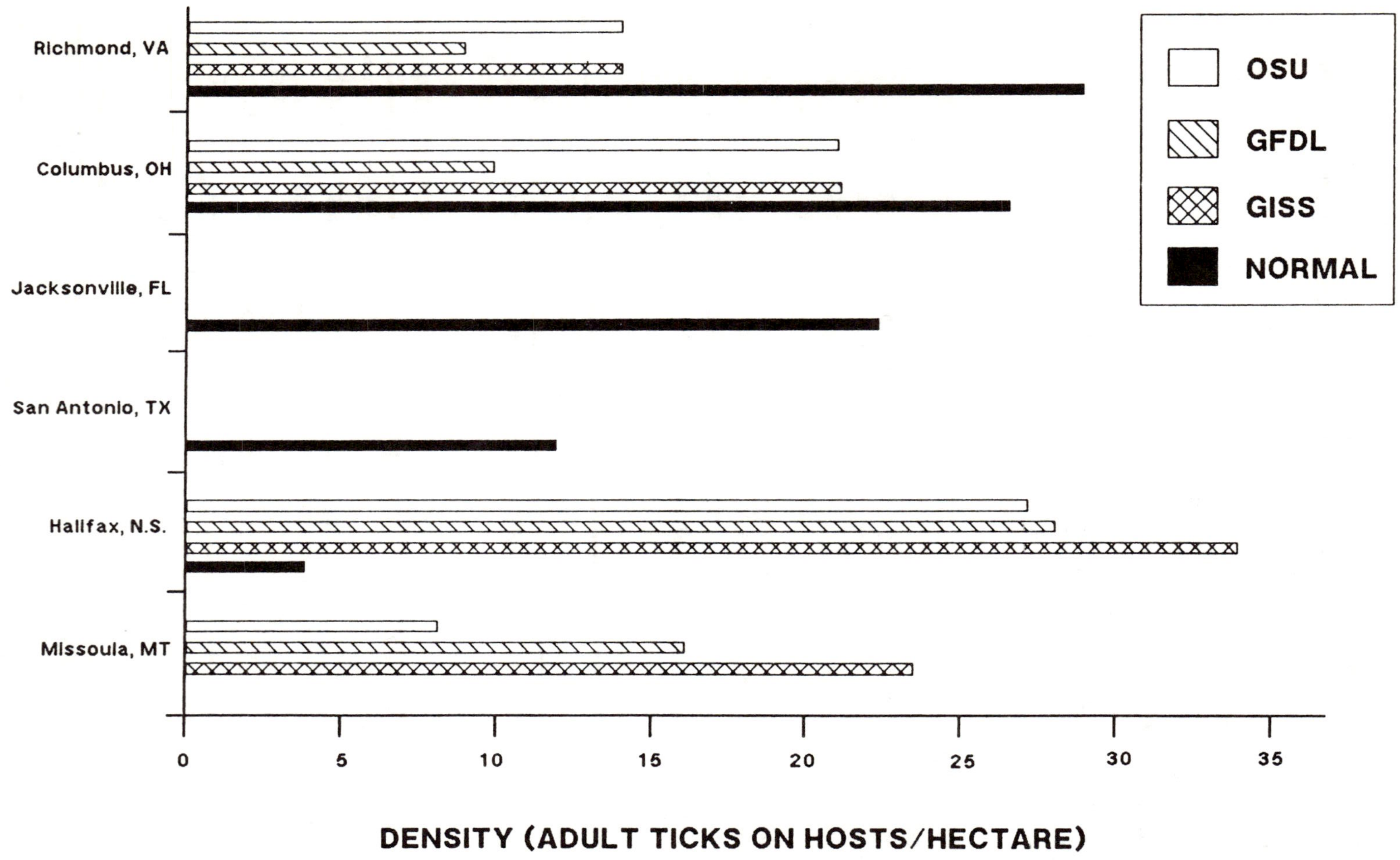

Figure 14-7. Simulated tick densities for selected cities under various scenarios of climate change.

Source: Haile. Volume __.

extremely important to these predictions. Keeping them constant, as was done in this analysis, could have underestimated the impact of climate change on the density of the American dog tick.

Mosquito-Borne Diseases

A second category of vector-borne diseases that can be affected by climate change consists of diseases carried by mosquitoes. Climate changes resulting in more days between 16 and 35°C, with humidity between 25 and 60%, are likely to favor the growth of mosquitoes (White and Hertz-Picciotto, 1985). Mosquito populations are also sensitive to the presence of standing water, which may or may not increase.

Worldwide, mosquito-borne diseases are associated with significant illness and mortality. In the United States, however, vector control programs and improved hygiene have virtually eliminated endogenously transmitted cases of these diseases, with the exception of sporadic outbreaks of arbovirus-encephalitis. (Imported cases are seen occasionally.) Numerous mosquito species are present in the United States, however. Recent restrictions on pesticide use, coupled with the influx of visitors and immigrants who can serve as sources of infectious agents, as well as the lack of available vaccines for many of the potential diseases, suggest the potential for reintroduction and establishment of these diseases in the United States -- particularly if global warming provides a more suitable climate for their growth and development (Wiseman and Longstreth, 1988).

Although there are numerous mosquito-borne diseases, at a recent workshop five were considered to pose a potential risk to U.S. populations if the status quo is disturbed by climate change (Wiseman and Longstreth, 1988). Malaria, dengue, and arbovirus-induced encephalitis were considered to be significant risks, and yellow fever and Rift Valley fever were considered to be possible risks.

Malaria

Malaria is an infectious disease transmitted by mosquitoes and induced by parasites (Plasmodia). The symptoms are highly variable, depending on the species of the agent. They include chills, sweats, and headache, and in severe cases, may progress to liver damage and even liver and renal failure.

Owing to effective vector control and treatment programs, malaria is no longer indigenous to the United States. However, imported cases occur regularly, and occasionally indigenous transmission has been documented (Wiseman and Longstreth, 1988). Current demographic trends in the United States, including a large number of legal and illegal immigrants from locations where malaria is endemic, could result in the presence of a sufficient pool of infected individuals that in conjunction with climate changes may create sufficiently favorable conditions for increased incidence of the disease.

Haile used the weather-dependent model MALSIM to evaluate the potential impact of climate change on malaria in an infected population living in an area where a competent carrier is present. The MALSIM model showed that several cities in the South (e.g., Miami, Key West, and Orlando) under current climate conditions are very favorable for malaria transmission. In general, however, very little change was estimated for the transmission potential of malaria in the United States (Figure 14-8). In a few cities, e.g., Richmond, Nashville, and Atlanta, the model estimated large increases in one scenario relative to those that would occur normally. However, the results varied with different climate scenarios and did not occur at all locations.

<u>Dengue Fever</u>

Dengue fever is an arbovirus-induced* illness characterized by fever, rash, and severe pain in the joints. Like malaria, it is not currently endemic in the United States, although potential carriers are present (see Figure 14-9) and the disease is imported here regularly by people who have traveled abroad.

The first infection by the virus generally is not fatal. However, a second infection, particularly by a type of virus different from the first, can result in much more serious forms of the disease that can be fatal, particularly in the very young and the elderly.

The ability of the vector to transmit the agent appears to depend on temperature, and current conditions do not appear to be favorable for this process. Climate changes that raise temperatures, however, may reduce the required incubation period

---

*An arbovirus is a virus transmitted by an arthropod. Arthropods are a group of animals that include insects and crustaceans. Examples of arthropods that transmit disease include mosquitoes and ticks.

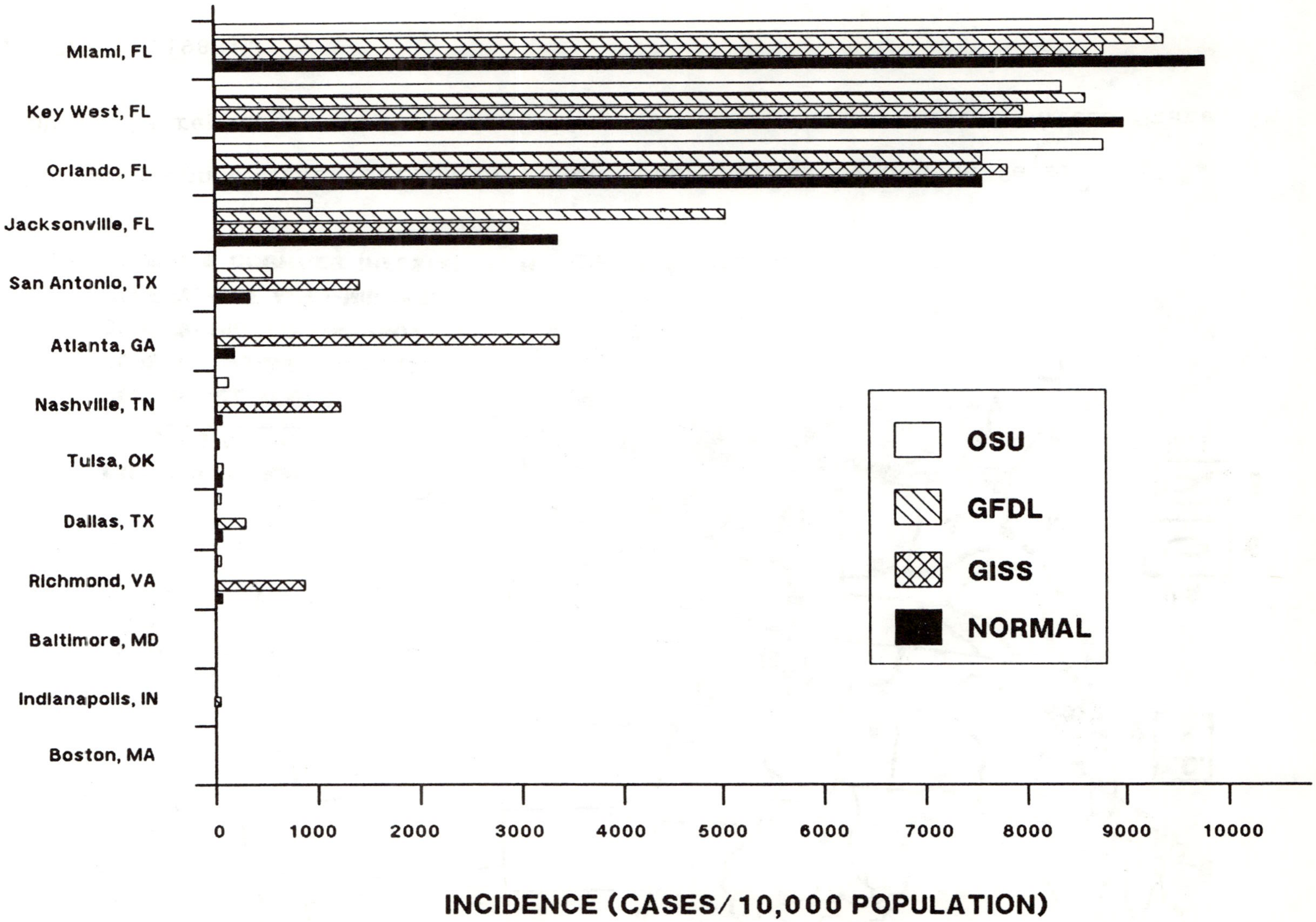

Figure 14-8. Simulated incidence of malaria for selected cities under various scenarios of climate change.

Source: Haile. Volume __.

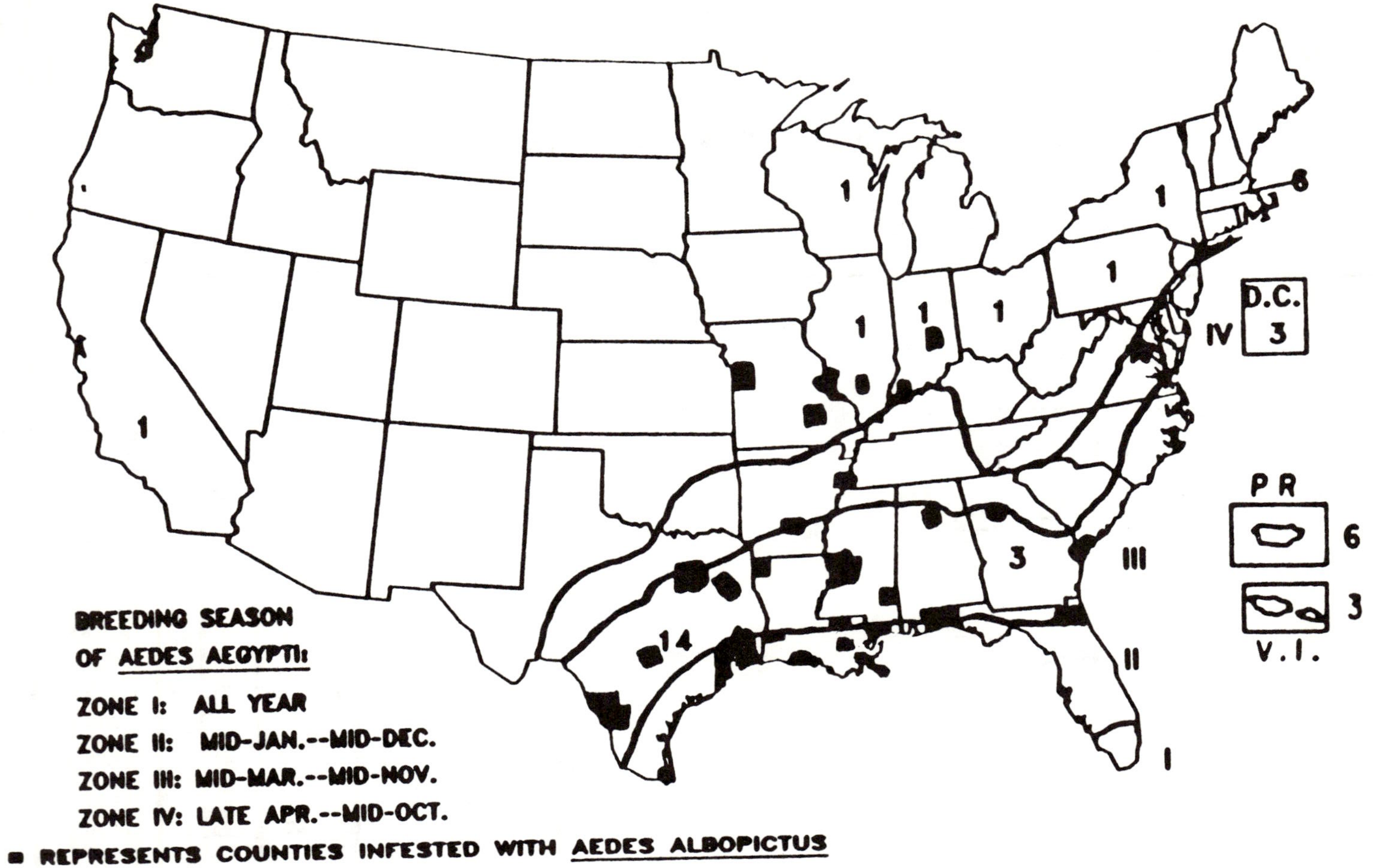

Figure 9. Geographic distribution of mosquitoes that carry dengue fever[a]

[a]Arabic numerals represent the number of imported cases found in the United States in 1984.

Source: Grant (1988).

and increase the infectivity of the carrier, increasing the potential transmission of the disease.

Arbovirus-Related Encephalitides

Arbovirus-related encephalitides are a group of acute inflammatory diseases that involve parts of the brain, spinal cord, and meninges. In mild cases, these infections result in feverish headaches or aseptic meningitis; in more severe cases, those symptoms can be accompanied by stupor, coma, convulsions (in infants), and occasionally spastic paralysis (APHA, 1985).

At least eight types of viruses causing encephalitis are present in the United States. Cases range in severity depending on the type of virus, with yearly fatality rates between 0.3 and 60%, These infections are rare. In 1984, 129 cases were reported to the Centers for Disease Control, which maintains an active surveillance program for them (CDC, 1986).

Outbreaks of encephalitis attributable to these viruses are normally limited to specific geographic locations and seasons for several reasons. First, summer temperatures are normally required for the viruses to multiply and to be transmitted to a new host. Increased temperatures may quicken the transmission process and promote epidemic disease. However, the extent of this effect depends largely on the particular virus. Some viruses require cooler weather and higher moisture conditions. Thus, higher temperatures may reduce their prevalence. Second, environmental conditions that favor the presence of carriers and hosts must prevail. For example, relative humidity may affect plant life necessary for the feeding of hosts.

Other Diseases

A variety of other U.S. diseases indicate a sensitivity to changes in weather. Higher humidity may increase the incidence and severity of fungal skin diseases (such as ringworm and athlete's foot), and yeast infections (candidiasis). Studies on soldiers stationed in Vietnam during the war indicated that outpatient visits for skin diseases (the largest single cause of outpatient visits) were directly correlated to increases in humidity, but showed a 4-month lag with relationship to temperature increases (Figure 14-10). In addition, excessively high temperatures can lead to such skin diseases as prickly heat and heat rash, which impair the ability of the skin to breathe and thus place additional stress on people already suffering from overexposure to heat from other causes.

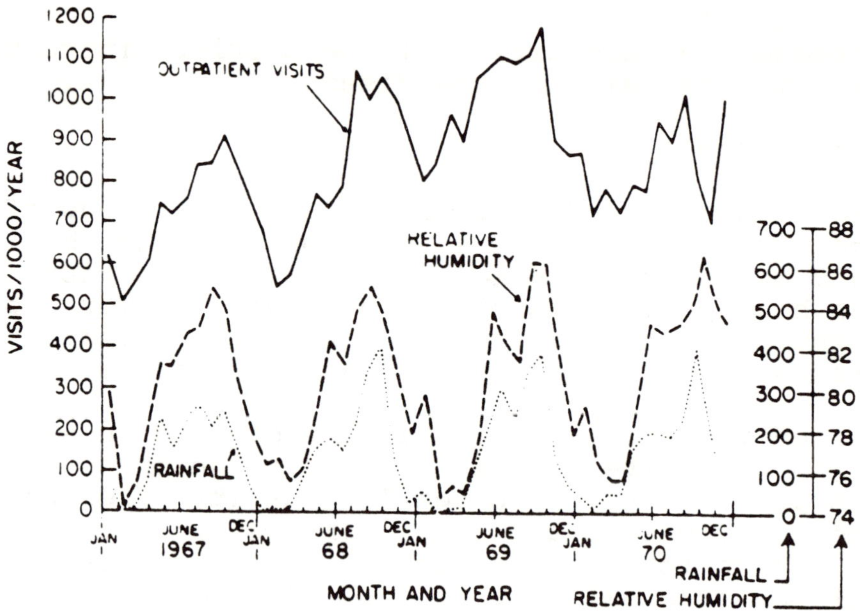

Outpatient visits for skin diseases in relation to mean monthly temperature index: U.S. Army personnel in Vietnam, 1967–70

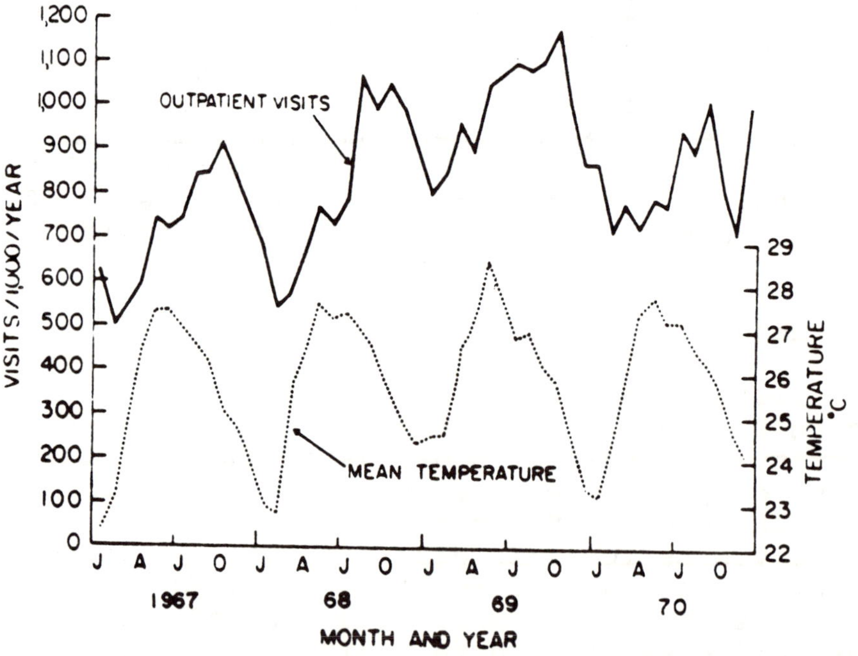

Figure 14-10. Relationship of Skin Infections to Humidity and Temperature.

Source: Longstreth and Wiseman, Volume __.

Several diseases appear to be associated with the acquisition of winter infections. If a reduction in winter severity is also accompanied by a decrease in wintertime infections, these diseases could be reduced under global warming.

For example, birth in cold winter months has been associated with a higher risk of schizophrenia in individuals whose schizophrenia is without an apparent genetic component (Kovelman and Scheibel, 1983). In addition, juvenile-onset diabetes, which has been reported to be increasing over the past several decades, has been shown to be associated with a seasonal variation in that the month of first admission peaks in the winter (Glatthaar et al., 1988; Patterson et al., 1988). It is a common clinical experience that a minor viral illness precedes the onset of symptoms.

NATIONAL IMPLICATIONS

Human societies possess considerable ability to adapt to change. The potential for climate to affect human health may be considerably modified by adaptive responses, such as immunizations, modification of the environmental temperature (e.g., use of air conditioners), and control of disease carriers.

In addition, several demographic and technological trends (the aging of the population, an influx of immigrants, advances in treatment techniques) make it difficult to analyze the potential impacts of climate change on human health. Although this chapter attempts to identify those human health effects at risk from climate change, the analyses were not designed to consider adaptive responses and should not be treated as absolute predictions of what will happen with climate change. Rather, the analyses presented here represent possible scenarios in the absence of consideration of demographic trends or adaptive responses that may either exacerbate or ameliorate the impact of climate change on human health.

Climate change may affect regional and national health care. For instance, the treatment requirements for asthma may increase or decrease as locations experience changes in the distribution and intensity of pollen concentrations. Increased resources may be needed to treat premature infants if there is an increase in the number of preterm births. If heart attacks, stroke, and respiratory problems increase, hospitalization costs and costs due to days lost from work may also increase. Higher health care costs might be particularly obvious in Medicaid and Medicare because those below the poverty line would be less able to take adaptive measures (e.g., air-conditioning), and the elderly are more susceptible to the ill effects of extreme weather conditions.

The United States is already dealing with an infant mortality problem higher than that of any other industrialized nation (World Bank, 1987). Infant mortality is higher in the summer; consequently, the increased temperatures expected with global warming may well exacerbate the problem.

The need for irrigation may increase in many regions of the U.S. (see Chapter 10, Agriculture). Irrigation may result in greater amounts of standing water and can therefore increase mosquito populations. Arbovirus encephalitis may become a greater problem than at present, and other mosquito-borne diseases, such as dengue or yellow fever, could be more easily spread if introduced.

POLICY IMPLICATIONS

The full impacts of climate change on human health will require more research. Agencies such as the Department of Health and Human Services should consider conducting studies a potential impact.

In the future a cadre of trained professionals may be needed to respond to outbreaks of diseases. A shift in the distribution of carriers of human disease may necessitate regional shifts in surveillance and eradication programs. States that do not have these programs may need to develop them in the future.

RESEARCH NEEDS

One of the largest potential impacts of climate change apparent in the literature is the increase in perinatal mortality and preterm births. An analysis similar to that performed by Kalkstein, but focusing strictly on deaths in the 0- to 1-year age group, might be warranted. In addition, some refinements to the Kalkstein analysis -- e.g., focusing on the distinct causes of death, and a more detailed analysis of the analogue modeling results -- are needed to provide additional insight into the robustness of the conclusions from that study.

The modeling that has been performed for vector-borne diseases did not assess the potential impact of changes in the host population (or habitat) that might result from climate change. Additional analyses are needed that evaluate what the changes in host populations and habitat are likely to be and how those changes could affect predictions from the ATSIM and MALSIM models.

REFERENCES

APHA. 1985. American Public Health Association. In: Berenson, A.S., ed. Control of Communicable Diseases in Man. Springfield, VA: John D. Lucas Printing Company.

CDC. 1986. Centers for Disease Control. Annual Summary 1984. MMWR 33:54.

Cooperstock, M., and R.A. Wolfe. 1986. Seasonality of preterm birth in the collaborative perinatal project: demographic factors. American Journal of Epidemiology 124:234-41.

Dukes-Dobos, F., 1981. Hazards of heat exposure. Scandinavian Journal of Work and Environmental Health 73-83.

Gill, J.S., P. Davies, S.K. Gill, and D.G. Beevers. 1988. Wind-chill and the seasonal variation of cerebrovascular disease. Journal of-Clinical Epidemiology 41:225-230.

Glatthaar, C., P. Whittall, T. Welborn, M. Giboon, B. Brooks, M.M. Ryan, and G. Byrne. 1988. Diabetes in Western Australian children: descriptive epidemiology. Medical Journal of Australia, 148:117-123.

Glenzen, W.P. 1982. Serious illness and mortality associated with influenza epidemics. Epidemiological Reviews 4:25-44.

Grant, L.D. 1988. Health effects issues associated with regional and global air pollution problems. Draft document prepared for World Conference on the Changing Atmosphere, Toronto.

Haile, D.G. 1988. Computer simulation of the effects of changes in weather patterns on vector-borne disease transmission. Report prepared for U.S. EPA, Office of Policy, Planning and Evaluation. U.S. EPA Project No. DW12932662-01-1.

Harris, R.E., F.E. Harrell, K.D. Patil, and R. Al-Rashid. 1987. The seasonal risk of pediatric/juvenile acute lymphocytic leukemia in the Unites States. Journal of Chronic Diseases 40:915-923.

Kalkstein, L.S., R.E. Davis, J.A. Skindlov, and K.M. Valimont. 1986. The impact of human-induced climate warming upon human mortality: a New York case study. Proceedings of the International Conference on Health and Environmental Effects of Ozone Modification and Climate Change. June 1986. Washington, DC.

Kalkstein, L.S. 1988. The impact of CO_2 and trace gas-induced climate change upon human mortality. Report prepared for the U.S. EPA, Office of Policy, Planning and Evaluation. U.S. EPA contract No. CR81430101.

Kalkstein, L.S., and K.M. Valimont. 1987. Effect on human health. In: Tirpak, D., ed. Potential Effects of Future Climate Changes on Forest and Vegetation, Agriculture, Water Resources, and Human Health, Vol. V, pp. 122-152. EPA 400/1-87/101E. Washington, DC.

Keller, C.A., and R.P. Nugent. 1983. Seasonal patterns in perinatal mortality and preterm delivery. American Journal of Epidemiology 118:689-98.

Kovelman, J., and A. Scheibel. 1986. Biological subtrates of schizophrenia. Acta Neurologica Scandinavica 73:1-32.

Kutschenreuter, P.H. 1959. A study of the effect of weather on mortality. New York Academy of Sciences 22:126-138.

Lopez, M., and J.E. Salvaggio. 1983. Climate-weather-air pollution. In: Middleton, E., and C.E. Reed, eds. Allergy, Chapter 54. St. Louis: C.V. Mosby Company.

Mount, G.A., and D.G. Haile. D.G. 1988. Computer simulation of population dynamics of the American dog tick, Dermacentor variabilis (Acari: ixodidae). Journal of Medical Entomology. In Press.

Patterson, C., P. Smith, J. Webb, M. Heasman, and J. Mann. 1988. Geographical variation in the incidence of diabetes mellitus in Scottish children during the period 1977-1983. Diabetic Medicine, Vol. 5, pp. 160-165.

Rogot, E., and S.J. Padgett. 1976. Associations of coronary and stroke mortality with temperature and snowfall in selected areas of the United States 1962-1966. American Journal of Epidemiology 103:565-575.

Sontaniemi, E., U. Vuopala, E. Huhta, and J. Takkunem. 1970. Effect of temperature on hospital admissions for myocardial infarction in a subartic area. British Medical Journal 4:150-1.

Vuori, I. 1987. The heart and the cold. Annals of Clinical Research 19:156-162.

White, M.R., and I. Hertz-Picciotto. 1985. Human health: analysis of climate related to health. In: White, M.R., ed. Characterization of Information Requirements for Studies of CO_2 Effects: Water Resources Agriculture, Fisheries, Forests, and Human Health. Department of Energy, DOE/ER/0236.

Wiseman, J., and J.D. Longstreth. 1988. The potential impact of climate change on patterns of infectious disease in the United States. Background paper and summary of a workshop. Report to USEPA Contract No. 68-10-7289.

World Bank. 1987. World Development Report 1987. New York: Oxford University Press.

CHAPTER 15

URBAN INFRASTRUCTURE

FINDINGS

Global climate change could require U.S. cities to make major changes in capital investments and operating budgets. Areas most likely to be affected include water supplies, roads, and bridges, storm sewers and flood control levees, and energy demand in municipal buildings and schools.

- Most urban infrastructure in the United States will turn over in the next 35 to 50 years. If changes in climate are considered, this turnover will allow cities to prepare for climate change at lower costs. In some cases, the risk of climate change should be incorporated into decisions beginning today, such as coastal drainage systems that are likely to last for 50 to 100 years.

Northern Cities

- Northern cities, such as Cleveland, may incur a change in the mix of their expenditures. In such locations, increased electricity costs for air conditioning could be offset by reductions in expenditures for heating fuel, snow and ice control, and road maintenance. Southern cities could see increases in operating budgets due to the demand for additional air conditioning.

Coastal Cities

- Coastal cities may be subject to more severe impacts. These include 12 of the 20 largest metropolitan areas. For example:

  - Sea level rise or more frequent droughts would increase the salinity of shallow coastal aquifers and tidal surface waters. Cities that rely on water from these sources need to examine water supply options. Such areas as Dade County, Florida, or New York City will probably be vulnerable.

  - As sea level rises, some coastal cities would require levees to hold back the sea or fill to raise the land surface area. In the case of Miami, the cost of these activities might be more than $500 million over the next 50 to 75 years.

Water Supply and Demand

- Climate change will influence the supply and demand for water in many cities. A lengthened summer season and higher temperatures will increase the use of water for air conditioners and lawns and gardens. Changes in rainfall patterns, runoff, and flood control measures may alter water supplies. In the Hudson River Basin, summer water demand could increase by 5 percent over the demand for water without climate change, while supplies might fall. Such a change would require new institutional and management approaches for both the Delaware and Hudson Rivers.

Policy Implications

- Climate change has implications for many national programs and policies. For example:

  - The National Flood Insurance Program may react to climate change by redrawing flood plain maps and adjusting insurance rates to account for sea level rise and changes in river flows. It may be prudent to examine whether this program should anticipate climate change by discouraging development that would be vulnerable in the future.

  - Because of the key role federal programs play in the development of cities, the Department of Housing and Urban Development should examine the implications of climate change on long-term policies. A minimum response might be to provide guidance on the certainties and uncertainties of climate change to groups such as the National League of Cities, the U.S. Conference of Mayors, and the American Planning Association.

  - Because water supply infrastructure may last for several centuries, improved planning is important. The U.S. Geological Survey should study the probable impacts of global climate change and sea level rise on the water supplies of major cities. The Corps of Engineers should factor climate change into the design of major projects.

  - Given the assumption that modest changes in the design and location of many transportation systems may facilitate an accommodation to climate change, the Department of Transportation should factor climate change into the design of roads, bridges, and mass transit facilities.

-- Voluntary standard organizations, such as the American Society of Civil Engineers, the Building Officials and Code Administrators, International and American Society of Heating and Refrigerating and Air Conditioning Engineers should examine the need for changes in existing standards to improve the use of energy and safety factors.

RELATIONSHIP BETWEEN URBAN INFRASTRUCTURE AND CLIMATE

Three-quarters of the U.S. population is concentrated in urban areas (Statistical Abstract, 1987). The majority of the nation's investment in water supply, wastewater transport and treatment facilities, drainage, roadways, airports, mass transit, electric power, solid waste disposal sites, and public buildings serves these urban areas. The current value of selected infrastructure nationwide, displayed in Table 15-1, provides insight into the aggregate investment at stake if climate changes. Most of these items could be considered part of urban infrastructure, and have been located and designed based on historic meteorologic information.

Eighteen of the 20 most populated U.S. cities have access to oceans, major lakes, or rivers* and have invested in additional infrastructure for port facilities and flood control. The expenditure required to construct coastal defense structures -- which prevent inundation by the sea, slow oceanfront erosion, control storm surges, slow saltwater advance up rivers, and reduce saltwater intrusion into aquifers -- is now minimal.

Although actual practice varies, the nominal replacement cycle for most infrastructure is 35 to 50 years (National Council on Public Works Improvement, 1988). Some water supply investments have 100-year cycles. Between planned replacement

*Of the 20 most populous cities in the U.S., twelve are tidal waterfront cities (Baltimore, Boston, Houston, Los Angeles, Miami, New York, Philadelphia/Wilmington, San Francisco/Oakland, San Diego, Seattle, Tampa/St. Petersburg, and Washington, D.C.), three are located on the Great Lakes (Chicago, Cleveland, and Detroit), three are on navigable rivers (Minneapolis, Pittsburgh, and St. Louis), and two are not on a navigable waterway (Atlanta and Dallas).

TABLE 15-1. Value of the Nation's Stock of Selected Infrastructure (Billions of 1984 Dollars)

Component	Value[a]
Water supply	$108
Wastewater	136
Urban drainage	60
Streets	470
Public airports	31
Mass transit	34
Electric power (private only)[b]	266
Public buildings	unknown
Total	$1,107+

[a]Based on a useful life of 35 to 50 years for most assets, and 10 to 20 years for transit vehicles.
[b]About 77% of electric power is privately produced.

Source: Statistical Abstract (1987); National Council on Public Works Improvement (1988).

cycles; however, sea level rise, temperature change, and changes in precipitation patterns all could alter the balance between water supply and demand. The nature and pattern of precipitation could affect drainage requirements as well as highway design and maintenance.

The heat wave of 1988 vividly illustrated some of the potential impacts. Hundred-degree weather distorted railroad tracks, forcing Amtrack to cut speeds from 200 to 128 km per hour between Washington and Philadelphia (Bruske, 1988) and possibly causing a train wreck that injured 160 people on a Chicago-Seattle run (The Washington Post, 1988). A U.S. Army Corps of Engineers contractor worked around the clock for 2 weeks to build a 170-m-wide, 9-m-high silt wall across the bottom 40% of the Mississippi River channel, 48 km below New Orleans (Sossaman, 1988a,b). This $2 million wall, designed to wash away when spring snow melt demands the full capacity of the channel, slowed an advancing wedge of saltwater that threatened the water supply in New Orleans and nearby parishes. In Manhattan, heat exacerbated the effects of long-standing leaks in 256 km of steam pipes, causing the asphalt to soften. As vehicles kneaded the soft asphalt, thousands of bumps formed on city streets, requiring extensive repairs (Hirsch, 1988). In the suburbs of Washington, D.C., steel expansion joints bubbled along a 21-km stretch of Interstate 66 (Lewis, 1988).

This chapter will examine such issues as the portions of the infrastructure that will be significantly affected, and anticipated costs and who will bear them.

PREVIOUS CLIMATE CHANGE STUDIES ON URBAN INFRASTRUCTURE

Available literature on the potential effects of global climate change on urban infrastructure is sparse. Rhoads et al. (1987) examined the potential impacts of sea level rise on water supply and flood protection in Dade and Broward Counties, Florida, and concluded that the effects might be severe. Hull and Titus (1986) analyzed the potential impact of sea level rise on water supply in the Philadelphia-Wilmington-Trenton area, and found that a rise of 0.3 m could require adding 140 million m^3 of reservoir capacity, about a 12% increase, to prevent saltwater from advancing past water intakes on the Delaware River. Additional investment would be required to prevent or respond to saltwater infiltration into underground aquifers.

Two recent studies illustrate the importance of considering sea level rise in urban coastal infrastructure planning and the uncertain nature of the decisions involved. Wilcoxen (1986) examined the impact of sea level rise on a portion of San Francisco's sewage transport system that is buried close to the shoreline. The study estimated that if sea level rose 0.6 m by the year 2100, an expenditure of roughly $70 million on beach nourishment might be required to prevent damage to a structure that cost $100 million to build in the late 1970s. The author suggested that consideration (at no additional cost) of sea level rise in siting the structure could have prevented these expenses. Titus et al. (1987) examined the impact of sea level rise on a proposed coastal drainage system in Charleston, South Carolina, and estimated that a 0.3-m sea level rise by 2025 would require almost $2.5 million in additional investments to maintain the target level of flood protection. The present value of these investments is $730,000. In contrast, only about $260,000, one-third of the cost of responding in 2025, would be required to add this level of protection at initial construction.

URBAN INFRASTRUCTURE STUDIES IN THIS REPORT

Several studies undertaken for this report examined some of the implications of climate change in relationship to urban infrastructure. One study comprehensively examined the impacts on infrastructure in several cities:

- Impact of Global Climate Change on Urban Infrastructure (Walker et al., The Urban Institute)

The following studies, referenced in this chapter, covered issues relating to urban infrastructure:

- Electric Utilities (Linder et al., 1987)
- The Impacts of Low Lake Levels on the Illinois Shoreline (Changnon, year)
- Case Studies of the Water Supply Systems of California and Atlanta, Georgia (Sheer, year)
- Defending Unprotected Shorelines (Leatherman, year)
- Defending Protected Shorelines (Weggel, year)

RESULTS OF THE INFRASTRUCTURE STUDIES

Impacts on Miami, Cleveland, and New York City

Walber et al. examined three cities distinctly affected by climate change to determine a range of impacts on urban infrastructure.

Study Design

The study was based on a critical review of existing infrastructure studies in the three cities, discussions of likely impacts with local infrastructure experts, analyses undertaken by these experts, and preliminary calculations of probable impacts. Experts were presented with GCM scenarios for CO_2 doubling, and scenarios were used to calculate effects on energy demand, roadways, and other systems. The study also derived conclusions based on experiences in other cities where current temperatures are analogous to temperatures projected for the cities under study, using the analogues identified by Kalkstein (Vol.).

Limitations

The principal limitation of the overall study is the limited use of hydrologic and other modeling. In addition, experts were asked to derive conclusions regarding conditions beyond their experience. Since only three cities are included, the full range of effects on urban infrastructure was not covered. The authors did not perform engineering analyses of cost-effective responses, and they did not assess the potential for reducing impacts through technological change.

Results and Implications

Miami's Infrastructure

Greater Miami is bounded by water on all sides during the rainy season. An extensive network of canals and levees has been built to control ocean and freshwater flooding and to recharge the aquifer beneath the area. Miami has one of the world's most porous aquifers, which lies less than 1.5 m below the surface in one-third of the developed area. Federal law requires that roughly 15% of Miami's freshwater be released into the Everglades National Park. The Dade County Development Master Plan predicts that by 2000, a water supply deficit will force use of desalinated water at three times current water costs.

The Miami case study examined the probable impacts of climate change and sea level rise on Dade County's water supply, water control and drainage systems, building foundations, roads, bridges, airports, solid waste disposal sites, and sewage transport and treatment systems, assuming that a gradual sea level rise would be managed through strategies such as raising the land in low-lying areas, upgrading levees and dikes with pumped outflows, retreat selectively from some areas, and increasing the freshwater head roughly in proportion to sea level rise to prevent saltwater infiltration into the aquifer.

As Table 15-2 shows, global climate change could require $500 million ($ ½ billion), in capital investment in Greater Miami over the next century. Because many systems could not be estimated and because a complete engineering analysis was not performed, these results should only be considered as rough estimates.

The south Florida aquifer is currently under pressure because of growth in the Miami area. Because the aquifer extends under the ocean, the typical urban response to a rising sea -- diking the water at the surface and pumping out the seepage from

TABLE 15-2. Probable Infrastructure Needs and Investment in Miami in Response to a Doubling of CO_2 (Millions of 1987 Dollars)

Infrastructure Need	Cost
Raising canals/levees	$60
Canal control structures	$50
Pumping	not estimated
Raising streets	$250 added to reconstruction cost
Raising yards and houses	not estimated
Pumped sewer connections	not estimated
Raising lots at reconstruction	not estimated
Drainage	$200-300
Airport	$30
Raising bridges	not estimated
Sewer pipe corrosion	not estimated
Water supply	more use of desalinated water at double
present	water prices

Source: Walker et al, Volume __.

ditches behind the dikes -- appears to be unworkable. Unless the dike extended down more than 45 m, rising seawater pressure would cause the sea to rush into the aquifer below the surface and push freshwater upward, almost to the surface.

The one-time capital costs for upgrading existing canals and levees in response to a 1-m sea level rise would be about $12 million. However, almost $50 million in new control structures, including extensive pumping capacity, might be required for the canals used to maintain the freshwater head. Large-scale pumping for canvas also could involve substantial operating costs, but these have not been estimated. Storm sewers and drainage would cycles; however, sea level rise, temperature change, and changes in precipitation patterns all could alter the balance between water supply and demand. The nature and pattern of precipitation need upgrading, requiring investment of several hundred million dollars above normal replacement costs.

Building foundations generally should remain stable if the freshwater head rises 1 m because houses are built on concrete slabs, most buildings in newer areas already are built on raised lots to meet Dade County's flood control ordinance, and the foundations of many larger buildings already extend into the water table.

Conversely, the water table could infiltrate the base of about a third of Dade County streets, which would have to be raised or risk collapse. If sea level rose gradually, thereby permitting raising of streets and related sewer mains during scheduled reconstruction, the added public cost might be approximately $250 million. Building owners would incur substantial costs to improve drainage, raise yards, raise lots at reconstruction, and pump sewage to mains. Miami's airport also would need better drainage, requiring an approximate $30 million investment.

A 1-m rise in sea level would require raising most bridges to ensure adequate clearances and reduce vulnerability to storm surges during hurricanes.

Global climate change could force Dade County to greatly expand its planned use of desalinated water. This may also be needed by the Everglades National Park, if it is protected from the rising sea.

Analysis of Miami's coastal defense and water supply options provides insight into the likely impacts of sea level rise on cities built on coral reefs, but not into the likely response of most mainland cities on the U.S. coastline. Dade County is unusual because readily extracted fill is extensively available on public lands having easy access to a canal system that can be navigated by flat-bottomed barges. Nevertheless, this case study suggests that global climate change could force large coastal cities to invest billions of dollars to add and upgrade infrastructure.

<u>Cleveland's Infrastructure</u>

The Cleveland case study examined impacts of climate change on snow and ice control costs, road construction and maintenance, heating and cooling costs and equipment needs, water supply, and storm and wastewater transport. The study also includes a preliminary analysis of the effects of a drop in the level of Lake Erie as estimated by Croley (see Great Lakes chapter of this report). The impact on the snow and ice control budget was estimated by analogy to the budget in Nashville, Tennessee.

Results are displayed in Table 15-3, which shows that the net impact of climate change on Cleveland's annual infrastructure costs could be negligible, although expenditures probably would shift between categories. In addition to the costs shown in Table 15-3, a one-time capital expenditure of $68-80 million could be required to add air-conditioners in public buildings. Also, many private residences probably would install air-conditioners.

Walker et al. estimated that global climate change could cause annual snowfall in Cleveland to drop from 1.25 to roughly 0.2 m, reducing annual snow and ice control costs by about $4.5 million. Decreased frost damage to roads and bridges could yield further savings estimated at $700,000 per year. An estimated drop of $2.3 million per year in heating costs for public buildings also was projected. Conversely, annual public air-conditioning costs seemed likely to rise by $6.6-9.3 million. The impacts on the transit operating budget seemed likely to mirror the impacts on the general budget, with reduced mishaps and traffic delays in ice and snow offsetting increased fuel costs for vehicle cooling.

The study suggested Cleveland might spend about $65-80 million to add air-conditioning to older schools and large nonoffice spaces such as gyms and repair garages. Much of this expenditure would occur as buildings were replaced or refurbished and might have occurred even without climate change.

The rise in winter temperatures associated with a doubling of CO_2 might allow Cleveland to use thinner pavement resulting in possible savings of about 3% in road resurfacing costs and 1% in reconstruction costs. The net savings could average about $200,000 per year or 1.3% of the city's current capital budget. Standard engineering estimates (AASHTO, 1987) suggested that the rate of pavement deterioration probably also should decline as winter temperatures rise, saving roughly $500,000 per year.

A climate-induced drop in the level of Lake Erie probably would not adversely affect Cleveland, although some dredging might be required in the Cuyahoga River. Upgrading of the city's combined storm and wastewater collection system appeared to be unnecessary, although this would depend upon rainfall variability.

TABLE 15-3. Estimated Impacts of a Doubling of CO_2 on Cleveland's Annual Infrastructure Costs (Millions of 1987 Dollars)

Cost Category	Annual Operating Costs
Heating	-$2.3
Air-conditioning	+$6.6-9.3
Snow and ice control	-$4.5
Frost damage to roads	-$0.7
Road maintenance	-$0.5
Road reconstruction	-$0.2
Mass transit	summer increase offsets winter savings
River dredging	less than $0.5
Water supply	negligible
Storm water system	negligible
Total	-$1.6 to +$1.1

Source: Walker et al. (Volume __); Weggel et al. (Volume).

If temperature rises several degrees, most northern cities probably could anticipate savings in snow and ice control, heating, and roadway construction and maintenance costs similar to those described for Cleveland. These savings might approximately offset the increase in air-conditioning costs. Communities south of Cleveland probably would experience modest budget increases. Cleveland could become a more attractive location for water-intensive industry if water supplies in other areas become less reliable. Resulting in-migration could bring further growth-related infrastructure costs. (See Great Lakes chapter for a further discussion of this issue.)

New York City's Water Supply

New York City's infrastructure may be affected in many ways by global climate change. Temperature change could affect the same capital expense categories in both New York City and Cleveland. In addition, the city may have to invest heavily to protect underground infrastructure from seawater infiltration.

The most pressing, and perhaps largest, problem facing the city may be the effects of global climate change on the adequacy of the city's water supply. The New York City study focused on that issue.

The New York metropolitan area draws water from the adjoining Hudson and Delaware River Basins and from underground aquifers serving coastal New Jersey and Long Island. Figure 15-1 shows the region and its water supply sources.

The water supply network is in deficit. The Mayor's Task Force (1987) has recommended remedying New York City's portion of the deficit through better management of water demand and detailed study of the possibility of reactivation of a water intake at Chelsea, a $223-391 million investment that would yield 375 to 750 million liters of water daily.

Walker et al. estimated changes in water demand using design standards for commercial cooling-tower demand, changes in electricity demand estimated by Linder et al. (1987), historic residential summer water use, and estimates by Hull and Titus (1986) of saltwater advance up the Delaware River. This last estimate is an order-of-magnitude estimate made by analogy. Water supply was also estimated by analogy, using a Great Lakes water balance model. Walker et al. assumed that baseline demand would not increase above projected demand in 2030, potentially underestimating the increased demand for water and saltwater advance.

Walker et al. also estimated that a rise in temperatures consistent with the GISS and GFDL scenarios would mean about a 20% increase in cooling degree days.* In response, average daily demand for water used in cooling large buildings could increase by 190 million liters during the summer and increased lawn watering could raise demand by 110 million liters per day, thereby generating a 5% rise in annual demand.

Higher temperatures could increase evaporation and evapotranspiration, decreasing the ability to store water efficiently in surface impoundments. The water balance model indicated the supply loss could range from 10 to 24%.

---

*The number of degrees by which the average daily temperature exceeds 65°F.

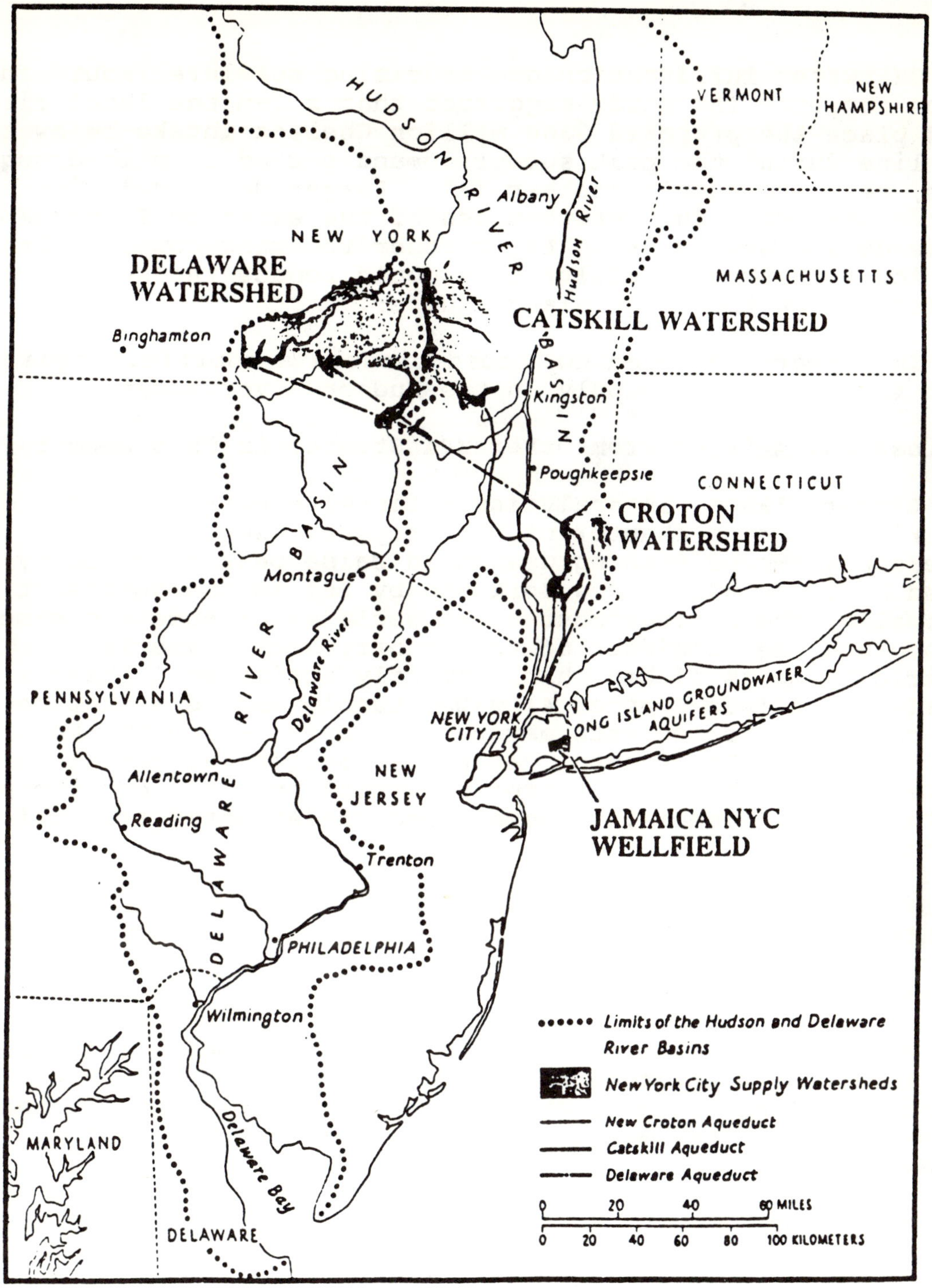

Figure 15-1. The Sources of New York City's Water Supply.

Source: Citizen's Union Foundation of the City of New York, 1986.

Saltwater infiltration due to rising sea level would further reduce supply. The study suggested that a 1-m sea level rise could place the proposed $300 million Chelsea intake below the salt line during the peak summer demand period in mild drought years, reducing supply another 13%. Larger sea level rise or greater droughts might prevent use of the existing Poughkeepsie intake during severe droughts, further reducing supply. In addition, subsurface infiltration could reduce the supply available from the Long Island aquifer.

In summary, doubled CO_2 could produce a shortfall equal to 28 to 42% of planned supply in the Hudson River Basin.

Implications Arising From Other EPA Studies in This Report

Linder (1987), in analyzing national electricity demand (see Electricity Demand chapter of this report), suggests that increased air-conditioning use could raise peak electricity demand by 10 to 20% in New York and by 20% in the Southeast. Nationally, the northernmost areas would experience decreased demand, and the remainder of the country would experience electricity needs higher than currently estimated. Sheer's study of California (see Chapter 4) water supply suggests that new surface water impoundments may be needed to meet urban water and other demands. The coastal defense strategies suggested in the Sea Level Rise chapter (Chapter 9) would apply to most urban coastal areas, especially those along the Atlantic and Gulf coasts.

RESULTS OF RELATED STUDIES

Metropolitan Water Supply

Schwarz and Dillard (1988) conducted telephone interviews with local infrastructure managers to identify the probable impacts of global climate change on water supply and drainage in several metropolitan areas. Results from some cities are discussed here.

Washington, DC

Longer hot spells could warm the Potomac River and cause trihalomethane formed during chlorination to rise above allowable limits. Remedying this could require a capital investment of roughly $50-70 million and could increase treatment costs. Also, lawn watering probably would increase during long spells of hot, dry weather. Although a substantial decrease in runoff could reduce supply in parts of the system, the availability of additional storage capacity would make a shortage unlikely.

New Orleans

Sea level rise could necessitate moving the water intakes considerably farther up the Mississippi and replacing cast iron water mains that would corrode if exposed to salt water. Reduced river flow also could increase settling and treatment requirements. Rising sea level could increase saltwater infiltration into the water system and could require increased pumping capacity.

New York City

This study raised many of the same concerns regarding water supply and demand as the study by Walker et al. (Vol. X) and indicated that even a 0.25-m sea level rise would mean the proposed Chelsea intake was too far downstream. The sanitary and storm sewage system capacity and design probably would not need revision. Nevertheless, in a few low-lying areas, higher sea level could increase sewer backups, ponding, and basement flooding when high tides coincided with high runoffs.

Tucson

Tucson is depleting its aquifer despite substantial conservation efforts and lawn watering with treated wastewater. Higher temperatures would increase demand and tighten supply, possibly jeopardizing the city's ability to draw on water from the Central Arizona Project on the already strained Colorado River. While modest savings might be achieved through stricter conservation measures and more wastewater use, purchase of water in the regional market most likely would be the only practical response to climate-related shortfalls.

IMPLICATIONS FOR THE NATION'S URBAN INFRASTRUCTURE

The implications of climate change for urban America vary spatially. Some localities, especially those along the Great Lakes, might benefit. Others, especially those along the coastlines and in water-short areas, could face major infrastructure costs. The costs would be especially high if changes came through abrupt "sawtooth" shifts, making it difficult to adapt infrastructure primarily during normal repair and replacement. The likely impacts of an effective doubling of atmospheric CO_2 could affect a wide range of infrastructure.

Water

Hotter temperatures could cause faster evaporation of groundwater and raise the demand for water to support commercial air-conditioning systems and lawn watering. Larger winter rains in the West could force a lowering of dam levels to ensure availability of enough capacity to control flood waters. At the same time, sea level rise could cause saltwater to advance up rivers and to infiltrate into coastal aquifers. In droughts, many existing water intakes might deliver brackish water.

The solution to these problems could be expensive: strong conservation measures, miles of aqueducts from new water intakes at higher river elevations, new reservoirs, sewage effluent recycling systems to support commercial cooling or lawn watering, and perhaps desalinization efforts along the coasts. The solution for the New York-Philadelphia corridor alone is likely to cost $3-7 billion. Communities in the Delaware River Basin, northern New Jersey, the lower Hudson, and Long Island might well form a multi-State water supply and management district of unprecedented size and complexity to handle financing and capital construction.

Drainage and Wastewater Systems

Increased storm size and intensity could tax many storm sewer systems. Sea level rise also could reduce coastal flood protection levels in low-lying areas. The resulting increases in flooding and releases of untreated waste into watercourses from combined storm and wastewater systems probably would motivate new sewer investments. In Dade County alone, costs to maintain flood protection at existing levels could be $200-300 million if sea level rose 1 m.

Temperature rise could increase hydrogen sulfide formation in sewer pipes, leading to internal corrosion and eventual failure. In coastal areas, increased ocean flooding could increase how frequently storm sewers must carry corrosive saltwater. Sea level rise also could cause more pipes in coastal areas to face the external risk of corrosive seawater. More frequent inspection and earlier replacement of much existing pipe, as well as a gradual shift to more corrosion-resistant pipe with plastic lining, might be required.

Coastal Defenses

Protection from a rising sea could require tremendous investment in many major coastal communities. In urban areas, a common approach might be the New Orleans solution, where extensively developed coastal areas are protected by dikes, and covered drainage ditches behind the dikes are pumped to keep out the saltwater.

Roads

Rising temperatures could reduce the costs of road construction and maintenance. Snow and ice control costs might drop dramatically. A decrease in deep freezes and freeze-thaw cycles also would mean fewer potholes. Warmer temperatures and the improved drainage resulting from higher evaporation rates could permit use of thinner pavements in many areas, but require enhanced expansion capabilities.

Bridges

Sea level rise and increased storm intensity could require upgrading of many bridges either through costly retrofit or as part of normal reconstruction. The range of temperature accommodated by expansion joints also might need to be increased. The costs might be modest if bridge planners upgraded in anticipation of climate change.

Mass Transit

In the North, buses and railcars could experience fewer snow-related delays. Conversely, slight increases in fuel costs could result from increased use of air-conditioners.

Electricity and Air-Conditioning

Hotter temperatures could increase air-conditioning use. Consequently, peak load capacity to generate electric power might have to increase in response to global climate change. Fortunately, air-conditioning equipment is replaced fairly frequently, so increased loads on existing equipment could be accommodated incrementally. Some houses and public buildings in northern climates might need to add air-conditioning, but such retrofitting has been performed since the first window air-conditioners were introduced.

NATIONAL POLICY IMPLICATIONS

The uncertain, yet potentially imminent, impact of global climate change has already increased the risks of infrastructure investment. Application of design standards and extrapolation from historical data still may not provide reasonable assurance that water and power supply, dam strength and capacity, bridge clearances, or storm sewerage capacity will be adequate for the 35-, 50-, and 100-year design cycles of these facilities. For example, the National Flood Insurance Program's maps identifying the historical 100-year floodplain and 500-year floodway may no longer provide a reliable basis for local building and zoning ordinances designed to minimize flood losses to life and property.

Investment Analysis Methods

Especially in coastal areas, the possibility of accelerating global climate change may soon require careful decisions regarding how and when to adapt the infrastructure. A strong emphasis on life-cycle costing and upgrading during reconstruction in anticipation of future changes could yield large, long-term cost savings. To accomplish this goal such institutions as the Department of Housing might work with the American Public Works Association, the National League of Cities, the U.S. Conference of Mayors, the American Planning Association, and similar groups to educate their constituencies regarding the uncertainties and ways to incorporate them into the decisionmaking process.

Water Supply

Water supply is of particular concern because decades are required to plan and complete projects, which then might last 100 years. Dams, reservoirs, and water intakes currently being planned and built could become obsolete or inadequate as a result of global climate change. Elsewhere, communities might be allowing development of land needed for reservoirs to meet the water shortages that would result from climate change.

Such federal agencies as the U.S. Geological Survey, U.S. Army Corps of Engineers, and EPA may wish to work with States and municipalities to study the probable impacts of climate change scenarios on the water supply of major metropolitan areas.

Water supply investments frequently affect multi-State areas, creating a need for Federal coordination. The Supreme Court has been forced to settle previous water rights disputes concerning many major rivers, and global climate change might well generate new disputes. Cost-effective response to climate change also might require new multi-State water projects. For example, a major project on the Hudson River that allowed New York City to reduce its use of Delaware River water might be the least costly way to increase water supply in Philadelphia. The upcoming State debates over water supply financing should be informed by the lesson of past infrastructure crises: water piping and pumping costs resulting from global climate change should be fully recovered from the water users to avoid stimulating artificial demand for bargain water.

Infrastructure Standards

Voluntary standards organizations, such as the American Society of Civil Engineers, the Building Officials and Code Administrators International, and the American Association of State Highway and Transportation Officials, may wish to educate their committees on global climate change. Growing uncertainty concerning future temperature, precipitation, and sea levels might dictate a reassessment of existing standards and safety factors for ventilation, drainage, flood protection, facility siting, thermal tolerances, resistance to corrosion, and so forth. Conversely, prompt detection of lasting changes could allow adjustment of geographically based standards -- for example, on roadbed depth and home insulation levels -- and provide significant savings. Thus, the standard-making organizations might beneficially establish policies concerning how and when their committees should account for global climate change or educate their committees about the prospects.

Voluntary standard organizations, such as the American Society of Civil Engineers, the Building Officials and Code Administrators, International and American Society of Heating and Refrigerating and Air Conditioning Engineers should examine the need for changes in existing standards to improve the use of energy and safety factors.

BIBLIOGRAPHY

AASHTO. 1987. American Association of State Highway and Transportation Officials. Manual for the Design of Permanent Structures, Appendix A. Treatment of Roadbed Swelling and/or Frost Heave in Design.

Bruske, E. 1988. 104 (phew!) degrees hottest in 52 years. The Washington Post 111(225):A1, A6. July 17.

Hirsch, J. 1988. As streets melt, cars are flummoxed by hummocks. The New York Times 137(47599):B1, B5. August 16.

Hull, C.H.J., and J.G. Titus, ed. 1986. Greenhouse Effect, Sea Level Rise, and Salinity in the Delaware Estuary. Washington, DC: U.S. Environmental Protection Agency. Publication no. 230-05-86-010.

Lewis, N. 1988. Two more heat records fall as summer of 1988 boils on. The Washington Post 111 (257):A1, A10, A11. August 18.

Linder, K.P., M.J. Gibbs, and M.R. Inglis. ICF Incorporated. 1987. Potential Impacts of Global Climate Change on Electric Utilities. New York State Energy Research and Development Authority, Albany, NY. Publication no. 824-CON-AEP-86.

Mayor's Intergovernmental Task Force on New York City Water Supply Needs, "Managing for the Present, Planning for the Future", December 1987.

Metropolitan Dade County Planning Department. 1987. Comprehensive Development Master Plan for Metropolitan Dade County, Florida. July 1979, Amended July 1985 and June 1987.

National Council on Public Works Improvement. 1988. Fragile Foundations - Final Report to the President and Congress, Washington, DC. February 1988.

Rhoads, P.B., G.C. Shih, and R.L. Hamrick. 1987. Water resource planning concerns and changing climate: A Florida perspective. In: Proceedings of the Symposium on Climate Change in the Southern United States: Future Impacts and Present Policy Issues. Norman: University of Oklahoma, pp. 348-363.

Schwarz, H.E., and L. Dillard. 1988. (Draft Report) Urban water. Chapter III-D. In: AAAS Global Climate Change Report. Washington, DC.

Sossaman, B.A. 1988a. News Release, U.S. Army Corps of Engineers, New Orleans District. June 28.

Sossaman, B.A. 1988b. News Release, U.S. Army Corps of Engineers, New Orleans District. July 15.

Statistical Abstract of the United States - 1987, U.S. Government Printing Office, 1988.

Titus, J.G., C.Y. Kuo, M.J. Gibbs, T.B. LaRoche, M.K. Webb, and J.O. Waddell. 1987. Greenhouse effect, sea level rise, and coastal drainage systems. Journal of Water Resources Planning and Management 113(2):216-227.

The Washington Post. 1988. Warped rails checked in Amtrak wreck. 111(246):A5. August 7.

Wilcoxen, P.J. 1986. Coastal erosion and sea level rise: Implications for ocean beach and San Francisco's westside transport project. Coastal Zone Management Journal 14(3):173-191.

CHAPTER 16

NATIONAL IMPACTS OF CLIMATE CHANGE ON ELECTRIC UTILITIES

FINDINGS

Global warming would increase electricity demand, generating capacity requirements, annual generation, and fuel costs. The impacts could be significant within a few decades, and would increase substantially over time if global warming continues.

- By 2010, new generating capacity requirements induced by climate change are estimated to increase by 25 to 55 gigawatts (GW) or 9 to 19 percent above estimated new capacity requirements, assuming no change in climate. Between 2010 and 2055, climate change impacts could accelerate, increasing new capacity additions by 200 to 400 GW (14 to 23 percent) above what would be needed in the absence of climate change. These capacity increases would require investments of approximately $200 to $300 billion. These results are sensitive to assumptions about the rates of economic growth, technological improvements, and the relationship between energy use and climate. The savings in other fuels (gas and oil) in northern regions and the impacts on hydroelectric supplies were not analyzed.

- Estimated increases in annual electricity generation and fuel use induced by climate change represent several thousand gigawatt-hours by 2055. The estimated increases are 1 to 2 percent in 2010 and 4 to 6 percent in 2055. Annual fuel, operation, and maintenance cost would be hundreds of millions of dollars in 2010 and several billion dollars in 2055.

- Estimated regional impacts differ substantially for the scenarios studied. The largest increases would occur in the Southeast and Southwest, where air conditioning costs are large relative to heating. Northern border states may have a net reduction in electricity generation relative to base over increases in generation capacity.

Policy Implications

- Utility executives and planners should begin to consider climate change as a factor in planning new capacity. The estimated impacts of climate change is similar to other uncertainties and issues utility planners may need to consider over the 20-30 year period. Additional climate and utility analyses are needed to develop refined risk assessments and risk management strategies.

- The increased demand for electricity induced by climate change also could offset other environmental problems, such as the implementation of "acid rain" strategies, adherence to the international NO_x treaty, state implementation plans for ozone control, and thermal pollution control permit requirements. The Environmental Protection Agency should analyze the impacts of climate change on long-range policies and should include climate change as an explicit criterion in making risk management decisions when appropriate.

- The increased demand for electricity could make policies to stabilize the atmosphere through energy conservation more difficult to achieve. The estimated increases in electricity generation induced by climate change alone could increase annual CO_2 emissions by 40-65 million tons in 2010 and by 250-500 million tons in 2055. This does not consider reduced demand for oil and gas in residential furnaces for heating. Future analyses of national and international strategies to limit greenhouse gases should include the additional energy demand created by global warming as a positive feedback.

CLIMATE CHANGE AND ELECTRICITY DEMAND

While climate change could affect a wide range of energy sources and uses, its implications with regard to the demand and production of electricity could be significant. Many electrical end uses vary with weather conditions. The principal weather-sensitive end uses are space heating and cooling and, to a lesser degree, water heating, cooking, and refrigeration. These applications of electricity may account for up to a third of total sales for some utilities and may contribute an even larger portion of seasonal and daily peak demands. (Electric energy used for air-conditioning exceeds that used for space heating nationwide, and the temperature-sensitivity associated with cooling is higher than that associated with heating.)

Changes in weather-sensitive demands for electricity can affect both the amount and the characteristics of generating capacity that a utility must build and maintain to ensure reliable customer service. These changes also can affect fuel requirements and the characteristics of efficient utility system operations, particularly the scheduling and dispatching of the utility's generating capacity.

Similarly, utilities in most regions experience their peak demands in the summer. A rise in air-conditioning and other temperature-sensitive summer loads could significantly increase peak loads and, as a result, would step up utility investments in new generating capacity needed to meet additional demands and to maintain system reliability.

PREVIOUS CLIMATE CHANGE STUDIES

A number of utilities conduct analyses relating short-term variations in weather conditions with a need to "weather-normalize" historical demand data and to test the sensitivity of system reliability and operations to these short-term variations. Furthermore, some researchers have speculated regarding the potential effects of longer term climate changes on electricity demand (e.g., Stokoe et al., 1987).

However, only one previous study has estimated the potential implications of longer term, global warming-associated temperature changes on electricity demands and on utility investment and operating plans. Linder et al. (1987) used General Circulation Model (GCM) results to estimate the potential impacts of temperature change on electricity demand and on the supply of hydropower for selected case study utility systems in two geographical areas: a utility located in the southeastern United States and the major utilities in New York State, disaggregated into upstate and downstate systems.

Linder et al. found that temperature increase could significantly heighten annual and peak electricity demands by 2015, and that a temperature rise would require construction of new generating capacity and increases in annual generation. The southeastern utility had higher estimated increases in electricity demand, generation, and production costs than the New York utilities because of greater electricity demands for air-conditioning. In addition, streamflow used to generate hydropower in New York could be reduced, requiring increased use of fossil fuel generation to meet customer demands for electricity.

Chapter 16

CLIMATE CHANGE STUDY IN THIS REPORT

Study Design

Linder and Inglis[1] expanded on the regional case studies of the sensitivity of electricity demand to climate change by conducting a national analysis of electricity demand. The national analysis used the same basic approach as that used in the regional case studies. Relevant regional results from Linder's national studies are discussed in the regional chapters of this report.

The analytic approach developed by Linder et al. formed the basis for estimating the regional and national impacts described in this report. The principal steps in the approach are summarized in Figure 16-1 (see Volume X for more details). Estimated impacts were developed for the relatively near term (from the present to 2010, within electric utility long-range resource planning horizons of 20-30 years) and over the longer term (to 2055), when the magnitude of temperature changes is expected to approach equilibrium levels representative of a doubling of atmospheric concentrations of CO_2. Linder and Inglis used Goddard Institute for Space Studies (GISS) A and B transient estimates of temperature change in 2010 and 2055 in his calculations. The changes in annual temperatures for the utility regions ranged from about 0.6 to 1.6°C in 2010 and from about 3.1 to 5.3°C by 2055.

Linder and Inglis used actual utility demand and temperature data from the case study utilities, and from five other large, geographically dispersed utility systems to develop a set of weather-sensitivity parameters for utility areas. On a weighted-average basis (weighted by electricity sales), utility peak demands were estimated to increase by about 3.1% per change in degree Celsius (ranging from -1.35 to 5.40% across utility areas), and annual energy demands were estimated to increase by about 1.0% per change in degree Celsius (ranging from -0.54 to 2.70%).

A number of uncertainties associated with the data and assumptions used to develop these weather-sensitivity relationships suggested that the relationships may understate customer response to climate change, particularly at higher ,

---

[1]The Potential Impact of Climate Change on Electric Utilities: Regional and National Estimates -- Linder and Inglis, ICF, Inc.

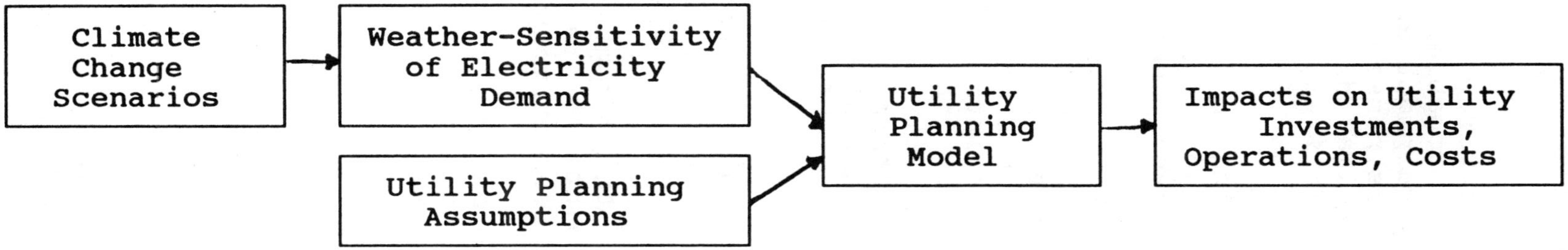

Figure 16-1. Analytic Approach

Source: Linder and Inglis, Volume __.

temperature change levels occurring in the future. For example the approach did not explicitly account for probable increases in the market saturation of air-conditioning equipment as temperatures rise over time. To address this possibility, an alternate case was designed in which the estimated weather-sensitivity values were increased by 50%. This was designated as the "higher sensitivity" case.

Since this study is focused on estimating how climate change may affect key utility planning factors, Linder used a planning scenario assuming no change in climate (a "base case") to serveas a basis for comparison with planning scenarios under alternative assumptions of climate change for 2010 and 2055. Thus, base case utility plans were developed for 2010 and 2055, using assumptions regarding future demands for electricity in the absence of climate change, generating technology option performance and costs, fuel costs, and other utility characteristics. Linder and Inglis assumed that future capacity and generation requirements will be met by investments either in new coal-fired baseload capacity or in oil- and natural gas-fired peaking capacity. Other sources, such as nuclear energy and renewables or innovative fossil fuel-fired technologies (e.g., fluidized bed combustion) were not considered.*

Demands for electricity in the absence of climate change can be related to the overall level of economic activity as represented by the gross national product (GNP). Because of the critical nature of economic growth, alternative GNP growth rates were assumed in developing the base cases; these ranged from 1.2 to 2.1% per year.** These alternative assumptions are referred to as "lower growth" and "higher growth," respectively.

These assumptions served as inputs to a regional planning model called the Coal and Electric Utilities Model (CEUM). CEUM output includes the amount and characteristics of new generating capacity additions, electricity generation by fuel type, and electricity production costs.

---

*Innovative technologies will penetrate future power markets. However, the analysis for this focused on distinction between requirements for peaking capacity vs. baseload capacity generally and not on utility investments in specific types of technologies.

**These GNP growth rates are relatively conservative, but they are comparable with GNP growth rates used by EPA in its Report to Congress on atmospheric stabilization.

Limitations

The study did not include analysis of temperature sensitivity for all utility regions of the United States. It focused narrowly on impact pathways, considering only the potential effects of temperature change on changes in electricity demand. Neither the potentially significant impacts of climate change on hydropower availability nor the impacts of reduced water supplies for cooling were included.

Furthermore, the study did not evaluate the sensitivity of the results to different GCM climate scenarios, although the use of the GISS transient experiment results for 2010 and 2055 indicates relative sensitivities to small and large temperature changes.

Very limited climate change information important for utility planning was available (e.g., variations in temperature changes and the occurrence of extreme events, which affect powerplant dispatch and determinations of peak demands, respectively).

Many uncertainties exist regarding the concepts, methods, and assumptions involved in developing and applying estimates of the temperature sensitivity of demand. For example, a key assumption is that the estimated sensitivities of demand to historical, short-term variations in temperature are adequate representations of future relationships between electricity demand and long-term changes in mean temperatures.

Uncertainties also exist regarding market, regulatory, technological, and other conditions that will face the utility industry in the future. For example, technological changes that improve the energy efficiency of weather-sensitive end-use equipment or electricity-generating equipment could reduce the estimated impacts of climate change.

Results

The potential national impacts for 2010 and 2055 are summarized in Table 16-1. The table presents base case values (i.e., assuming no change in climate) for each year and estimated impacts represented by changes from the base case values. The impacts for 2055 are presented for both the lower growth GNP and the higher growth GNP cases. Also, where ranges of impacts are presented, they summarize the estimates under alternative climate change scenarios (GISS A and GISS B) and assumptions of the

TABLE 16-1. The Potential National Impacts of Climate Change on Electric Utilities

	2010		2055			
			Lower GNP		Higher GNP	
	Base	**Increase**	**Base**	**Increase**	**Base**	**Increase**
Peak demand (GW)	774	20-44	1,355	181	1,780	238-357
New capacity reqirements (GW)[a]						
Peaking	50	13-33	176	118	254	182-286
Baseload	226	11-22	1,011	67	1,423	74- 98
Total	376	24-55	1,187	185	1,677	227-384
Annual sales (bkWh)	3,847	39-67	6,732	281	8,848	370-555
Annual generation[b] (bkWh)						
Oil/gas	287	(12)-(29)	221	2	308	27- 51
Coal	2,798	54-103	6,242	305	8,295	381-560
Other	1,092	1- (1)	846	(2)	1,003	(7)- 0
Total	4,177	43- 72	7,309	305	9,607	401-611
Cumulative capital costs[c] (billions of 1986 $)	669	25-48	1,765	173	2,650	222-328
Annual costs (billions of 1986 $)	162	3-6	474	33	655	48-73

[a]Includes reserve margin requirements; does not include "firm scheduled" capacity.
[b]Includes transmission and distribution losses.
[c]"Base" values include regional capital expenditures for utility-related equipment in addition to new generating capacity (e.g., new transmission facilities).

Abbreviations: GW = gigawatts; bkWh = billion kilowatthours.
Source: Linder and Inglis, Volume __.

weather sensitivity of demand ("estimated sensitivity" and "higher sensitivity").

Estimated increases in peak demand on a national basis range from 2 to 6% by 2010. Changes in estimated annual energy requirements by 2010 are more modest, ranging from 1 to 2%. In 2055, peak national demands are estimated to increase by 13 to 20% above base case values, and annual energy requirements are estimated to increase by 4 to 6%. By 2010, new climate change-induced generating capacity requirements increase by 6 to 19%, or about 24 to 55 GW, representing an average increase of up to 1 GW per State (approximately the capacity of a large nuclear or coal-fired baseload powerplant). The majority of the capacity increase is for peaking capacity rather than baseload capacity. The investment associated with these capacity increases is several billion dollars (in constant 1986 dollars). By 2055, the change in new capacity requirements increases in percentage terms and represents several hundred GW. Under high GNP and higher weather-sensitivity assumptions, the estimated increase attributable to climate change is almost 400 GW, or 23%.

Annual generation increases for the United States are not as large in percentage terms as those estimated for new generating capacity requirements, but nonetheless, they account for several hundred billion kWh by 2055. In the near term (i.e., to 2010), increased levels and changing patterns of climate change-induced electricity demand permit utilities in some areas having excess generating capacity to serve the growing needs of utilities in other areas through substitution of lower cost coal-fired generation for higher cost oil- and natural gas-fired generation. On net, oil- and gas-fired generation would be lower owing to climate change in 2010.* In 2055, oil- and gas-fired generation is projected to increase along with coal-fired generation, because all the excess capacity that had existed in 2010 either would have been fully used by growing demands to 2055 or would have been retired. The estimated impacts of climate change on national new generating capacity requirements and annual generation are illustrated in Figure 16-2.

Table 16-1 also indicates that the increase in annual costs for capital, fuel, and O&M associated with climate change-induced modifications in utility investments and operations are a few

---

*This result is confirmed and explained in more detail in the case study report (Linder et al., 1987).

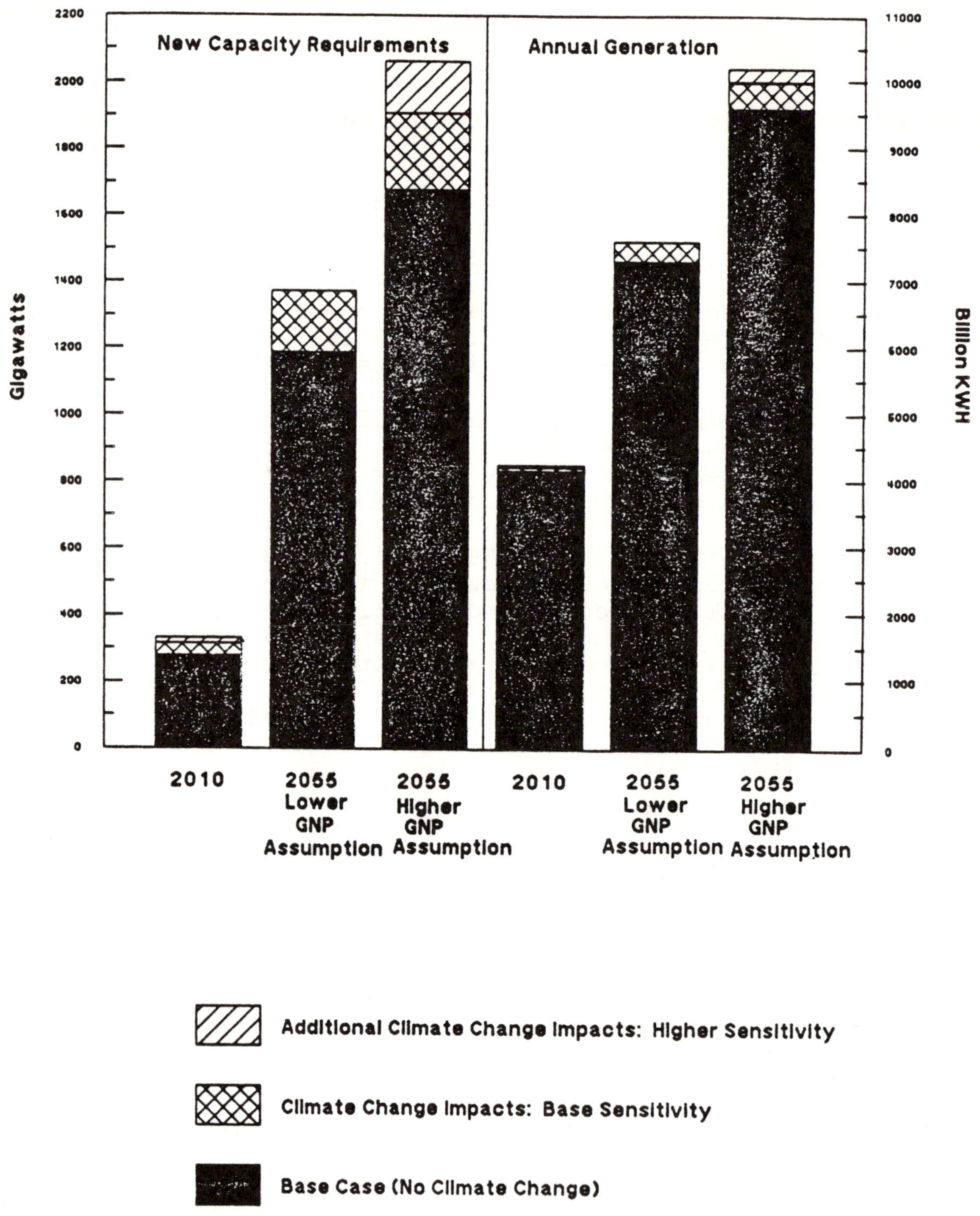

Figure 16-2. Potential impacts of climate change on electric utilities, United States

Source: Linder and Inglis, Volume __.

billion dollars in 2010 and are $33-73 billion by 2055, a 7 to 15% increase over base case values for 2055.

Figures 16-3 and 16-4 illustrate the diversity of the estimated results for generating capacity on a state-by-state basis. The State and regional differences reflect differences in current climate conditions (e.g., seasonal temperature patterns), assumed future climate changes, and electricity end-use and utility system characteristics (e.g., market saturation of weather-sensitive appliances and equipment).

Figure 16-3 shows that estimated reductions in new capacity requirements induced by climate change are limited to the winter-peaking regions of the extreme Northeastern and Northwestern States. The Great Lakes, Northern Great Plains, and Mountain States, as well as California, are estimated to experience increased new capacity requirements by 2055 in the range of 0 to 10%. Increases greater than 20% are concentrated in the Southeast, Southern Great Plains, and Southwest.

Figure 16-4 shows a somewhat similar geographic pattern of impacts for electricity generation in 2055. Reductions in generation are estimated in the Northern States, and the greatest increases are concentrated in the Southwest. Despite substantial use of air-conditioning in the Southeast, the estimated increases in generation are only in the 5 to 10% range. There is a relatively high market saturation of electric heat in the region, and the increase in cooling is partly offset by a decrease in heating as a result of warmer winters.

Because regions are affected differently, the results indicate potential changes in the patterns of interregional bulk power exchanges and capacity sales overtime and as climate change. For example, under the assumption of increasing temperatures, some regions may require significant amounts of additional generating capacity to meet increased demands reliably during peak (cooling) seasons, but may experience lower demands in other (heating) seasons. As a result, the region's needs may be for power plants that are utilized heavily during only part of the year. Low annual utilization in the region would not justify construction of high capital and low fuel cost baseload power plants that can produce electricity more cheaply (per kwh) than low capital and high fuel cost peaking units. However, when considered across several regions, the least-cost plan may be to construct baseload power plants in certain regions, utilize them to an extent greater than required by the region, and sell the "excess" electricity from these plants into other regions. The

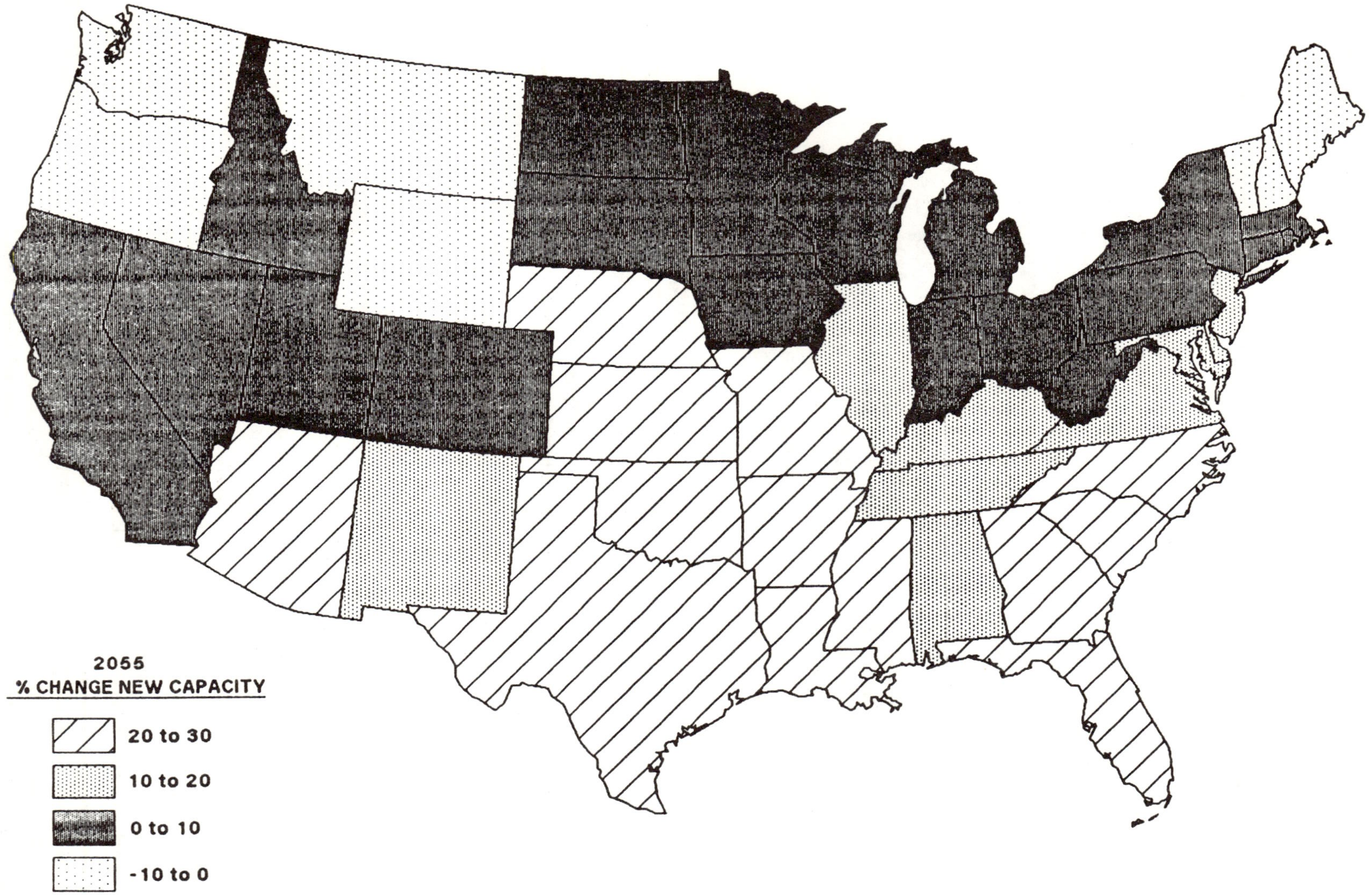

Figure 16-3. Changes in electric utility capacity additions by state: induced by climate change to 2055.

Source: Derived from Linder and Inglis, Volume __.

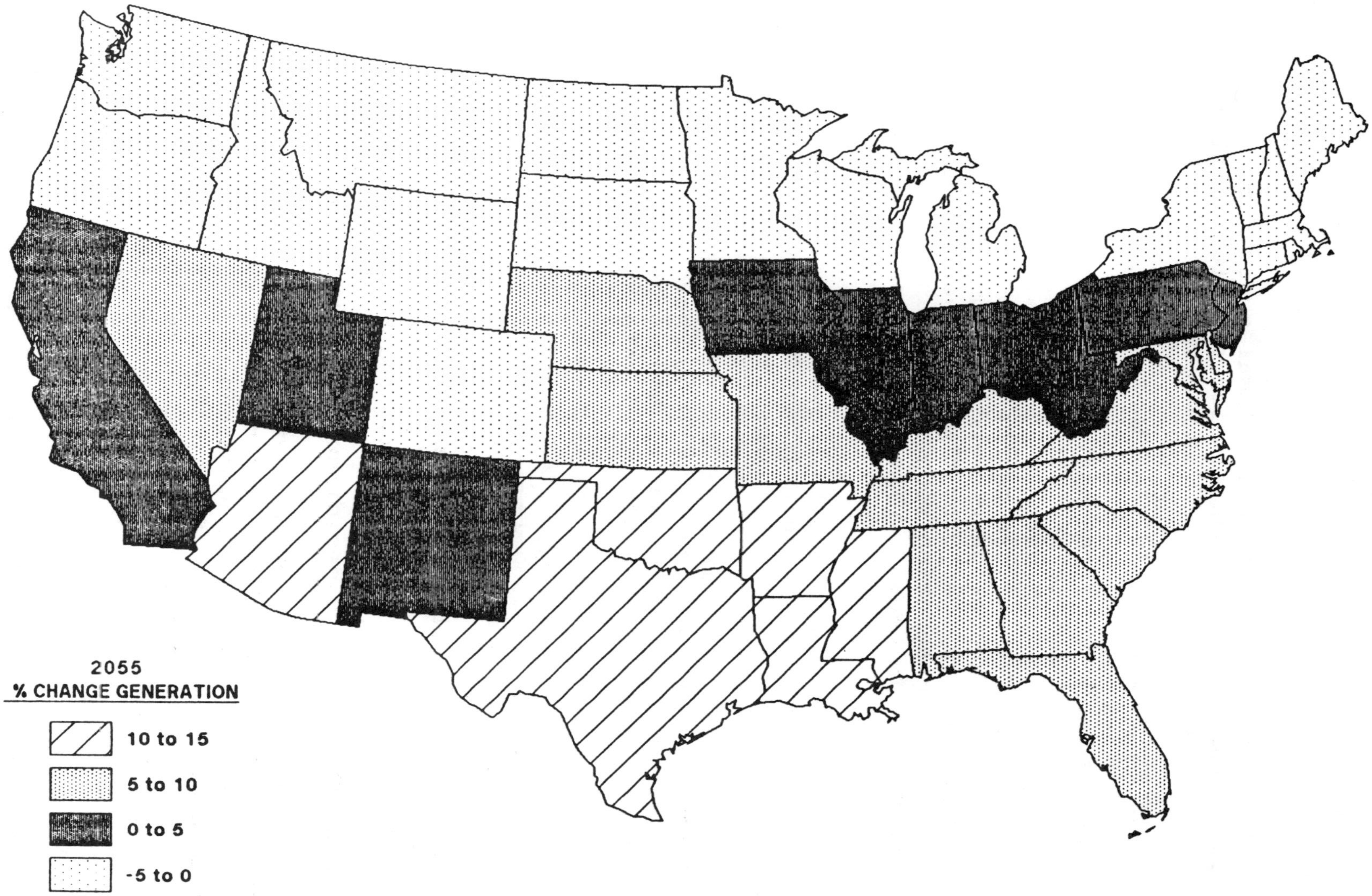

Figure 16-4. Changes in electricity generation by state: induced by climate change in 2055.

Source: Derived from Linder and Inglis, Volume __.

location and amount of these interregional sales would be subject to the transfer capabilities of transmission capacity in place.

SOCIOECONOMIC AND ENVIRONMENTAL IMPLICATIONS

Despite the limitations of the analysis and the need for more research to refine the data and methods used, the results are judged to be reasonable estimates of potential climate change impacts associated with temperature change and its effects on electricity demand. Key socioeconomic and environmental implications of the results stem from the increases in electric generating capacity and generation requirements associated with climate-induced changes in demand. The implications include the following:

- Climate change could result in overall fuel mixes for electricity generation that differ from those expected in the absence of climate change.

- Regional differences in capacity and generation requirements suggest that important, new opportunities for interregional bulk power exchanges or capacity sales may arise as a result of climate change.

- The impacts of uncertain climate conditions over the long term could pose significant planning and economic risks. Because of long lead times required to plan and build economic baseload generating capacity, the ability of utility planners to correctly anticipate climate change could result in lower electricity production costs. The magnitude of these risks in some regions (e.g., the Southeast and the Southern Great Plains) could be similar to other uncertainties that utility planners and decisionmakers must face.

- If the result is confirmed that the majority of new capacity requirements in response to climate change are for peaking capacity, a new technological and market focus would be directed at this type of generating plant. Related to this would be increased R&D on electricity storage technologies, which would allow lower cost, more efficient powerplants to generate, at off-peak times, electricity for use during peak periods.

- Because increases in customer demands for electricity may be particularly concentrated in certain seasons and at peak periods, conservation and especially load management programs that improve the efficiency or change the patterns of customer uses of electricity could be more cost-effective when considered in the context of potential changes in climate.

- o Increased demands could increase electricity imports, adversely affecting the U.S. balance of payments.

- o Increased electricity generation implies increased adverse environmental impacts associated with the following:

  - -- air quality (e.g., emissions of sulfur dioxide, NO_x, and other pollutants);

  - -- land use for new powerplant sites, fuel extraction, fuel storage, and solid waste disposal;

  - -- water quality and use (e.g., for powerplant cooling and fuel processing); and

  - -- resource depletion, especially of such nonrenewable fuels as natural gas, which may be in short supply.

  Of particular concern would be additional water withdrawal and consumption requirements in areas where water supplies may be reduced by climate change.*

- o Increased electricity generation also implies increased emissions of CO_2 and other greenhouse gases. For example, if the estimated increases in climate change-induced generation reported in Table 16-1 were met by conventional technologies, CO_2 emissions could increase

*For example, increased electricity generation induced by climate change in Northern California could increase requirements for water withdrawal by 600-1,200 million ft^3 and for water consumption by 200-400 million ft^3 in 2055. Comparable figures for the Southern Great Plains in 2055 would be water withdrawal of 5,800-11,500 million ft^3 and consumption of 1,800-3,500 million ft^3.

by 40-65 million tons per year by 2010 and by 250-500 million tons per year by 2055.**

POLICY IMPLICATIONS

In general, the study results suggest that utility planners and policymakers should begin now to assess more fully and to consider climate change as a factor affecting their planning analyses and decisions. If more complete and more detailed analyses support the socioeconomic and environmental implications of the climate change effects described above, they should be explicitly addressed in planning analyses and decisions.

Specific policy implications related to the findings include the following:

- In formulating future National Energy Plans, the Department of Energy may wish to consider the potential impacts of climate change on utility demands.

- The interactions of climate change and the current efforts of the Federal Energy Regulatory Commission (FERC) to restructure the electric utility industry are difficult to assess. For example, the industry's response to FERC policies could either accelerate or reduce the rate of emissions of greenhouse gases, depending upon changes in the mix of generating fuels and effects on the efficiency of electricity production. The possible, alternative responses should be assessed, and FERC policies should be considered with respect to their potential implications related to climate change issues.

- Increases in electricity demands induced by climate change will make achievement of energy conservation goals more difficult. For example, the draft conference statement from "The Changing Atmosphere: Implications for Global Strategy" (Environment Canada, 1988) calls for reductions in CO_2 emissions to be achieved in part through increased efforts in energy efficiency and other conservation measures. An initial

---

**Note, however, that these increases in emissions from electricity production could be offset at least in part by reduced demand for space heating provided by natural gas and oil furnaces or by other direct uses of fossil fuels.

goal for wealthy, industrialized nations set by the conference is a reduction in CO_2 emissions through conservation of approximately 10% of 1988 emissions levels by 2005. The impacts of climate change to increase electricity demand should be factored into the policies and plans designed to achieve this conservation goal.

- Similarly, climate change impacts may exacerbate the difficulties or costs associated with implementing acid rain mitigation strategies being considered by the Congress. However, these strategies center primarily on near-term solutions focusing on emissions reductions from existing powerplants, and the impacts of climate change may not be large within that time frame.

RESEARCH NEEDS

Important areas for further climate change research include improved methods for developing and disseminating climate change scenarios, with particular emphasis on (1) estimates of variables (in addition to temperature) relevant to impact assessment (e.g., hydrologic factors, winds); (2) estimates of the possible impacts of global warming on variations in weather conditions and the occurrence of extreme events; (3) continued attention to estimates of the rate of climate change over time; and (4) estimates of climate change at a more disaggregated regional or local level.

Follow-on research suggestions on the utility side include (1) refinement of the analytical approach, in part through lessons learned from additional utility-specific analyses; (2) more detailed and complete analyses of the weather sensitivity of customer demand for electricity; (3) extension of the approach to consider other pathways (including indirect and secondary effects) in which climate change could affect utility investments and operations; and (4) an assessment of the value of improved climate change information to utility planners and managers.

REFERENCES

Environment Canada. 1988. The Changing Atmosphere: Implications for Global Security Conference Statement. Environment Canada, Government of Canada. Toronto. June 27-30, 1988.

Linder, K.P., M.J. Gibbs, and M.R. Inglis. ICF Incorporated. 1987. Potential Impacts of Climate Change on Electric Utilities. Report 88-2. Albany, New York: New York State Energy Research and Development Authority. (Note: Publication of the report by the Electric Power Research Institute, Palo Alto, California, is forthcoming.)

Stokoe, P.K., and M. LeBlanc. P. Lane and Associates, Ltd., and Discovery Consultants, Ltd. 1987. Socio-economic Assessment of the Physical and Ecological Impacts of Climate Change on the Marine Environment of the Atlantic Region of Canada, Phase I. Halifax, Nova Scotia, Canada: School for Resource and Environmental Studies, Dalhousie University.

CHAPTER 17

ANALYSIS OF CLIMATE VARIABILITY AND EXTREME EVENTS

FINDINGS

Climate variability is an important factor needed for detailed climate impact studies. Analysis of changes in variability estimated by two GCMs is inconclusive. Inconsistencies are obtained in comparing the changes in climatic variability with a changing climate calculated by two different global circulation models (GCMs) in four regions of the United States.

- -- For interannual variability the GISS model produces reduced variability for temperatures for January - April but varied results for the other months. The interannual variability of precipitation tends to increase, although not consistently.

- -- The GISS model suggests reduction in the variability of daily temperature with warming, but these results are not statistically significant on a regional-by-regional basis. The NCAR model (Washington version) produces both increases and decreases in daily temperature variability (although most of these changes are not statistically significant). The daily variability of precipitation with a warming climate tends to increase in the majority of cases (months and locations).

- -- The amplitude of the diurnal cycle (GISS results) tends to decrease in summer with warming. In the other seasons, both increases and decreases occur.

- o On the basis of the mixed model results on how climatic variability will change with a changed climate, the assumption of no change in variability in the scenarios used for this report must be recognized as reasonable, given the current state of knowledge.

- o Comparison of observed and model results for the current climate for the two GCMs for selected regions of the United States reveals both interesting contrasts and similarities regarding the reproduction of climate variability.

  - -- Regarding mean statistics for temperature and precipitation, there are significant discrepancies in both models, compared to observed values in all four regions examined. The GISS model is in general too cool in summer and fall, whereas the NCAR model is either too warm or too cool, depending on the region and the time of year. Errors for both models are in general on the order of several degrees C. The GISS model and the NCAR model (Chervin version) consistently overpredict precipitation in the four regions.

-- Additional comparisons of mean values of precipitation and temperature with two additional versions of the NCAR model revealed both under and over estimations of mean precipitation, whereas mean annual temperature was consistently overestimated.

-- The interannual variability of relative humidity and absorbed radiation (NCAR results) are fairly well reproduced by the model, but tend to be slightly lower than the observed. The daily variance of relative humidity is much lower than the observed at all locations examined.

-- The GISS model overestimated interannual variability of temperature, whereas the NCAR model (Chervin version) tends to underestimate it. Both models reproduce the variability "reasonably well" at some locations some of the time.

-- Both models reproduce the relative variability of annual precipitation fairly well at all locations, although the GISS model overpredicts the absolute variability. There are indications that both models fail to properly reproduce the daily variability of precipitation. Both underestimate the frequency of light rain days, and overestimate the frequency of high-intensity rain days.

-- Explanations for some discrepancies, such as why the daily variances of temperature are too high, relate to how the surface hydrology is modeled in both models. More investigations of model results are necessary in order to improve understanding of future changes in climate variability.

NATURE AND IMPORTANCE OF CLIMATE EXTREMES

The impacts of climate change on society accrue not necessarily from the relatively slow trends in the mean of a climate variable, but rather from the attending shifts in the frequency of extreme events. This issue has already received some attention in the literature (e.g., Schwarz, 1977; Parry, 1978; Mearns et al., 1984) where the nonlinear relationship between changes in the mean and extreme events has been underlined. However, less is known about this factor than about most other aspects of climate change.

For the purposes of climatic impact analysis, Heathcote (1985) states that extreme climatic events may be considered "short term perturbations of the energy flows which provide magnitudes outside the normal spectrum or range of the typical period." What constitutes normal is, of course, a central issue in defining extremes as well as what constitutes the averaging period.

Extreme events relevant to impact analysis function in different time scales, depending upon the climate variable involved and the impact area of interest. Thus, events can range from the length of time (in minutes and hours) that minimum temperatures in Florida remain below a critical value, resulting in damage to citrus crops, to the length of time (in months and years) that precipitation is particularly low in California, resulting in serious water shortages for industry and agriculture. The probability of extreme events can also vary considerably, for example, a range from recurrence intervals for extreme snowfall in the Buffalo, New York, area of a magnitude similar to that of the 1976-77 winter (1/20,000 or 0.00005), to heat waves (temperatures above 100°F for 5 consecutive days) in Dallas, Texas (0.38).

What defines an event as extreme is not only a certain statistical property (for example, likely to occur less than 25% of the time), but also how prepared a particular system is to cope with an event of such magnitude. Hence very few extreme events have a fixed absolute value independent of particular response systems at a particular location. This implies that what constitutes an extreme event can also change over time due to changes in the relevant response system (Heathcote, 1985).

It is thus very difficult to comprehensively review all climatic extreme events of importance to society, and what is presented here is far from an exhaustive catalogue. Because one of the purposes of this review is to highlight the extreme events of importance that can serve as guides for choosing the extreme events that should be quantitatively analyzed in general circulation model (GCM) doubled CO_2 experiments, priority is given to events related to relatively easily analyzable variables.

This review considered the two most important climate variables and their extremes (maxima and minima) and one major meteorological disturbance: temperature, precipitation, and severe storm effects. Extremes in these variables affect the areas of energy use and production, human mortality and morbidity, agriculture, and unmanaged ecosystems (although not all areas are discussed under each climatic extreme event). It should be noted that although the research on extreme climate events not as extensive for natural ecosystems as for some of the other are as covered here, in present research on climatic change impacts on forest ecosystems, the primary importance of variability changes has been underlined (e.g., by Soloman and West, 1985) who assert that climate variance controls geographic ranges of mature tree species, and the rates and outcomes of plant succession; and in chapter 11 of this report where limitations of the EPA forest studies since they did not consider change in climate variability, are discussed).

Temperature

Given the scientific consensus that higher atmospheric concentration of greenhouse gases will raise average global temperatures, extreme temperature effects are given priority in this analysis.

<u>Maximum Temperatures</u>

The area of extreme temperature effects that has received the most attention in the scientific literature is the effects on human mortality and morbitidity (e.g., Kalkstein, 1988; Becker and Wood, 1986; Jones et al., 1982; Bridger et al., 1976; and Ellis, 1972). This is partly because the relevant climatic factors (i.e., maximum daily temperatures and relative humidity) are readily available for analysis.

Heat waves consist of a series of days with abnormally high temperatures. Examples include the 1980 heat wave in the United States when Kansas City had 17 consecutive days above 102°F, St. Louis had 16 days above 100°F (Jones et al., 1982), and Dallas had 42 consecutive days with temperatures above 100°F (Becker and Wood, 1986). The death toll that year was several times above normal (1,265 lives). Major heat waves of at least regional extent in the United States in this century occurred in 1901 (9,058 deaths in the Midwest), 1934, 1935, and 1936 (4,768 deaths nationwide in 1936 alone), 1954, 1955, 1966, and 1980.

Studies have specifically tried to pinpoint the most significant meteorological factors associated with heat-related death and illness. Oeschli and Buechley (1970) identified the maximum temperature with a 1-day time lag for three Los Angeles heat waves. Jones et al. (1982) determined that high maximum temperatures, the number of days that the temperature is elevated, high humidity, and low wind velocity contributed to excess mortality in Kansas City and St. Louis in the 1980 heat wave. Kalkstein et al. (1987) established for New York City that runs of days with high minimum temperatures, low relative humidities, and maximum temperatures above 92°F contributed to heat-related deaths. In a more comprehensive study (reviewed in Chapter 14, Health), Kalkstein (1988) established statistical models relating weather and mortality to heat intensity.

Increases in heat waves are virtually certain, assuming global warming. But how they increase (longer or greater departure from the mean) very much depends on changes in variability that would affect the persistence of high temperatures.

Such crops as corn, soybeans, wheat, and sorghum are sensitive to high temperatures during their bloom phases. For example, Shaw (1983) reported that severe temperature stress during a 10-day period around silking will result in crop failure. Corn silking is a very critical period, during which the number of kernels on the ear are determined. Barnell and Eftmescu (1973) reported that maximum temperatures exceeding 32°C around tasseling and silking resulted in high rates of kernel abortion. Thompson (1968) developed a statistical model to determine corn yields and found that the accumulation of daily maximum temperatures greater than 92°F in July and August was negatively correlated with yield. McQuigg (1981) reported that the corn crop was severely damaged in July 1980 as a result of temperatures exceeding 100°F. The destructive effects of runs of hot days on corn yields was particularly in evidence in 1983 in the U.S. Corn Belt. Although the damage from high temperatures is best documented for corn, it has also been noted in wheat and soybean yields (e.g., Neild, 1982; Mederski, 1983).

Although not as much research has been performed on the effects of temperature extremes on natural ecosystems, some research has been done on forest responses to temperature extremes. Solomon and West (1985) indicate in their summary of climate effects on forests that the frequency, intensity, and lengths of heat waves under climate change conditions are important factors influencing seedling survival, and can contribute to the loss of a species from an ecosystem. A run of warm years can affect the location of tree lines. Schugart et al. (1986) established that a period of warm summers at high altitudes during the 1930s, when the mean annual temperature was no more than 1°C higher than average, resulted in a burst of regeneration in boreal forest trees near polar and altitudinal limits in North America.

High temperatures have their most immediate impact on energy by causing an increase in demand for electricity for air-conditioning. In a report prepared for New York State, the potential impacts of climate change on electric utilities were investigated. Using climate scenarios similar to those used by EPA (see Chapter 3), Linder et al. (1987) found that energy demand would significantly increase in summer (on the order of 3% for an average August day in 2015 for downstate New York). Net energy demand would increase, even though winter heating demand would fall. The study was limited by the way the climate change scenarios were developed. For example, it assumed that the distributions of daily, monthly, and seasonal temperatures all change consistently by the same amount. Since energy demand is affected by changes in extremes, the omission of potential changes in variability limited the study. This is obviously the case for electricity demand is concerned, since capacity is determined in part, by peak demand, and peak demand is determined, in part, by heat waves.

Minimum Temperatures

Since it is presumed that extreme minimum temperatures will be less of a problem with CO_2-induced climate change, they receive less attention here. However, changes will most likely occur in the growing areas of certain crops, where risks of frost damage, for example, may not be clearly known.

The best example of frost damage to crops is the effect of low minimum temperatures on citrus trees; this issue has been studied in depth for the citrus crop in Florida. (See Glantz, Volume X, for a discussion of the Florida citrus industry's responses to freezes in the early 1980s.) The most striking aspect of these freezes is the very short freezing time necessary for damage to occur. New citrus growth (i.e., bloom buds) can be completely killed during a 30-minute exposure to 26°F or a 3-hour exposure to 28°F. The effect of freezes is exacerbated if the crops have not hardened with the cold. Thus, if a freeze follows a warm period (i.e., indicating high daily temperature variability) when dormancy has been broken, more damage will occur at less extreme temperatures. A recent example of a freeze that affected the Florida citrus crop was the 1983 freeze, December 24-26, which resulted in a 30% reduction in citrus yield compared with the previous year (Mogil et al., 1984).

Extreme lows on a seasonal basis tend to most directly affect winter energy use for heating. The difference in heating fuel use in the United States between a warm and a cold winter can vary by as much as 400 million gallons of oil. During the extremly cold winter of 1976-77, heating degree days (calculated on a base of 65°F) were 10% greater than normal for the nation as a whole (Dare, 1981).

Precipitation

Anticipated changes in precipitation through climate change are not well known at this point. However, it is likely that geographic shifts in rainfall patterns will occur, with increases in annual total precipitation, decreases in different U.S. regions, and changes in the seasonal distribution of precipitation. Changes in the frequencies of extremes of both droughts and floods must be considered. Drought is of particular interest at the time of this writing because of the current drought in the United States and the energetic speculations being made concerning its possible connection with CO_2-induced climate change (Wilford, 1988).

Drought

The most basic definition of drought may be lack of sufficient water to meet essential needs (Gibbs, 1984). It is a condition determined relative to some long-term average condition of balance between rainfall and evapotranspiration in a particular region (Wilhite and Glantz, 1987). Different types of drought are recognized, such as meteorological drought (a departure of precipitation from normal), agricultural drought (insufficient soil moisture based on crop growth needs), or hydrological drought (based on departures from normal or relevant hydrologic parameters, such as streamflow). These "types" of drought are not completely independent, but can show up at different time lags one from the other.

Drought of any kind is anomolous as an extreme climatological event in that it is a "creeping" phenomenon; neither its onset nor its end is clearly punctuated in time. It is difficult to measure drought severity, since drought is a combination of factors: duration, intensity, and areal extent. Drought also can be one of the longer lived extreme events in that it can be measured in terms of seasons or, more frequently, years.

In the United States, major droughts have usually been defined in terms of several years, and the rate of occurrence is most strongly influenced by interannual variability of precipitation. Major drought episodes in the United States in the 20th century occurred in 1910, 1911, 1913 (serious but short lived), the 1930s (particularly 1934-36), the 1950s (1953-1957, primarily in the southern Great Plains), 1961-66 (northeastern United States), 1976-77 (primarily California and sections of the northern Great Plains), and the current drought (1988), which has been compared to drought conditions in 1936.

Although much research has been aimed at forecasting and planning for drought under the current climate (e.g., Wilhite and Glantz, 1987; Rosenberg, 1978, 1979) we still remain highly vulnerable to its impacts, as is being vividly demonstrated by current reports of agricultural disaster in areas of the northern Great Plains and forecasts of serious economic ramifications from the ongoing drought.

The effect of drought on crop production is perhaps the impact of drought that has received most research attention. The occurrence of droughts has been a major cause for yearly variability in crop production in the United States (Newman, 1978). During the 1930s, drought yields of wheat and corn in the Great Plains dropped as much as 50% below normal, whereas the drought in the 1950s brought less dramatic declines in yields (Warrick et al., 1975). The reduction in vegetative cover associated with drought also brings about severe soil erosion, which will diminish future crop productivity. Low yields of forage crops result in inadequate food for livestock and premature selling off of livestock.

Soil moisture deficits affect natural vegetation as well as crops. Again, much of the research in natural ecosystems has been on forests. Soloman and West (1985) identify drought as the cause for death of seedlings and for slowed or stopped growth of mature trees.

Aside from the direct effects of insufficient moisture on unmanaged ecosystems, indirect effects also result from increased incidence of fires. In the Australian drought of 1982-83, for example, brush fires destroyed over 8,000 km^2 of land; most of this area was forested (Gibbs, 1984). During the drought of 1988, forest fires have broken out across the country.

The effects of drought on U.S. energy resources are most obvious with regard to the generation of hydroelectric power. In the study described above on the impacts of climate change on electric utilities (Linder et al., 1987), the effect of decreased streamflow due to drought on the production of hydroelectric power in New York was discussed. In New York, 25% of electricity is supplied by hydrogeneration, mainly from the Niagara and St. Lawrence projects. Historical fluctuations have been extreme at times, as seen in 1964 during the northeastern drought when the Niagara Project generated 36% less energy than normal (Linder et al., 1987). Other regions of the country are even more dependent on hydrogeneration of electricity. Two-thirds of the total electricity in the Pacific Northwest is generated by hydropower. In this region, large variations in energy capacity result from variability in precipitation (Northwest Power Planning Council, 1983).

Floods

On average, 200 people die each year from flooding; flash floods account for most of these deaths (AMS, 1985). Floods also destroy property, crops, and natural vegetation, and disrupt organized social systems.

Floods result from a combination of meteorological extremes (heavy precipitation from severe storms, such as hurricanes and thunderstorms), the physical characteristics of particular drainage basins, and modifications in drainage basin characteristics made by urban development. Loss of life and property is increasing as use of vulnerable floodplains increases.

The recurrence interval of flooding is most important in applying effective control and protection mechanisms. These include building dams, reservoirs, and levees, and improving channels and floodways (White et al., 1975). For example, flood control reservoirs are designed to operate at a certain level of reliability, and the reliability is determined by a certain flood magnitude that the reservoir can handle, such as a 100-year flood.* Hence, the statistics of flooding are key for designing for protection and are based on a certain climatic variability determined from the historical record. As that variability changes, the reliability of the protection system will change.

---

*A 100-year flood is a flood of a certain magnitude specific to a drainage basis, which has a 1% probability of occurring in a year.

This is currently seen in systems where longer precipitation and streamflow records change the recurrence interval of particular magnitudes of flooding, such as in the case of Folsom Dam in northern California (Reibsame, 1987). Overtopping of dams accounted for 33% of total flood loss during 1903-58. During 1900-79, 26% of dam failures were due to overtopping (Clark, 1987). In 1972, overtopping of levees accounted for at least 40% of losses from tropical storm Agnes, which was one of the largest natural catastrophes in terms of property damage. With changing variability in climate, overtopping could become more common, and losses from floods could increase.

Major recent floods include the following: (1) Rapid City, South Dakota (June 1972), 231 deaths, more than $100 million in damage; (2) northeastern United States (June 1972), 120 deaths and about $4 billion damage -- inundation from hurricane Agnes; (3) Big Thompson Canyon, Colorado (July 1976), 139 deaths, $50 million in damage -- a result of a stalled thunderstorm system that delivered 12 inches of rain in less than 6 hours (Henz and Sheetz, 1976); (4) Johnstown, Pennsylvania (July 1977), 76 deaths and $200 million in property damage -- a result of slowly moving thunderstorms that deposited 11 inches of rain in 9 hours.

Floods in the 1980s have been less serious in terms of loss of life, but changing frequencies of severe storms, such as thunderstorms and hurricanes as well as general shifting of precipitation patterns could result in unprecedented losses from floods in a climate-changed world.

<u>Severe Storms - Hurricanes</u>

Three important kinds of climatic extremes are present in hurricanes: strong winds, intense and high precipitation amounts, and extreme storm surges. A hurricane is an extreme form of a tropical cyclone, characterized by torrential rains, typically as much as 5 to 10 inches in one storm; high windspeeds, which can exceed 100 mph; very steep pressure gradients, with pressure at the center as low as 915 millibars (mb) and with diameters of 100 to 400 miles.

The deadliest hurricane to hit the United States was the Galveston hurricane of 1900, which killed 6,000 people. However, the deadliest hurricane worldwide struck Bangladesh in 1970, resulting in 300,000 deaths. Hurricane Agnes in 1972 caused over $4 billion in damages to Florida and the northeastern United States. (Much of this damage resulted from flooding inland in the Northeast. Damages from high winds and surges were confined to coastal areas.) Hence, the frequency and intensity of hurricanes is of major concern for the inhabitants of U.S. regions where they commonly occur -- i.e., the gulf and the South Atlantic coasts.

Hurricanes are classified according to their severity on the Saffir/Simpson Scale (1-5), taking into account the central pressure, windspeed, and surge. Major hurricanes are considered all those of categories 3-5 wherein central pressure is less than 945 mb, windspeeds exceed 110 mph, and the surge is greater than 8 feet (Herbert and Taylor, 1979).

From 1900 through 1978, 53 major hurricanes (categories 3-5) directly hit the United States (averaging two major hurricanes every 3 years). Overall, 129 hurricanes of any strength hit the United States, which averages to approximately two per year. In recent decades, the number of major hurricanes has declined. From 1970 to 1978, only three hurricanes occurred compared with six or more in earlier decades. The last hurricane of category 4 or 5 to strike the United States was hurricane Camille in 1969. (Hurricane Allen (1980), which at one time reached force 5, weakened before it struck a relatively unpopulated segment of the Texas coast (Oliver, 1981).) Since then, the population of the south coastal regions of the United States has grown tremendously, and most inhabitants have never experienced a major force hurricane. Building in coastal areas has also increased with population, which raises the potential for high property damage. Thus, the population may be more vulnerable and less prepared to handle this particularly devastating extreme event (Sanders, 1982). Any increase in the frequency and/or intensity of these storms, which could result from climate change, would be of great concern to south coastal regions of the United States.

Coleman (1988) has found some limited evidence for increased frequency on the basis of examining the historical record for the number of storms formed in the North Atlantic during years of warmer than average sea surface temperatures. Emmanuel (1988a) has found through a hurricane modify experiment that the intensity of hurricanes increase under warmer conditions. The extreme intensity of Hurricane Albert this past summer is consistent with the findings. Emmanuel (1988b) also asserts the importance of establishing a general theory of hurricane development independent of current atmospheric conditions, so that scientists can predict changes in frequency and intensity of storms with climate change.

PRIOR STUDIES ON THE MODELING OF CLIMATE VARIABILITY

Studies comparing variability statistics of observed time series with variability statistics of general circulation model (GCM)-generated time series of climate variables relevant to climate impacts are not numerous in the atmospheric sciences literature, although studies first appeared in the early 1980s (e.g., Manabe and Hahn, 1981; Chervin, 1981). Such studies are critical if climate change research is to determine if the variability statistics of doubled CO_2 experiments with GCMs are valid. To accomplish this, the ability of GCMs to reproduce the variablity statistics of the present-day climate must be examined, and a thorough understanding of descrepancies attained. Five of the more recent and relevent works that address comparison of variability are briefly reviewed in this section. Very often, these studies also involve a study of the change in variability under doubled CO_2 conditions.

Chervin (1986) investigated interannual climate variability and climate prediction. He designed a special version of the NCAR, CCMO (National Center for Atmospheric Research Community Climate Model) to eliminate external variability of interannual time series of climate variables, so that discrepancies between modeled and observed variability would reflect the external component of variability present in the observed data. The additional variability attributed to external boundary conditions is considered the potential predictability of the climate. The model was modified such that external boundary conditions remained the same each year of the 20-year model run. This entailed the prescription of monthly sea surface temperatures and sea ice distributions. The variability of mean sea level pressure and 700-mb geopotential height (which is related to large-scale wind patterns) were analyzed in the Northern hemisphere, with particular focus on the United States. Results indicated no significant differences between modeled and observed variabilities of mean sea level pressure over the United States and only limited areas of differences in the variability of 700-mb geopotential height. It should be noted, however, that these results are based on the assumption that processes responsible for the internal variation of the atmosphere are correctly incorporated in the model.

Bates and Meehl (1986) also used the CCM to investigate changes in the frequency of blocking events on a global scale under doubled CO_2 conditions. Their version of the CCM included a seasonal cycle, computed hydrology, and a simple mixed layer ocean. The statistics of the 500-mb height field (the height above the surface at which atmospheric pressure equals 500 mb) are examined, as are blocking events defined as persistent positive height anomalies (which indicate high pressure). Blocking events are strongly related to persistent surface temperature anomalies, such as heat waves in the summer. Hence, changes in the frequencies of blocking events have significant implications for changes in the frequencies of extreme temperature events. In comparing modeled data with observed data for a 10-year period, they found that the model does a "reasonable" job of simulating 500-mb height standard deviations, particularly in winter in both hemispheres. The model, however, produces too few extreme blocking events. Under doubled CO_2 conditions, standard deviations of 500-mb height and blocking activity were found to decrease in all seasons (i.e., the variability of blocking events decreased).

Two studies were recently conducted on local or regional scales using the U.K. Meterological Office 5-layer GCM. Reed (1986) analyzed observed versus model control run results for one gridpoint in eastern England, and addressed the problem of creating the correspondence between the model grid square and the spatial average of several observation points. He analyzed the mean and variance of temperature for a 3-year integration of model runs. Compared to observations, the model tended to produce temperatures that were too cool and variability that was too high as measured by the standard deviation. For precipitation, the model produced too many rain days but did not successfully simulate extreme rain events of greater than 20 mm/day.

Most recently, Wilson and Mitchell (1987) examined the modeled distribution of extreme daily climate events over Western Europe, using the same model. Comparisons were made of minimum surface temperatures and precipitation. Again, the model produced temperatures that were too cold, and hence, extreme minimum temperatures were overestimated. This problem was most pronounced in grid boxes away from the coasts. The model also produces too much precipitation in general, but does not successfully reproduce observed highest daily totals. The number of rain days is overestimated. Wilson and Mitchell examined changes under qradrupled CO_2 conditions and found that variability of temperature generally decreased, which given the general warming, indicates a marked reduction in the occurrence of freezing temperatures. The authors compared the changes in intra-annual temperature variability using the more rigorous procedures of Katz (1984) as well as using the more standard F-test. Both tests indicate a signicant decrease in winter temperature variability. The authors emphasized the importance of examining variables of importance to climate impact analysis, but also of using a model that faithfully reproduces the present-day climate in terms of means and extremes.

Hansen et al. (1988) used the GISS model II general circulation model to simulate the global climate effects of time-dependent variations of atmospheric trace gases and aerosols. Several different scenarios of trace gas increase from 1958 to the present were used. A one hundred year control run was also produced. From this run it was determined that globally the model only slightly underestimates the observed interannual variability. However, the model's variability tends to be larger than observed over land. Among the calculations made with output from the transient (i.e. progressive climate change) run were changes in the frequencies of extreme temperature events. This was accomplished by adding the change in temperature with climate warming predicted by the model to observed local daily temperatures. Hence the authors assumed no change in variability in making these calculations. Results indicate that predicted changes in the frequency of extremes beyond the 1900s at such locations as New York, Washington, and Memphis, become quite large, and would have serious impacts.

The studies reviewed above indicate some important shortcomings of GCMs with regard to their ability to faithfully reproduce observed variability statistics. More research is clearly needed to further determine the sensitivity of the models to changes in physics, resolution, and so forth, with regard to the determination of variability. Moreover, only one of these studies explicitly concerns variables of importance to climate impact analysis. Studying the higher moments (e.g. variance) of climate variable statistics, and carefully verifying the models' ability to reproduce observed variability on regional scales, are the necessary prerequisites to rigorously analyzing possible changes in these statistics under doubled CO_2 conditions.

STUDIES FOR THIS REPORT

Two research efforts were undertaken for this report to attempt to increase knowledge concerning how climate variability will change. The climate change scenarios for use in the climate change impact studies reviewed in this report excluded consideration of changes in variability (see Chapter 3). The following two studies on GCM estimates of current and future variability were performed for this report:

1. Variability and the GISS Model - Rind, Goldberg, and Ruedy, Goddard Institute for Space Studies.

2. Variability and the NCAR Model - Mearns, Schneider, Thompson and McDaniel, National Center for Atmospheric Research.

It should be recalled that scenarios of climate change generated by the GISS GCM are used in most of the impact studies summarized in this report. The results of these two studies are directly compared in a later section.

The GISS Study

Rind et al. (1988) examined how well the GISS GCM simulates the observed variability of climate by comparing the model and the observed interannual and daily variations of temperature and precipitation. They describe the model assessment of changes in variability of these two major climate variables under climate change using the GISS doubled CO_2 run and a transient climate change experiment in which trace gases are increased gradually. The analysis is conducted in four regions of the United States, which are the four regions upon which this report is focused: the Great Plains, the Southeast, the Great Lakes region, and the West Coast. The locations of the four model grids representing these areas are indicated in Figure 17-1.

The GISS model, a grid general circulation mode, is run at the 8 by 10° resolution, and the results are analyzed for four different grid boxes corresponding to the regions listed above. Observed data consist of the average of observations at 9 different stations per grid box. Grid boxes are indicated in Figure 17-1. First, mean conditions are compared for actual weather observations with the GCM control run (or single CO_2), the doubled CO_2 run, and the transient run. The model values for mean temperatures for 4 months in the four regions are generally cooler than observations (particularly in summer and fall), they are always within several degrees Celsius of the observations. Model precipitation values are fairly close to observed values in the Great Lakes and Southeast grid boxes, but model values are higher than observed for the other two regions (e.g., January in the southern Great Plains: model = 2.1 mm/day, observed = 0.46 mm/day). Under the doubled CO_2 scenarios, temperatures increase over the control run by 4-6°C in the winter and 3-4°C in the summer. Warming in the transient scenarios is progressive, but temperature change occurs more gradually than with simply doubling the CO_2 amount. Winter warms more than summer, and so the annual seasonal cycle is reduced under climate change. Precipitation changes are not statistically significant at individual grid, but there is an overall tendency for increased precipitation.

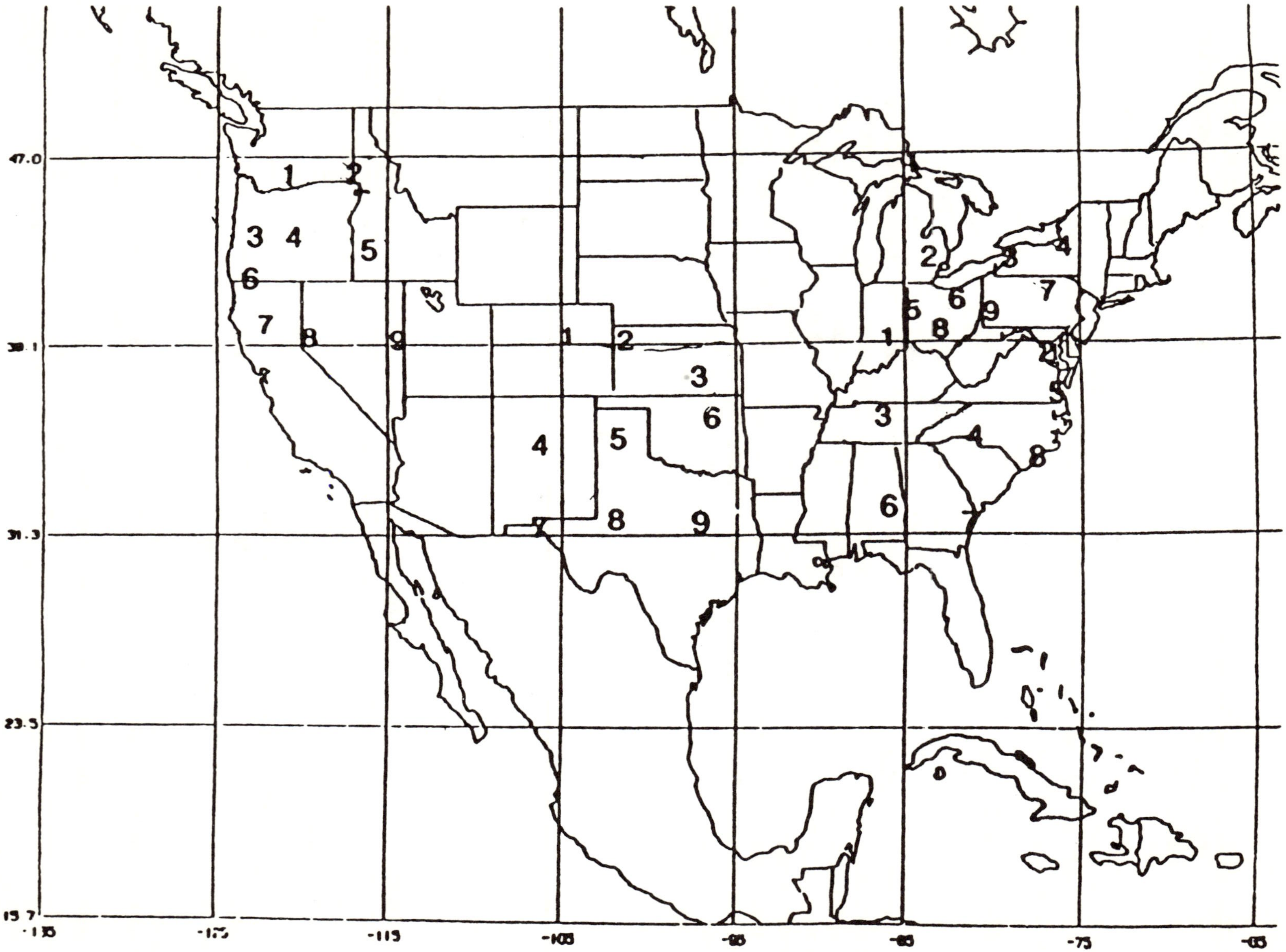

Figure 17-1. The Locations of the Four Model Grids.

Interannual Variability

Standard deviations of temperature and precipitation of observed and modeled data were compared for all months. In most months, the model year to year temperature variability is similar to the observed variability in the four regions, but in summer the variability was overestimated by 0.3-0.6°C. Precipitation variability is overestimated in half of the cases where precipitation amount is also overestimated. The relative annual variability of precipitation (that is the standard deviation relative to the mean) of the model is generally in agreement with observations.

Under conditions of climate change (doubled CO_2) in comparing control versus climate change interannual standard deviations (a sample size of 10 years in each case) for each month in all regions, there is generally reduced variability of temperature from January through April. (These results were further substantiated by examining the changed variability at other tropical grid boxes in the United States.) Results for other seasons of the year are more ambiguous. For precipitation, the doubled CO_2 climate resulted in increased variability in most months at the four grids (in 31 of 48 cases), but was particularly striking at the southeast grid. These changes, however, were often of the same order as the model's natural variability (from examination of the 100-year control run). The sign of the change in mean value and the sign of change in interannual variability are highly correlated.

Daily Variability

Daily variability of temperature was analyzed by taking the daily departures from monthly means and comparing the resulting model distribution with the distribution formed in the same manner from the observational data. Two statistical tests were used.

Ten years of control run for the transient experiment for 4 months (January, April, July, and October) were compared with 30 years of observations. Distributions of observed versus model daily temperature data were, in general, not significantly different. Comparisons were also made by calculating the standard deviations of the departures from the mean for the four months (Table 17-1a). These results indicate that the model's values are significantly greater than the observed, which demonstrates that the model is producing too many extremes.

Table 17-1a. Daily Temperature Standard Deviations (°C) (Rind et al.)

Month	Location	Observed SD	Current SD	2010s SD	2030s SD	2060 SD
January	Southern Great Plains	4.81	8.15	0.61	-1.19	-0.83
	Southeast	4.53	6.90	-0.14	-1.14	-0.23
	West Coast	3.63	5.86	-0.61	0.05	-0.16
	Great Lakes	4.97	5.79	0.44	-0.33	-0.44
April	Southern Great Plains	3.72	5.77	-0.57	-0.27	-0.80
	Southeast	3.71	5.50	-0.65	-1.61	-1.24
	West Coast	2.59	4.29	0.77	0.60	0.33
	Great Lakes	4.65	6.15	-0.51	-0.26	-1.39
July	Southern Great Plains	1.74	2.56	0.54	-0.19	0.18
	Southeast	1.50	2.34	0.14	-0.22	-0.24
	West Coast	2.40	3.56	0.03	0.54	0.28
	Great Lakes	2.38	3.02	-0.48	-0.84	-0.14
October	Southern Great Plains	3.79	5.16	1.16	0.97	1.35
	Southeast	3.59	5.21	-0.54	-0.25	-0.73
	West Coast	3.15	6.51	-0.55	-0.30	-0.80
	Great Lakes	4.09	5.46	-0.37	0.91	-0.06

Under climate change, comparing temperature distributions of the control run with a series of decades from the transient experiment reveals that the distributions did not significantly differ. Results comparing standard deviations (presented in Table 17-1a) indicate that, although changes with time are not strictly progressive, most cases by the end of the climate change experiment show reductions in the standard deviation (although these reductions are not statistically significant). Hence, a decrease of daily temperature variability is not demonstrated by these results.

For precipitation, comparisons are more complex. For example, the number of observation stations used to represent a grid box does affect the results. Model rainfall distributions differ significantly from observed in half of the cases (in three seasons for the West Coast and the southern Great Plains). The model also produces less days of light rain in general and more extreme values in the winter in all four regions. These results are also confirmed by comparison of standard deviations (Table 17-1b).

As the climate warms, again considering the various decades of the transient experiment, the precipitation distributions differ from the control climate in about one-fourth of the time with no general progression over the decades. Figure 17-2 presents a sample set of distributions for precipitation during several decades of warming for the West Coast in April. In comparing standard deviations (Table 17-1b), in half of the cases, the warmest time period exhibits increases in standard deviations. These results are again consistent with those for interannual variability.

Variability of the Diurnal Cycle

It would be expected that the diurnal cycle would decrease under changed climate as the additional greenhouse gases could limit nighttime cooling. Comparisons of control model results with observations are reasonable in the four regions. Under doubled CO_2 conditions, it was found that the amplitude of the diurnal cycle very definitely decreases in summer but changes inconsistently in the other seasons. The reason for this is the dominance of radiative heating in the summer and of other forms of heating and cloud cover change in other seasons.

Table 17-1b. Daily Precipitation Standard Deviations (mm d^{-1}) (Rind et al.)

Month	Location	Observed SD	Current SD	2010s SD	2030s SD	2060 SD
January	Southern Great Plains	1.08	2.80	0.05	0.05	1.68
	Southeast	4.35	4.62	-1.20	-1.35	-0.85
	West Coast	3.23	4.55	-0.18	0.34	0.13
	Great Lakes	2.23	4.06	-1.07	-0.94	-0.50
April	Southern Great Plains	2.51	3.26	0.94	1.99	1.17
	Southeast	4.35	3.85	0.95	-0.15	0.81
	West Coast	1.41	2.76	0.07	1.02	-0.12
	Great Lakes	3.85	3.29	-0.43	-0.31	0.44
July	Southern Great Plains	2.79	3.08	-0.10	-0.09	0.36
	Southeast	4.13	3.31	0.28	0.29	0.11
	West Coast	0.57	1.53	0.44	0.24	0.71
	Great Lakes	3.68	2.48	-0.06	0.72	0.35
October	Southern Great Plains	2.75	1.79	0.52	0.34	0.00
	Southeast	3.77	3.88	0.72	-0.15	-0.28
	West Coast	1.86	2.69	1.20	-0.63	1.34
	Great Lakes	3.58	2.26	0.52	0.76	0.95

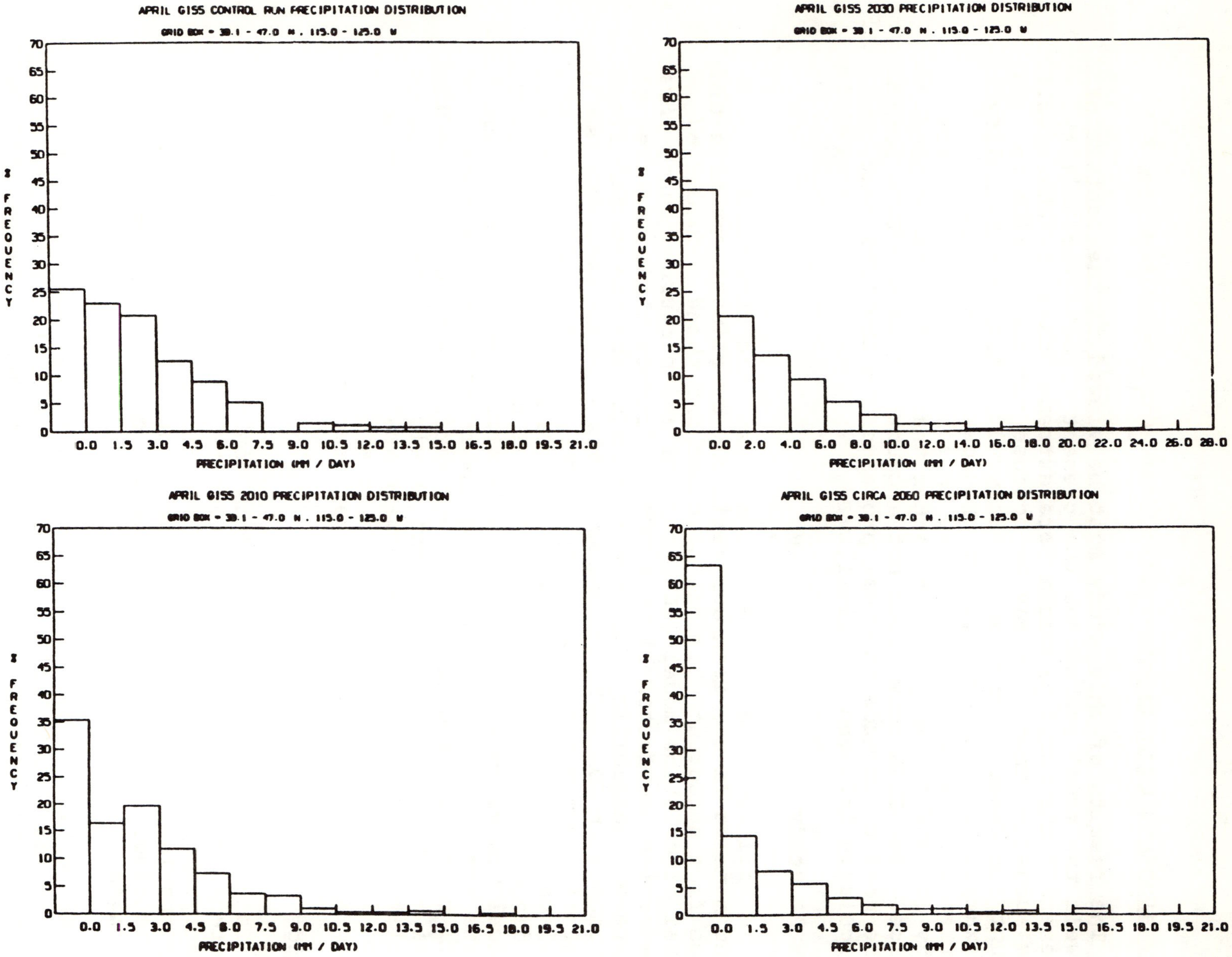

Figure 17-2. Sample Set of Distributions

Source: Rind et al., Volume _.

Discussion and Conclusions

The results of this study must be viewed in the context of how well the model reproduces the present climate variability. In general, the model produces more extremes than are observed. It is suggested that this may be because the model fails to account for feedback processes that would minimize variability. For example, the conservation of groundwater limits the occurrence of temperature extremes in summer, as some surface energy is used for evaporation rather than heating the ground surface. The GISS model does not account for this. This points to the need to include physical processes in the GCMs that particularly influence variability.

The NCAR Study

In this study, Mearns et al. analyzed the first and second moments (i.e., mean and variance) of climate variable time series from selected empirical stations and those produced by general circulation model (GCM) control and perturbed runs, first to determine how faithfully the GCMs reproduce these measures of the present variability and then to examine how the variability will change in CO_2 perturbed cases. This type of research is essential because the variability performance of a GCM not only is relevant to the credibility of its forecasts but also serves as a diagnostic of the validity of model physics, since it is the physical processes within the model's structure that give rise to climate variability. By comparing the relative performance (i.e., model versus observations) of various versions of the NCAR CCM (i.e., versions with different physical parameterizations or formulations), they help to determine what formulations may be needed for forecasting certain measures of variability and how much credibility to assign to those forecasts.

Methods

This study used the output from control runs of three different versions of the NCAR Community Climate Model (CCM). However, the version of Chervin (1986) is the primary one used for comparison of observed and model control output (i.e.,model runs to simulate the actual present-day climate), since it has the longest time integration (20 years).

The version of the CCMO(A) model used is documented in Chervin (1986). The CCMO is a spectral general circulation model originally developed by Bourke and collaborators (Bourke, 1974; Bourke et al., 1977), which has been modified by the incorporation of radiation and cloud parameaterization schemes.

The model has a resolution for physical processes of approximately 4.5° in latitude and 7.5° in longitude, and has nine levels in the vertical. This version includes an annual solar cycle featuring identical, seasonally varying forcing each year. Identical forcing each year was accomplished by prescribing for each month ocean surface temperatures and sea ice distributions.

The other two versions of the CCM used are the Washington version of the CCMO(A) (Washington and Meehl, 1984), which includes an interactive thermodynamic ocean and surface hydrology, and the Dickinson version (Dickinson et al., 1986), a version of the more sophisticated CCM1(B) containing a diurnal cycle and a very sophisticated land surface package, the Biosphere-Atmosphere Transfer Scheme (BATS).

The same four regions of the United States were chosen for investigation as were chosen for the GISS study: the Great Plains (represented by three grid boxes), the Southeast, the Great Lakes and the West Coast. The locations of the grid boxes are indicated on Figure 17-3. Note that the NCAR model grids for the Great Plains are further north than the GISS grid for the southern Great Plains.

Real data corresponding to each grid box were generated from the average of six observation stations for time series of temperature and precipitation. Solar radiation and relative humidity real data were taken from one Solmet observation station located in each grid square. Figure 17-3 shows the location of all observation stations and grid boxes.

Four variables deemed particularly relevant to climate impact analysis were chosen for this analysis: daily mean temperature, daily total precipitation, mean daily relative humidity, and mean daily absorbed solar radiation.

Comparison of Observed versus Chervin Control Run

Temperature

Figure 17-4 displays the time series of daily average temperature for modeled and observed data for the four regions investigated. The model successfully simulates the annual cycle for the four regions, which represents the variability within the year. There are regional differences, however. In the Great Plains region, the model annual range is slightly smaller than that of the observed time series (Figure 17-4a), whereas in the Great Lakes region (Figure 17-4b) the model annual range is

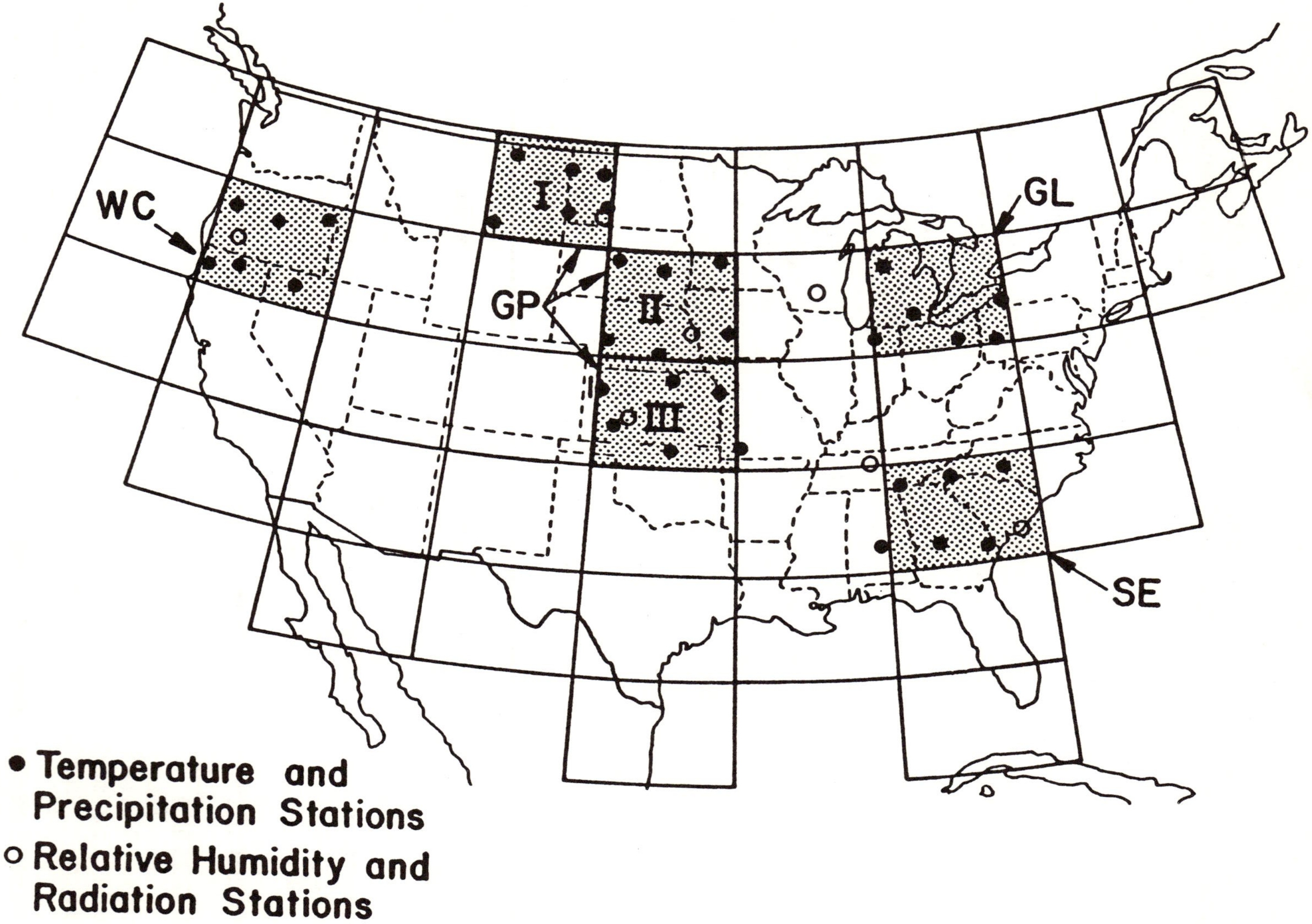

Figure 17-3. Model Grid Cells and Station Locations.

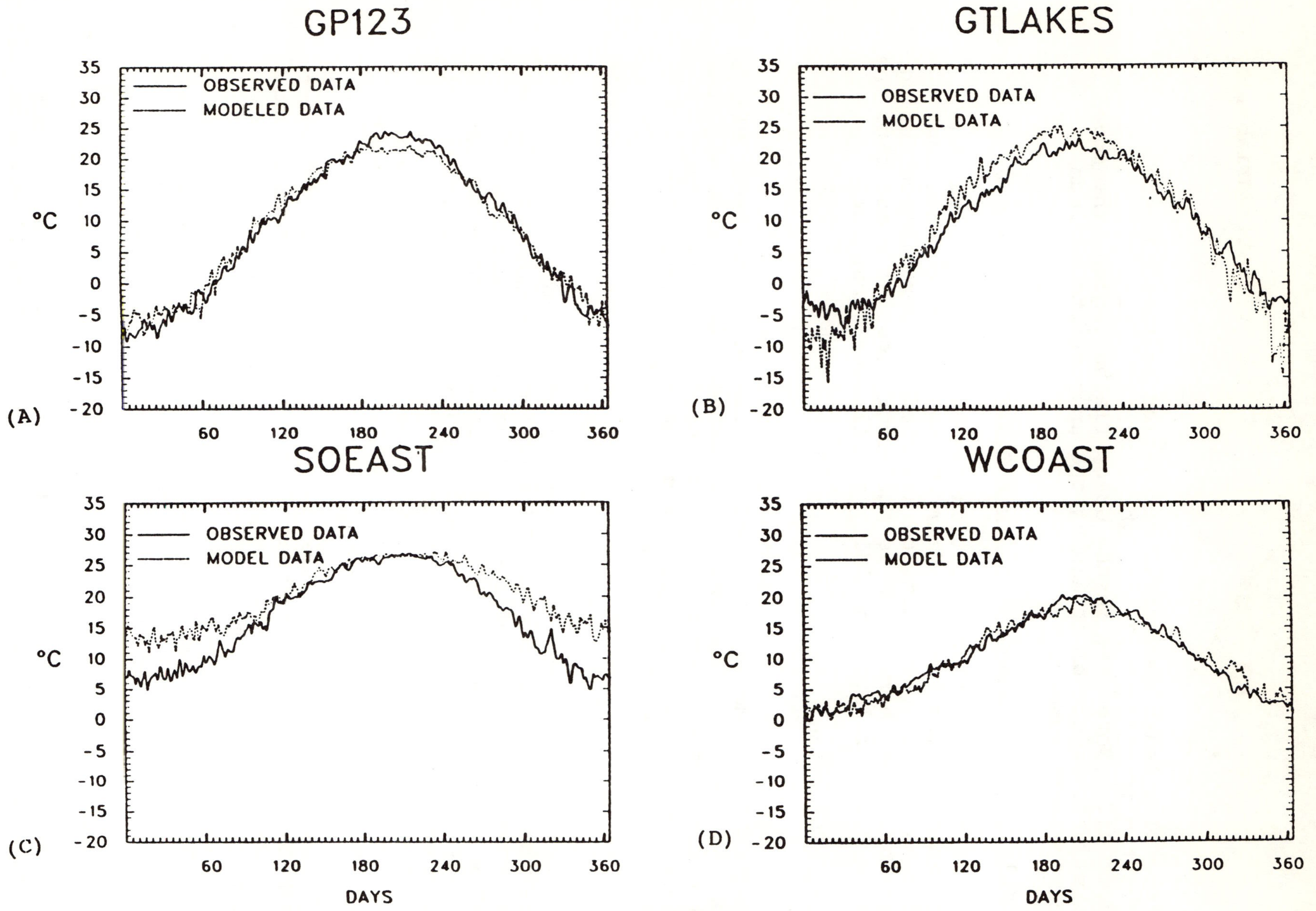

Figure 17-4. Average Temperature for a 20-year Average Year.

Source: Mearns, et al., Volume _.

greater. The least successful simulation is that for the Southeast grid (figure 17-4c), where winter minima are too high by about 8°C. Note that in the Great Plains region, the model falls slightly short of attaining the appropriate maximum temperatures in summer.

Solar Radiation and Relative Humidity

Simulation of solar radiation ranges from very good (the Great Plains region) to only fair at the Southeast where the model consistently overestimated absorbed solar radiation during all months. The Chervin CCM is strikingly poor at simulating the annual cycle of relative humidity at all four locations.

Precipitation

The Chervin CCM consistently overestimates precipitation, although the seasonal cycle is well simulated in the Great Plains region and the West Coast grid. The authors do not know why the model overestimates precipitation but speculate that it may partially be a result of a precipitation parameterization criterion of 80% relative humidity.

Variability Comparisons of the Chervin CCM

Interannual variability of temperature is generally underestimated by the Chervin CCM in all four regions. Interannual variability of precipitation (i.e., relative variability, the standard deviation relative to the mean) is generally in reasonable agreement with observed data, although it is occasionally overestimated. This is a particularly encouraging result for the credibility of predicting climate changes given how inaccurate the control precipitation results are in terms of absolute values.

In terms of daily variance, model relative humidity tends to be much less variable than observed values at all locations and in most months. Results for temperature for January and July indicate that the Chervin model generally overestimates daily temperature variance (on the basis of statistical tests on the innovation variance, i.e., the variance that remains when the effect of autocorrelation is removed). Solar radiation daily variance comparisons are not particularly interesting.

Intercomparisons of Three CCM Versions and Observed Data

Given that all GCMs fail in certain respects to accurately reproduce observed climate, Mearns et al. felt that it would be useful to compare different model versions' simulations of present-day climate to be aware of the possible ranges of errors and to understand more fully the effect of model structural differences. The present-day climate runs of models incorporating physics different from those of the CCMO(A) version of Chervin (1986) investigated in the third section are compared. Both the Washington and Dickinson runs consist of 3-year integrations. Table 17-2 presents comparisons of annual and seasonal mean conditions simulated by the three models and observations for the Southeast (SE) grid. Statistics of interannual variability cannot be analyzed because there are too few years of simulated data.

There is considerable variability in how well the models reproduce mean total precipitation for the SE grid, ranging from the very good results of Dickinson's model, to the overestimation of Chervin's model, the underestimation of Washington's model. On the basis of mean annual and seasonal comparisons, no one model is clearly superior to the other two in accurately reproducing mean climate at the four locations.

The Dickinson model also produced much lower daily variability of temperature than the other two models or in some cases, than the observations. This result is graphically illustrated in the temperature histograms (three models and observed) for two key months for the SE grid (Figure 17-5).

The reasons for these discrepancies have yet to be explored in depth, but are likely related to different land surface packages in the models. A possible explanation for the lowered daily temperature variability of the Dickinson model concerns the more sophisticated surface energy balance used, which includes consideration of soil heat capacity.

Control versus CO_2 Perturbed Runs

The authors included a preliminary analysis of changes in precipitation and temperature, under a scenario of doubled CO_2, using the output from Washington's control and doubled CO_2 runs for the four regions. Interannual variability could not be analyzed because the time series are too short. However, examination of daily variance of temperature is presented.

Table 17-2. Southeast Grid -- Multicomparisons of CCM Model Version and Observed Mean Temperature and Total Precipitation Statistics (Mearns et al.)

	Obs.	Chervin	Wash.	Dick.
Annual				
Mean T (°C)	16.6	20.0	20.1	21.5
Total precipitation (mm)	1137.2	3085.5	589.1	1401.8
Winter				
Mean T (°C)	7.4	14.1	14.8	16.0
Total precipitation (mm)	738.9	295.3	393.4	293.2
Summer				
Mean T (°C)	25.5	25.9	24.2	26.6
Total precipitation (mm)	317.0	786.4	75.7	346.9

Obs. = Observations.
Chervin = Chervin CCM version (20-year run).
Wash. = Washington CCM version (3-year run).
Dick. = Dickinson CCM version (3-year run).

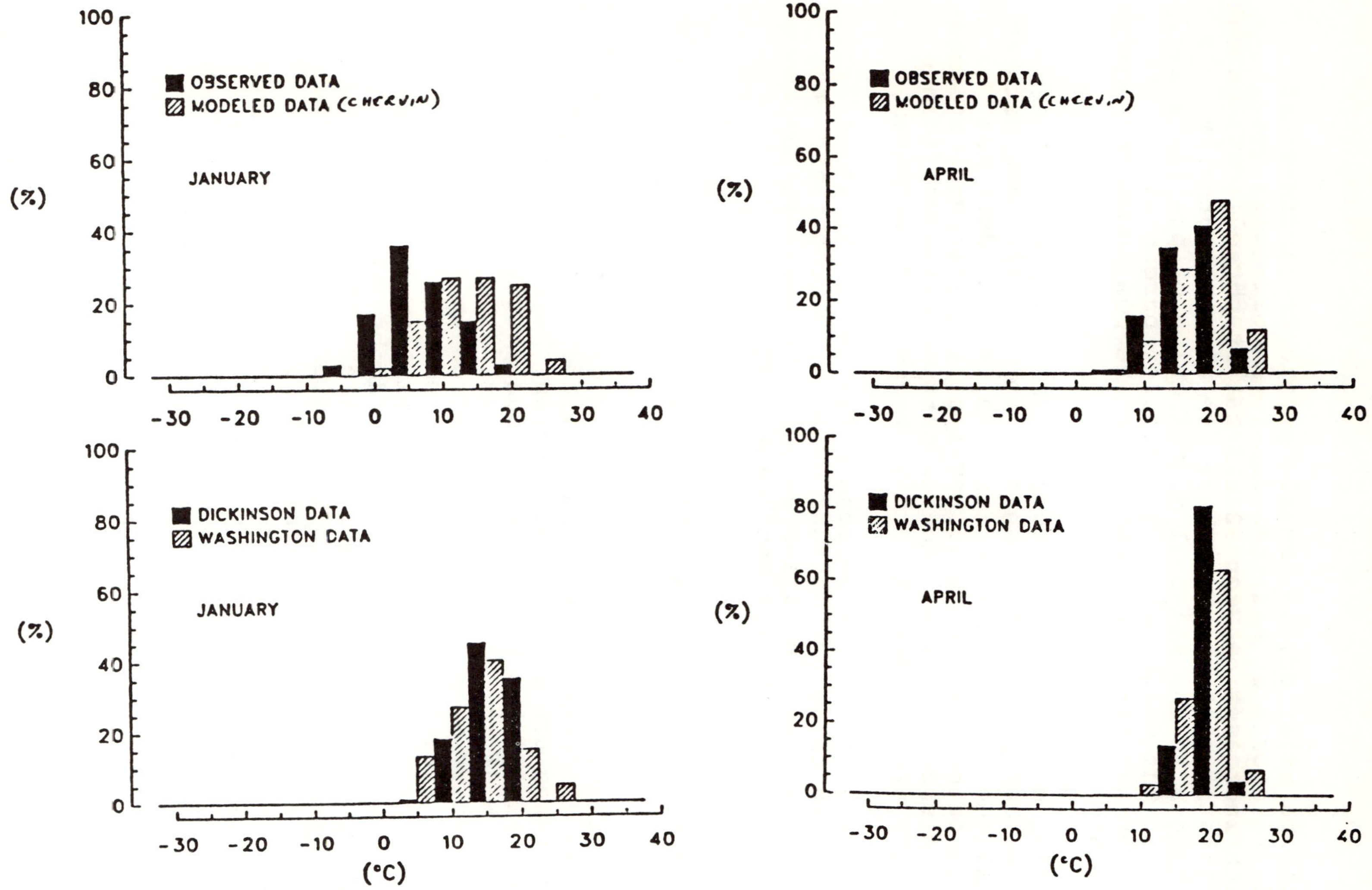

Figure 17-5. The percentage of events for the intervals ± 2.5°C from the given temperature.

Source: Mearns et al., Volume _.

As would be expected, an annual temperature increase of about 2 and 3°C occurs at all locations. Annual total precipitation increases between 22 and 26% at three locations but decreases slightly (2%) at SE. There are also potentially important changes in the seasonal distribution of precipitation. For example, at the Southeast grid a smaller percentage of the annual total occurs during the summer in the CO_2 perturbed case (from 13% to 6%).

Statistics comparing daily temperature variance of the control and perturbed runs for January, April, July, and October indicate that the temperature variance in general does not significantly change (at the 0.05 level of significance) at these four grids. Without consideration of statistical significance levels, results are mixed with both increases and decreases.

Percentage of rain days decreases in the summer under climate change in three of the four grids. Interestingly, the occurrence of small precipitation amounts decreases in the summer in the perturbed case for the Southeast grid.

COMPARISON OF GISS AND NCAR RESULTS

It is difficult to compare the two studies. The modeling experiments were conducted partially with different purposes in mind using two different models (which differ not only in how physical processes are modeled but also in their spatial resolutions). They also use different qualitative and statistical methods for making comparisons. The GISS experiment was aimed primarily at examining the changes in variability with climate change, whereas the immediate purpose of the NCAR experiment was primarily to examine and explain discrepancies in variability between model control runs and observations. Since the spatial resolutions of the models differ, the grid boxes of the models do not coincide and so the regions analyzed differ. These are only some of the problems that would affect these comparisons. Nevertheless, an attempt is made here to compare some of their results that roughly coincide. Some regions coincide fairly well (such as the Great Lakes grids, see Figures 17-1 and 17-3), and some similar analyses were conducted.

First a brief comparison is made of how the models reproduce the observed mean climate. In general, the GISS model is too cool and the NCAR model(s) too warm. Precipitation is overestimated at two grids by the GISS model and is overestimated at all grid boxes by the Chervin version of the NCAR model (although this is not true of two other versions of the NCAR CCM).

The following sections compare the observed and control runs of interannual and daily variability of temperature and precipitation, and then comparisons of climate change variabilities are made. It must be emphasized again that comparison of control runs with observed data is an essential step in the process of trying to evaluate the models' forecasts of climate change.

Interannual Variability

The easiest and most straightforward comparison can be made of the two studies' results for interannual variability of temperature and precipitation control runs. Relative variability values (standard deviation relative to the mean) for the GISS study were provided by its authors (Rind, personal communication). Rind et al. use a 100-year control run for interannual variability calculations. Their observational data set consists of 30 years (1951-80). The NCAR study uses a 20-year control run of Chervin (1986) and a 20-year observational data set (1949-68). The differences in sample size should be noted. Table 17-3a and 17-3b present the relevant results, winter and summer standard deviations for temperature, and annual coefficients of variation (i.e., a measure of relative variability) for precipitation for the four regions for both studies. The overlap for the GISS southern Great Plains grid box and the NCAR Great Plains III grid (see Figures 17-1 and 17-3) is considerably less than in the case of the other three regions (Great Lakes, Southeast, and West Coast). In comparing the model temperature standard deviations with the observed for each study, there is little agreement in the direction of error from model to observed. Both models overestimate the variability of the Great Plains region in winter. (However, the difference in the NCAR study was deemed statistically insignificant.) Both models underestimate the temperature variability (but the NCAR model much more so than the GISS) for the West Coast winter. In all cases the GISS study shows an overestimate and the NCAR study an underestimate by the model, of variability.

Regarding the relative variability of precipitation, the results for the two models are rather similar, although the GISS values tend to be absolutely higher. The differences between observed and model values are very close (from 1 to 6 percentage points) in each study. The NCAR model slightly underestimates the variability at each location, whereas the slight errors in the GISS results are mixed.

Table 17-3. Interannual Standard Deviations. Temperature and Precipitation, GISS and NCAR Control Runs (Rind, personal communication and Mearns et al.)

		Temperature (°C)		Precipitation
		Dec.-Feb.	June-Aug.	Annual (%)
a. GISS (n = 100)				
SGP	Model	1.65	1.05	15
	Obs	1.20	0.75	21
SE	Model	1.65	1.05	22
	Obs	1.65	0.70	18
WC	Model	1.35	1.35	18
	Obs	1.45	0.75	23
GL	Model	1.35	1.25	18
	Obs	1.50	0.70	18
b. NCAR (n = 20)				
GP III	Model	1.3	0.14	17
	Obs	1.1	1.2	22
SE	Model	1.0	0.38	10
	Obs	1.8	0.74	12
WC	Model	2.2	0.71	10
	Obs	1.6	0.88	11
GL	Model	0.8	0.17	17
	Obs	1.45	0.81	17

SGP = Southern Great Plains; SE = Southeast; WC = West Coast; GL = Great Lakes.

The reasons for the lack of agreement in the two studies are far from obvious, and speculation can only be rough. Certainly the difference in how the atmosphere ocean interaction is modeled may play a role (i.e., the NCAR model uses fixed sea surface temperatures, whereas the GISS model computes sea surface temperatures from a simple ocean mixed-layer model). In fact, the number of differences in the two studies largely precludes any straightforward analysis of discrepancies.

Daily Variability

Daily variability of temperature can be compared for two season months (January and July) at the four locations using the standard deviations (Table 17-4). Each study also examines changed daily variability using some other statistical measure (chi-squared test of distribution similarities in the GISS study and innovation variance comparison in the NCAR study), but these are not comparable. Comparisons on the basis of standard deviations alone cannot be statistically tested because of problems encountered in meeting the assumptions required for the tests, (although in a later version Rind et al. does apply tests to the standard deviation after correcting for dependence in the sample size). Hence, these comparisons must be viewed strictly qualitatively.

In seven of the eight cases, the studies agree (that the model overestimates daily temperature variability). In the case where there is disagreement (Great Plains, July), the NCAR difference is not statistically significant on the basis of tests on the innovation variance (the variance after the effect of autocorrelation is removed). However, the GISS discrepancy is fairly large, suggesting that there is an overestimation by the GISS model. Still, the overall agreement is that the models tend to overestimate daily variability at these four locations. This tendency is graphically illustrated in Figures 17-6a and 17-6b, where the observed and model histograms of daily temperature for April at the Great Lakes (GL) grid for the two studies are compared.

In both studies, explanations for the overestimations are related to the modeling of surface hydrology, i.e., that both models fail to completely account for important surface-atmosphere interactions that would tend to reduce daily temperature variability.

Table 17-4. Daily Temperature Standard Deviations (°C) (Rind et al. and Mearns et al.)

	GISS		NCAR	
	Obs.	Model	Obs.	Model
January				
Great Plains	4.81	8.15	4.81	8.15
Southeast	4.53	6.90	5.41	5.92
Great Lakes	4.97	5.79	5.5	11.20
West Coast	3.63	5.86	4.1	5.0
July				
Great Plains	1.74	2.56	2.90	2.79
Southeast	1.50	2.34	1.55	1.70
Great Lakes	2.38	3.02	2.67	2.82
West Coast	2.40	3.56	2.18	3.52

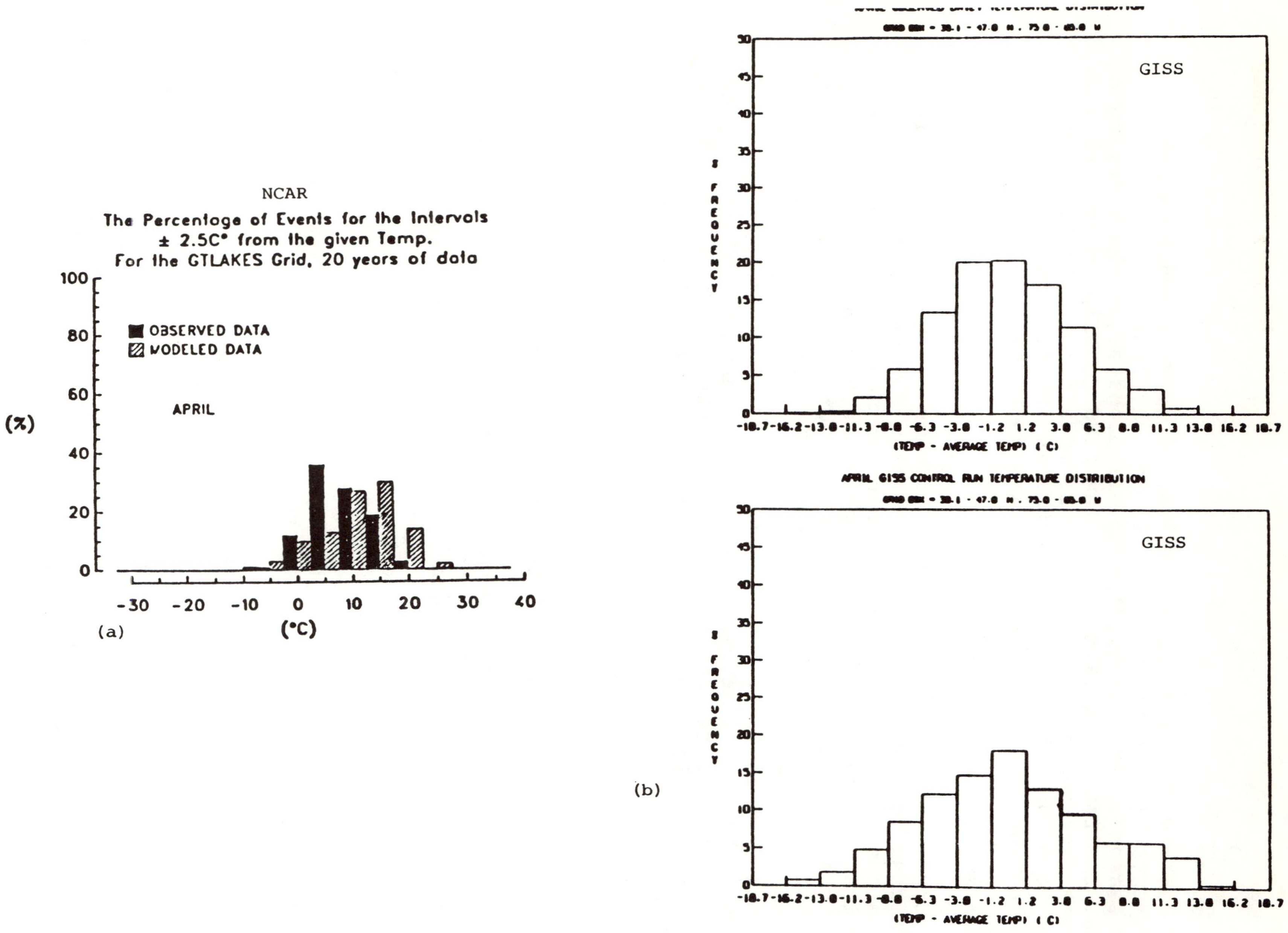

Figure 17-6. The percentage of events for the intervals ± 2.5°C from the given temperature.

Source: Rind et al., Volume _; Mearns et al., Volume _.

Daily variability of precipitation is compared by summary statements made in the two studies for the four locations for four seasons. In general, the models produce, in the majority of cases, too few light rain days. The GISS model produces too many extreme rain events in winter at all locations. The NCAR model tends to produce too many high extremes in all four seasons. Neither study accounts for these discrepancies.

Comparison of Climate Change

Comparison of climate change results of the two models is restricted to changes in daily temperature variance for four months for the four locations, since the NCAR study includes a quantitative analysis of only temperature variability change. The two studies do not agree on the direction of change of daily temperature variability. The NCAR results are mixed showing both increases and decreases, although most of these changes are statistically insignificant. Rind et al. conclude that in general, there is a decrease in daily temperature variability on the basis of changes in standard deviations (but the changes are not statistically significant). On the basis of the two research reports, no clear statement may be made about changes in daily temperature variability under CO_2 warming conditions.

Limitations of the Two Studies

It should be noted that both studies underline the importance of viewing climate change results of the models in the context of how well they reproduce the present climate. Model deficiencies can be expected to limit the reliability climate change results, and that faith in quantitative results are probably misplaced. However, model results give crude estimates as to the importance of physical processes responsible for variability and what must be done to improve them. There is a need for further testing to determine how the models' deficiencies in reproducing present-day climate affects predictions for a CO_2-warmed future climate.

IMPLICATIONS FOR STUDIES OF CLIMATE CHANGE IMPACTS

As indicated in the second section of this chapter, virtually all systems are affected by climate variability, although some more than others. The relative importance of climate variability and changes in variability, as a result of climate change, to particular impact areas is reflected in the results and limitations of some of the studies summarized in this report.

Of greatest concern was the lack of information regarding changes in the variability of temperature and precipitation that would attend climate change. The lack of this information resulted in the formation of climate scenarios wherein the temporal variability of both precipitation and temperature were not changed (see Chapter 3 on scenario development). This was considered a limitation or concern in many studies, some of which are discussed below.

In the Johnson et al. study on agricultural runoff and leaching (reviewed in Chapter 10, Agriculture), the results were considered to be limited by the failure to consider changes in storm frequency and duration that would result from climate change. Such changes would obviously affect the results of the Pesticide Root Zone Model (PRZM), which determines the distribution and transport of pesticides in soils. Also, the frequency of precipitation was left unchanged from the present-day climate and, hence, estimates of total annual erosion could be underestimated. Furthermore, the magnitude of extreme rainfall events could be exaggerated, since the number of days of rainfall is not changed, but different (usually larger) amounts of rainfall are assumed. This could result in overestimation of pesticide losses through soil erosion and leaching in a particular year. The results of this study could be vastly different from those presented, depending upon assumptions concerning precipitation duration, frequency, and intensity, all of which would change if a changed daily variability were assumed.

Several studies on hydrology summarized in this report also are highly dependent upon assumptions about precipitation variability. These include the Lettenmaier et al. study on the hydrology of catchments in the Central Valley and the Sheer and Randall study on the impact of climate scenarios on water deliveries, both reviewed in Chapter 6. The hydrologic inputs for the models used in these studies are interdependent. Lettenmaier et al. assumed that the number of days of rainfall remains the same under the climate change. Model results in terms of predicting runoff amounts would be quite different if more rainfall events of lower intensity were assumed compared to the same number of rainfall events with (generally) higher intensity. Since the Sheer and Randall study uses the output of the Lettenmaier study as hydrologic inputs to their water delivery model, their results are also dependent upon the assumptions about unchanged variability and, hence, their results could also change considerably if precipitation variability changes were considered.

The studies for the Southeast (Chapter 6) did not consider changes in the frequency of droughts or severe storms such as hurricanes, which could certainly affect the likelihood of flooding for some coastal communities. However, these concerns are considered secondary to changes in sea level that would dominate in terms of changing the likelihood of floods.

Crop yields are very dependent on daily variability. For example, heat waves during grain filling lower wheat yields. Whether a drought occurs early or late in the growing season has differential effects on yields. Changes in variability were not considered in the Rosenzweig, Peart et al., Ritchie, and Dudek studies.

Changes in the frequencies of extreme events are considered to be of great importance to potential forest disturbance, as discussed in Chapter 11. Here, an important distinction is made between the ability of forests to adapt to changes in mean climate conditions versus changes in extremes. The possibility of increases in the frequencies of events such as wind, ice, or snow storms, and droughts and flooding may be of greater significance to forest survival than the gradual mean change in climate that has been studied so far.

The Kalkstein study, which is reviewed in Chapter 14 on Human Health effects, is strongly dependent upon the determination of certain maximum temperature threshold values beyond which human mortality increases. In applying the death/weather effects statistical models to scenarios of climate change, temperature variability is held constant, so that temperatures that exceed the threshold values are determined unrealistically. Of course, there is greater accumulation beyond the thresholds in the warmer scenarios, but not to the same extent as would be the case if the both variance and mean of the series are changed. This problem is explicitly stated in the section on limitations.

Changes in the variability of temperature both seasonally and daily are important to studies concerned with the effect of temperature change on the change in electricity demand discussed in Chapter 16. Although new generating capacity requirements for the nation for 2010 and beyond are calculated assuming climate change, the numbers generated could be considerably different for any particular year depending mainly on air-conditioning needs, which would be the major use increase for electricity. Such needs are sensitive to extremes in daily maximum temperatures and the persistence of such temperatures (i.e., heat waves).

Potential change in extreme events is identified as a research need in this chapter.

It would be impossible to quantitatively or even qualitatively estimate how different the results of these studies would be if changes in climate variability had formed part of the climate scenarios made available as input for the various climate impact models used. Primarily, it is impossible because the variability changes are not known; second, it is impossible because most of the studies are of such complexity that the effect of a change in one variable (a complex change at that) is not intuitively obvious in most cases. Sensitivity analyses of the impact models involved to changes in variability would be required to provide specific answers. What can be said at this point is that the lack of information on climate variability has limited a number of studies in this report and has limited the completeness of the answers they could provide. The magnitude of change of some impacts could be affected by the inclusion or exclusion of variability information.

FUTURE RESEARCH NEEDS

The research reported above clearly indicates that research of changes in climate variability with climate change is truly in its infancy. Much needs to be done. Future research needs may be broken into three categories.

1. Further investigation of variability in presently existing GCM control and perturbed runs. Results summarized here represent only an initial effort at looking at variability in GCMs. The first research priority is to examine at many more grid boxes the daily and interannual variability of many climate variables in addition to temperature and precipitation, such as relative humidity, solar radiation, and severe storm frequency. Moreover, the most sophisticated statistical techniques must be used or where needed, developed, such that uniform quantitative indicators are available to evaluate both how well the current models reproduce present variability and how they forecast the change in variability under climate conditions. The causes for discrepancies in present-day climate variability and control run variability must be better understood, so that a clearer understanding of future climate changes can be attained.

2. Improvements in GCMs geared toward better reproduction of climate variability. From the results of Rind et al. and Mearns et al., some indications are given that oversimplifications in the land surface packages of GCMs contribute to overpredictions of daily temperature variability. This possibility is further underlined by the better results obtained with Dickinson's model, which includes a more sophisticated land surface package. More detailed analyses of current GCMs are necessary to confirm this speculation as well as to determine the causes of other errors in variability such as for precipitation. The next step involves altering the GCMs so that variability is properly simulated. Only then can much faith be put in GCM forecasts of variability changes with a perturbed climate.

3. Sensitivity analyses of climiate change impact models to changes in climate variability. It also must be determined how important changes in variability will be to different areas of impact. Since the variability of climate variables produced from GCMs cannot be "trusted" or even easily analyzed at this point, these sensitivity analyses of impact models should be performed with statistically simulated time series of climate variables, as has been performed by Schwarz (1976) and Mearns et al. (1984). By simulating time series, different levels of autocorrelation and variance in the time series may be controlled for and systematically varied. By this means, important thresholds of variability change for different variables as they affect the output of impact models can be determined. Moreover, ranges of possible impacts of variability change can be determined and can serve as guides until better information is available on how variability will change in a CO_2-warmed world.

REFERENCES

American Meteorological Society. 1985. Flash floods: a statement of concern by the AMS. Bulletin of the American Meteorological Society 66(7):858-859.

Barbecel, O., and M. Eftimescu. 1973. Effects of agrometeorological conditions on maize growth and development. Institute of Meteorology and Hydrology, Bucharest, Romania. pp. 10-31.

Bates, G.T., and G.A. Meehl. 1986. The Effect of CO_2 concentration on the frequency of blocking in a general circulation model coupled to a simple mixed layer ocean model. Monthly Weather Review 114:687-701.

Becker, R.J., and R.A. Wood. 1986. Heatwave. Weatherwise 39(4):195-6.

Bourke, W. 1974. A multi-level spectral model. I. Formulation and hemispheric integrations. Monthly Weather Review 102:687-701.

Bourke, W., B. McAvaney, K. Puri, and R. Thurling. 1977. Global modeling of atmospheric flow by spectral methods. Methods in Computational Physics 17:267-324.

Bridger, C.A., F.P. Ellis, and H.L. Taylor. 1976. Mortality in St. Louis, Missouri, during heat waves in 1936, 1953, 1954, 1955, and 1966. Environmental Research 12:38-48.

Chervin, R.M. 1986. Interannual variability and seasonal predictability. Journal of the Atmospheric Sciences 43:233-251.

Chervin, R.M. 1981. On the comparison of observed GCM simulated climate ensembles. Journal of the Atmospheric Sciences 38:885-901.

Clark, R.A. 1987. Hydrological design criteria and climate variability. In: Soloman, S.I., M. Beran, and W. Hogg, eds. The Influence of Climate Change and Climatic Variability on the Hydrologic Regime and Water Resources. International Association of Hydrological Sciences Publ. No. 168. Oxfordshire: IAHS Press.

Coleman, J.E. 1988. Climatic warming and increased summer aridity in Florida, U.S.A. Climatic Change 12:164-178.

Dare, P.M. 1981. A study of the severity of the midwestern winters of 1977 and 1978 using heating degree days determined from measured and wind chill temperatures. Bulletin of the American Meteorological Society 62(7):674-682.

Dickinson, R.E., A. Henderson-Sellers, P.J. Kennedy, and M.F. Wilson. 1986. Biosphere-Atmosphere Transfer Scheme (BATS) for the NCAR Community Climate Model. NCAR Technical Note 275, NCAR, Boulder.

Ellis, F.P. 1972. Mortality from heat illness and heat-aggravated illness in the U.S. Environmental Research 5(1):1-58.

Emmanuel, K.A. 1988a. The dependence of hurricane intensity on climate. Nature 326:483-485.

Emmanuel, K.A. 1988b. Toward a general theory of hurricanes. American Scientist 76(4):370-379.

Gibbs, W.J. 1984. The great Australian drought: 1982-1983. Disasters 8/2/84 89-104.

Hansen, J., I. Fung, A. Lacis, S. Lebedeff, D. Rind, R. Ruedy, G. Russell, P. Stone. 1988. Global climate changes as forecast by the Goddard Institute for Space Studies three-dimensional model. Journal of Geophysical Research 93(D8): 9341-9364.

Heathcote, R.L. 1985. Extreme event analysis. In: Kates et al. Climate Impact Assessment, SCOPE 27. Chichester: John Wiley and Sons.

Henz, J.F., and Scheetz, V.R. 1976. The big Thompson flood of 1976 in Colorado. Weatherwise 29(5):278-285.

Herbert, P.J., and Taylor, G. 1979. Everything you always wanted to know about hurricanes. Weatherwise 32(2):61-67.

Jones, T.S. et al. 1982. Morbidity and mortality associated with the July 1980 heat wave in St. Louis and Kansas City, MO. Journal of the American Medical Association 247:3327-3331.

Kalkstein, L.S. 1988. The Impact of CO_2 and Trace Gas-Induced Climate Changes Upon Human Mortality. Report prepared for U.S. EPA, contract No. 68-01-7033.

Kalkstein, L.S., R.E. Davis, J.A. Skindlov, and K.M. Valimont. 1987. The impact of human-induced climate warming upon human mortality: a New York case study. Unpublished? Pre-report for the EPA. 26 pp.

Katz, R.W. 1984. Procedures for Determining the Statistical Significance of Changes in Variability Simulated by an Atmospheric General Circulation Model. Climate Research Institute Report No. 48. Oregon State University, Corvallis.

Linder, K.P., M.J. Gibbs, and M.R. Inglis. 1987. Potential Impacts of Climate Change on Electric Utilities. New York State Energy Research and Development Authority, Report 88.2 Debany, New York.

Manabe, S., and D.G. Hahn. 1981. Simulation of atmospheric variability. Monthly Weather Review 109:2260-2286.

McQuigg, J.D. 1981. Climate variability and crop yield in high and low temperature regions. In: Back W., J. Pankrath, and S.H. Schneider, eds. Food-Climate Interactions. Reidel, 121-138.

Mearns, L.O., S.H. Schneider, S.L. Thompson, and L.R. McDaniel. 1988. Analysis of climate variability in general circulation models (in preparation).

Mearns, L.O., R.W. Katz, and S.H. Schneider. 1984. Extreme high-temperature events: Changes in their probabilities with changes in mean temperature. Journal of Climate and Applied Meteorology 23:1601-1613.

Mederski, H.J. 1983. Effects of heater and temperature stress on soybean plant growth and yield in humid temperature climates. In: Raper, C.D., and P.J. Kramer, eds. Crop Reactions to Water and Temperature Stresses in Humid, Temperate Climates. Boulder, CO: Westview Press, pp. 35-48.

Mogil, H.M., et al. 1984. The great freeze of '83: Analyzing the causes and the effects. Weatherwise 37(6):304-308.

Neild, R.E. 1982. Temperature and rainfall influences on the phenology and yield of grain soybean and maize. Agricultural Meteorology 27:79-88.

Newman, J.E. 1978. Drought impacts on American agricultural productivity. In: Rosenberg, N.J., ed. North American Droughts. Boulder, CO: Westview Press, pp. 43-63.

Northwest Power Planning Council. 1983. Regional Conservation and Electric Power Plan 1983. Portland, OR: Northwest Power Planning Council.

Oeschsli, F.W., and R.W. Buechley. 1970. Excess mortality associated with three Los Angeles September hot spells. Environmental Research 3:277-284.

Oliver, J. 1981. The nature and impact of Hurricane Allen - August 1980. Journal of Climatology 1:221-235.

Parry, M.L. 1978. Climatic Change, Agriculture and Settlement. Dawson, Folkstone.

Quirk, W.J. 1981. Climate and energy emergencies. Bulletin of the American Meteorological Society 62(5):623-631.

Reed, D.N. 1986. Simulation of time series of temperature and precipitation over eastern England by an atmospheric general circulation model. Journal of Climatology 6:233-257.

Reibsame, W.E., and D.J. Smith. 1988. Sensitivity and adjustment to climatic fluctuation in water resource management. Climatic Change (13) (in press).

Rind, D., R. Goldberg, and R. Ruedy. 1988. Change in climate variability in the 21st century. Submitted to Climate Change.

Rosenberg, N.J., ed. 1978. North American Droughts. Boulder, CO: Westview Press.

Rosenberg, N.J. (ed). 1979. Drought in the Great Plains: Research on Impacts and Strategies. Water Resources Publications, Littleton, Co.

Sanders, J.F. 1982. The hurricane dilemma. Weatherwise 35(4):174-178.

Schugart, H.H., M. Ya. Antonovsky, P.G. Jarvis, and A.P. Sandford. 1986. CO_2, climatic change, and forest ecosystems. In Bolin, B., B.R. Doos, J. Jager and R.A. Warrick (eds.), The Greenhouse Effect, Climatic Change, and Ecosystems. SCOPE 29, John Wiley and Sons, Chichester, U.K. pp. 475-521.

Schwarz, H.E. 1977. Climatic change and water supply: How sensitive is the Northeast? In: Climate, Climate Change and Water Supply. Washington, DC: National Academy of Sciences, pp. 111-120.

Shaw, R.H. 1983. Estimates of yield reductions in corn caused by water and temperature stress. In: Raper, C.D., and P.J. Kramer, eds. Crop Reactions to Water and Temperature Stresses in Humid, Temperate Climates. Boulder, CO: Westview Press, pp. 49-66.

Soloman, A.M., and D.C. West. 1985. Potential responses of forests to CO_2-induced climate change. In: M.R. White, ed. Characterization of Information Requirements for Studies of CO_2 Effects: Water Resources, Agriculture, Fisheries, Forests, and Human Health. U.S. Department of Energy, DOE/ER-0236. pp. 145-170.

Thompson, L.M. 1968. Weather and technology in the production of corn. In: Purdue Top Farmer Workshop, Corn Production Proceedings. West Lafayette, IN: Purdue University, pp. 3-19.

Warrick, R.A., et al. 1975. Drought Hazard in the United States: A Research Assessment. University of Colorado, Boulder. IBS Monograph No. NSF-RA-E-75-004.

Washington, W.M. and G.A. Meehl 1984. Seasonal cycle experiment on the climate sensitivity due to a doubling of CO_2 with an atmospherice general circulation model coupled to a simple mixed-layer ocean model. Journal of Geophysical Research 89(D6):9475-9503.

White, G.F. et al. 1975. Flood Hazard in the U.S.A. Research Assessment. University of Colorado, Boulder. IBS Monograph No. NSF-RA-e-75-006.

Wilhite, D.A., and M.H. Glantz. 1987. Understanding the drought phenomenon: The role of definitions. In D.A. Wilhite and W. Easterling (eds.), Planning for Drought. Westview, Boulder. pp. 11-27.

Wilford, J.N. 1988. Vest Persistent Pattern Spreading Heat Wave. The New York Times, p. 1; July 19.

Wilson, C.A., and J.F.B. Mitchell. 1987. Simulated climate and CO_2-induced climate change over Western Europe. Climate Change 10:11-42.

CHAPTER 18

RESEARCH NEEDS

The Environmental Effects report has summarized the results of over 50 individual research projects interpreted from regional and national viewpoints. As described in previous chapters, this report is organized around a series of regional case studies performed for the Great Plains, the Southeast, the Great Lakes region, and north-central California, as described in Chapter 1. In addition, several larger environmental issues, such as impacts on agriculture, forest resources, coastal zones, and water demand, were analyzed from a national perspective. In each regional and national study, an attempt was made to make use of the highest quality and most recent science available. This process has resulted in two outcomes:

1. The results provide insight about the sensitivities of different sectors and regions to climate change.

2. The results themselves raise new scientific questions and issues that cannot now be resolved. For example, the importance of certain climate data for impact analyses has been recognized. The following partial list of examples may be helpful to circulation modelers:

   - the importance of winds in analyses of Great Lakes levels;
   - the need for information on major storms and hurricanes to fully assess impacts on coastal regions;
   - the role of water vapor, clouds, boundary layer, and windspeed and direction for air quality analyses;
   - the impact of precipitation variability on pesticide runoff studies;
   - the effect of changes in freezing conditions on ice cover in the Great Lakes and on snow pack in California;
   - the importance of soil moisture and temperature on forest growth;
   - the effect of temperature and precipitation variability on agriculture;

- o the role of water runoff for future estimates of hydroelectric power, flood control, and water quality; and

- o the possible importance of synoptic weather patterns on health.

This chapter summarizes the major scientific research issues on the environmental effects of climate change that the Environmental Protection Agency has identified while preparing this report. It is not our purpose to make recommendations to the U.S. Government or to the full scientific research community. However, we have, where appropriate, suggested areas where strong research coordination should occur and have attempted to describe where such research coordination is already occurring.

Detailed research needs have been identified and justified in each chapter. Table 18-1 summarizes these research needs for several major topics. The assumption that underlies both this table and the overall research strategy is that focusing on regional-scale effects from global atmospheric change is an effective way to address scientific issues relevant to environmental policymakers. This report has upheld that initial assumption, and we expect that future research on impacts in other parts of the globe and on other resources will do so as well. In addition, however, it will be important to understand broad national consequences of the regional impacts of climate change. We propose that this can be accomplished by giving careful attention to both aspects in research, and by conducting scientific studies and policy studies in close cooperation, as has been done in this report.

This chapter contains four main sections:

1. A description of the research role that EPA will play in the U.S. Government's overall research program in global climate change;

2. A description of the main research thrusts to be pursued in the environmental consequences of global climate change;

3. A description of the supporting atmospheric science research that will be necessary from the standpoint of understanding environmental effects; and

4. A short summary of the research topics implicit in integrating regional and national-scale research issues in such a way that they are relevant to the development of environmental policy.

THE ENVIRONMENTAL PROTECTION AGENCY'S RESEARCH ROLE

EPA's main responsibility is to develop, implement, and enforce policies to protect human health and the environment. In recent years, the concept of risk assessment and risk-based decisionmaking has been used as a model for developing policy-relevant scientific information in a logical fashion that takes into account the inherent uncertainties in data and understanding (Ballou et al., 1981; Barnthouse et al., 1982; Harwell and Harwell, 1988). While this concept has its roots in health and toxicological research and regulatory activities in EPA (North and Yosie, 1987), it can also be extended in concept to other, more ecological issues.

Global climate change presents a challenge to this concept, as EPA, other Federal agencies, and international research organizations have discovered. Since probable losses due to climate change cannot be quantified in a truly predictive fashion, a strong research emphasis must be placed on scientific assessment: we must identify not only what we do and do not know, but how well we understand the possible consequences of change. The role of EPA's research program and other research programs should thus clearly be to provide understanding and scientific assessments necessary to develop policy options and to analyze their consequences for the environment.

EPA and the U.S. Government are moving strongly in this direction. The Global Climate Protection Act of 1987 directs EPA and the State Department to coordinate the development of national policy for global climate change. This coordination involves many other agencies with essential policy roles, such as the Department of Energy. In addition, the act directs EPA, in cooperation with other agencies, to prepare a scientific assessment of climate change with international scope and representation. EPA has been designing the scientific assessment in cooperation with several Federal departments and research agencies: the National Aeronautics and Space Administration, the National Oceanic and Atmospheric Administration, the National Science Foundation, the U.S. Geological Survey, the Department of Energy, the Office of Naval Research, and the U.S. Department of Agriculture. These agencies, along with the State Department, have been organized under the auspices of the Committee on Earth

Sciences in the Office of Science and Technology Policy and in cooperation with the National Climate Program Office. The scientific assessment will address the scientific information that is important for policymakers to consider as they move toward developing policy options.

Within the U.S. Government, EPA's overall role encompasses both policy development and scientific research. In scientific research, its main responsibility will be to understand the potential consequences of global atmospheric change, the anthropogenic contribution of global change and options to minimize it, and the development of adaptation responses.

That responsibility and the research in this report have led to the formulation of several important scientific questions that are also relevant to EPA's policy mission. These questions should be thought of as overriding themes, rather than as a list of all the potential issues:

- What is the likelihood of a change in the global climate system, occurring over the next several decades to a century, that is clearly attributable to anthropogenic influences?
- How will climate and atmospheric chemistry change in different regions?
- What is the extent and magnitude of ecological and environmental change associated with changes in regional chemistry and climates?
- Over what time spans are regional atmospheric change and subsequent ecological changes likely to occur?
- What are the potential impacts of climate change on environmental resources?

These policy-oriented questions are viewed as the beginning stages for analyzing possible environmental changes resulting from climate change and eventually analyzing the possible avenues of managing risks. They also define the specific areas in which EPA will conduct and sponsor research, e.g., emissions of trace gases, global atmospheric chemical modeling, regional atmospheric consequences of change, regional environmental consequences of change, and monitoring of sensitive ecosystems for change.

Goals and Objectives

The major research goals and objectives of EPA's research program have been structured to address the policy-oriented questions. They reflect the need to establish scientific direction that is consistent with major policy uncertainties. In so doing, the goals and objectives are more technical statements of research needed to address the policy questions.

The goals of EPA's proposed research are the following:

- To assess the probability and magnitude of changes in the composition of the global atmosphere, and the anthropogenic contributions to those changes, for the purpose of evaluating the likelihood and magnitude of subsequent climate change.

- To assess the likely extent, magnitude, and rate of regional environmental changes as a function of variability in climate, for the purpose of evaluating the risks associated with changes in the climate system.

Since this report has focused on the effects of climate change, and not on the processes leading to climate change, we will focus here on the second goal. The following objectives relate to this goal:

- To relate global scale changes in climate and chemistry to regional scale changes by constructing a series of regional atmospheric scenarios.

- To predict environmental responses to climatic variation and to test the processes that control those responses.

- To document the spatial covariation of regional climate change with regional ecological change to establish comprehensive ecological monitoring in selected locations, cooperatively with other EPA and Federal programs.

- To produce periodic scientific assessments in conjunction with other Federal agencies and international research organizations, and to perform research to evaluate the consequences of adaptation policies.

ENVIRONMENTAL CONSEQUENCES OF CLIMATE CHANGE

Background

The long-term goal of the environmental research proposed is to create quantitative regional models of environmental responses to climate change for decisionmakers. These models need to explain landscape-level and multi-resource interactions, e.g., links between air quality, forest health, and climate change, that are beyond the scope of this report. Before such models can be created, however, many gaps in knowledge remain to be addressed.

Research still needs to be done that will (1) identify regions and resources sensitive to global climate change; (2) establish quantitative, causal relationships between climate and selected ecological indicators and attributes of concern; (3) predict the impact of changing climate on selected systems and regions; (4) refine the assessments and methodologies to include quantitative, multi-resource interactions; and (5) develop the ability to address multiple stressors, integrated at the regional scale.

The process consists of four generic steps:

- Classify and characterize the environmental resources of concern;
- Determine the resources' current status, recent trends in condition, and possible associations with climatic variability;
- Use mathematical models to estimate future status under different climatic scenarios; and
- Characterize the possible changes in each system by sensitivity studies, and by formally considering the uncertainty inherent in the mathematical models and underlying data.

Objectives

In the first phase of research, one must continue to identify resources and regions sensitive to climate change through review and synthesis studies (e.g., NAS, 1982; MacCracken and Luther, 1985; Trabalka, 1985; Bolin et al. 1986; Shands and Hoffman, 1987; White, 1985). These studies will proceed from the identification of extant data through exploratory analyses and model enhancement to regional characterizations of environmental changes associated with global climate change. The second phase will build on the first to initiate experimental studies investigating causal relations among climatic stressors and ecological responses at the ecosystem, landscape, and regional scales. In the third phase, several years away, research should emphasize trends detection, long-term experimental manipulation, and model testing and refinement. In each phase, close cooperation with the U.S. land management and research agencies will be necessary for success.

The following objectives are identified for each of the above phases.

Short-Term Objectives

- Define environmental sensitivity qualitatively, and identify sensitive climatic, biotic, and aquatic regions for study using extant data.

- Produce conceptual models of expected climate-induced responses, including changes in biogenic sources of radiatively important trace gases.

- Prioritize the regions and systems, as well as the most appropriate temporal and spatial scales, for in-depth observational and experimental (field and laboratory) studies.

Mid-Term Objectives

- Establish quantitative, causal relationships between climate and selected environmental attributes of concern, and begin to quantify the uncertainties.

- Proceed from qualitative to quantitative, predictive models.

- Produce preliminary predictions of changing climatic impacts on selected systems and regions, and quantify the uncertainties.

Long-Term Objectives

- Refine the assessments and methods produced in the first two phases to include quantitative, multi-resource interactions.

- Develop and validate capabilities to address multiple stressors and resources, integrated at regional scales for decisionmaking.

Research Strategy

This report has clearly shown that concerns over the adaptability and fate of natural ecosystems are well founded. Forested ecosystems, aquatic and marine biota, wildlife in refuges, water quality in small lakes, and certainly other resources may be unresistant to rapid climate change. The strategies for mitigating changes in these systems are likely to be large and exceedingly difficult to implement. Consequences for environmental attributes such as biodiversity, primary productivity, and cycling of nutrients, while difficult to quantify, may be equally difficult, if not impossible, to reverse. Federal agencies with both stewardship and research responsibilities for managing natural resources such as the U.S. Forest Service, the National Park Service, the Fish and Wildlife Service, the U.S. Geological Survey, the Bureau of Land Management, and the National Oceanic and Atmospheric Administration, should recognize that data, models, and continued research on these environmental resources and attributes are vital. This report has additionally shown that while intensively managed ecosystems, especially agroecosystems, may also be affected by a climate change, there seem to be more opportunities for human intervention to mitigate or adapt to the response. Thus, the critically important questions become whether or not the capacity for human intervention can keep pace with the rate of change induced by changing climate. Areas of major interest, therefore, are historical changes in the distribution and yield of agroecosystems as a function of climate, interactive effects of climate change and carbon dioxide increases on crop yields, and the adaptation rate of management practices. The Agricultural Research Service and Department of Energy have been, and should continue to be, important contributors to these research areas. Whether the same generalizations will hold for agriculture abroad remains to be investigated, and EPA will cooperate with U.S. and international organizations to explore this issue further.

Other managed environmental resources, such as air and water quality, or the preservation of coastal wetlands, are already subject to regulation, but may well be sensitive to climate change, as shown in Chapters 8, 9, and 13. EPA will investigate the possible consequences of climate change for these resources, examine their influence on ecosystems, and analyze the implications for regulatory policies. Other already regulated resources, such as drinking water and groundwater, may also be investigated.

Chapter 18

Possible impacts on human health, as emphasized in this report, are both troubling and very difficult to quantify. The correlative studies reported in Chapter 14 need to be substantiated with more targeted research on the mechanisms of impact. While EPA is not likely to be a major contributor in this area, other organizations, such as the National Institutes of Health or the National Institute for Environmental Health Studies, may be logical sponsors for further research. The possibilities that disease vectors may change their distributions as a function of climate change, while speculative, should be pursued more completely. To the extent that these are as much ecological questions as they are health questions, contributors will include EPA as well as organizations such as the Centers for Disease Control.

It is clearly not possible to study all potential effects. Because of the concerns raised in this report, and EPA's responsibilities for understanding the consequences of climate change, initial focus will be on the possibilities of irreversible change as well as when changes may occur rapidly. Therefore, early work will concentrate on providing data on the sensitivities of environmental systems to climate change, constructing conceptual models, and providing input to future regional and national-scale assessments. Later research will lead to increasingly quantified understanding of environmental response to climate change in several research areas: regional environmental attributes, feedbacks, scaling issues, disturbance regimes, and landscape interactions.

To conceptually organize research, attributes of systems will be investigated as a function of ecological and biotic regions (Solomon and West, 1985; Braun, 1950; Brown and Lowe, 1980; Fowells, 1965; Vankat, 1979; Wells, 1979). Then the structural and functional aspects of each attribute will be investigated. Ecosystems and regions will be defined by extant environmental conditions and existing data, with boundaries along maximum rates of change in system attributes. Ecosystems and regions will therefore be well defined near the center of their range or distribution. The boundaries will necessarily be less well defined, but they will be more precisely delineated as a function of continuing analyses. Examples of similar approaches to resource definition are shown in Figures 18-1 through 18-3.

The different scales of spatial patterns on the ground can impart great difficulties in the "scaling up" of local processes. Therefore, a high priority is to develop the ability to interpolate from large spatial scales to small and extrapolate from local patterns and processes to regional generalizations.

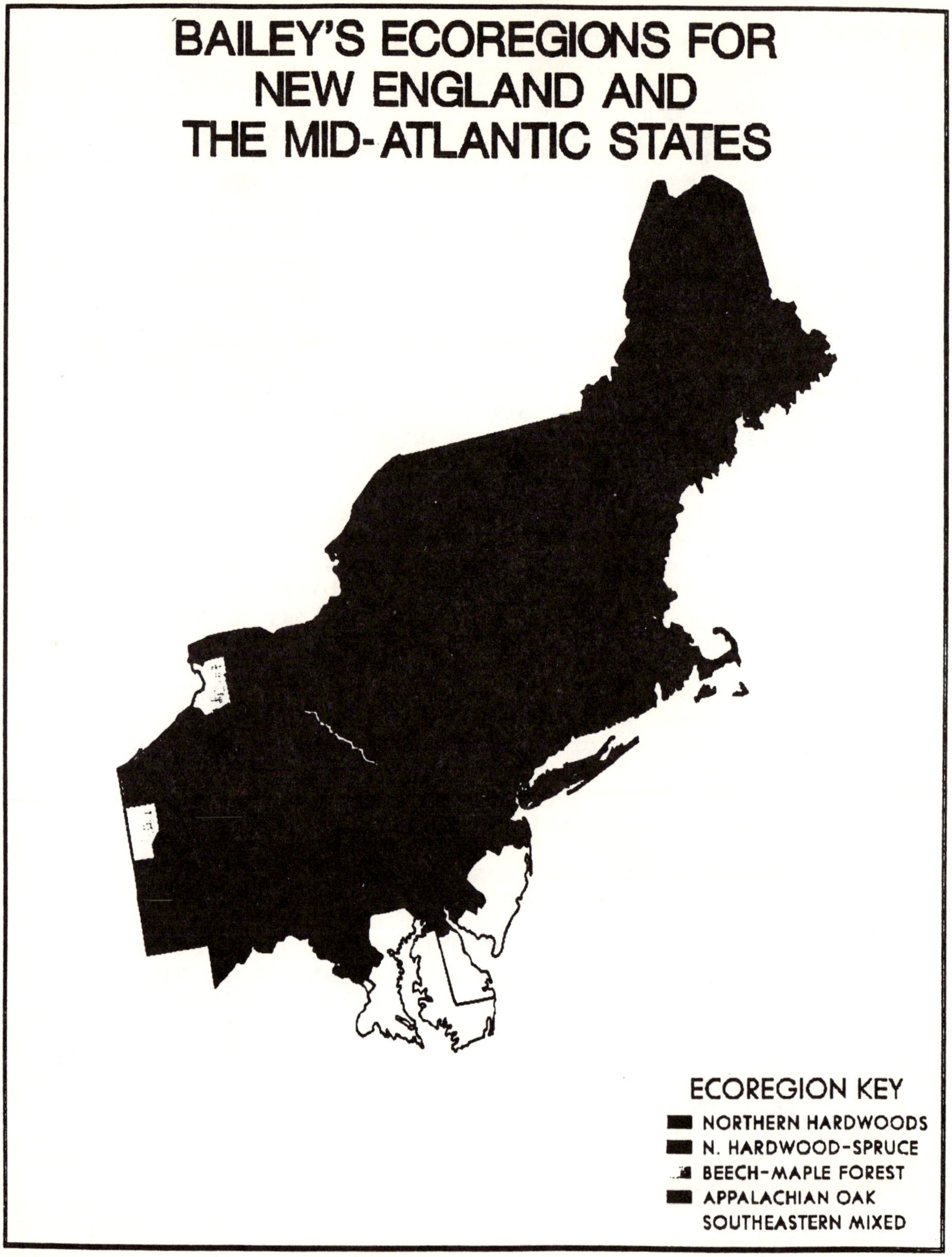

Figure 18-1. Bailey's ecoregions for New England and the Mid-Atlantic States.

Source:

Figure 18-2. Kuchler's ecoregions for New England and the Mid-Atlantic States.

Source:

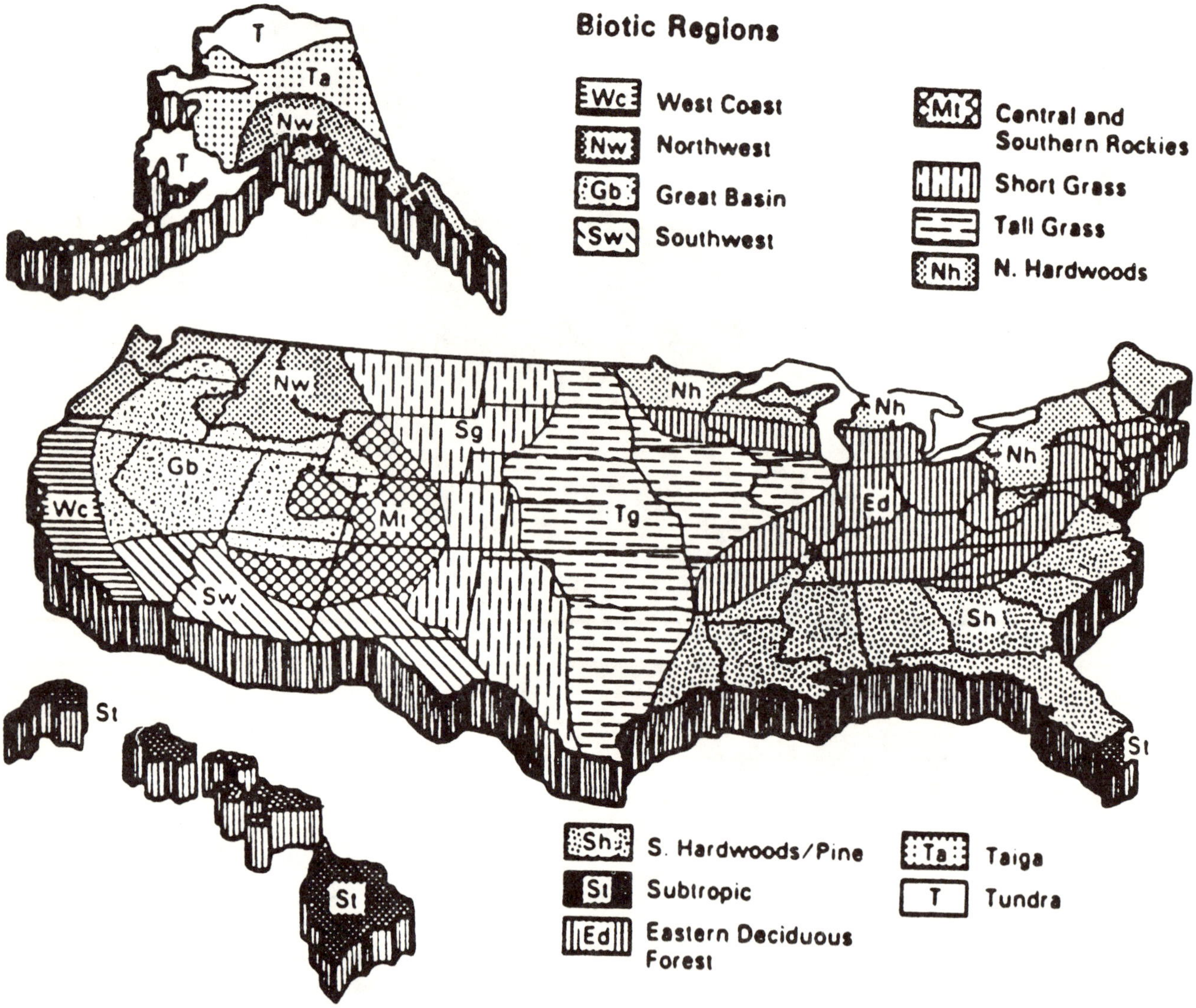

Figure 18-3. Major biotic regions of the United States.

Source: Compiled from Braun (1950); Fowells (1965); Vankat (1979); Wells (1979); Brown and Loew (1980).

Intercalibration of measurement approaches will improve confidence in results and should allow low-cost regional estimates from satellite measurements. The National Science Foundation, the National Aeronautics and Space Administration, and the National Oceanic and Atmospheric Administration will be key contributors to this research. However, a significant investment in methods development of sensors as well as ground-truthing must be done before this goal becomes a reality. Analysis of the effects of changes in climatic variability is critical to understanding potential environmental changes. Variations in the frequency of weather events are important for many environmental processes, and the rates of those processes may vary as a function of both spatial scales and temporal scales. Changes in climatic variability will be critical as the problem to be addressed becomes one of attempting to predict the occurrence of large-scale, rapid changes in the biosphere.

Critical feedbacks in ecosystem distributions result from (1) interactions between bordering systems, such as the forest-prairie ecotone; (2) differential lags, such as the different migratory and developmental rates of different kinds of organisms and physical components within any given system; (3) interactions between biogeochemical processes and population processes; and (4) boundary layer effects.

The differing lifespans and dynamic processes in the soil-plant-atmosphere continuum create lags that will cause the response time of systems to differ substantially from the rate of climate change. Changes in the biogeography or distribution of species will change biogeochemical balances, which will then change the size of reservoirs of carbon and other elements, thus changing the rates of production of some radiatively important trace gases.

Physical and biological disturbances on any scale exercise considerable influence over the patterns and processes of ecological systems on both larger and smaller scales. Frequency, extent, and magnitude of disturbances will likely change as a function of global climate changes. The relationships between disturbances and ecological responses will therefore be critical to any modeling that attempts to predict the appearance and rate of change of large areas of landscape.

The initial stages of research should involve the following: (1) setting priorities for systems and regions, based on initial concepts of sensitivity and results from studies in this report and others; (2) constructing conceptual models; and (3) providing preliminary information for national-scale assessments of environmental effects of climate change. The major indicators of regional sensitivity to be used should include natural variability of weather and concomitant variability in the growth and distribution of major plant species, based on historical records and proxy data.

Early scientific assessments should provide information on (1) location of sensitive regions, and (2) conceptual models that include causal processes. To evaluate the potential sensitivity of regions, landscapes, and ecosystems, the temporal variability in the spatial distribution of both weather patterns and ecological resources must be analyzed systematically. This is difficult, because increased resolution on one scale generally results in decreased resolution on the other. However, the results will provide the conceptual models of regional responses to global climate change. A similar approach is applicable to aquatic biota and water resources, wetlands, agriculture, and other environmental resources.

The general procedure is to identify climatic regions, as determined by both temporal (seasonal) and spatial variability in weather patterns (Borchert, 1959; Bryson, 1966; Mitchell, 1969; Wendland and Bryson, 1981). Climatic regions so identified are then associated with attributes of corresponding environmental and ecological systems that might be sensitive to these changing weather patterns. The zones of seasonal weather patterns and interannual variability can then be analyzed for covariation with corresponding data on major natural resources, such as vegetation distribution and growth (Braun, 1950; Fowells, 1965; Vankat, 1979; Wells, 1979; Brown and Lowe, 1980). Characteristics of the plant distributions, specifically species boundaries or transition zones, are related through causal hypotheses to variability in weather patterns. These characteristics and model results are then used to indicate the expected future covariation between climatic and natural resources regions, ecotonal areas, and ecosystems potentially sensitive to large-scale shifts in the climate system.

The critical areas to be identified during these analyses include (1) links between indicators of climate change and upper atmospheric weather patterns; and (2) links between upper atmospheric weather patterns, surface weather patterns, and resource variability (Neilson, 1986); thus providing (3) links between indicators of climate change and resource variability. This information can then be integrated to evaluate the expected responses in resource indicators to a change in the climate system. The entire process yields mechanistic, conceptual models of regional ecosystem responses to global climate change.

This approach can also be used to identify potentially susceptible water resources and surface water regions of the country (Figure 18-4) (USGS, 1986). This will produce a depiction of the processes that control seasonal runoff patterns and their year-to-year variability, as well as a direct relationship between rainfall and runoff at a regional scale. The interactions between regional rainfall, runoff, vegetation cover, soil moisture, and evapotranspiration will then need to be addressed. Integrating hydrologic flows, land-use patterns, and ecosystem dynamics will allow characterization of water quality patterns within regions. These associations can then be used to characterize the effects of climate change on water resources on regional spatial scales.

The biotic and hydrologic studies then must be integrated to identify regions and sites for future experimental study. This is critical as terrestrial, aquatic, wetland, and coastal systems do not exist in isolation; their interactions affect each ecosystem both directly and indirectly.

Experimentation, Modeling, and Monitoring

Field and laboratory experiments will be required to address directly some of the proposed mechanisms and ecological processes that mediate landscape and regional scale responses. Experimentation will assist in model development, model calibration, and identification of baselines against which change can be measured.

Experiments and observations should assess quantitatively the changes in resources as a function of position along regional environmental gradients. Attributes to be assessed include (1) past variability of resources to climate change; (2) establishment and mortality of plant species; (3) decomposition and productivity patterns; (4) successional and biodiversity patterns; (5) biogenic emissions of trace gases; and (6) runoff and water quality patterns.

Monitoring studies must be initiated for the purposes of (1) resolution of ecological patterns at different temporal and spatial scales; (2) regional detection of trends; and (3) calibration and validation of models. The locales, systems, and attributes to monitor should be based on the extant data analyses and field studies outlined above.

Modeling experiments should be performed to reduce uncertainties in our capabilities to predict ecological responses. Sources of uncertainties include (1) dispersal

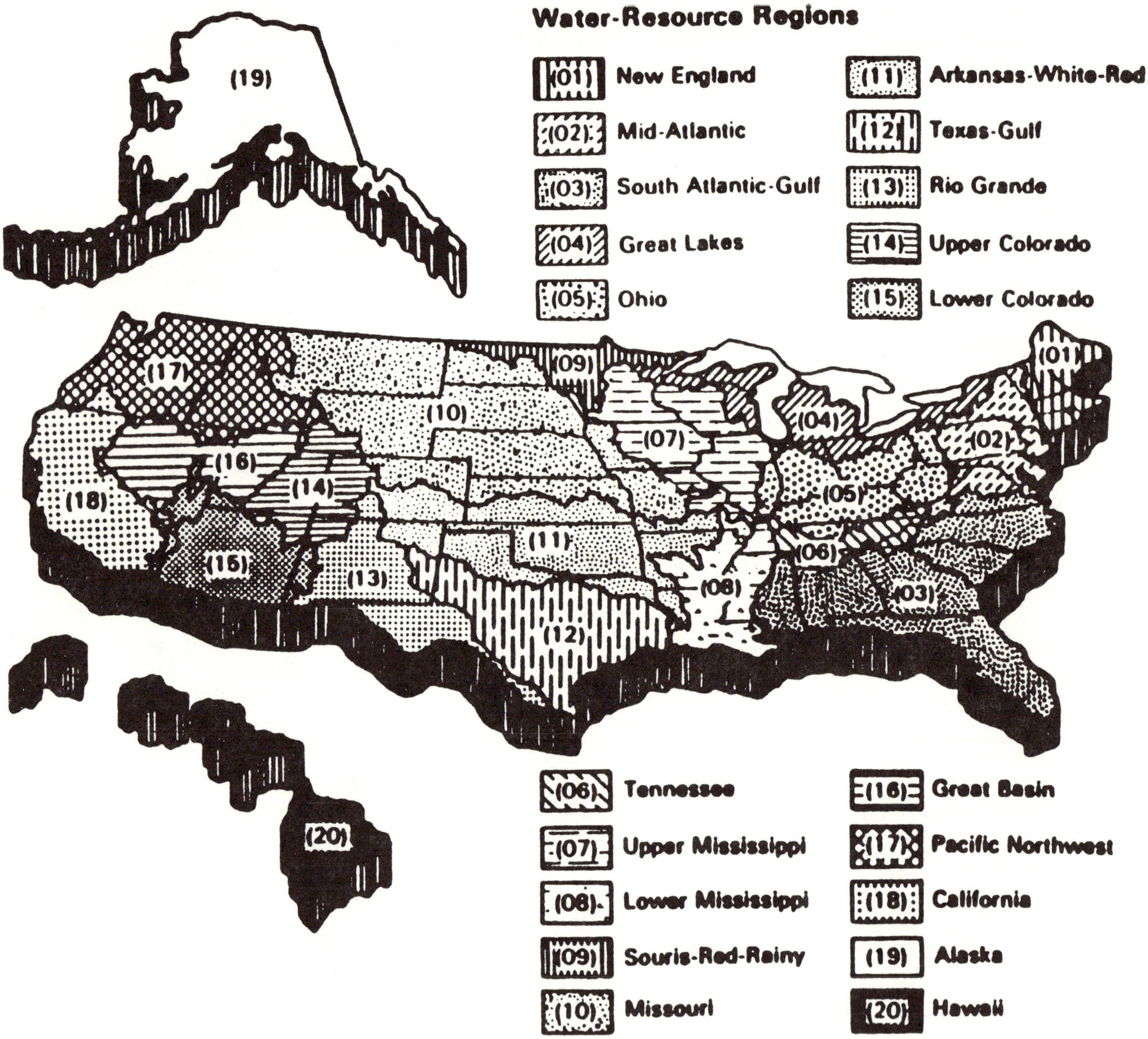

Figure 18-4. Water resource management regions based on Water Resources Council and USGS delineation.

Source: USGS (1986).

through a heterogeneous and changing landscape; (2) lags imparted by life cycles of dominant organisms and ecological processes; (3) rate changes imparted by feedbacks from direct effects of CO_2 on plant processes; (4) landscape transfers of sediment, nutrients, organic matter, energy, water, and propagules among terrestrial, aquatic, wetland, and near-coastal systems; (5) geomorphic processes; and (6) possible threshold effects or synergistic effects in any of these processes, including the possibility of exotic or weedy species overrunning newly created habitats.

The modeling strategy will be most effective in a phased approach, focusing initially on developing conceptual models and critically assessing already available techniques and quantitative models. This report has begun to contribute to this phase due to its use of existing models for analysis. Components of existing models will then be evaluated for use in regional models or for direct application. Conceptual models will then provide a framework for integrated landscape and regional simulation models. Modeling exercises should initially parallel the field experiments outlined above.

Mid-term research tasks will be a direct outgrowth of earlier work and also will address major feedback loops and quantifying temporal and spatial scale relationships.

The feedback loops fall into three general categories: (1) landscape interactions; (2) direct effects of CO_2; and (3) biogenic emissions of radiatively important trace gases. Scaling issues to be resolved in this phase involve relating global scale circulation patterns to regional scale weather and then to landscape scale effects consistent with ecosystem processes.

Experimental approaches will be initiated to improve the conceptual models developed earlier and to permit more quantitative modeling and understanding. New research areas of importance include (1) remote sensing to detect areas of plant stress, to inventory possible biogenic emission sources, and to monitor ecological status; (2) ground-based inventories of possible biogenic emissions sources; (3) quantification of ecosystem-landscape processes and interactions; and (4) continued hierarchical modeling of system and landscape level responses to climate change.

This phase will make possible a more quantitative assessment of the potential ecological effects of climate change, including identification of uncertainties and directions for research to further reduce critical uncertainties. The major change is that assessments at this time will have the choice of reflecting a national scope with regional resolution or of focusing on quantitative potential effects within one or a few sensitive regions.

The next stages of model enhancement should be to formulate landscape-scale models. This will involve linking models of different resources and extending models from point or site representations to include spatial variability. Remote sensing will play a key role in calibrating and validating such models.

Simplifying modeling approaches to describe resource links will be a major challenge. One approach is "top-down" modeling, similar to the approach first used to identify sensitive regions (Landsberg, 1986). Conceptual models will be constructed from this perspective with process-based complexities added as needed. While it will be necessary to account for additional variables at individual sites, this may not be necessary for describing regional patterns.

Major emphases must be placed on linking models or model results across ecosystem boundaries. One approach will include searching for response surfaces that can directly link inputs and outputs, bypassing internal model complexities. These response surfaces could be used to represent model behavior in linked systems. The original, process-based models would be retained for testing and evaluation purposes. Linking models at both the landscape and regional scales should then provide information on the important processes and inputs at that scale, and as well as on what climatic signals are transferred to other scales.

Regional Estimates and Trends Detection

A long-term requirement for environmental effects research is an inventory of ecosystems and landscape units on a regional basis. Statistical methods for trends detection over time and space will also be needed. The land management agencies, the National Science Foundation, the Department of Energy, and EPA all have existing and planned activities in these areas. Resource inventories and ecological monitoring data must continue to be high priorities for funding, if we ever hope to be able to validate our current or planned models of ecosystem response to climate changes.

ATMOSPHERIC SCIENCES RESEARCH

The goal in the proposed strategy for atmospheric sciences research is to better understand the physical and chemical dynamics of the atmosphere to assess the impact of climate and climate change on the environment. Research is needed to produce scenarios of both physical and chemical atmospheric conditions that will support ecological assessments and policy analyses of the effect of global climate change on regional environmental resources (Kellogg and Schware, 1981; Wigley et al., 1986; NRC, 1982). We will refer to these more comprehensive scenarios simply as "climate scenarios," with reference to both the physical and chemical components of the atmosphere as climate.

Objectives

Four objectives must be achieved to accomplish the overall goal:

1. To produce regional climate scenarios that address biological and ecological responses, incorporating appropriate meteorological and chemical dynamics that are related to the expected biological effects.

2. To produce regional climatic scenarios that allow assessments of future tropospheric and boundary layer air quality for regulatory and policy analysis.

3. To enhance understanding of the atmospheric phenomena controlling impacts by producing models that describe the physical and chemical processes and their interactions on subregional to global scales.

4. To examine secular air quality and climate changes to identify important current trends and to allow evaluation of model predictions and the testing of mechanistic hypotheses.

Major research efforts are needed to link anthropogenic changes in atmospheric chemistry, changes in climate, and the effects of the new climate on environmental systems. The National Science Foundation, National Oceanic and Atmospheric Administration, the National Aeronautics and Space Administration, and the Department of Energy have traditionally been the major contributors to research in these aspects of the atmospheric sciences because of their support for development and analysis of general circulation models (GCMs). Their current and planned research programs are vital to continued progress and success in linking the atmospheric and ecological sciences. EPA, because of its existing expertise in meteorology and atmospheric chemistry, and its emphasis on environmental effects, will become an increasingly large contributor as well.

Research Strategy

Process models offer the potential for quantitative, deterministic understanding but suffer from known (and perhaps some unknown) deficiencies and limitations due to generalizing, parameterizing, or distorting physical and chemical relationships. Such deficiencies are unavoidable, so models must be tested for both sensitivity and accuracy. The scale of the model is chosen based upon the nature of the problems being studied. As a consequence, some processes are important factors on one scale of space and time but not on another. While the level of confidence in individual models varies widely from one use to another, it is clearly possible, as shown in this report, to produce early versions of atmospheric scenarios. These early efforts are clearly not entirely satisfactory, but they have provided useful directions for additional work. The approach from here on must be iterative in nature. Because models will not be able to provide all the needed information, analogue and statistical techniques must also be used.

Scenario Production

Scenario production in the future should be a three-step process involving the following: defining the needs of the users for scenarios; choosing, combining, and integrating the various methods of producing scenarios; running the various process and analogue models to produce sets of numbers, with guidance on their use; and distributing the scenarios to the research community. This will be an iterative process, producing new scenarios tailored to user needs and changing as models are developed or refined.

Because no one scenario will meet all the needs of ecological research groups concerned with different effects, a large number of scenarios tailored to the needs of each research group will have to be produced. For all the users, however, scenarios must address the frequency of occurrence of weather events; the significance of the environmental consequences; the degree of understanding connecting weather events and their consequences; and the knowledge of whether or not the proper atmospheric information can be provided (Robinson and Hill, 1987).

Scenario production teams will require statements of data needed from their clients. One potential problem is that many user needs for climatic data are poorly understood and defined, due in part to a lack of interaction between climatologists and other scientific disciplines.

General circulation models, as have been used for this report, are only partially suited for these purposes; regional and continental-scale simulations of precipitation, cloudiness, and soil moisture, and resolution of regional-climate detail on coarse computational grids, are all problems that need to be addressed. GCMs also do not include many atmospheric processes, such as chemistry and boundary layer meteorology. Since much of the necessary ecological research will be on the regional scale, GCM output must be augmented for scenario development. Promising approaches include using large-scale two-dimentional chemical models, regional air quality models, and analogue and statistical techniques. These various approaches need to be developed in a manner that will allow their inputs and boundary conditions to be taken, directly or indirectly, from GCMs.

Once regional models, large-scale models, and analogue techniques have been developed separately, they should be linked where desirable to produce the optimal tools for scenario production. One example might be to use analogue methods to modify the output of a GCM, then to use the output as input boundary conditions for regional climate and air quality models, then to use other methods to change their output from episodic to long-term means, and finally to use weather generators to find exceedances of air quality standards, drought probability, or other climatic indicators of interest in a small region.

The components mentioned above that make up atmospheric scenarios each need development programs. They are described in subsequent sections.

General Circulation Models

The major strength of general circulation models is that they are strictly process-oriented, and are thus capable of indicating the nature, magnitude, and causes of climate change. Not all processes can be treated in full detail; many must be incorporated through parameterization schemes that relate, either statistically or empirically, the processes to the output variable of concern. The major weakness of general circulation models is that their prediction capabilities are well characterized primarily for temperature and precipitation, which may not be sufficient variables for particular assessments. Furthermore, GCMs have difficulty predicting changes in regional climate. Therefore, techniques are needed to translate the significant portions of large-scale climate change available from these models into the smaller scales on which ecosystem effects usually take place. This spatial and temporal detail can be provided using either local models or historical analogue approaches.

Local Climate Models

Numerous process-oriented models have been developed on the scale of individual point-sources. While the basic model formulations are site-independent, the results depend on local conditions, and extrapolation is therefore difficult or impossible. Similarly, air quality models have characterized the urban boundary layer and considered detailed processes, but it is presently extremely difficult to produce a detailed model applicable to anything other than the specific location for which the parameterization was developed.

Mesoscale models are being developed, however, and the process orientation of mesoscale meteorological models suggests that they have the potential to be incorporated explicitly as subgrid scale processes within GCMs. One of the best developed of these models is MM-4, a numerical model designed to simulate or predict atmospheric circulation systems for short time periods on a spatial scale ranging from 20 to 2,000 km. MM-4 has been used to drive the Regional Acid Deposition Model (RADM) and can itself be driven by large-scale GCMs. It is not, however, a climate model and cannot presently be used as such.

One area where major improvements of local climate models can be expected is cloud processes. Clouds are strong players in potential nonlinear climate feedbacks, particularly on a subregional scale, as they influence radiative transfer, chemical reactions, transport of pollutants, and precipitation. Current models assume clouds exist when certain saturation or near-saturation conditions are met; however, real clouds form when supersaturation conditions exist relative to cloud condensation nuclei. There is a strong potential for feedback through changes in the ultraviolet part of solar radiation that affects the formation of cloud condensation nuclei. Before any local climate models can be built, linkages between atmospheric chemistry, cloud condensation nuclei, and the formation of all types of clouds must be understood.

Chemical Models

EPA is one of the national leaders in the construction, testing, and operation of regional-scale, three-dimensional meteorological and chemical transformation process models. These models have improved the understanding of the complex and nonlinear interactions of regional emissions, long-range transport, and free-radical gas phase and cloud (aqueous) phase chemistry of trace gases and aerosols. A natural extension of the present EPA mission is to expand these process-based tools to incorporate elements needed to address the effects of global climate change on the atmospheric environment.

Scale differences, both temporal and spatial, are a major impediment to simply merging modeling techniques between the climate model and regional air quality communities. Likewise, simply adding chemistry to a GCM is not presently feasible for the following reasons: (1) even the simplest chemistry model is presently very computer-intensive; and (2) due to the large spatial scales in GCMs, much of the critical concentration-dependent chemistry of the urban and industrial sources would be lost as the emissions were "smeared" over the whole grid cell volume. The manner in which emissions or sources affect global climate change must be understood through an understanding of how many of the trace gases are changed by urban and regional chemical systems. At the same time, changes in global climate will modify how the urban and regional chemistry and climate systems behave. Improvements in basic computer science and technology may help solve some of the strictly computational problems. However, the other issues require new science.

More developmental work is needed to improve our understanding of chemical mechanisms, especially for aromatic and biogenic species, before these mechanisms can reliably predict the fate of much of the mass of material emitted into the atmosphere by human sources.

Linking Available Models

Chemical and climate models have been developed largely in isolation from each other, so the process of linking them is not simple. In addition to technical problems associated with computer limitations, model structures, and basic formulation, a major problem area exists: merging the differing temporal and spatial scales used.

From the perspective of regional chemical modeling, the major research question must be: "How are the existing regional models, which are all episodic (a few days at most) in design and operation, and which need boundary conditions and externally specified weather streams, to be used to estimate the impacts of global climate change?" Clearly, we will need to incorporate new data sources, perhaps especially remote sensing of vegetation types, atmospheric chemical composition (Whitten, 1986), and climate into our modeling systems.

Model Evaluation

Performing model evaluations for models as complex as those discussed here requires a concerted group effort involving many scientists with both observational and theoretical skills and experience. A range of models will often be needed to understand these complex systems. To test such models, data will be needed that are associated with the operational aspects of the models (their ability to predict the items of interest) and the diagnosis of model performance (their ability to give the right answer for the right reason). Several problems exist in model evaluation, including observations taken at a different scale from the model's predictions; no available technology that can measure what the model calculates; model outputs that must be inferred from secondary data; models with incomplete descriptions of processes; biases of groups developing models; and different assumptions and scenarios used by different modeling groups. Accounting for these limitations will require field observations, laboratory data, or the use of simpler models emphasizing the processes being investigated.

Observational Analysis

Analyses are needed that provide information about the current state of the climate with emphasis on secular trends and their regional variations. These are needed for (1) the evaluation of process models, and (2) the production of scenarios for ecological and air quality risk assessments. There are many variables and time and space scales of interest. Two general needs, then, are to (1) develop methods of analysis capable of spanning several scales and variables, and (2) to develop priorities for particular variables and scales for specific purposes.

In general, the traditional physical elements of climate have been measured at numerous sites for many years. Other elements may have long records at a fairly restricted number of sites. In addition, a long tradition of paleoclimatic research using proxy data has provided a very long record for some regions and variables.

In contrast, the chemical variables of climate are routinely measured at few sites, and even those tend to have short record lengths. Few paleoclimatic investigations have considered chemical variables; however, some chemical data are becoming available from tree ring and ice core elemental analysis.

Quality control of observations of physical climate variables must be a major concern prior to any detailed climatological analysis. Many observations were originally obtained for operational weather forecasting purposes, so quality control was a relatively minor concern. There may have been changes in observational practices and equipment or in the surrounding area that may have created spurious climate changes. Adjusting climatic data sets is a difficult and highly skilled task; however, several data sets are available or are being developed (Karl et al., 1987).

The majority of quality-controlled data sets being developed emphasize factors needed for comparison with GCMs. However, for a particular scenario, these may not be the appropriate variables or time period. Quality controlled information on other variables, such as humidity or extreme winds, may be required. Another troublesome potential stems from the general failure to rigorously address the question of quality control of satellite-derived information.

Current Climate Statistics

The development of current climate statistics must be based on analytical techniques capable of addressing various time and space scales for different variables, requiring the use of appropriate statistical techniques combined with physical and chemical insight. The selection of appropriate summarizing statistics is complicated when the record is a continuously varying function on several time scales. Choosing stations representative of regions and scaling up from a series of points to a value representative of an entire area introduces further problems.

These questions have been addressed in the most depth on the global scale; however, apart from temperature, no physical climate variable has been assessed on a global scale (Mitchell, 1984; Jones et al., 1986). Chemical variables that have been assessed globally include gases such as CO_2 and CH_4. Regional variations, such as for tropospheric ozone concentrations, have also received considerable attention. Subglobal physical analyses have emphasized the continental scale. The regional scale has also been the one at which many of the long-term data sets have been developed (e.g., Karl et al., 1984; Lamb, 1987).

These analyses provide a wealth of detail about the current climate and its recent variability. However, identifiable deficiencies in the present state of knowledge remain, including the following: (1) no common statistical rationale to allow comparisons, and (2) an emphasis on surface phenomena without linkage to causal atmospheric phenomena, thus making linkage with process models or use in scenario development difficult.

A high priority is the development of a statistical framework for future analyses. The framework will need to allow flexibility when considering different variables in different regions, while allowing comparisons between analyses and providing a straightforward means of comparing results with process models.

Chemical elements of climate must also be studied to develop appropriate statistical methods of spatial and temporal analyses, development of techniques to link the various scales of interest, and development of techniques to relate the near-surface climate to the overall atmospheric processes causing them. A severe handicap for current work stems from the relative sparseness of the data.

<u>Analogue Scenarios</u>

Analogue scenarios are of two major types: those based on paleoclimatic data from proxy evidence (e.g., tree rings or lake sediments), and those based on the modern historical record. They have been developed primarily as complements to GCM results, to contribute to the understanding of how altered processes are reflected in different climates (Butzer, 1980; Williams and Wigley, 1983). Because of their connection to process models, currently available scenarios have emphasized the variables given by the models, mainly precipitation and temperature. As before, these may not be what the user of a scenario needs (Gates, 1985). A potential exists, however, for the development of very detailed analogue scenarios with specific user orientations.

A major research challenge is the development of new, scientifically valid techniques of analogue scenario production, the creation or assembly of appropriate data sets, the assessment of specific user needs, and the testing of the physical reality of scenarios produced. The analogue approach may provide the detail of a scenario, while process models provide the boundary conditions. However, the analysis of current trends in regional climate and the development of user-oriented analogue regional scenarios are likely to indicate the need for additional observational information.

RESEARCH INTEGRATION TOPICS

This report has treated each regional case study as a vehicle for illustrating potentially important environmental effects that may result from a change in the climate system. However, the capacity for truly integrating the research results of individual projects within each region has been limited. The purpose of integrating diverse aspects of science in a regional or subregional context is to ensure that scientific assessments derived from a research program have an internally consistent set of assumptions and relationships. Future regional case studies will need to address this problem, if regional and local concerns of environmental decisionmakers are to be reflected in scientific research and assessment.

Spatial Data Bases for Integration

Research in both the ecological and atmospheric sciences should produce early, qualitative estimates of continental and regional sensitivities based on the current understanding of environmental resources and the processes that control the responses of each ecosystem to climatic variability. The beginning stages of this process have clearly been taken in this report and will be added to, as noted earlier in this chapter.

However, once the early definitions of sensitivity have been made, there still remains the daunting task of determining the possible relationships between climatic variability and the underlying systems in the atmosphere and biosphere in a way that will enable future regional studies to best investigate sensitivities to climate change. This can be done only if current results and understandings are portrayed in a manner that will take into account both the actual results from modeling and experimental studies and their location on the landscape, i.e., in spatially structured data bases, analyzed with geographic information system (GIS) technologies.

Such spatially structured data bases can be used to produce maps, but their use goes far beyond traditional two-dimensional representations of geographic data. In particular, their main advantage may be the potential to track not only the results of experiments but also the uncertainty associated with both underlying measurements and with modeling algorithms. Thus, the spatially oriented data bases, if used in conjunction with geographic information system technology, can play both an assessment role and a quality assurance role.

Spatially oriented data bases therefore have three major purposes: (1) they provide a method of prioritizing research efforts within funding constraints; (2) they provide a ready-made method for validating projections, especially if temporal scales can be portrayed adequately; and (3) they provide a direct relationship to the regional, national, and international policy issues that are of most concern to environmental decisionmakers.

The production of such data bases should be thought of as the first stage in an ongoing iterative procedure. The initial state of the science will both qualitatively, and in some cases quantitatively, limit the utility of maps. However, the research priorities outlined here and pursued by other Federal and international organizations should enable us to improve the underlying information base and thereby improve the next estimates of resources' responses to climate change.

Scenario-Assessment Approach to Integration

Integrated modeling has been attempted in large-scale environmental issues many times before and may be useful for policy analysis or for heuristic purposes. However, the goal of a model adequate to simulate the dynamics of geophysical, chemical, and biological processes on global scales is clearly achievable only after decades of research (ESSC, 1988).

Given that achieving such a goal is at least several decades in the future, the question of integrating diverse aspects of science in global climate change and its potential effects still must be addressed now. A promising approach for integrating research results is to treat the entire cycle of information flow (Figure 18-5a) as a series of two-stage or three-stage processes (Figures 18-5b and 18-5c).

Within each two-stage process, research results should be treated as follows. The first part of the process will be the creation of a set of scenarios, where we define a scenario as a possible and plausible combination of variables derived from a set of internally consistent assumptions. The second part of the process will evaluate the range of changes that are potentially due to each scenario and will evaluate the sensitivity of the underlying systems to different aspects of the scenarios. Thus, scenarios of land-use change can be used to evaluate the possible changes in emissions; scenarios of emissions can be used to evaluate the possible changes in atmospheric composition; scenarios of atmospheric composition can be used to evaluate changes in climate; climatic scenarios can be used to evaluate the possible changes in ecosystems; and scenarios of ecosystem and land-use change can in turn be used to evaluate possible changes in emissions.

The use of a scenario-assessment approach as an integrating tool has several advantages. It will provide clear priorities for research on the sensitivities of important environmental processes in each scientific area. It maintains an holistic view of the problems of global change. It preserves the information on the uncertainty of model results and data, in both qualitative and potentially quantitative fashion. Each pair of

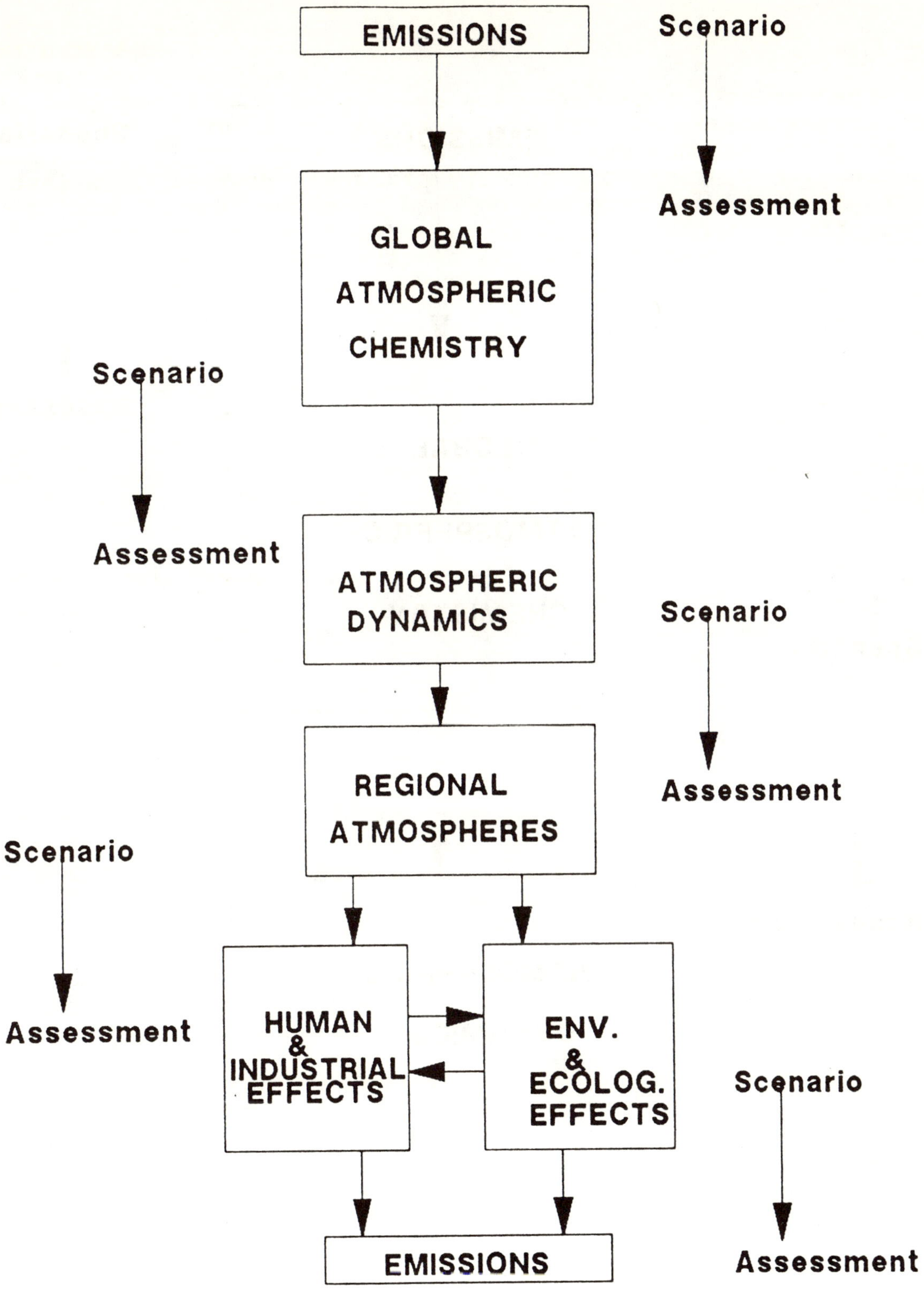

Figure 18-5a. Scenario approach to integration.

Source:

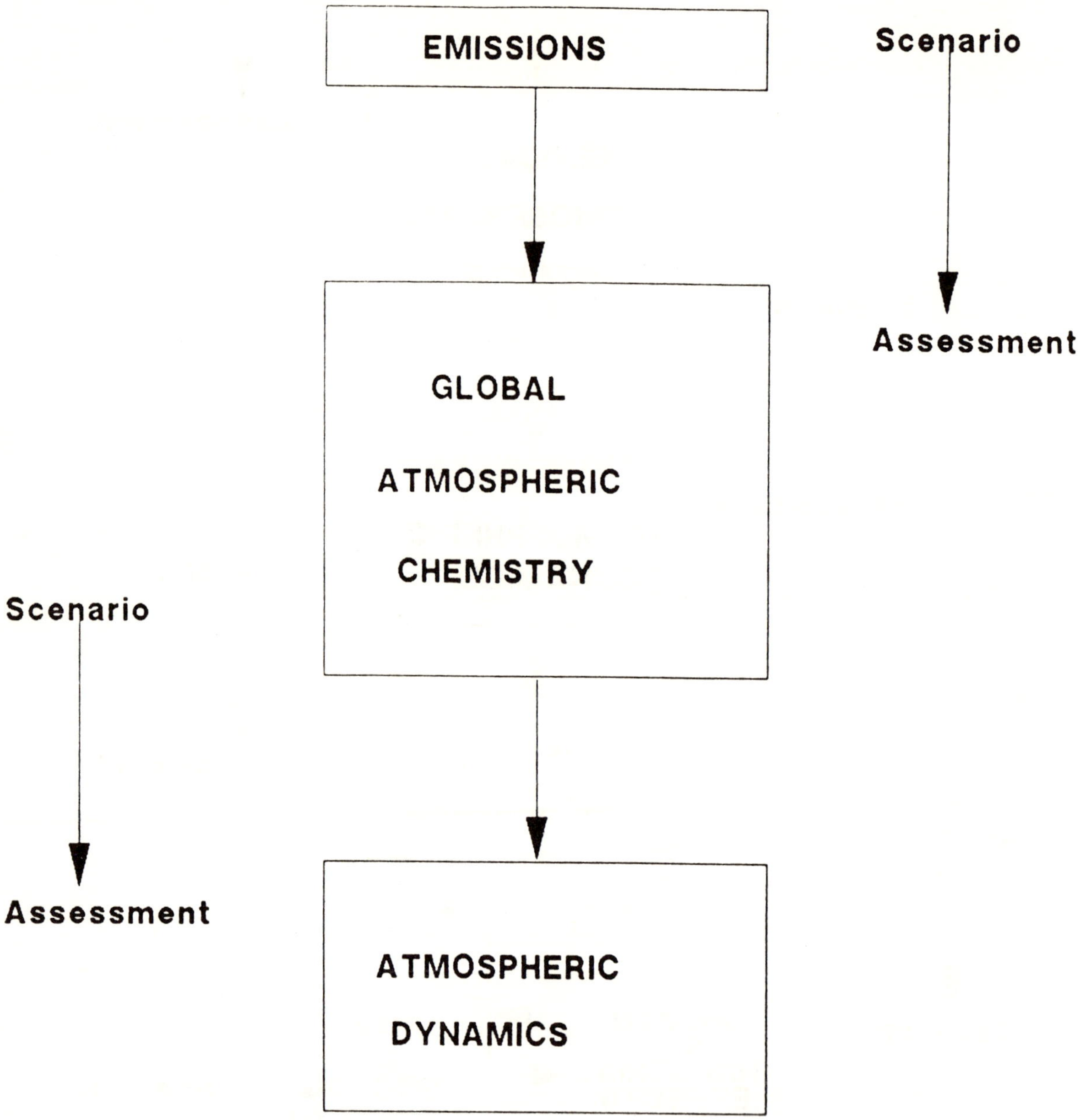

Figure 18-5b. Scenario approach to integration.

Source:

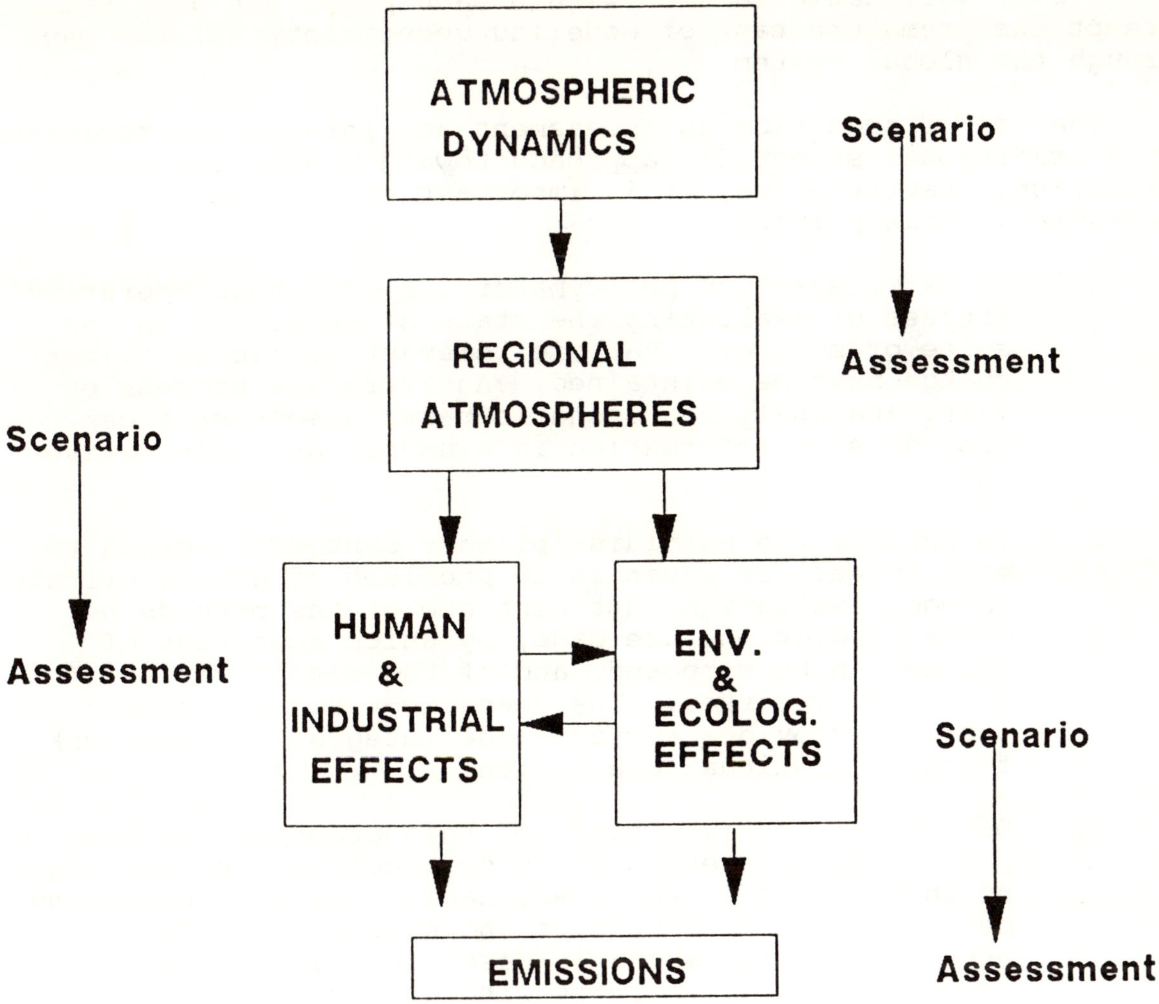

Figure 18-5c. Scenario approach to integration.

Source:

scenario-response steps is explicitly decoupled from other pairs, while remaining consistent with them. Thus, this approach can indicate both ranges and sensitivities of responses in potentially verifiable fashion within each pair, but does not attempt the premature task of modeling uncertainty all the way through the global system.

The use of scenarios as assessment and integrative tools is not a traditional scientific approach toward prediction and validation. Nevertheless, it is important from three programmatic standpoints:

1. To be relevant to policymakers, a continued iterative process of evaluating the state of knowledge in the suite of sciences that are relevant to global climate change must be maintained. An iterative process of using and analyzing scenario-based assessments can provide such information in a usable and informative way.

2. To achieve the multidisciplinary syntheses needed to make scientific advances in problems of global climate change, evaluation must continue of the methods by which predictions are made, by which scenarios of change can be composed, and of the sensitivities of affected processes. The scenario-based assessment approach provides a ready-made integrating framework for such continual evaluation.

3. Because of the importance of this proposed research in public policy arenas, it is critical not to lose sight of what is and is not predictable. By distinguishing between a set of scenarios and actual verifiable predictions, the scenario-based approach can best illustrate the difference without becoming a morass of hedged bets.

CONCLUSION

In this chapter, we have attempted to lay out some of the most important strategic areas for scientific research in deriving an understanding of the environmental consequences of global atmospheric change. The recommendations have not been limited to the ecological sciences but have addressed the atmospheric sciences as well. Throughout, the focus has been on the research topics in the environmental sciences that will have the best chance of producing information that will be of use to environmental decisionmakers and policymakers.

In so doing, we have made suggestions that transcend research efforts that the Environmental Protection Agency will be able to pursue on its own. Our intent in doing so is to underscore the institutional requirements for cooperation among U.S. and international research organizations; close coordination with efforts such as the International Geosphere Biosphere Programme (NAS, 1986); and a continuing need for productive dialog between scientists and policy analysts. Our hope is that this report, and the research needs outlined here, can be seen as a beginning step in this direction.

Table 18-1: Major Research Topics From This Report

Research Area	Role of climate change
Water Resources	Analyses of impacts on small lakes and streams Better estimates of rainfall, runoff relationships, and regional runoff patterns Analyses of relationship between waterflow and water quality Analyses of changes in pesticide and fertilizer management practices on water quality
Agriculture	Interactive effects of carbon dioxide, climate change, air pollution, and ultraviolet-B (UV-B) radiation Development of drought-resistant varieties Field studies to validate crop yield models
Forest Ecosystems	Processes and rates of migration in critical regions Changes in land-use practices Interactive effects of climate change, air pollution, UV-B radiation, and carbon dioxide Ecosystem-level processes quantified for different landscape types Modeling and sampling strategies for regional effects
Human Health	Methods for assessing morbidity, mortality patterns Influence on disease vectors
Air Quality	Influence on urban, mesoscale air quality including acid deposition and zones
Fishery Resources	Impacts in sensitive estuaries Shifts of distributions in near-coastal species Ecological impact on freshwater species
General Circulation Models	Climate variability for $1xCO_2$ and $2xCO_2$ Simulations of climate on regional and local scales Simulations of major storms, hurricanes, and blocking events

REFERENCES

Ballou, S.W., Levenson, J.B., Robek, K.E., and Gabriel, M.H. 1981. Regional Ecological Assessments; Concepts, Procedures, and Application. Argonne, IL: Argonne National Laboratory.

Barnthouse, L.W., DeAngelis, D.L., Gardner, R.H., O'Neill, R.V., Powers, C.D., Suter, G.W., and Vaughn, D.S. 1982. Methodology for Environmental Risk Analysis. Oak Ridge, TN: Oak Ridge National Laboratory, Environmental Science Division. Publication No. 2679.

Bolin, B., Doos, B.R., Jager, J., and Warrick, R.A. 1986. The Greenhouse Effect, Climate Change, and Ecosystems: SCOPE 29. New York: John Wiley and Sons.

Borchert, J.R. 1959. The climate of the Central American grassland. Annals AAG. 15(1).

Braun, E.L. 1950. Deciduous Forests of Eastern North America. New York: Hafner Press.

Brown, D.E., and Lowe, C.H. 1980. Biotic Communities of the Southwest. USDA Forest Service General Technical Report RM-78.

Bryson, R.A. 1966. Air masses, streamlines, and the boreal forest. Geographical Bulletin. 8(3).

Butzer, K. 1980. Adaption to global environmental change. Prof. Geogr.

Carbon Dioxide Assessment Committee. 1983. Changing Climate. Washington, DC: National Academy Press.

Fowells, H.A. 1965. Silvics of Forest Trees of the United States. Agriculture Handbook, No. 271. Washington, DC: USDA Forest Service.

Gates, W.L. 1985. The use of general circulation models in the analysis of ecosystem impacts of climate change. Climate change.

Hansen, J., Russell, G., Rink, D., Stone, P., Lacis, A., and Lebedeff, S. 1983. Efficient three-dimensional global models for climate studies: models I and II. Monthly Weather Review.

Harwell, M.A., and Harwell, J.S. 1988. Environmental decision-making in the presence of uncertainties. In: Levin, Harwell, Kelly, and Kimball, eds. Ecotoxicology: Problems and Approaches. New York: Springer-Verlag.

Karl, T.R., and Riebsame, W.E. 1984. The identification of 10- and 20-year temperature and precipitation fluctuations in the contiguous United States. Journal of Climatology and Applied Meteorology.

Kellogg, W.W., and Schware, R. 1981. Climate Change and Society: Consequences of Increasing Atompheric Carbon Dioxide. Boulder, CO: Westview Press.

Lamb, P.J. 1987. On the development of regional climatic scenarios for policy-oriented climatic-impact assessment. Bulletin of the American Meteorologic Society.

Landsberg, J.J. University of Sussex. 1986. In: Sutcliff, J.F., ed. Physiological Ecology of Forest Production from Applied Botany: A Series of Monographs. London, England: Academic Press.

MacCracken, M.C., and Luther, F.M. 1985. Projecting the Climatic Effects of Increasing Carbon Dioxide. Washington, DC: U.S. Department of Energy. DOE/ER-0237.

Mitchell, J.M. Jr. 1984. The effects of pollutants on global climate. Meterological Magazine.

Mitchell, V.L. 1969. The Regionalization of Climate in Montane Areas. Ann Arbor, MI: University Microfilms, Inc.

NAS. 1982. National Academy of Sciences. Carbon Dioxide and Climate: A Second Assessment. National Academy of Sciences, Climate Board.

National Academy of Sciences. 1986. Global Change in the Geosphere-Biosphere: Initial Priorities for an IGBP. National Research Council. Washington, DC: National Academy Press.

Neilson, R.P. 1986. High-resolution climatic analysis and Southwest biogeography. Science 232.

North, W., and Yosie, T.F. 1987. Risk Assessment: What It Is; How It Works. EPA Journal, U.S. Environmental Protection Agency, 13(9).

Robinson, P.J., and Hill, H.L. 1987. Toward a Policy for Climate Impacts. Bulletin of the American Meteorological Society.

Shands, W.E., and Hoffman, J.S. 1987. The Greenhouse Effect, Climate Change, and U.S. Forests. Washington, DC: The Conservation Foundation.

Solomon, A.M., and West, D.C. 1985. Potential Responses of Forests to CO_2-Induced Climate Change. Characterization of Information Requirements for Studies of CO_2 Effects: Water Resources, Agriculture, Fisheries, Forests and Human Health. Washington, DC: U.S. Department of Energy. DOE/ER-0236.

Trabalka, J.R., ed. 1985. Atmospheric Carbon Dioxide and the Global Carbon Cycle. Washington, DC: U.S. Department of Energy. DOE/ER-0239.

USGS. 1986. U.S. Geological Survey. National Water Summary 1985-Hydrologic Events and Surface-Water Resources. Washington, DC: U.S. Government Printing Office.

Vankat, J.L. 1979. The Natural Vegetation of North America. New York: John Wiley and Sons.

Wells, P.V. 1979. An equable glaciopluvial in the West: pleniglacial evidence of increased precipitation on a gradient from the Great Basin to the Sonoran and Chihuahuan Deserts. Quaternary Research 12.

Wendland, W.M., and Bryson, R.A. 1981. Northern hemisphere airstream regions. Monthly Weather Review 109.

White, M.R., ed. 1985. Characterization of Information Requirements for Studies of CO_2 Effects: Water Resources, Agriculture, Fisheries, Forests and Human Health. Washington, DC: U.S. Department of Energy. DOE/ER-0236.

Whittier, T.R., Hughes, R.M., and Larsen, D.P. 1988. The correspondence between ecoregions and spatial patterns in stream ecosystems in Oregon. Canadian Journal of Fisheries and Aquatic Sciences. In press.

Wigley, T.M.L., Jones, P.D., and Kelly, P.M. 1986. Empirical climate studies: warm world scenarios and the detection of a CO_2-induced climate change. In: Bolin, B., Doos, B.R., Jager, J., and Warrick, R.A., eds. The Greenhouse Effect, Climate Change and Ecosystems: SCOPE 29. New York: John Wiley and Sons.

Williams, L.D., and Wigley, T.M.L. 1983. A Comparison of Evidence for late Holocene summer temperature variations in the northern hemisphere. Quaternary Research.

AUTHORS

Joseph J. Bufalini	USEPA - Atmospheric Research and Exposure Assessment Laboratory - Research Triangle Park, North Carolina
Lauretta M. Burke	The Bruce Company
Margaret M. Daniel	The Bruce Company
Robert L. DeVelice	USEPA - Environmental Research Laboratory - Corvallis, Oregon
Eugene C. Durman	USEPA Office of Policy, Planning and Evaluation
Peter L. Finkelstein	USEPA - Atmospheric Research and Exposure Assessment Laboratory - Research Triangle Park, North Carolina
Anthony Janetos	USEPA - Office of Research and Development
Roy Jenne	National Center for Atmospheric Research
Ross A. Kiester	USEPA - Environmental Research Laboratory - Corvallis, Oregon
George A. King	USEPA - Environmental Research Laboratory - Corvallis, Oregon
Kenneth P. Linder	ICF, Inc.
Janice A. Longstreth	ICF, Inc.
Linda O. Mearns	National Center for Atmospheric Research
Ted R. Miller	The Urban Institute
Mark W. Mugler	Apogee Research, Inc.
Ronald P. Neilson	USEPA - Environmental Research Laboratory - Corvallis, Oregon
Cynthia Rosenzweig	Columbia University/Goddard Institute for Space Studies

William E. Riebsame	University of Colorado
Michael C. Rubino	Apogee Research, Inc.
Joel B. Smith	USEPA Office of Policy, Planning and Evaluation
Dennis A. Tirpak	USEPA Office of Policy, Planning and Evaluation
James G. Titus	USEPA Office of Policy, Planning and Evaluation
Jack K. Winjum	USEPA - Environmental Research Laboratory - Corvallis, Oregon
Robert C. Worrest	USEPA - Environmental Research Laboratory - Corvallis, Oregon

CONTRIBUTING INVESTIGATORS AND PROJECTS

Authors: Adams, Richard M., J. David Glyer, and Bruce A. McCarl
Institution: Oregon State University
Title: The Economic Effects of Climate Change on U.S. Agriculture: A Preliminary Assessment.

Authors: Allen, Richard C., and Francis N. Gichuki
Institution: Utah State University
Title: Effects of Projected CO_2-Induced Climate Changes on Irrigation Water Requirements in the Great Plains States.

Author: Assel, Raymond, A.
Institution: Great Lakes Environment Research Lab
Title: Impact of Global Warming on Great Lakes Ice Cycles.

Author: Baldwin, Malcolm F.
Institution: Baldwin Associates
Title: Applicability of Federal Long-Range Planning and Environmental Impact Statement Processes to Global Climate Change Issues.

Authors: Blumberg, Alan F., and Dominic M. DiToro
Institution: HydroQual, Inc.
Title: The Effects of Climate Warming on Lake Erie Water Quality.

Authors: Botkin, Daniel B., Robert A. Nisbet, and Tad E. Reynales
Institution: University of California, Santa Barbara
Title: Effects of Climate Change on Forests of the Great Lakes States.

Authors: Byron, Earl R., Alan Jassby, and Charles R. Goldman
Institution: University of California, Davis
Title: The Effects of Global Climate Change on the Water Quality of Mountain Lakes and Streams.

Authors: Changnon, Stanley A., Steven Leffler, and Robin Shealy
Institution: University of Illinois
Title: Impacts of Extremes in Lake Michigan Levels along Illinois Shorelines: Low Levels.

Authors: Croley, Thomas F., and Holly C. Hartmann
Institution: Great Lakes Environment Research Lab
Title: Effects of Climate Changes on the Laurentian Great Lakes Levels.

Author: Davis, Owen K.
Institution: University of Arizona
Title: Ancient Analogs for Greenhouse Warming of Central California.

Author: Dudek, J. Daniel
Institution: Environmental Defense Fund
Title: Climate Change Impacts upon Agriculture and Resources: A Case Study of California.

Author: Easterling, William E.
Institution: University of Illinois
Title: Farm-Level Adjustments by Illinois Corn Producers to Climate Change.

Author: Glantz, Michael
Institution: National Center for Atmospheric Research
Title: Societal Responses to Regional Climate Change: Forecasting by Analog.

Author: Haile, Daniel G.
Institution: U.S. Department of Agriculture, Agriculture Research Service - Gainesville
Title: Computer Simulation of the Effects of Changes in Weather Pattern on Vector-Borne Disease Transmission.

Author: Hains, David K.
Institution: C.F. Hydrologist, Inc.
Title: Impacts on Runoff in the Upper Chattahoochee River Basin.

Authors: Johnson, Howard L., Ellen J. Cooter, and Robert J. Sladewski
Institution: University of Oklahoma
Title: Impacts of Climate Change on the Fate of Agricultural Chemicals Across the U.S.A. Great Plains and Central Prairie.

Author: Josselyn, Michael, and John Callaway
Institution: San Francisco State University
Title: Ecological Effects of Global Climate Change: Wetland Resources of San Francisco Bay.

Author: Kalkstein, Laurence S.
Institution: University of Delaware
Title: The Impact of CO_2 and Trace Gas-Induced Climate Changes upon Human Mortality.

Author: Keith, Virgil F., Carlos DeAvila, and Richard M. Willis
Institution: Engineering Computer Optecnomics, Inc.
Title: Effect of Climatic Change on Shipping within Lake Superior and Lake Erie.

Author: Leatherman, Stephen P.
Institution: University of Maryland
Title: National Assessment of Beach Nourishment Requirements Associated with Accelerated Sea Level Rise.

Authors: Lettenmaier, Dennis P., Thian Yew Gan, and David R. Dawdy
Institution: University of Washington
Title: Interpretation of Hydrologic Effects of Climate Change in the Sacramento-San Joaquin River Basin, California.

Authors: Linder, Kenneth P., and Mark R. Inglis
Institution: ICF, Inc.
Title: The Potential Impacts of Climate Change on Electric Utilities: Regional and National Estimates.

Author: Livingston, Robert J.
Institution: Florida State University
Title: Projected Changes in Estuarine Conditions Based on Models of Long-Term Atmospheric Alteration.

Authors: Longstreth, Janice, and Joseph Wiseman
Institution: ICF, Inc.
Title: The Potential Impact of Climate Change on Patterns of Infectious Disease in the United States.

Authors: Magnuson, John J., Henry A. Regier, Brian J. Shuter, David K. Hill, John A. Holmes, and J. Donald Meisner
Institution: University of Wisconsin
Title: Potential Responses of Great Lakes Fishes and their Habitat to Global Climate Warming.

Author: McCormick, Michael J.
Institution: Great Lakes Environment Research Lab
Title: Potential Climatic Changes to the Lake Michigan Thermal Structure.

Author: Mearns, Linda O.
Institution: National Center for Atmospheric Research
Title: Analysis of Climate Variability in GCM Control Runs.

Authors: Meo, Mark, and Steve Ballard
Institution: University of Oklahoma
Title: Policy Implications of Global Climatic Change Impacts upon the Tennessee Valley Authority Reservoir System, Apalachicola River, Estuary, and Bay, and South Florida.

Authors: Miller, Barbara, and W. Gary Brock
Institution: Tennessee Valley Authority
Title: Potential Impacts of Climatic Change on the Tennessee Valley Authority Reservoir System.

Authors: Morris, Ralph E., Michael W. Gery, Mei Kao Liu, Gary E. Moore, Chris Daly, and Stanley M. Greenfield
Institution: Systems Applications, Inc.
Title: Examination of the Sensitivity of a Regional Oxidant Model to Climate Variations.

Authors: Overpeck, Jonathan T., and Patrick J. Bartlein
Institution: Lamont-Doherty Geological Observatory
Title: Assessing the Responses of Vegetation to Future Climate Change: Ecological Response Surfaces and Paleological Model Validation.

Authors: Park, Richard A., Manjit S. Trehan, Paul W. Mausel, and Robert C. Howe
Institution: Butler University
Title: The Effects of Sea Level Rise on U.S. Coastal Wetlands.

Authors: Peart, Robert M., James W. Jones, and R. Bruce Curry
Institution: University of Florida
Title: Impact of Climate Change on Crop Yield in the Southeastern U.S.A.

Authors: Penner, Joyce E., Peter S. Connell, Donald J. Wuebbles, and Curtis C. Covey
Institution: Lawrence Livermore National Laboratory
Title: Climate Change and Its Interactions with Air Chemistry: Perspectives and Research Needs.

Authors: Ray, Daniel K., Kurt N. Lindland, and William J. Brah
Institution: The Center for the Great Lakes
Title: Effects of Global Warming on the Great Lakes: The Implications for Policies and Institutions.

Author: Riebsame, William E.
Institution: University of Colorado
Title: Some Policy Implications of Climate Change, with Emphasis on Natural Resources Management and the Role of Climate Perception on Policy Response.

Authors: Riebsame, William E., and Jeffrey W. Jacobs
Institution: University of Colorado
Title: Some Policy Implications of Climate Change in the Sacramento-San Joaquin Region of California.

Author: Rind, David
Institution: Goddard Institute for Space Studies
Title: Change in Climate Variability in the 21st Century.

Authors: Ritchie, Joe T., Brian D. Baer, and T.Y. Chou
Institution: Michigan State University
Title: Effect of Global Climate Change on Agriculture: Great Lakes Region.

Author: Rose, Elise
Institution: Consultant
Title: Direct Effects of Increasing CO_2 on Plants and Their Interactions with Direct Climatic Effects.

Author: Rosenzweig, Cynthia
Institution: Columbia University/Goddard Institute for Space Studies
Title: Potential Effects of Climate Change on Agricultural Production in the Great Plains: A Simulation Study.

Authors: Schmidtmann, Edward T., and J.A. Miller
Institution: U.S. Department of Agriculture, Agriculture Research Service - Beltsville, Maryland
Title: Effect of Climate Warming on Population of the Horn Fly.

Author: Schuh, G. Edward
Institution: University of Minnesota
Title: Agricultural Policies for Climate Changes Induced by Greenhouse Gases.

Authors:	Sheer, Daniel P., and Dean Randall
Institution:	Water Resources Management Inc.
Title:	Methods for Evaluating the Potential Impacts of Global Climate Change: Case Studies of the Water Supply Systems of the State of California and Atlanta, Georgia.
Authors:	Stem, Edgar, Gregory A. Mertz, J. Dirck Strycker, and Monica Huppi
Institution:	Tufts University
Title:	Changing Animal Disease Patterns Induced by the Greenhouse Effect.
Authors:	Stinner, Benjamin R., Robin A.J. Taylor, Ronald B. Hammond, Foster F. Purrington, David. A. McCartney, Nick Rodenhouse, and Gary Barrett
Institution:	Ohio State University
Title:	Potential Effects of Climatic Change on Plant-Pest Interactions.
Authors:	Titus, James G., and Michael Greene
Institution:	U.S. Environmental Protection Agency
Title:	An Overview of the Nationwide Impacts of Sea Level Rise.
Authors:	Urban, Dean L., and Herman H. Sheer
Institution:	University of Virginia
Title:	Forest Response to Climate Change: A Simulation Study for Southeastern Forests.
Authors:	Walker, Christopher J., Ted R. Miller, G. Thomas Kingsley, and William A. Hyman
Institution:	The Urban Institute
Title:	Impact of Global Climate Change on Urban Infrastructure.
Authors:	Weggel, J. Richard, Scott Brown, Juan Carlos Escajadillo, Patrick Breen, and Edward L. Doheny
Institution:	Drexel University
Title:	The Cost of Defending Developed Shorelines Along Sheltered Waters of the United States from a Two Meter Rise in Mean Sea Level.
Author:	Williams, Philip B.
Institution:	Philip Williams & Associates
Title:	The Impacts of Climate Change on the Salinity of San Francisco Bay.

Authors:	Woodman, James N., and Cari L. Sasser
Institution:	North Carolina State University
Title:	Potential Effects of Climate Change on U.S. Forests: Case Studies of California and the Southeast.

Author:	Yohe, Gary W.
Institution:	Wesleyan University
Title:	The Cost of Holding Back the Sea: Phase 1, Economic Vulnerability.

Author:	Zabinski, Katherine and Margaret Davis
Institution:	University of Minnesota
Title:	Hard Times Ahead for Great Lakes Forests: A Climate Threshold Model Predicts Responses to CO_2-Induced Climate Change.